T0348573

Plant Virus–Host Interaction
Molecular Approaches and Viral Evolution

Plant Virus–Host Interaction

Molecular Approaches and Viral Evolution

R.K. Gaur
Associate Professor and Head, Department of Science, FASC,
Mody Institute of Technology and Science,
Lashmangarh, Sikar, India

Thomas Hohn
Professor Emeritus, Botanical Institute, University of Basel,
Basel, Switzerland

Pradeep Sharma
Senior Scientist (Biotechnology), Directorate of Wheat Research,
Karnal, India

AMSTERDAM • BOSTON • HEIDELBERG • LONDON
NEW YORK • OXFORD • PARIS • SAN DIEGO
SAN FRANCISCO • SINGAPORE • SYDNEY • TOKYO

Academic Press is an imprint of Elsevier

Academic Press is an imprint of Elsevier
The Boulevard, Langford Lane, Kidlington, Oxford, OX5 1GB, UK
225 Wyman Street, Waltham, MA 02451, USA

First published 2014

British Library Cataloguing in Publication Data
A catalogue record for this book is available from the British Library

Library of Congress Cataloguing in Publication Data
A catalogue record for this book is available from the Library of Congress

ISBN: 978-0-12-411584-2

For information on all Academic Press publications
visit our website at store.elsevier.com

Printed and bound by CPI Group (UK) Ltd, Croydon, CR0 4YY
14 15 16 17   10 9 8 7 6 5 4 3 2 1

Working together
to grow libraries in
developing countries

ELSEVIER   Book Aid International

www.elsevier.com • www.bookaid.org

Contents

3.   Host–virus interactions in banana-infecting viruses

R. Selvarajan and V. Balasubramanian

4.   Recent advances on interactions between the whitefly *Bemisia tabaci* and begomoviruses, with emphasis on *Tomato yellow leaf curl virus*

Pakkianathan Britto Cathrin and Murad Ghanim

5.   Hosts and non-hosts in plant virology and the effects of plant viruses on host plants

András Takács, József Horváth, Richard Gáborjányi,
Gabriella Kazinczi, and József Mikulás

6.   Interference with insect transmission to control plant-pathogenic viruses

María Urizarna España and Juan José López-Moya

7. Transmission and host interaction of *Geminivirus* in weeds

Avinash Marwal, Anurag Kumar Sahu, and R.K. Gaur

8. *Tombusvirus*-induced multivesicular bodies: origin and role in virus–host interaction

L. Rubino, M. Russo, and G.P. Martelli

9. *Papaya ringspot virus-P*: overcoming limitations of resistance breeding in *Carica papaya* L.

Sunil Kumar Sharma and Savarni Tripathi

10. Synergism in plant–virus interactions: a case study of CMV and PVY in mixed infection in tomato

Tiziana Mascia and Donato Gallitelli

11. Methods of diagnosis, stability, transmission, and host interaction of *Papaya lethal yellowing virus* in papaya

J. Albersio A. Lima, Aline Kelly Q. Nascimento, Roberto C.A. Lima, Verônica C. Oliveira, and Geórgia C. Anselmo

12. Establishment of endogenous pararetroviruses in the rice genome

Ruifang Liu and Yuji Kishima

13. Volatile organic compounds and plant virus–host interaction

Y. L. Dorokhov, T.V. Komarova, and E.V. Sheshukova

14. Diversity of latent plant–virus interactions and their impact on the virosphere

K.R. Richert-Pöggeler and J. Minarovits

15. Viroid–insect–plant interactions

Noémi Van Bogaert, Guy Smagghe, and Kris De Jonghe

16. Engineering crops for resistance to geminiviruses

Akhtar Jamal Khan, Sohail Akhtar, Shahid Mansoor, and Imran Amin

17. *Cauliflower mosaic virus* (CaMV) upregulates translation reinitiation of its pregenomic polycistronic 35S RNA via interaction with the cell's translation machinery

Mikhail Schepetilnikov and Lyubov Ryabova

18. Molecular mechanism of *Begomovirus* evolution and plant defense response

T. Vinutha, Om Prakash Gupta, G. Rama Prashat, Veda Krishnan,
and P. Sharma

19. Impact of the host on plant virus evolution

Xiao-fei Cheng, Nasar Virk, and Hui-zhong Wang

20. Virus-induced physiologic changes in plants

András Takács, József Horváth, Richard Gáborjányi, and
Gabriella Kazinczi

21. Virus–Virus interactions

*András Takács, Richard Gáborjányi, József Horváth, and
Gabriella Kazinczi*

Contributors

Sohail Akhtar   Department of Crop Sciences, College of Agricultural and Marine Sciences, Sultan Qaboos University, Sultanate of Oman

Imran Amin   National Institute of Biotechnology and Genetic Engineering, Faisalabad, Pakistan

Geórgia C. Anselmo   Federal University of Ceará, Laboratory of Plant Virology, Fortaleza, Brazil

V. Balasubramanian   National Research Centre for Banana (ICAR), Tiruchirappalli, India

Pakkianathan Britto Cathrin   Department of Entomology, Agricultural Research Organization, The Volcani Center, Bet Dagan, Israel

Xiao-fei Cheng   College of Life and Environmental Science, Hangzhou Normal University, Hangzhou, Zhejiang, P R China

Mirosława Chrzanowska   Młochów Research Center, Plant Breeding and Acclimatization Institute—National Research Institute, Młochów, Poland

Kris De Jonghe   Laboratory of Virology, Plant Sciences Unit—Crop Protection, Institute for Agricultural and Fisheries Research (ILVO), Merelbeke, Belgium

Y.L. Dorokhov   A N Belozersky Institute of Physico-Chemical Biology, Moscow State University, Moscow, Russia, and N I Vavilov Institute of General Genetics, Russian Academy of Science, Moscow, Russia

Richard Gáborjányi   University of Pannonia, Georgikon Faculty of Agricultural Sciences, Institute for Plant Protection, Keszthely, Hungary

Donato Gallitelli   Dipartimento di Scienze del Suolo, della Pianta e degli Alimenti, Università degli Studi di Bari Aldo Moro, Bari, Italy

R.K. Gaur   Department of Science, Faculty of Arts, Science and Commerce, Mody Institute of Technology and Science, Lakshmangarh, Rajasthan, India

Murad Ghanim   Department of Entomology, Agricultural Research Organization, The Volcani Center, Bet Dagan, Israel

Om Prakash Gupta   Division of Quality and Basic Sciences, Directorate of Wheat Research, Karnal, India

József Horváth   University of Pannonia, Georgikon Faculty of Agricultural Sciences, Institute for Plant Protection, Keszthely, Hungary, and Kaposvár University, Department of Botany and Plant Production, Kaposvár, Hungary

Gabriella Kazinczi   Kaposvár University, Department of Botany and Plant Production, Kaposvár, Hungary

Akhtar Jamal Khan Department of Crop Sciences, College of Agricultural and Marine Sciences, Sultan Qaboos University, Sultanate of Oman

Yuji Kishima Laboratory of Plant Breeding, Research Faculty of Agriculture, Hokkaido University, Sapporo, Japan

T.V. Komarova A N Belozersky Institute of Physico-Chemical Biology, Moscow State University, Moscow, Russia, and N I Vavilov Institute of General Genetics, Russian Academy of Science, Moscow, Russia

Veda Krishnan Division of Biochemistry, IARI, New Delhi, India

J.Albersio A. Lima Federal University of Ceará, Laboratory of Plant Virology, Fortaleza, Brazil

Roberto C.A. Lima BioClone, Eusébio—CE, Brazil

Ruifang Liu Laboratory of Plant Breeding, Research Faculty of Agriculture, Hokkaido University, Sapporo, Japan

Juan José López-Moya Centre for Research in Agricultural Genomics CRAG, CSIC-IRTA-UAB-UB, Campus UAB—Bellaterra, Barcelona, Spain

Shahid Mansoor National Institute of Biotechnology and Genetic Engineering, Faisalabad, Pakistan

G.P. Martelli Dipartimento di Scienze del Suolo, della Pianta e degli Alimenti, Università di Bari Aldo Moro, Bari, Italy, and Istituto di Virologia Vegetale del CNR, UOS Bari, Bari, Italy

Avinash Marwal Department of Science, Faculty of Arts, Science and Commerce, Mody Institute of Technology and Science, Lakshmangarh, Rajasthan, India

Tiziana Mascia Dipartimento di Scienze del Suolo, della Pianta e degli Alimenti, Università degli Studi di Bari Aldo Moro, Bari, Italy

Krystyna Michalak Młochów Research Center, Plant Breeding and Acclimatization Institute—National Research Institute, Młochów, Poland

József Mikulás Corvinus University of Budapest Institute of Viticulture and Oenology Research Station, Kecskemét, Hungary

J. Minarovits University of Szeged, Faculty of Dentistry, Department of Oral Biology and Experimental Dental Research, Szeged, Hungary

Thomas Mitchell Department of Plant Pathology, Ohio Agriculture Research and Development Center, The Ohio State University, Wooster, OH, US

Aline Kelly Q. Nascimento Federal University of Ceará, Laboratory of Plant Virology, Fortaleza, Brazil

Verônica C. Oliveira Federal University of Ceará, Laboratory of Plant Virology, Fortaleza, Brazil

G. Rama Prashat Division of Genetics, IARI, New Delhi, India

Feng Qu Department of Plant Pathology, Ohio Agriculture Research and Development Center, The Ohio State University, Wooster, OH, US

K.R. Richert-Pöggeler Institute for Epidemiology and Pathogen Diagnostics, Julius Kühn-Institut, Braunschweig, Germany

L. Rubino Istituto di Virologia Vegetale del CNR, UOS Bari, Bari, Italy

M. Russo Istituto di Virologia Vegetale del CNR, UOS Bari, Bari, Italy

Lyubov Ryabova Institut de Biologie Moléculaire des Plantes du CNRS, Université de Strasbourg, France

Anurag Kumar Sahu Department of Science, Faculty of Arts, Science and Commerce, Mody Institute of Technology and Science, Lakshmangarh, Rajasthan, India

Mikhail Schepetilnikov Institut de Biologie Moléculaire des Plantes du CNRS, Université de Strasbourg, France

R. Selvarajan National Research Centre for Banana (ICAR), Tiruchirappalli, India

P. Sharma Division of Crop Improvement, Directorate of Wheat Research, Karnal, India

Sunil Kumar Sharma Indian Agricultural Research Institute, Regional Station, Agricultural College Estate, Shivajinagar, Pune, India

E.V. Sheshukova N I Vavilov Institute of General Genetics, Russian Academy of Science, Moscow, Russia

Jasleen Singh Department of Plant Pathology, Ohio Agriculture Research and Development Center, The Ohio State University, Wooster, OH, US

Guy Smagghe Department of Crop Protection, Faculty Bioscience Engineering, Ghent University, Ghent, Belgium

Lucy R. Stewart Department of Plant Pathology, Ohio Agriculture Research and Development Center, The Ohio State University, Wooster, OH, US, and USDA-ARS Corn, Soybean and Wheat Quality Research Unit, Wooster, OH, US

András Takács University of Pannonia, Georgikon Faculty of Agricultural Sciences, Institute for Plant Protection, Keszthely, Hungary

Savarni Tripathi Indian Agricultural Research Institute, Regional Station, Agricultural College Estate, Shivajinagar, Pune, India

María Urizarna España Centre for Research in Agricultural Genomics CRAG, CSIC-IRTA-UAB-UB, Campus UAB—Bellaterra, Barcelona, Spain

Noémi Van Bogaert Department of Crop Protection, Faculty Bioscience Engineering, Ghent University, Ghent, Belgium, and Laboratory of Virology, Plant Sciences Unit—Crop Protection, Institute for Agricultural and Fisheries Research (ILVO), Merelbeke, Belgium

T. Vinutha Division of Biochemistry, IARI, New Delhi, India

Nasar Virk Atta-ur-Rahman School of Applied Biosciences, National University of Sciences and Technology, Islamabad, Pakistan

Hui-zhong Wang College of Life and Environmental Science, Hangzhou Normal University, Hangzhou, Zhejiang, P R China

Zhimin Yin Młochów Research Center, Plant Breeding and Acclimatization Institute—National Research Institute, Młochów, Poland

Xiuchun Zhang   Department of Plant Pathology, Ohio Agriculture Research and Development Center, The Ohio State University, Wooster, OH, US, and Institute of Tropical Bioscience and Biotechnology, Chinese Academy of Tropical Agricultural Sciences, Key Laboratory of Biology and Genetic Resources of Tropical Crops, Ministry of Agriculture, Haikou, Hainan, P R China

Ewa Zimnoch-Guzowska   Młochów Research Center, Plant Breeding and Acclimatization Institute—National Research Institute, Młochów, Poland

Preface

Plant viruses have evolved as combinations of genes whose products interact with cellular components to produce progeny virus throughout the plant. Some viral genes, particularly those involved in replication and assembly, tend to be relatively conserved, whereas other genes that have evolved for interaction with a specific host, for movement, and to counter host-defense systems, tend to be less conserved.

The ability of the virus to move from the initially infected cell throughout the plant appears to be one of the major selective forces for the evolution of plant viruses. Successful systemic infection of plant viruses result from replication in initially infected cells, followed by two distinct processes: cell-to-cell and long-distance movement. Cell-to-cell movement is a process that allows the virus to pass to adjacent cells by successful interactions between virus-encoded movement proteins and host factors. Long-distance movement is a multistep process that allows the virus to enter the sieve element from an adjacent cell, followed by passive movement of the virus through the phloem to a distal region of the plant by exiting into a cell adjacent to the phloem. Further cell-to-cell movement from the phloem-associated cells allows the virus to invade most of the cells at a distal region of the plant. Viral proteins, and host factors that are involved in the cell-to-cell movement of plant viruses, have been widely examined. However, the host factors that are involved in long-distance transport of plant viruses and the mechanisms of long-distance movement—such as factors that are involved in virus entry into phloem tissue and virus exit at a distal region of the plant—are less well understood. Additionally, plants have host-defense mechanisms, including RNA silencing, that must be overcome by the virus for effective movement within the plant. Viruses have evolved gene products to suppress these defense mechanisms.

The long-term research of the group focuses on understanding the emergence of new viral diseases. Plant viral diseases have a high socioeconomic impact, as they affect crop and forest productivity as well as ecosystem composition and dynamics. The highest impact of diseases in host populations is often caused by emerging diseases, defined as those whose incidence in a host population is increasing as a result of long-term changes in their underlying epidemiology. Major factors favoring disease emergence are genetic change in pathogen and host populations and changes in host ecology and environment. Hence, we proposed an edited book covering the research interests of the virus group and organized the chapters in and around plant virus evolution and the mechanisms of plant–virus interaction at a molecular level.

Editors

Dr Rajarshi Kumar Gaur is presently working as Associate Professor and Head, Department of Science, FASC, Mody Institute of Technology and Science, Lashmangarh, Sikar, India. He did his PhD on molecular characterization of sugarcane viruses of India. He had partially characterized three sugarcane viruses, viz., sugarcane mosaic virus, sugarcane streak mosaic virus, and sugarcane yellow luteovirus. He received a MASHAV fellowship in 2004 from the government of Israel for his postdoctoral studies, joined the Volcani Centre, Israel, and then moved to the Ben Gurion University, Negev, Israel. In 2007 he received the Visiting Scientist Fellowship from the Swedish Institute for 1 year to work at Umeå University in Sweden. He was also a recipient of the Italian ICGEB Post Doctoral fellowship in 2008. He worked on the development of marker-free transgenic plants against cucumber viruses. He has made significant contributions on sugarcane viruses, published 85 national/international papers, and presented some 45 papers at national and international conferences. He has also edited 10 books published by internationally reputable publishers. He was awarded a Fellowship of the Linnean Society, London, UK. He has also visited Thailand, New Zealand, the UK, Canada, the US, and Italy to attend conferences/workshops. Currently, he is handling different projects funded by the Government of India on plant viruses and disease management.

Thomas Hohn is Professor emeritus at the Botanical Institute, University of Basel. He is Austrian, has studied at the Max-Planck Institute of Tübingen, and has performed postdoctoral studies in Stanford, California. He was junior group leader at the Bicocenter of the University of Basel, and group leader at the Friedrich Miescher Institute, Basel. His interests are in virology, originally in bacteriophages; he has studied self-assembly of small bacteriophages as well as morphogenesis of bacteriophage λ. The latter work, performed together with his wife, led to DNA-packaging. Later he shifted to plant viruses, where he recognized the first plant pararetrovirus (CaMV) and detected special viral translation strategies. Recently he became interested in the topic of RNA-interference and transgenesis. He has published more than 200 papers during his career. For several years he has been involved in the Indo-Swiss Collaboration in Biotechnology project, working together with Indian scientists to apply biotechnology for the improvement of pulses (leguminosae) and cassava for use by subsistence farmers.

Dr Pradeep Sharma now works as Senior Scientist (Biotechnology) at the Directorate of Wheat Research in Karnal, India. He did his PhD in plant pathology (molecular virolgy) at Haryana Agricultural University, Hisar, on cotton leaf curl begomovirus. Dr Sharma has made a significant contributions in geminiviruses from South East Asia and published more than 80 national and international research papers, invited chapters, and reviews and has edited three books in biotechnology and geminivirology. He has studied the role of suppressors encoded by monopartite Tomato leaf curl Java and Ageratum yellow vein begomoviruses. Dr Sharma also has the distinction of receiving numerous honors, international fellowships, and national awards in recognition of his excellent academic and research contributions. These include: Young Scientist's Award (biannual 2005–2006) of the National Academy of Agricultural Sciences, the PranVohra award (2008–2009) of the Indian Science Congress Association, etc. As a recipient of the JSPS fellowship, he worked in Tohoku University, Japan, during 2006–2008, and received a postdoc fellowship from the Ministry of Agriculture and Foreign Affairs, Israel; he also joined the Volcani Centre, ARO, Israel, in 2005–2006. He was elected as Fellow of the Indian Virological Society, Delhi, India. He has worked and visited many pioneer laboratories in the US, UK, Japan, France, China, the Netherlands, Indonesia, Turkey, and Israel. He is member and reviewer of several national and international scientific societies. Currently he is working on functional genomics, small RNAs, and bioinformatics of wheat.

Chapter 1

Role of double-stranded RNA-binding proteins in RNA silencing and antiviral defense

Jasleen Singh
Department of Plant Pathology, Ohio Agriculture Research and Development Center, The Ohio State University, Wooster, OH, USA

Xiuchun Zhang
Department of Plant Pathology, Ohio Agriculture Research and Development Center, The Ohio State University, Wooster, OH, USA, and Institute of Tropical Bioscience and Biotechnology, Chinese Academy of Tropical Agricultural Sciences, Key Laboratory of Biology and Genetic Resources of Tropical Crops, Ministry of Agriculture, Haikou, Hainan, P R China

Lucy R. Stewart
Department of Plant Pathology, Ohio Agriculture Research and Development Center, The Ohio State University, Wooster, OH, USA, and USDA-ARS Corn, Soybean and Wheat Quality Research Unit, Wooster, OH, USA

Thomas Mitchell and Feng Qu
Department of Plant Pathology, Ohio Agriculture Research and Development Center, The Ohio State University, Wooster, OH, USA

INTRODUCTION

RNA silencing is an ancient genome surveillance system conserved in most eukaryotic organisms ranging from fission yeast to human beings. It is primarily triggered by the intracellular presence of double-stranded RNA (dsRNA), with the final outcome being down-regulation of the expression of genes that share substantial sequence homologies with the dsRNA trigger (Voinnet 2009, Ding 2010). The signaling cascade begins with the perception of the silencing-inducing dsRNAs by the RNA silencing machinery, which then uses a dsRNA-specific nuclease designated Dicer or Dicer-like (DCL) nuclease in plants to process these dsRNAs into short RNA duplexes of 21–25 nucleotides (nt) referred to as small interfering RNAs or siRNAs (Hamilton & Baulcombe 1999). siRNAs are key sequence specificity determinants of RNA silencing, as one strand of the siRNA duplex is recruited by another nuclease, referred

Plant Virus-Host Interaction.

to as Argonaute (AGO or AGL in some animal models such as *Caenorhabditis elegans*), and guides the later to single-stranded RNAs (ssRNAs) that contain sequences complementary to the siRNAs. Another important role of siRNAs is that, at least in plants, they also act as partners for RNA-dependent RNA polymerases (RDRs). RDRs use siRNA-complementary ssRNAs as templates to synthesize more dsRNAs, which are again processed by DCLs, thus amplifying the RNA silencing process (Fig. 1.1).

In addition to DCLs, a family of double-stranded RNA-binding proteins (DRBs) have also been found to be required for the processing of dsRNA substrates. Therefore, the plant RNA silencing machinery primarily consists of four different families of proteins: DCLs, DRBs, AGOs, and RDRs. The genome of the model plant *Arabidopsis thaliana* encodes four DCLs, five DRBs, ten AGOs, and six RDRs, which participate in a number of RNA silencing pathways (Hammond 2005, Brodersen & Voinnet 2006, Vaucheret 2006) to regulate diverse developmental and physiologic processes, to mediate responses to biotic as well as abiotic stresses, to interfere with virus infections, and to ensure genome integrity. The different RNA silencing pathways are typically distinguishable by the sources of dsRNAs, the proteins involved in the silencing cascade, and the nature of their targets, although significant functional redundancy and crosstalk exists between some of the pathways (Table 1.1).

The *microRNA* (*miRNA*) pathway is probably the best understood RNA silencing pathway in plants (Voinnet 2009). miRNAs are encoded in the plant genome by *MIR* genes, which are transcribed by DNA-dependent RNA polymerase II (PolII) to generate primary miRNA (pri-miRNA). Pri-miRNAs form partially double-stranded hairpins through extensive intra-molecular base pairs (Fig. 1.1) and are processed by DCL1 sequentially to produce mature miRNAs (Kurihara et al 2006). miRNA-programmed RNA-induced silencing complexes (RISCs)

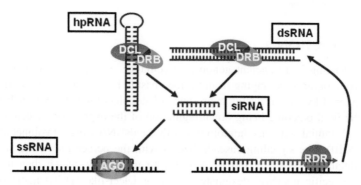

FIGURE 1.1 A simplified representation of the RNA silencing cascade in plants. Both long dsRNA and hairpin RNA with a significant length of double-stranded region (hpRNA) can be processed by the DCL/DRB complex into siRNAs, which, in turn, mediate the degradation or translational repression (not shown) of ssRNAs, or the remodeling of chromatin (not shown), In addition, siRNAs could also serve as primers to prime RDR-mediated synthesis of new dsRNAs.

are directed to mRNAs with complementary sequences to mediate cleavage as well as translational repression. At least three AGOs (AGO1, 7, and 10) have been associated with miRNAs (Qi et al 2005, Baumberger & Baulcombe 2005, Montgomery et al 2008). Other plant proteins critical for miRNA biogenesis include HYL1, a DRB; HEN1, an enzyme that methylates the 2′OH of the 3′ end nucleotide of miRNAs, and SERRATE, a zinc finger protein (Han et al 2004, Yu et al 2005, Kurihara et al 2006). Functions of miRNAs include regulation of developmental processes, and response to abiotic as well as biotic stresses (Sunkar et al 2006, Navarro et al 2006, 2008).

Similar to miRNAs, *transacting siRNAs* (tasiRNAs) are derived from their respective genes (*TAS* genes). However, the *TAS* transcripts are first processed by miRNA-mediated cleavage to become templates for RDR6. dsRNAs generated by RDR6 are then processed by DCL4 to generate tasiRNAs (Allen et al 2005, Yoshikawa et al 2005). The biogenesis and functionality of tasiRNAs in *Arabidopsis* depend on at least six plant proteins: DCL1, DCL4, RDR6, DRB4, HEN1 and AGO1, and, in some cases, AGO7 (Peragine et al 2004,

TABLE 1.1 Partial genetic requirements for various small RNA pathways

miRNA pathway	DCL1
	AGO1, 7, 10
	HYL1
tasiRNA pathway	DCL1, 4
	RDR6
	AGO1, 7
	DRB4
casiRNA pathway	DCL3
	RDR2
	AGO4
	PolIV, V
nat-siRNA pathway	DCL1, 2
	RDR6
	AGO1
	PolIV
lsiRNA pathway	DCL1, 4
	RDR6
	AGO7
	PolIV, V
Viral siRNA pathway	DCL1, 2, 3, 4
	RDR1, 2, 6
	AGO1, 4, 7
	DRB4
	PolIV, V

Vazquez et al 2004, Adenot et al 2006, Hunter et al 2006, Montgomery et al 2008).

Cis-acting siRNAs (casiRNAs) are by far the most abundant class of endogenous siRNAs. They are derived from transposons and other highly repeated sequences in plant genomes, and function to silence these repetitive elements by promoting DNA methylation and heterochromatin formation (Henderson et al 2006, Kasschau et al 2007). Their biogenesis and functionality depend on DCL3, RDR2, AGO4, AGO6, and newly discovered DNA-dependent RNA polymerases PolIV and PolV (Herr et al 2005, Zheng et al 2007, Wierzbicki et al 2008). In addition, Raja and colleagues (2008) suggested that the same set of plant proteins are responsible for defending plants against infections by geminiviruses with single-stranded DNA genomes, revealing an antiviral role of this pathway.

Natural antisense siRNAs (nat-siRNAs) arise from partially overlapping ends of mRNA pairs transcribed in opposite directions. These siRNAs are often detected when plants are exposed to abiotic stresses or incompatible bacterial infections, which induces the expression of one of the paired mRNAs (Borsani et al 2005, Katiyar-Agarwal et al 2006). dsRNA formed by the paired ends then triggers a pathway that produces nat-siRNAs using DCL1, DCL2, RDR6, AGO1, PolIV, and other proteins. Another class of small RNAs of 30–40 nt, called *long siRNAs* (lsiRNAs), is also generated from similar natural antisense pairs (Katiyar-Agarwal et al 2007; Table 1.1). Both nat-siRNAs and lsiRNAs play important roles in plant stress responses.

Antiviral defense is another primary function of RNA silencing in plants. For RNA viruses, DCL4 and DCL2 have been shown to be the primary Dicers of dsRNAs of virus origins, which could be intermediates of viral RNA replication, or hairpin RNA formed by intra-molecular base-pairing of viral RNAs. For example, *Arabidopsis* plants infected with *Turnip crinkle virus* (TCV) accumulate low levels of DCL4-generated 21 nucleotide (nt) viral siRNAs (vsRNAs), but moderate levels of 22 nt vsRNAs (Deleris et al 2006, Qu et al 2008). However, once the suppressor of RNA silencing encoded by this virus is abolished, 21 nt siRNAs become the dominant viral siRNA class (Cao et al 2010, Zhang et al 2012). These observations suggest that the activity of DCL4 is suppressed by TCV infection, whereas DCL2 acts as a surrogate in the absence of DCL4.

Several recent examples have further revealed the role of at least one DRB protein in antiviral defense in plants. In this chapter we attempt to first summarize the current understanding of the functional mechanisms of DRB proteins in both animal and plants. We then focus our discussions on DRBs in plants and their participation in antiviral defense. As it will become clear as we progress, both available literature and our unpublished data suggest that DCL function is mostly dependent on the presence of an accompanying DRB. It will be interesting to find out how plant DCLs and DRBs interact biochemically to facilitate the biogenesis of miRNAs and siRNAs.

DRBs IN HUMANS AND ANIMALS

Because Dicers in animals and DCLs in plants possess dsRNA-binding motifs (dsRBMs), it was a surprise to discover that the functionality of a *Drosophila* Dicer (DCR-2) requires another dsRBM-containing protein, R2D2 (Liu et al 2003). These authors first purified an RNAi-competent protein complex from *Drosophila* cells and found that it contained both DCR-2 and R2D2. They further demonstrated that DCR-2 and R2D2 physically interact with each other, and that the interaction is indispensable for the functionality of the purified protein complex. In addition, they also mapped the R2D2 action to a step after DCR-2 processing, thus serving as the bridge between the production of siRNAs and the slicing of siRNA targets. This study unequivocally implicated R2D2 in *Drosophila* RNA silencing. It was then quickly discovered that the *Drosophila* genome encodes two more dsRBM proteins, Pasha and Loquacious (Loqs), that are needed for two different RNAi pathways. Like *Arabidopsis*, *Drosophila* also has an miRNA pathway responsible for producing miRNAs that play key roles in controlling proper developmental transitions. In *Drosophila* the processing of primary miRNA (the original transcripts) and precursor miRNA (the hairpin RNA with the tails trimmed) is accomplished by two different Dicers, namely Drosha and DCR-1. Accordingly, it has been established that both Drosha and DCR-1 require their own DRB partners, with Pasha pairing with Drosha, Loqs with DCR-1 (Denli et al 2004, Saito et al 2005; also see Fig. 1.2). In addition, Loqs is also needed for DCR-2 mediated processing of long dsRNAs, at a step prior to R2D2 (Han et al 2004a, Marques et al 2010).

While only one Dicer is encoded by the human genome, it pairs with three different DRB partners to exert specified functions. For example, DGCR8 is needed for the primary miRNA processing to produce precursor miRNA (Gregory et al 2004, Han et al 2004a). By contrast, both TRBP paralogs and PACT are involved in the subsequent processing of precursor miRNA, as well as the biogenesis of siRNAs in human cells (Chendrimada et al 2005, Lee et al 2006). Further highlighting the importance of DRBs in RNAi processes, two DRBs, PASH-1 and RDE4, have been found to participate in RNAi in *C. elegans*, another excellent model animal that is considered evolutionarily more ancient than *Drosophila* (Tabara et al 2002, Lehrbach et al 2012). PASH-1 is the *C. elegans* ortholog of DGCR8 in human and Pasha in *Drosophila*, and has been shown to be required for miRNA biogenesis (Lehrbach et al 2012). Conversely, RDE-4 is required for the processing of long dsRNA by DCR-1 to produce silencing-mediating siRNAs (Tabara et al 2002, Parker et al 2006, 2008). Although initially thought to contain two dsRBMs, RDE-4 was found to contain a third, more degenerate dsRBM at its C-terminus with *in vitro* dsRNA-binding assays (Parker et al 2008).

Compared with human and *Drosophila* DRBs, which were identified biochemically, both PASH-1 and RDE-4 of *C. elegans* were identified through genetic screens of mutated worm populations for mutants defective in various

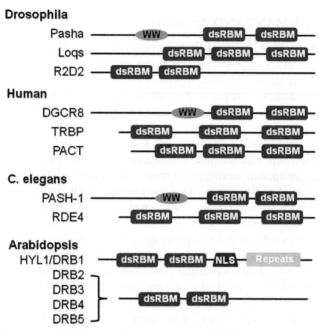

FIGURE 1.2 DRBs in animals and plants. dsRBM: dsRNA-binding motif. WW: a domain with two conserved, consecutive tryptophan residues (Tomari & Zamore 2005). NLS: nuclear localization signal. Repeats: six incomplete repeats of a 28 amino acid region.

RNA silencing pathways. It is remarkable that both approaches succeeded in discovering similar sets of DRB proteins that play critical roles in the RNAi processes. This again highlights the highly conserved nature of the RNA silencing machinery. As we discuss next, similar DRB proteins have also been identified from the model plant *Arabidopsis* using a combination of genetic screening and genomic data mining.

DRBs IN PLANTS

The first evidence for the involvement of plant DRBs in RNA silencing came from the characterization of *hyl1*, an *Arabidopsis* mutant that displayed pleiotropic development and growth defects (Lu & Fedoroff 2000). The *hyl1* plants have narrow rosette and cauline leaves, stunted stature with a reduced hypocotyl elongation rate, delayed flowering, reduced fertility, and aberrant responses to different plant hormones such as abscisic acid, cytokinins and auxins. The HYL1 protein contains two dsRBMs in the N-terminal half, and six near complete repeats of 28 amino acids in the C-terminal half (Fig. 1.2). The expression level of certain endogenous growth and development-related genes was elevated in *hyl1* plants, suggesting that HYL1 is either directly or indirectly involved in the negative regulation of these genes at either the transcriptional or post-transcriptional level. The capability of HYL1 to negatively regulate the transcription of other genes, coupled with broad-spectrum growth and development defects of the mutant plants provided circumstantial evidence for the involvement of HYL1 in RNA silencing-based negative regulation.

In two subsequent studies, HYL1 was implicated in the biogenesis of miRNAs by acting as a partner of DCL1, the plant Dicer for the miRNA pathway (Han et al 2004b, Vazquez et al 2004). Consistent with this conclusion, the mRNAs up-regulated in *hyl1* plants are targets of a number of miRNAs. These studies not only established a definitive role for HYL1 in miRNA biogenesis but also showed that the long dsRNA mediated post-transcriptional gene silencing is unaffected in *hyl1* plants. Separately, *in vitro* experiments by Hiraguri and colleagues (2005) found that HYL1 specifically binds dsRNA, and that its first dsRBM alone was sufficient for the dsRNA binding activity. These *in vitro* experiments also revealed a highly specific interaction between recombinant HYL1 and DCL1 proteins, mediated by both dsRBMs of HYL1. These experiments shed light on the differences between the two seemingly similar dsRBMs in DRBs and paved the way for *in vivo* experiments to determine the biologic significance of these interactions.

The *Arabidopsis* genome encodes five DRBs: HYL1/DRB1, and DRB2 to DRB5 (Hiraguri et al 2005). In addition to HYL1/DRB1, recent reports suggest that DRB2 may play a role in regulating the biogenesis or stability of a subset of miRNAs in shoot apical meristems, and DRB3 and DRB5 may assist DRB2 in meristematic tissues at steps downstream of miRNA biogenesis (Eamens et al 2012a,b). Finally, DRB4 participates in the biogenesis of some families of endogenous tasiRNAs by functioning as a partner of DCL4 (Adenot et al 2006, Nakazawa et al 2007, Curtin et al 2008). It is interesting to note that while disruption of DCL4 function resulted in the loss of the 21-nt tasiRNAs of all families, disruption of DRB4 function is associated with the loss of only tasiRNAs of the *TAS3* family (Adenot et al 2006). These results suggest that other DRB(s) might also act as partners of DCL4. Finally, DRB4 has also been found to function in the production of 21-nt siRNAs from short hairpin RNAs expressed from transgenes (Bouche et al 2006, Deleris et al 2006, Fusaro et al 2006).

STRUCTURAL ANALYSES OF *ARABIDOPSIS* DRB PROTEINS

All five *Arabidopsis* DRBs contain two dsRBMs in their N-terminal halves (Fig. 1.2). The C-terminal half of HYL1/DRB1 contains six near-complete repeats of 28 amino acids, the function of which remains to be resolved as truncated HYL1 with the complete repeat domain deleted was able to fully complement the defect of *hyl1* knockout mutant plants (Hiraguri et al 2005, Han et al 2004b, Wu et al 2007). No conserved motifs could be discerned from the C-terminal regions of DRB2 to DRB5. By contrast, the two N-terminal dsRBMs are very important for the proper function of DRBs, serving as platforms for both protein–protein interaction with DCLs, and protein–dsRNA interactions

(Hiraguri et al 2005, Kurihara et al 2006, Nakazawa et al 2007, Fukudome et al 2011). This was most convincingly demonstrated by Wu and colleagues (2007), who examined the functional requirements of HYL1 by attempting to complement the loss of HYL1 function in the *hyl1* knockout *Arabidopsis* plants by transgenically expressing an array of HYL1 deletion mutants. They found that while a HYL1 variant containing only the two dsRBMs almost completely eliminated the *hyl1* defects, another HYL1 variant containing only the first dsRBM did not. Thus, both HYL1 dsRBMs are needed for its functionality. Conversely, the C-terminal half of HYL1 protein, including the nuclear localization signal (NLS), can be deleted without significantly compromising the function of HYL1 (Wu et al 2007). Consistent with these genetic analyses, the dsRBMs of HYL1 were shown to physically interact with their DCL1, the miRNA Dicer (Hiraguri et al 2005).

DRB4 has been examined in a similar manner to determine its functional requirement. It was shown that the portion of DRB4 containing both dsRBMs binds to dsRNA with high affinity (Hiraguri et al 2005). Also, the DRB4-DCL4 interaction has been demonstrated *in vitro*, and confirmed *in vivo* through co-immunoprecipitation, and was found to correlate with the level of certain tasiRNAs (Hiraguri et al 2005, Nakazawa et al 2007). Adenot and colleagues (2006) first showed that tasiRNA biogenesis is impaired in *drb4* plants. One interesting finding in this study was that, among the three *TAS* genes known at that time, only *TAS3* associated tasiRNA levels were severely diminished by the loss of DRB4 function. Furthermore, the reduction of *TAS3* tasiRNA levels was much less pronounced in the flowers of *drb4* plants. The fact that levels of *TAS1* and *TAS2* associated tasiRNAs were unaffected by the loss of DRB4 function suggests the involvement of other DRBs in tasiRNA biogenesis, although it is also possible that DCL4 does not require the activity of any DRB protein in order to generate tasiRNAs from *TAS1* and *TAS2* loci. These results were somewhat contradicted by another study that showed that the phenotype of *drb4* plants overlaps with that of *dcl4* plants and that indeed *drb4* plants are impaired in tasiRNA accumulation from the *TAS1* and *TAS2* loci (Nakazawa et al 2007). Interestingly, the phenotype of *drb4 dcl4* double mutant plants was more severe than either of the single mutant, and *dcl4* plants had a more penetrating phenotype than *drb4* plants. This is consistent with the idea that other DRBs may act in absence of DRB4 to aid DCL4 to some extent in the tasiRNA biogenesis pathway.

In addition to being involved in the biogenesis of tasiRNAs, DRB4 is required for the processing of other long dsRNAs by DCL4 *in vitro* to produce 21 nt siRNAs (Fukudome et al 2011). It was found that protein extracts of wild-type *Arabidopsis* seedlings, but not of *drb4* or *dcl4* seedlings, were able to process long dsRNA into 21 nt siRNAs. Furthermore, adding recombinant DRB4 or DCL4 proteins into the *drb4* or *dcl4* extracts, respectively, restored long dsRNA processing activity to the extracts. Consistently, mutant DRB4 proteins that lost DCL4-interaction capabilities were unable to complement the *drb4* defect. These results clearly demonstrated that both DRB4 and DCL4 are necessary for dsRNA processing activity.

An interesting recent development is the demonstration of self-interaction of DRB4 (Marrocco et al 2012). These authors found that DRB4 interacts with itself in yeast as well as plant cells, and this interaction could be mapped to the second dsRBM. This interaction has also been independently observed in our laboratory, with the notable difference being that, in our experiments, the entire N-terminal half consisting of both dsRBMs are needed for this interaction, as splitting these two dsRBMs resulted in the complete loss of DRB4 self-interaction (Singh, unpublished data). We further delineated the amino acid residues essential for this interaction by mutating the highly conserved amino acid residues within the two dsRBMs. Mutating both the histidine and lysine residues, at position 32 and 133, respectively, led to a complete loss of DRB4 self-interaction in yeast. Interestingly, these amino acids have been previously shown to be critical for the DRB4-DCL4 interaction. These data suggest that DRB4 likely exists as a dimer in cells and that this dimerization is necessary for its interaction with DCL4 and its function. Alternatively, it is also possible that DRB4 self-interaction plays a unique role in the RNA silencing cascade that is in addition to the role of DRB4-DCL4 interactions.

Subcellular localization of DRB proteins

The intracellular location of proteins is often indicative of the cellular processes they participate in or regulate. Accordingly, there have been several efforts to map the localization of DRB family proteins (Lu & Fedoroff 2000, Hiraguri et al 2005, Wu et al 2007). Owing to the presence of a bipartite NLS, the HYL1 protein is highly nuclear localized. Nuclear localization of HYL1 is consistent with its function in miRNA processing which takes place in the nucleus. Notably, Wu and colleagues (2007) found that the NLS in HYL1 is not required for its function. The dispensability of the NLS could be attributed to the fact that HYL1 exists in complex with DCL1 which is also nuclear localized and hence can ride with DCL1 into the nucleus. Indeed, the N-terminal fragment of HYL1 containing only two dsRBMs (without NLS) fused with GFP failed to localize to the nucleus of bombarded onion epidermal cells (Wu et al 2007). Interestingly, none of the other DRBs contains a NLS. Nevertheless, transiently expressed DRB4 was still found to localize to cell nuclei (Hiraguri et al 2005). The most likely explanation for the nuclear localization of DRB4 could be its partnership with DCL4, which contains a known NLS (Hiraguri et al 2005). Thus, the subcellular localization of DRBs could be modified by the proteins they interact with. In contrast to HYL1 and DRB4, DRB2 is cytoplasmically localized as GFP-DRB2 fusion protein localizes to cytoplasm (Hiraguri et al 2005). The subcellular localization of DRB3 and DRB5 has yet to be determined.

FUNCTIONAL DIVERSIFICATION OF DRBs IN ANIMALS AND PLANTS

It is clear from the studies we reviewed so far that the proper function of all Dicers or DCLs is dependent on one or more DRB partners. Nevertheless, DRB proteins seemed to have continued their adaptive evolution after animals and plants parted their ways. This is evidenced by the fact that, while all plant DRBs contain two dsRBMs, many animal DRBs, including Loqs of *Drosophila*, TRBP and PACT in humans, as well as RDE4 in *C. elegans*, contain three of them. DRBs likely play more pivotal roles in sorting Dicers to different types of dsRNAs in animals, as animal genomes typically encode fewer Dicers than plants. Indeed, both *C. elegans* and human genomes encode only two dsRNA-processing nucleases, with the first one, Drosha, exclusively committed to the processing of primary miRNAs. Thus, the remaining one, Dicer, has to rely on various DRBs to differentiate between precursor miRNAs and fully paired dsRNAs (Tabara et al 2002, Lehrbach et al 2012). Furthermore, biochemical evidence suggests an additional level of complexity in human systems (Chendrimada et al 2005, Lee et al 2006). Both TRBP and PACT are present in the same protein complex with Dicer and human Ago2 (hAgo2). However, they do not appear to be needed for the biogenesis of miRNAs from the partially double-stranded pre-miRNAs. Rather, they appear to participate in the target repression step of miRNA-mediated translational repression. Another twist of TRBP and PACT functionality is that, with fully paired dsRNA, both TRBP and PACT are needed at the siRNA biogenesis step, but are unnecessary for the degradation of siRNA targets (Kok et al 2007).

Pasha is the *Drosophila* ortholog of DGCR8 in humans, which is solely committed to Drosha-mediated primary miRNA processing (Denli et al 2004). On the other hand, Loqs was found to interact with both *Drosophila* Dicers, Dcr-1 and Dcr-2, to ensure the proper processing of precursor miRNAs as well as of dsRNAs (Forstemann et al 2005, Saito et al 2005, Liu et al 2006, Czech et al 2008, Marques et al 2010). Notably, R2D2, the third *Drosophila* DRB, appears to be needed only for bridging Dcr-2 and AGO2 to allow the translocation of siRNAs (Liu et al 2003). These organism-specific differences in functional mechanisms of DRB proteins strongly suggest that DRB proteins have undergone additional adaptive changes more recently.

Plants have evolved to encode several DCL proteins. The small *Arabidopsis* genome has four of them, whereas the rice genome encodes at least six different ones. Nevertheless, plant DRBs have also undergone diversifying evolution as *Arabidopsis* encodes even more DRBs than DCLs. In addition to the obvious functional differences between HYL1 and DRB4 of *Arabidopsis*, a most striking evidence of continuous evolution of DRB proteins lies in HYL1 orthologs of different plant species. While *Arabidopsis* HYL1 and its tomato ortholog share a substantial level of homology at their N-terminal half containing the two dsRBMs, their C-terminal halves differ drastically. In contrast with the *Arabidopsis* HYL1 that contains six nearly perfect repeats of a 28 amino acid section, tomato HYL1 contains four repeats of 24 amino acids, with entirely different sequences. Even more strikingly, no similar repeats can be identified in the rice HYL1 (Qu data not shown). This level of diversity is a strong indication of recent diversifying evolution. Thus, while RNA silencing is considered to be an ancient defense mechanism highly conserved in all eukaryotes, the specific components of various RNA silencing pathways have been clearly under selection pressure.

DRBs AND ANTIVIRAL DEFENSE

Because one of the primary functions of RNA silencing is to defend host cells against virus invasion, it is not surprising to discover that some of the DRBs, both in animals and in plants, participate in antiviral defense. Indeed, Lu and colleagues (2009) showed that *C. elegans* containing a loss-of-function mutation in the RDE4 DRB was much more susceptible to virus infection and accumulated substantially higher levels of viral RNA. Similarly, mutant *Drosophila* containing a loss-of-function mutation in the R2D2 gene also permitted much elevated levels of viral RNA accumulation (Han et al 2011). Interestingly, Loqs of *Drosophila* appears to participate in the production of viral siRNAs without affecting the levels of viral genomic RNAs (Han et al 2011).

In *Arabidopsis*, DRB4 has been implicated in defense against a number of viruses, including a DNA virus. Qu and associates (2008) infected *drb4* mutant plants with a suppressor-less TCV and found that the accumulation of viral RNA was significantly increased in *drb4* plants. However, the level of viral siRNAs was only slightly decreased. This indicates that DRB4 might not be directly involved in the biogenesis of viral siRNAs but can help in their stabilization and subsequent loading into AGOs to constitute the RISC. Nevertheless, the partial involvement of DRB4 in antiviral defense is analogous to its partial involvement in the tasiRNA biogenesis pathway (Adenot et al 2006, Nakazawa et al 2007). An additional study reported that DRB4 is targeted and inactivated by P6 of *Cauliflower mosaic virus*, the suppressor of RNA silencing encoded by this double-stranded DNA virus (Haas et al 2008). These findings clearly demonstrated the importance of DRB4 in antiviral silencing.

More recently, in a further attempt to investigate the antiviral role of DRB4 and DCL4, Jakubiec and colleagues (2012) examined the *drb4* and *dcl4* mutants for their response to TYMV infection. The TYMV-infected *dcl4* plants showed more severe virus symptoms and elevated viral RNA levels with a simultaneous decrease in TYMV-derived siRNAs. Surprisingly, the *drb4* mutant plants did not show increased viral RNA levels despite increased severity in symptoms. Instead, the TYMV-infected *drb4* plants accumulated higher levels viral coat protein (CP) without a simultaneous increase in CP subgenomic RNA. This observation is of significant interest as the RNA silencing-based regulation of viral protein levels has never been reported in plants before. It was further demonstrated that DRB4 interacts with a highly structured (+) RNA region of TYMV known to form a transfer RNA-like structure responsible for translational enhancement (Dreher 2009). More interestingly, a high percentage of TYMV siRNAs was derived from this region. These results suggest that DRB4 might preferentially target viral RNA through translational repression.

CONCLUSION

RNA silencing plays important roles in safeguarding the integrity of plant and animal genomes, maintaining the temporal and spatial orders of developmental progression, conditioning swift responses to environmental stresses, and defending the hosts against viruses and other microbial pathogens. In all organisms studied so far, Dicers or DCLs are key enzymes as they are responsible for processing the silencing-triggering dsRNAs into silencing-guiding siRNAs. Recent studies revealed the importance of DRBs as partners of Dicers or DCLs to function at various steps of the RNA silencing cascade. Despite this general theme, DRBs from different plant and animal species have abundant differences in their structures and functional modes, strongly suggesting continuous adaptive evolution.

In plants, while the role of HYL1 and DRB4 is relatively better understood, the functions of other DRB family members await further investigations. A recent report appears to suggest that DRB2 antagonizes the function of DRB4 in the production of Pol IV-dependent siRNAs in flowers (Pelissier et al 2011). This implies that DRB2 might act in the RdDM pathway to regulate specific genes in flowers and thus modulate growth and reproduction. Other studies implicated DRB2, DRB5, and DRB3 in miRNA biogenesis and/or stability in meristem tissues (Eamens et al 2012a,b). Although most animal DRBs have been found to partner with Dicer to participate in various steps of siRNA/miRNA biogenesis, the DCL partner(s) of *Arabidopsis* DRB2, 3 and 5 have yet to be identified. While HYL1 and DRB4 are known to partner with DCL1 and DCL4, respectively, the function of C-terminal halves, especially the repeat domain of HYL1, are yet to be elucidated. We expect that future investigations of DRB functions in plants will be geared toward biochemical characterization of protein complexes containing various DRBs, which should reveal more mechanistic details of these proteins.

ACKNOWLEDGMENT

We thank Ohio Agricultural Research and Development Center for a SEEDS grant.

REFERENCES

Adenot, X., Elmayan, T., Lauressergues, D., et al., 2006. DRB4-dependent TAS3 trans-acting siRNAs control leaf morphology through AGO7. Current Biology 16, 927–932.
Allen, E., Xie, Z., Gustafson, A.M., Carrington, J.C., 2005. microRNA-directed phasing during trans-acting siRNA biogenesis in plants. Cell 121, 207–221.

Baumberger, N., Baulcombe, D.C., 2005. *Arabidopsis* ARGONAUTE1 is an RNA Slicer that selectively recruits microRNAs and short interfering RNAs. Proceedings of the National Academy of Sciences of the USA 102, 11928–11933.

Borsani, O., Zhu, J., Verslues, P.E., et al., 2005. Endogenous siRNAs derived from a pair of natural cis-antisense transcripts regulate salt tolerance in *Arabidopsis*. Cell 123, 1279–1291.

Bouche, N., Lauressergues, D., Gasciolli, V., Vaucheret, H., 2006. An antagonistic function for Arabidopsis DCL2 in development and a new function for DCL4 in generating viral siRNAs. EMBO J 25, 3347–3356.

Brodersen, P., Voinnet, O., 2006. The diversity of RNA silencing pathways in plants. Trends in Genetics 22, 268–280.

Cao, M., Ye, X., Willi, K., et al., 2010. The capsid protein of Turnip crinkle virus overcomes two separate defense barriers to facilitate systemic movement of the virus in *Arabidopsis*. Journal of Virology 84, 7793–7802.

Chendrimada, T.P., Gregory, R.I., Kumaraswamy, E., Norman, J., Cooch, N., Nishikura, K., Shiekhattar, R., 2005. TRBP recruits the Dicer complex to Ago2 for microRNA processing and gene silencing. Nature 436, 740–744.

Curtin, S.J., Watson, J.M., Smith, N.A., Eamens, A.L., Blanchard, C.L., Waterhouse, P.M., 2008. The roles of plant dsRNA-binding proteins in RNAi-like pathways. FEBS Letters 582, 2753–2760.

Czech, B., Malone1, C.D., Zhou, R., Stark, A., Schlingeheyde, C., Dus, M., Perrimon, N., Kellis, M., Wohlschlegel, J.A., Sachidanandam, R., Hannon, G.J., Brennecke, J., 2008. An endogenous small interfering RNA pathway in *Drosophila*. Nature 453, 798–802.

Deleris, A., Gallego-Bartolome, J., Bao, J., Kasschau, K.D., Carrington, J.C., Voinnet, O., 2006. Hierarchical action and inhibition of plant Dicer-like proteins in antiviral defense. Science 313, 68–71.

Denli, A.M., Tops, B.B.J., Plasterk, R.H.A., Ketting, R.F., Hannon, G.J., 2004. Processing of primary microRNAs by the microprocessor complex. Nature 432, 231–235.

Ding, S. W., 2010. RNA-based antiviral immunity. Nature Reviews in Immunology 10, 632–644.

Dreher, T.W., 2009. Role of tRNA-like structures in controlling plant virus replication. Virus Research 139, 217–229.

Eamens, A.L., Kim, K.W., Curtin, S.J., Waterhouse, P.M., 2012a. DRB2 is required for miRNA biogenesis in *Arabidopsis thaliana*. PLoS One 7, e35933. doi:10.1371/journal.pone.0035933.

Eamens, A.L., Kim, K.W., Waterhouse, P.M., 2012b. DRB2, DRB3 and DRB5 function in a non-canonical microRNA pathway in *Arabidopsis thaliana*. Plant Signaling & Behavior 7, 1224–1229.

Forstemann, K., Tomari, Y., Du, T., Vagin, V.V., Denli, A.M., Bratu, D.P., Klattenhoff, C., Theurkauf, W.E., Zamore, P.D., 2005. Normal microRNA maturation and germ-line stem cell maintenance requires Loquacious, a double-stranded RNA-binding domain protein. PLoS Biology 3, e236.

Fukudome, A., Kanaya, A., Egami, M., Nakazawa, Y., Hiraguri, A., Moriyama, H., Fukuhara, T., 2011. Specific requirement of DRB4, a dsRNA-binding protein, for the in vitro dsRNA-cleaving activity of *Arabidopsis* Dicer-like 4. RNA 17, 750–760.

Fusaro, A.F., Matthew, L., Smith, N.A., Curtin, S.J., Dedic-Hagan, J., Ellacott, G.A., Watson, J.M., Wang, M.B., Brosnan, C., Carroll, B.J., Waterhouse, P.M., 2006. RNA interference-inducing hairpin RNAs in plants act through the viral defence pathway. EMBO Reports 7, 1168–1175.

Gregory, R.I., Yan, K.-P., Amuthan, G., Chendrimada, T., Doratotaj, B., Cooch, N., Shiekhattar, R., 2004. The microprocessor complex mediates the genesis of microRNAs. Nature 432, 235–240.

Haas, G., Azevedo, J., Moissiard, G., Geldreich, A., Himber, C., Bureau, M., Fukuhara, T., Keller, M., Voinnet, O., 2008. Nuclear import of CaMV P6 is required for infection and suppression of the RNA silencing factor DRB4. EMBO Journal 27, 2102–2112.

Hamilton, A.J., Baulcombe, D.C., 1999. A species of small antisense RNA in posttranscriptional gene silencing in plants. Science 286, 950–952.

Hammond, S.M., 2005. Dicing and slicing: the core machinery of the RNA interference pathway. FEBS Letters 579, 5822–5829.

Han, J., Lee, Y., Yeom, K.-H., Kim, Y.-K., Hua, J., Kim, V.N., 2004a. The Drosha–DGCR8 complex in primary microRNA processing. Genes & Development 18, 3016–3027.

Han, M.H., Goud, S., Song, L., Fedoroff, N., 2004b. The *Arabidopsis* double-stranded RNA-binding protein HYL1 plays a role in microRNA-mediated gene regulation. Proceedings of the National Academy of Sciences of the USA 101, 1093–1098.

Han, Y.-H., Luo, Y.-J., Wu, Q., Jovel, J., Wang, X.-H., Aliyari, R., Han, C., Li, W.-X., Ding, S.-W., 2011. RNA-based immunity terminates viral infection in adult *Drosophila* in the absence of viral suppression of RNA interference: characterization of viral small interfering RNA populations in wild-type and mutant flies. Journal of Virology 85, 13153–13163.

Henderson, I.R., Zhang, X., Lu, C., Johnson, L., Meyers, B.C., Green, P.J., Jacobsen, S.E., 2006. Dissecting *Arabidopsis thaliana* DICER function in small RNA processing, gene silencing and DNA methylation patterning. Nature Genetics 38, 721–725.

Herr, A.J., Jensen, M.B., Dalmay, T., Baulcombe, D.C., 2005. RNA polymerase IV directs silencing of endogenous DNA. Science 308, 118–120.

Hiraguri, A., Itoh, R., Kondo, N., Nomura, Y., Aizawa, D., Murai, Y., Koiwa, H., Seki, M., Shinozaki, K., Fukuhara, T., 2005. Specific interactions between Dicer-like proteins and HYL1/DRB-family dsRNA-binding proteins in *Arabidopsis thaliana*. Plant Molecular Biology 57, 173–188.

Hunter, C., Willmann, M.R., Wu, G., Yoshikawa, M., de la Luz Gutierrez-Nava, M., Poethig, S.R., 2006. Trans-acting siRNA-mediated repression of ETTIN and ARF4 regulates heteroblasty in *Arabidopsis*. Development 133, 2973–2981.

Jakubiec, A., Yang, S.W., Chua, N.H., 2012. *Arabidopsis* DRB4 protein in antiviral defense against Turnip yellow mosaic virus infection. Plant Journal 69, 14–25.

Kasschau, K.D., Fahlgren, N., Chapman, E.J., Sullivan, C.M., Cumbie, J.S., Givan, S.A., Carrington, J.C., 2007. Genome-wide profiling and analysis of *Arabidopsis* siRNAs. PLoS Biology 5, e57.

Katiyar-Agarwal, S., Gao, S., Vivian-Smith, A., Jin, H., 2006. A pathogen-inducible endogenous siRNA in plant immunity. Proceedings of the National Academy of Sciences of the USA 103, 18002–18007.

Katiyar-Agarwal, S., Morgan, R., Dahlbeck, D., Borsani, O., Villegas, A., Zhu, J.-K., Staskawicz, B.J., Jin, H., 2007. A novel class of bacteria-induced small RNAs in *Arabidopsis*. Genes & Development 21, 3123–3134.

Kok, K.H., Ng, M.-H.J., Ching, Y.-P., Jin, D.-Y., 2007. Human TRBP and PACT directly interact with each other and associate with Dicer to facilitate the production of small interferin RNA. Journal of Biological Chemistry 282, 17649–17657.

Kurihara, Y., Takashi, Y., Watanabe, Y., 2006. The interaction between DCL1 and HYL1 is important for efficient and precise processing of pri-miRNA in plant microRNA biogenesis. RNA 12, 206–212.

Lee, Y., Hur, I., Park, S.-Y., Kim, Y.-K., Suk, M.R., Kim, V.N., 2006. The role of PACT in the RNA silencing pathway. EMBO Journal 25, 522–532.

Lehrbach, N.J., Castro, C., Murfitt, K.J., Abreu-Goodger, C., Griffin, J.L., Miska, E.A., 2012. Post-developmental microRNA expression is required for normal physiology, and regulates aging in parallel to insulin/IGF-1 signaling in *C. elegans*. RNA 18, 2220–2235.

Liu, Q., Rand, T.A., Kalidas, S., Du, F., Kim, H.E., Smith, D.P., Wang, X., 2003. R2D2, a bridge between the initiation and effector steps of the *Drosophila* RNAi pathway. Science 301, 1921–1925.

Liu, X., Jiang, F., Kalidas, S., Smith, D., Liu, Q., 2006. Dicer-2 and R2D2 coordinately bind siRNA to promote assembly of the siRISC complexes. RNA 12, 1514–1520.

Lu, C., Fedoroff, N., 2000. A mutation in the *Arabidopsis* HYL1 gene encoding a dsRNA binding protein affects responses to abscisic acid, auxin, and cytokinin. Plant Cell 12, 2351–2366.

Lu, R., Yigit, E., Li, W.-X., Ding, S.-W., 2009. An RIG-I-like RNA helicase mediates antiviral RNAi downstream of viral siRNA biogenesis in *Caenorhabditis elegans*. PLoS Pathogens 5, e1000286. doi:10.1371/journal.ppat.1000286.

Marques, J.T., Kim, K., Wu, P.H., Alleyne, T.M., Jafari, N., Carthew, R.W., 2010. Loqs and R2D2 act sequentially in the siRNA pathway in *Drosophila*. Nature Structural and Molecular Biology 17, 24–30.

Marrocco, K., Criqui, M.-C., Zervudacki, J., et al., 2012. APC/C-mediated degradation of dsRNA-binding protein 4 (DRB4) involved in RNA silencing. PLoS ONE 7, e35173.

Montgomery, T.A., Howell, M.D., Cuperus, J.T., Li, D., Hansen, J.E., Alexander, A.L., Chapman, E.J., Fahlgren, N., Allen, E., Carrington, J.C., 2008. Specificity of Argonaute7-miR390 interaction and dual functionality in TAS3 trans-acting siRNA formation. Cell 133, 128–141.

Nakazawa, Y., Hiraguri, A., Moriyama, H., Fukuhara, T., 2007. The dsRNA-binding protein DRB4 interacts with the Dicer-like protein DCL4 *in vivo* and functions in the trans-acting siRNA pathway. Plant Molecular Biology 63, 777–785.

Navarro, L., Dunoyer, P., Jay, F., Arnold, B., Dharmasiri, N., Estelle, M., Voinnet, O., Jones, J.D.G., 2006. A plant miRNA contributes to antibacterial resistance by repressing auxin signaling. Science 312, 436–439.

Navarro, L., Jay, F., Nomura, K., He, S.Y., Voinnet, O., 2008. Suppression of the microRNA pathway by bacterial effector proteins. Science 321, 964–967.

Parker, G.S., Eckert, D.M., Bass, B.L., 2006. RDE-4 preferentially binds long dsRNA and its dimerization is necessary for cleavage of dsRNA to siRNA. RNA 12, 807–818.

Parker, G.S., Maity, T.S., Bass, B.L., 2008. dsRNA binding properties of RDE-4 and TRBP reflect their distinct roles in RNAi. Journal of Molecular Biology 384, 957–979.

Pelissier, T., Clavel, M., Chaparro, C., Pouch-Pelissier, M.N., Vaucheret, H., Deragon, J.M., 2011. Double-stranded RNA binding proteins DRB2 and DRB4 have an antagonistic impact on polymerase IV-dependent siRNA levels in *Arabidopsis*. RNA 17, 1502–1510.

Peragine, A., Yoshikawa, M., Wu, G., Albrecht, H.L., Poethig, R.S., 2004. SGS3 and SGS2/SDE1/RDR6 are required for juvenile development and the production of trans-acting siRNAs in *Arabidopsis*. Genes & Development 18, 2368–2379.

Qi, Y., Denli, A.M., Hannon, G.J., 2005. Biochemical specialization within *Arabidopsis* RNA silencing pathways. Molecular Cell 19, 421–428.

Qu, F., Ye, X., Morris, T.J., 2008. *Arabidopsis* DRB4, AGO1, AGO7, and RDR6 participate in a DCL4-initiated antiviral RNA silencing pathway negatively regulated by DCL1. Proceedings of the National Academy of Sciences of the USA 105, 14732–14737.

Raja, P., Sanville, B.C., Buchmann, R.C., Bisaro, D.M., 2008. Viral genome methylation as an epigenetic defense against geminiviruses. Journal of Virology 82, 8997–9007.

Saito, K., Ishizuka, A., Siomi, H., Siomi, M.C., 2005. Processing of pre-microRNAs by the Dicer-1-Loquacious complex in *Drosophila* cells. PLoS Biology 3, e235.

Sunkar, R., Kapoor, A., Zhu, J.-K., 2006. Posttranscriptional induction of two Cu/Zn superoxide dismutase genes in *Arabidopsis* is mediated by downregulation of miR398 and important for oxidative stress tolerance. Plant Cell 18, 2051–2065.

Tabara, H., Yigit, E., Siomi, H., Mello, C.C., 2002. The dsRNA binding protein RDE-4 interacts with RDE-1, DCR-1, and a DExH-box helicase to direct RNAi in *C. elegans*. Cell 109, 861–871.

Tomari, Y., Zamore, P.D., 2005. Perspective: machines for RNAi. Genes & Development 19, 517–529.

Vaucheret, H., 2006. Post-transcriptional small RNA pathways in plants: mechanisms and regulations. Genes & Development 20, 759–771.

Vazquez, F., Gasciolli, V., Crete, P., Vaucheret, H., 2004. The nuclear dsRNA binding protein HYL1 is required for microRNA accumulation and plant development, but not posttranscriptional transgene silencing. Current Biology 14, 346–351.

Voinnet, O., 2009. Origin, biogenesis, and activity of plant microRNAs. Cell 136, 669–687.

Wierzbicki, A.T., Haag, J.R., Pikaard, C.S., 2008. Noncoding transcription by RNA polymerase PolIVb/PolV mediates transcriptional silencing of overlapping and adjacent genes. Cell 135, 635–648.

Wu, F., Yu, L., Cao, W., Mao, Y., Liu, Z., He, Y., 2007. The N-terminal double-stranded RNA binding domains of *Arabidopsis* HYPONASTIC LEAVES1 are sufficient for pre-microRNA processing. Plant Cell 19, 914–925.

Yoshikawa, M., Peragine, A., Park, M.Y., Poethig, R.C., 2005. A pathway for the biogenesis of trans-acting siRNAs in *Arabidopsis*. Genes & Development 19, 2164–2175.

Yu, B., Yang, Z., Li, J., Minakhina, S., Yang, M., Padgett, R.W., Steward, R., Chen, X., 2005. Methylation as a crucial step in plant microRNA biogenesis. Science 307, 932–935.

Zhang, X., Zhang, X., Singh, J., Li, D., Qu, F., 2012. Temperature-dependent survival of Turnip crinkle virus-infected *Arabidopsis* plants relies on an RNA silencing-based defense that requires DCL2, AGO2, and HEN1. Journal of Virology 86, 6847–6854.

Zheng, X., Zhu, J., Kapoor, A., Zhu, J.K., 2007. Role of *Arabidopsis* AGO6 in siRNA accumulation, DNA methylation and transcriptional gene silencing. EMBO Journal 26, 1691–1701.

Chapter 2

Alteration of host-encoded miRNAs in virus infected plants—experimentally verified

Zhimin Yin, Mirosława Chrzanowska, Krystyna Michalak, and Ewa Zimnoch-Guzowska

Młochów Research Center, Plant Breeding and Acclimatization Institute—National Research Institute, Młochów, Poland

INTRODUCTION

Plant microRNAs (miRNAs) are a near-ubiquitous class of 20–24 nucleotides (nt) long RNA molecules that regulate eukaryotic gene expression post-transcriptionally (Bartel 2004, Voinnet 2009, Chen 2010). Research evidence also indicated the possibility that miRNAs might guide transcriptional silencing in plants (Bao et al 2004). In 2002, efforts to clone small RNAs (sRNAs) from *Arabidopsis* by several groups led to the first discovery of miRNAs from plants (Llave et al 2002, Park et al 2002, Reinhart et al 2002). Up to June 2013, there were 7300 mature miRNA sequences registered for 71 plant species, representing *Coniferophyta*, *Embryophyta* and *Magnoliophyta*, among others (miRBase 2013). miRNAs negatively regulate gene expression by targeting specific messenger RNAs (mRNAs) for cleavage or translational inhibition (Carrington & Ambros 2003, Bartel 2004, Lanet et al 2009). In plants, an *MIR* gene is first transcribed to primary miRNAs (pri-miRNA) by RNA polymerase II; the primary miRNAs are processed by RNAse III-like Dicer-like I endonuclease (DCL1) to generate miRNA precursors (pre-miRNAs) (Schauer et al 2002, Xic et al 2003). Further cleavage of the pre-miRNA by DCLI releases a miRNA/miRNA* duplex. The duplex is then translocated into the cytoplasm by HASTY (Bollman et al 2003), and the canonical mature miRNA of 20–22 nt is selectively incorporated into the RNA-induced silencing complex (RISC) associated with Argonaute 1 (AGO1). In the RISC complex, miRNAs bind to mRNA and inhibit gene expression through perfect or near-perfect complementarity between the miRNA and the mRNA (Bartel 2004).

Note. A list of abbreviations, and their meanings, is given at the end of this chapter.

Plant Virus-Host Interaction.

Plant miRNAs are widely recognized as major players in gene regulation that affect almost all aspects of plants (Jones-Rhoades et al 2006, Chen 2010). They play essential roles in plant growth and development as well as in the regulation of miRNA and small interfering RNA (siRNA) biogenesis and function (Jones-Rhoades et al 2006, Zhang et al 2006a). Besides, miRNAs are also important components in the plant's response to various biotic and abiotic stresses, including infection by viral pathogens (Shukla et al 2008, Sunkar 2010, Khraiwesh et al 2012, Sunkar et al 2012).

RNA silencing is a eukaryotic surveillance mechanism against invasive nucleic acids, including plant viruses (Waterhouse et al 2001). To counteract the antiviral RNA silencing, most plant viruses have evolved silencing suppressor proteins that block one or more steps in the RNA silencing pathway (Voinnet et al 1999, Vaucheret et al 2001). Systemic infection of a plant by a viral pathogen frequently results in symptoms that include a wide range of developmental abnormalities (Hull 2009), in contrast to the normal plant development which is tightly controlled by miRNAs and their targets (Pasquinelli & Ruvkun 2002). Therefore, the basis for virus-induced disease in plants may be explained at least in part by interference with miRNA-controlled developmental pathways that share components with the antiviral RNA silencing pathway (Kasschau et al 2003).

A recent bioinformatics study indicated that plant miRNAs have a strong potential for antiviral activity (Pérez-Quintero et al 2010). Moreover, *in planta,* experimental evidence revealed a number of modifications in miRNA-based gene regulation—such as increased or reduced miRNA accumulation, null or reduced miRNA-guided cleavage and subsequent alteration of host target mRNA in virus-infected plants (Bazzini et al 2007, Cillo et al 2009, Du et al 2011). This chapter summarizes the experimentally evidenced changes in host-encoded miRNAs and their targets following viral infection in a susceptible plant, as well as the interaction between plant miRNA, target mRNA, and the viral silencing suppressor.

miRNAs RESPONSIVE TO VIRUS INFECTION

Although a common set of host miRNAs is induced by diverse RNA or DNA viruses in infected plants, some others exhibited virus-, strain-, plant-, or tissue-specific expression. Novel miRNAs and certain miRNA*s (antisense miRNAs) were identified in virus-infected plants. However, currently, no miRNAs have been identified in plant virus genomes. miRNAs responsive to virus infection in diverse plant species are summarized in Table 2.1.

miRNAs commonly induced

Expression levels of many miRNAs are altered by diverse DNA or RNA viruses in different plant species (Table 2.1). Out of them, induction of miR168 is ubiquitous in plant–virus interactions (Várallyay et al 2010, Lang et al 2011a; Table 2.1).

TABLE 2.1 Experimentally validated miRNAs responsive to virus infection in diverse plant species

Infection by *Begomovirus, Phytoreovirus, Tenuivirus, Polerovirus*

| | DNA virus | | | | | RNA virus | | | |
| | *Begomovirus* | | | | | *Phytoreovirus* | *Tenuivirus* | *Polerovirus* | |
miRNA	ACMV	CblCuV	TYLCV	CLCuMV	ToLCNDV	RDV	RSV	CLRDV	References[a]
miR156	Nbe↓	Nbe↓	Nbe↑	Nbe↑	nd	nd	Osa↓	nd	[6,8]
miR159	Nbe↑	Nbe↑	Nbe↑	Nbe↑	Sly↑	nd	Osa↓	nd	[6,8,14]
miR159*	nd	nd	nd	nd	nd	nd	Osa↑	nd	[14]
miR319	nd	nd	nd	nd	Sly↑	nd	nd	nd	[14]
miR160	Nbe↓	Nbe↑	Nbe↑	Nbe↑	Sly↓	nd	Osa−	nd	[6,8,14]
miR160*	nd	nd	nd	nd	nd	nd	Osa↑	nd	
miR162	nd	nd	nd	nd	Sly↑	nd	nd	Ghi↑	[5,14]
miR164	Nbe↑	Nbe↑	Nbe↑	Nbe↑	Sly↑	nd	Osa↓	nd	[6,8,14]
miR165	Nbe↑	Nbe↑	Nbe↑	Nbe↑	nd	nd	nd	nd	[6]
miR166	Nbe↑	Nbe↑	Nbe↑	Nbe↑	nd	nd	Osa↑	nd	[6,8]
miR166*	nd	nd	nd	nd	nd	nd	Osa↑	nd	
miR167	Nbe↑	Nbe↑	Nbe↑	Nbe↑	nd	Osa↓	Osa↓	nd	
miR167*	nd	nd	nd	nd	nd	nd	Osa↑	nd	

Continued

TABLE 2.1 Experimentally validated miRNAs responsive to virus infection in diverse plant species—cont'd

Infection by Begomovirus, Phytoreovirus, Tenuivirus, Polerovirus

miRNA	DNA virus Begomovirus					RNA virus Phytoreovirus	Tenuivirus	Polerovirus	References[a]
	ACMV	CbLCuV	TYLCV	CLCuMV	ToLCNDV	RDV	RSV	CLRDV	
miR168	Nbe↑	Nbe↑	Nbe↑	Nbe↑	Sly↑	Osa-	Osa↑	nd	[6,8,14]
miR168*	nd	nd	nd	nd	nd	nd	Osa↑	nd	
miR169	Nbe↓	Nbe↑	Nbe↑	Nbe↑	Sly↓	nd	nd	nd	[6,14]
miR171	nd	nd	nd	nd	Sly↓	Osa↓	Osa↑/-	nd	[8,14]
miR171*	nd	nd	nd	nd	nd	nd	Osa↑	nd	
miR172	nd	nd	nd	nd	Sly↑	Osa-	Osa-	nd	
miR172*	nd	nd	nd	nd	nd	nd	Osa↑	nd	
miR390	nd	nd	nd	nd	nd	nd	Osa↑	nd	[8]
miR390*	nd	nd	nd	nd	nd	nd	Osa↑	nd	
miR391	nd	nd	nd	nd	Sly↓	nd	nd	nd	[14]
miR393	nd	nd	nd	nd	nd	nd	Osa↓	nd	[8]

Continued

miRNA									
miR396	nd	nd	nd	nd	SlyT	nd	OsaL	nd	[8,14]
miR396*	nd	nd	nd	nd	nd	nd	OsaT	nd	
miR397	nd	nd	nd	nd	SlyT	nd	nd	nd	[14]
miR398	nd	nd	nd	nd	SlyT	nd	nd	nd	
miR408	nd	nd	nd	nd	SlyT	nd	nd	nd	
miR444*	nd	nd	nd	nd	nd	nd	OsaT	nd	[8]
miR447	nd	nd	nd	nd	SlyT	nd	nd	nd	[14]
miR528	nd	nd	nd	nd	nd	nd	OsaL	nd	[8]
miR528*	nd	nd	nd	nd	nd	nd	OsaT	nd	
miR535	nd	nd	nd	nd	nd	nd	OsaT	nd	
miR1318*	nd	nd	nd	nd	nd	nd	OsaT	nd	
miR1425	nd	nd	nd	nd	nd	nd	Osa-	nd	
miR1425*	nd	nd	nd	nd	nd	nd	OsaT	nd	
miR1863	nd	nd	nd	nd	nd	nd	OsaL	nd	
miR1884	nd	nd	nd	nd	nd	nd	OsaL	nd	

TABLE 2.1 Experimentally validated miRNAs responsive to virus infection in diverse plant species—cont'd

Infection by Tobamovirus, Potyvirus, Cucumovirus, Potexvirus

| | RNA virus | | | | | | | | | | |
| | Tobamovirus | | | Potyvirus | | | | Cucumovirus | | Potexvirus | |
miRNA	TMV	ToMV	ORMV	TEV	PVY	ZYMV	MV	CMV	TAV	PVX	References
miR156	Nta↓/↑Ath↑	Nta↑	nd	Nta↑	Nta↓Nbe↑	Cpe↑	nd	Nta-Syl↑&-	nd	Nta↓Nbe↑	[1,2,3,6,7 9,13,15]
miR156*	nd	nd	nd	nd	nd	Cpe↑	nd	nd	nd	nd	
miR157	Nta↓/↑Ath↑	Nta↑	nd	Nta↑	Nta↓	Cpe↑	nd	Nta-Sly-	nd	Nta↑	[1,9,13,16]
miR157*	nd	nd	nd	nd	nd	Cpe↑	nd	nd	nd	nd	
miR158	Bra-Ath↑	nd	nd	nd	nd		Bra↑	Bra-	nd	nd	[4,15]
miR159	Nta↓/↑Bra-Ath↑	Nta↑	nd	Nta↑	Nta↑	Cpe↑	Bra↑	Nta↓Bra-Sly↑/-	Sly↑	Nta↓Nbe↑	[1,2,3,4,6,9,10,13,15]
miR159*	Bra-	nd	nd	nd	nd	Cpe↑	Bra↑	Bra-	nd	nd	
miR319	Nta↓/↑	Nta↑	nd	Nta↑	Nta↑	nd	nd	Nta↓	nd	Nta↑	[1,2]
miR160	Nta↑Ath-	Nta↑	nd	Nta↑	Nta↓	nd	nd	Nta↓Sly↑	Sly↑	Nta↑/-Nbe↓	[3,6,9,11,15]
miR160*	nd	nd	nd	nd	nd	Cpe↑	nd	nd	nd	nd	
miR161	Nta↑	nd	nd	nd	nd	nd	nd	nd	nd	nd	[15]
miR162	nd	nd	nd	nd	nd	nd	nd	Nta-Sly↑&-	Sly↑	Nta-	[3,10,11,12,13]
miR162*	nd	nd	nd	nd	nd	nd	nd	Sly↑	nd	nd	
miR163	Ath↑	nd	nd	nd	nd	nd	nd	nd	nd	nd	[15]

Continued

miRNA											
miR164	Nta↓/↑Ath↑	Nta↑	nd	Nta↑	Ntal	nd	nd	Nta↑Sly↑	Sly↑	Nta↑Nbe↑	[1,2,3,6,10,11,12,13,15]
miR164*	nd	nd	nd	nd	nd	nd	nd	Sly↑	nd	nd	
miR165	Nta↓/↑	Nta↑	nd	Nta↑	Nta-	Cpe↑	nd	Nta-Sly↑	Sly↑	Nta↑/-Nbe↑	[1,2,3,6,9,10,11,12,13]
mR165*	nd	nd	nd	nd	nd	Cepa↑	nd	Sly↑	nd	nd	
miR166	Nta↓/↑Nta↑	Nta↑	Ath↑	Nta↑	Nta-/↑	Cpe↑	nd	Nta-Sly↑	Sly↑	Nta↑/-Nbe↑	[1,2,3,6,9,11,13,15,16]
miR166*	nd	nd	Ath↑	nd	nd	Cepa↑	nd	Sly↑	nd	nd	
miR167	Nta↓/↑Ath↑	Nta↑	nd	Nta↑	Nta↑	Cpe↑	nd	Nta-Sly↑	Sly↑	Nta↑/-Nbe↑	[1,2,3,6,9,10,11,13,15]
miR167*	nd	nd	nd	nd	nd	Cepa↑	nd	nd	nd	nd	
miR168	Nta↓Ath↑	nd	Ath↑	nd	Nbe↑	Cpe↑	nd	Nta↑Sly↑	Sly↑	Nta↑Nbe↑	[1,2,3,6,7,9,10,11,12,13,15,16]
miR168*	nd	nd	Ath↑	nd	nd	Cpe↑	nd	Sly↑	nd	nd	
miR169	Nta↓/↑Ath↑	Nta↑	nd	Nta↑	Ntal	nd	nd	Nta-Sly↑	Sly↑	Nta↑Nbe↑	[1,2,3,6,11,15]
miR171	Nta↓/↑Ath↑	Nta↑	nd	Nta↑	Nta↑Nbe↑	Cpe↑	nd	Nta-Sly↑&-	Sly↑	Nta↑/-Nbe↑	[1,2,3,7,9,10,13,15]
miR171*	Nta↑	Nta↑	nd	Nta↑	Nta↑	Cpe↑	nd	Sly↑	nd	Nta↑	
miR172	Nta↓Ath↑	nd	nd	nd	nd	nd	nd	nd	nd	nd	[1,2,15]
miR173	nd	nd	Ath↑	nd	nd	nd	nd	nd	nd	nd	[16]
miR173*	nd	nd	Ath↑	nd	nd	nd	nd	nd	nd	nd	
miR319	Ath↑	nd	nd	nd	nd	nd	nd	Sly↑&-	nd	nd	[12,13,15]
miR319*	nd	nd	nd	nd	nd	nd	nd	Sly↑	nd	nd	

TABLE 2.1 Experimentally validated miRNAs responsive to virus infection in diverse plant species—cont'd Infection by *Tobamovirus, Potyvirus, Cucumovirus, Potexvirus*

	RNA virus										
	Tobamovirus			Potyvirus				Cucumovirus		Potexvirus	
miRNA	TMV	ToMV	ORMV	TEV	PVY	ZYMV	MV	CMV	TAV	PVX	References
miR390	Nta*Atht	nd	nd	nd	nd	Cpef	nd	Syl^	nd	nd	[1,9,13,15]
miR390*	nd	nd	nd	nd	nd	Cpef	nd	nd	nd	nd	
miR391	NtatAtht	nd	nd	nd	nd	nd	nd	nd	nd	nd	[1,15]
miR395	Nta^	nd	nd	nd	nd	nd	nd	nd	nd	nd	[1]
miR396	Atht	nd	nd	nd	nd	Cpef	nd	Sly-	nd	nd	[9,13,15]
miR396*	nd	nd	nd	nd	nd	Cpef	nd	Slyt	nd	nd	
miR397	Nta˅	nd	nd	nd	nd	nd	nd	nd	nd	nd	[1]
miR398	Nta*Atht	nd	Atht	nd	Nbef	nd	nd	nd	nd	Nbef	[1,7,15,16]
miR398*	nd	nd	Atht	nd	nd	nd	nd	nd	nd	nd	
miR399	Ntat	nd	nd	nd	nd	nd	nd	nd	nd	nd	[1]
miR403	Ntal	nd	nd	nd	nd	nd	nd	Sly-	nd	nd	[1,13]

miRNA									
miR408	Nta↑	nd	nd	nd	nd	nd	nd	nd	[1]
miR415	Nta↓	nd	nd	nd	nd	nd	nd	nd	
miR472	Nta↑	nd	nd	nd	nd	nd	nd	nd	
miR474	Ntal	nd	nd	nd	nd	nd	nd	nd	
miR535	Nta↓	nd	nd	nd	nd	nd	nd	nd	
miR822	Atht	nd	nd	nd	nd	nd	nd	nd	[15]
miR823	Atht	nd	nd	nd	nd	nd	nd	nd	
miR824	Atht	nd	nd	nd	nd	nd	nd	nd	
miR1885	Bra-	nd	nd	Braf	Bra-	nd	nd	nd	[4]
miR1885* (AC189642)	nd	nd	nd	Braf	nd	nd	nd	nd	
miR1885* (AM391091)	nd	nd	nd	Bra-	nd	nd	nd	nd	

↑: up-regulated. ↓: down-regulated. ↑↓: initially up-regulated then reduced. ↓↑: initially down-regulated then up-regulated.
-: no change. ↑&↓ (±): some members were up-regulated, some were down-regulated (or no changes. nd: not determined.
↑/↓ (±): the same miRNA (family) was up-regulated or down-regulated (or no change) depending on the results published by different research groups.
Ath: *Arabidopsis thaliana* L. Bra: *Brassica rapa* L. Cpe: *Cucubita pepo* L. Ghi: *Gossypium hirsutum* L. Nbe: *Nicotiana benthamiana* L. Nta: *Nicotiana tabacum* L. Osa: *Oryza sativa* L. Sly: *Solanum lycopersicum* L.
ACMV: *African cassava mosaic virus*. CbLCuV: *Cabbage leaf curl virus*. CLCuMV: *Cotton leaf curl Multan virus*. CLRDV: *Cotton leafroll dwarf virus*. CMV: *Cucumber mosaic virus*. ORMV: *Oilseed rape mosaic virus*. PVX: *Potato virus X*. PVY: *Potato virus Y*. RDV: *Rice dwarf virus*. RSV: *Rice stripe virus*. TAV: *Tomato aspermy virus*. TEV: *Tobacco etch virus*. TMV: *Tobacco mosaic virus*. ToLCNDV: *Tomato leaf curl New Delhi virus*. ToMV: *Tomato mosaic virus*. TuMV: *Turnip mosaic virus*. TYLCV: *Tomato yellow leaf curl virus*. ZYMV: *Zucchini yellow mosaic virus*.

ªReferences. 1 (Bazzini et al 2011); 2 (Bazzini et al 2007); 3 (Lang et al 2011a); 4 (He et al 2008); 5 (Silva et al 2011); 6 (Amin et al 2011); 7 (Pacheco et al 2012); 8 (Du et al 2011); 9 (Shiboleth et al 2007); 10 (Feng et al 2009); 11 (Feng et al 2012); 12 (Cillo et al 2009); 13 (Lang et al 2011b); 14 (Naqvi et al 2010); 15 (Tagami et al 2007); 16 (Hu et al 2011).

Microarray profiling of the tomato miRNAs responsive to *Cucumber mosaic virus* (CMV) infection revealed temporal expression patterns (Lang et al 2011b). For example, the miR168 expression levels remained unaffected during 15 days post-inoculation (dpi), but were up-regulated at 20 dpi. The miR390, miR164, miR165 and miR166 levels showed little differences at 7 dpi, but exhibited more strongly induced expression at 14 dpi, while the expression levels were subsequently reduced at 20 dpi. Members of the miR159 family exhibited relatively constant expression levels up to 7 dpi, a sharp decrease at 14 dpi, and then a drastic increase at 20 dpi. The miR167 family showed lowered expression up to 7 dpi, and clear up-regulation at 14 and 20 dpi.

In *Tobacco mosaic virus* (TMV)-infected tobacco plants, a temporal alteration in the level of miRNAs showed two clear distinct stages (Bazzini et al 2011). In particular, miR415, miR156/7, miR390, miR398, miR168, miR167, miR171, miR397, miR535, miR165/6 and miR160 which form 'cluster A' were down-regulated at an early stage (5 dpi, no virus detectable) and up-regulated at a later stage (15 and/or 22 dpi, high level of viral accumulation) of infection (Bazzini et al 2011). The cluster A miRNAs were previously described as responsive to different biotic and abiotic stresses (Liu et al 2008, Trindade et al 2010, Khraiwesh et al 2012). However, the expression of another set of nine miRNAs, miR403, miR159/319, miR474, miR472, miR395, miR169, miR399, miR391 and miR408, was also altered, although without showing the biphasic trend (Bazzini et al 2011).

miRNAs induced by specific viruses

Some miRNAs may be induced by certain viruses, but not by others. Moreover, the same set of miRNAs may be differentially regulated upon infection by different viruses. In *Brassica napus* and *Brassica rapa*, miR158 and miR1885 were induced by *Turnip mosaic virus* (TuMV) but not by TMV or CMV (He et al 2008). In *Nicotiana benthamiana*, miR156 was down-regulated by the bipartite DNA begomoviruses *African cassava mosaic virus* (ACMV) and *Cabbage leaf curl virus* (CbLCuV), but up-regulated by the monopartite DNA begomoviruses *Cotton leaf curl Multan virus* (CLCuMV) and *Tomato yellow leaf curl virus* (TYLCV) and the RNA virus *Potato virus X* (PVX) (Amin et al 2011). CLCuMV, CbLCuV or TYLCV infection resulted in a slight increase, while ACMV and PVX infection resulted in down-regulation of miR160 (Amin et al 2011). In tobacco plants, miR159 and miR165/166 were inhibited by CMV but induced by PVX (Lang et al 2011a). In tomato, expression of miR168 was significantly induced by CMV at 10 dpi, whereas *Tomato aspermy virus* (TAV) infection induced miR162 only at 30 dpi (Feng et al 2009). Infection by *Rice stripe virus* (RSV), but not *Rice dwarf virus* (RDV), led to altered expression of selective rice miRNAs (Du et al 2011).

miRNA levels altered by specific virus strains

In CMV-infected tomato, the expression of tested miRNAs was altered with evident strain-specific differences (Cillo et al 2009, Feng et al 2012). Infection of tomato with the severe subgroup IA strain, CMV-Fny, severely altered leaf morphogenesis, inducing the typical reduction of leaflet blade size and of whole-plant growth. Infection of tomato with the mild subgroup II strain, CMV-LS, induced a moderate reduction of leaf blade size not associated with decrease of plant size. The accumulation levels of miR159, miR160, miR162, miR164, miR165/166, miR167, miR168, miR169 and miR319 were significantly enhanced by infection with the severe strain CMV-Fny compared to the mild strain CMV-LS.

Species- or tissue-specific induction of miRNAs upon viral infection

miRNAs may respond to the same virus in a different way depending on plant species or plant tissue. miR158 and miR1885 were greatly induced by TuMV in *B. rapa* but not in *Arabidopsis* (He et al 2008). In rice, miR41 and miR47 were increased in leaves and flowers of *Rice tungro* infected plants, with higher levels observed in flowers than in leaves; whereas miR29 was expressed only at a detectable level in both infected tissues (Sanan-Mishra et al 2009). In tomato, *Tomato leaf curl New Delhi virus* (ToLCNDV) induced the over-expression of miR159 and miR172 in leaves but not in flowers (Naqvi et al 2010).

miRNA*s

Some miRNA*s (antisense miRNAs) were detected only in virus-infected plants. In CMV-infected tomato, some members of miR168*, miR396* and miR166* were detected only in CMV-infected plants at 20 dpi, but not in the mock-infected and the 3, 7 and 14 dpi of CMV-infected ones (Moxon et al 2008, Lang et al 2011b).

In virus-infected plants, such as those infected with RSV, ZYMV or TuMV, the levels of many miRNA*s were higher in comparison to their corresponding miRNAs. These include miRNA*s for members of miR156*, miR157*, miR159*, miR160*, miR165*, miR166*, miR167*, miR168*, miR171*, miR172*, miR390*, miR396*, miR444*, miR528*, miR1318*, miR1425* and miR1885* (Shiboleth et al 2007, He et al 2008, Du et al 2011). In *Oilseed rape mosaic virus* (ORMV)-infected *Arabidopsis* plants, the relative amount of miRNA*s increased to 14% compared to that in mock infection (2%) (Hu et al 2011). Moreover, the miRNA* levels in severe strain ZYMV[FRNK]-infected squash plants were higher than those of attenuated strain ZYMV[FINK]-infected plants, and the average differences ranged from less than 2-fold to 20-fold (Shiboleth et al 2007).

miRNA*s, the passenger strands of mature miRNAs, are normally labile and degraded (Mallory et al 2002, Baumberger & Baulcombe 2005, Yu et al 2005). The detection of miRNA*s, on the other hand, may indicate dysfunction of the normal miRNA processing (Lang et al 2011b). Such observation cannot solely be explained by the ability of viral silencing suppressor to bind miRNA duplexes because this should lead to equal accumulation of both strands of the duplexes (Hu et al 2011). More complex biologic processes or some unknown mechanisms might be involved (Shiboleth et al 2007, He et al 2008, Du et al 2011, Hu et al 2011, Lang et al 2011b).

Novel miRNAs in virus infected plants

Virus-infected plants are good sources for identifying novel miRNAs (Tagami et al 2007). Some novel miRNAs were identified only in the virus-infected plants, e.g. plants infected by TMV, TuMV or RSV. The newly identified miR847(5′), miR822, miR823 and miR824 were more abundant in TMV-infected *Arabidopsis* than in mock infected plants (Tagami et al 2007). In *B. rapa*, the novel miR158 and miR1885 were greatly induced by TuMV infection, but not by CMV or TMV (He et al 2008). Studies of non-conserved *Arabidopsis* miRNAs support the 'inverted duplication' model of miRNA evolution, which proposes that *de novo* generation of miRNA genes arises from inverted duplications of their target genes (Kasschau et al 2003, Fahlgren et al 2007). BLASTN searches against the TAIR *Arabidopsis* gene database revealed that the novel *Brassica* miR158 and miR1885 may target the same family of the disease-resistant *R* proteins encoded by *Arabidopsis* mRNAs (He et al 2008). Both tandem gene duplications and ectopic duplication were found in *Arabidopsis R* genes (Richly et al 2002, Meyers et al 2003, Leister 2004). He et al (2008) speculated that the two hairpin-forming sequences were probably transitional loci derived from *Brassica* protein-coding genes homologous to *R* genes in *Arabidopsis*. One or more of these small RNAs might evolve into true and new miRNA(s) if their silencing activities were advantageous.

In rice, RSV infection induced expression of new phased miRNAs from conserved precursors, such as three-duplex phase forms of miRNA/miRNA* produced from precursors of miR159 and two-duplex phase forms derived from the precursors of miR394 (Du et al 2011). The new miRNA/miRNA* pairs were often detected at higher levels than the originally reported pairs. The finding that new phased miRNAs are induced during the infection of a plant virus significantly broadens the landscape of phased miRNA biogenesis during pathogen infection (Du et al 2011). The authors suggested that at least some viruses might have evolved mechanisms to induce expression of phased miRNAs from well-conserved cellular miRNA precursors.

miRNA precursors could produce additional sRNAs in *Arabidopsis* upon virus infection (Hu et al 2011). Some miRNA-like sRNAs (ml-sRNAs), which are rare or absent in non-infected *Arabidopsis*, accumulate to high levels upon

ORMV infection (Hu et al 2011). These miRNA precursor-derived ml-sRNAs are often arranged in phase and form duplexes (Zhang et al 2010, Hu et al 2011). Their production depends on the same biogenesis pathway as their sibling miRNAs (Zhang et al 2010). However, these ml-sRNAs are different from the new phased miRNA sequences reported by Du et al (2011). It appears likely that the ml-siRNAs accumulating in ORMV-infected *Arabidopsis* are caused by stabilization by VSR or by a yet unknown virus-induced effector complex (Hu et al 2011). The pool of miRNAs seems to be larger than was previously recognized, and miRNA-mediated gene regulation may be broader and more complex than previously thought (Zhang et al 2010).

miRNAs in synergistic infection

In comparison to a single virus infection, co-infection often results in increased systemic symptoms (synergism). Pacheco et al (2012) demonstrated that double infection by PVX and *Potato virus Y* (PVY) or *Plum pox virus* (PPV), that produced the most severe symptoms in *N. benthamiana*, alerted accumulation of miR156, miR171, miR398 and miR168 to a greater extent or in a different direction than the single infections that produced milder symptoms. This finding indicated a differential effect on miRNA metabolism of the combined infection by two unrelated plant viruses, which may account in part for the severe symptoms caused by PVX/potyvirus-associated synergisms.

Virus-encoded miRNAs

Virus-encoded miRNAs were first identified in the *Epstein–Barr virus* (EBV), a causative agent of infectious mononucleosis (Pfeffer et al 2004). Since then, virus-encoded miRNAs were found in at least 12 additional mammalian viruses (Sarnow et al 2006, Griffiths-Jones et al 2008). So far, no miRNAs have been identified in plant virus genomes (Lu et al 2008), although siRNAs derived from some plant viruses such as *Tobacco rattle virus* (TRV), CMV, geminiviruses or caulimoviruses had been characterized (Akbergenov et al 2006, Blevins et al 2006, Diaz-Pendon et al 2007). More recently, using next-generation sequencing platforms, the virus-specific siRNAs produced by a variety of plant viruses were profiled (Donaire et al 2009, Szittya et al 2010, Wang et al 2010). These viruses included *Cymbidium ringspot virus* (CymRSV), CMV, PVX, *Melon necrotic spot carmovirus* (MNSV), *Tobacco rattle virus* (TRV), *Pepper mild mottle tobamovirus* (PMMoV), *Watermelon mosaic potyvirus*, TuMV and *Tomato yellow leaf curl begomovirus*, a DNA virus.

EFFECT OF VIRAL INFECTION ON mRNA TARGETS OF miRNAs

Plant miRNAs perfectly or near perfectly complement their mRNA targets, leading to target cleavage and degradation (Bartel 2004). Thus, the miRNAs and their targets essentially show mutually antagonistic expression levels in a virus-infected plant (Naqvi et al 2010). In some cases, with unknown mechanisms, a parallel increase in expression of target mRNAs and the corresponding miRNA species was observed in plants infected by an RNA virus or occasionally by a DNA virus. In addition, regulation of some target mRNAs was virus-, strain- or tissue-specific. The miRNA-targets responsive to virus infection in diverse plant species are summarized in Table 2.2. The potential function of miRNA-targets responsive to virus infection is shown in Table 2.3.

Mutually antagonistic expression of mRNA targets and the corresponding miRNAs

In virus-infected plants, the up-regulated miRNAs were functional in down-regulating the respective mRNA targets, and *vice versa*. In tomato plants infected with the DNA begomovirus ToLCNDV, the enhanced expression of certain miRNAs resulted in significantly reduced expression of the corresponding mRNA targets in agroinfected leaves (Naqvi et al 2010). These altered mRNA targets included *AGO1* and *DCL1* involved in the regulation of miRNA pathways, affecting the transcription factor (TF) targets Apetala 2 (*AP2*), *MYB33* and lanceolate (*LA*), and the non-TF targets copper superoxide dismutase (*CSD*) and ubiquitin activating enzyme (*UAE1*) (Naqvi et al 2010). The other TF targets, Scarecrow-like TF (*SCL6*), NAC domain TF (*NAM*) and *CBF*, showed an expected increase in the infected leaves while the corresponding miRNAs decreased (Naqvi et al 2010). Tobacco plants infected with the RNA virus, *Sunn-hemp mosaic virus* (ShMV) or with TMV, showed a significant increase in HD-ZIP III homeobox 8 (*ATHB-8*) and *SCL6* mRNA, which negatively correlated with the reduction of miR165/6 and miR171 levels (Bazzini et al 2011). In RSV-infected rice plants, *HD-ZIP* and auxin response factor (*ARF8*) were up-regulated, in agreement with the down-regulation of miR166 and miR167 (Du et al 2011).

Parallel increase in expression of mRNA targets and the corresponding miRNAs

Unlike the mutually antagonistic expression, a parallel increase in expression of target mRNAs and the corresponding miRNA species was also observed in plants infected by an RNA virus or occasionally by a DNA virus, particularly at the later stage of infection. Enhanced mRNA levels of *AGO1* and *DCL1,* along with the corresponding miR168 and miR162, were observed in CMV- or TAV-infected tomato (Cillo et al 2009, Feng et al 2009, 2012). Várallyay et al (2010) demonstrated that induction of miR168 is ubiquitous in plant–virus interactions, and the increased miR168 accumulation is accompanied by *AGO1* mRNA induction. Enhanced expression of some TF targets together with the corresponding miRNAs were observed

TABLE 2.2 Experimental validated miRNA-targets responsive to virus infection in diverse plant species[a]

miRNA and targets[b]	DNA virus	RNA virus											References[c]
	Begomovirus	Tobamovirus			Potyvirus		Polerovirus	Potexvirus	Phytoreovirus	Tenuivirus	Cucumovirus		
	ToLC-NDV												
	NDV	TMV	ShMV	ORMV	TuMV	PVY	CLRDV	PVX	RDV	RSV	CMV	TAV	
miR156/7	ns	Nta↑	Nta-	Ath↑	ns	Nbe↑	ns	Nbe↑	ns	ns	ns	ns	[1,7,16]
SPL	nd	**Nta-**	**Nta-**	**Ath↑**	nd	**Nbe↑**	nd	**Nbe-**	nd	nd	nd	nd	
miR159	Sly↑	nd	nd	nd	nd	nd	nd	nd	nd	nd	Sly↑	Sly↑	[10,11,14]
MYB	**Sly↓**	nd	nd	nd	nd	nd	nd	nd	nd	nd	**Sly↑**	**Sly↑**	
miR160	ns	ns	ns	Ath↑	ns	ns	ns	ns	ns	Osa-	ns	ns	[8,16]
miR160*	ns	ns	ns	ns	ns	ns	ns	ns	ns	Osa↑	ns	ns	
Os11g38-140	nd	nd	nd	nd	nd	nd	nd	nd	nd	**Osa↓**	nd	nd	
Os02g49-240	nd	nd	nd	nd	nd	nd	nd	nd	nd	**Osa↓**	nd	nd	
ARF	nd	nd	nd	**Ath↑**	nd	nd	nd	nd	nd	nd	nd	nd	

Continued

TABLE 2.2 Experimental validated miRNA-targets responsive to virus infection in diverse plant species[a]—cont'd

| | DNA virus | RNA virus | | | | | | | | | | | |
| | Begomovirus | Tobamovirus | | | Potyvirus | | Polerovirus | Potexvirus | Phytoreovirus | Tenuivirus | Cucumovirus | | |
miRNA and targets[b]	ToLC-NDV	TMV	ShMV	ORMV	TuMV	PVY	CLRDV	PVX	RDV	RSV	CMV	TAV	References[c]
miR161 miR400	ns	ns	ns	Ath↑	ns	ns	ns	ns	ns	ns	ns	ns	[16]
PPR	nd	nd	nd	**Ath↑**	nd	nd	nd	nd	nd	nd	nd	nd	
miR162	Sly↑	ns	ns	ns	ns	ns	Ghi↑	ns	Osa-	Osa-	Sly↑	Sly↑	[5,11,12,14]
miR162*	ns	ns	ns	ns	ns	ns	ns	ns	nd	nd	Sly↑	ns	
DCL1	**Sly↓**	nd	nd	nd	nd	nd	**Ghi-**	nd	**Osa-**	**Osa-**	**Sly↑**	**Sly↑**	
miR164	Sly↓	ns	ns	ns	ns	ns	ns	ns	ns	ns	Sly↑	Sly↑	[11,12,14]
miR164*	ns	ns	ns	ns	ns	ns	ns	ns	ns	ns	Sly↑/-	ns	
NAC1	nd	nd	nd	nd	nd	nd	nd	nd	nd	nd	**Sly↑**	**Sly↑**	
NAM	**Sly↑**	nd	nd	nd	nd	nd	nd	nd	nd	nd	nd	nd	

Continued

miR165/6	ns	_Nta_↓	_Nta_↓	ns	ns	ns	ns	ns	_Osa_-	_Osa_↓	_Sly_↑	_Sly_↑	[1,8,11,12]
miR165*	ns	ns	ns	ns	ns	ns	ns	ns	ns	ns	_Sly_↑	ns	
ATHB-8	nd	**_Nta_↑**	**_Nta_↑**	nd	nd	nd	nd	nd	nd	nd	nd	nd	
HD-ZIP	nd	nd	nd	nd	nd	nd	nd	nd	_Osa_-	**_Osa_↑**	_Sly_↑	**_Sly_↑**	[8,10]
miR167	ns	ns	ns	ns	ns	ns	ns	ns	_Osa_-	_Osa_↓	_Sly_↑	_Sly_↑	
ARF8	nd	nd	nd	nd	nd	nd	nd	nd	_Osa_-	**_Osa_↑**	_Sly_↑	**_Sly_↑**	
miR168	_Sly_↓	ns	ns	_Ath_↑	ns	ns	ns	ns	_Osa_-	_Osa_↓	_Sly_↑	_Sly_↑	[8,10,11, 12,14,16]
miR168*	ns	ns	ns	ns	ns	ns	ns	ns	ns	ns	_Sly_↑	ns	
AGO1	**_Sly_↑**	nd	nd	**_Ath_↑**	nd	nd	nd	nd	**_Osa_-**	**_Osa_↑**	**_Sly_↑**	**_Sly_↑**	
miR169	_Sly_↓	ns	ns	ns	ns	ns	ns	ns	ns	ns	ns	ns	[14]
CBF	**_Sly_↑**	nd	nd	nd	nd	nd	nd	nd	nd	nd	nd	nd	
miR171	_Sly_↓	_Nta_↓	_Nta_↓	ns	ns	_Nbe_↑	ns	_Nbe_↑	ns	_Osa_↓	_Sly_↑	_Sly_↑	[1,7,8,10,14]
miR171*	ns	ns	ns	ns	ns	ns	ns	ns	ns	_Osa_↓	ns	ns	
SCL6	**_Sly_↑**	**_Nta_↑**	**_Nta_↑**	nd	nd	**_Nbe_↑**	nd	**_Nbe_↑**	nd	nd	**_Sly_↑**	**_Sly_↑**	
Os03g38-170	nd	nd	nd	nd	nd	nd	nd	nd	nd	**_Osa_↑**	nd	nd	
miR172	_Sly_↓	ns	ns	ns	ns	ns	ns	ns	_Osa_-	_Osa_-	ns	ns	[8,14]
AP2	**_Sly_↓**	nd	nd	nd	nd	nd	nd	nd	**_Osa_-**	**_Osa_↓**	nd	nd	

TABLE 2.2 Experimental validated miRNA-targets responsive to virus infection in diverse plant species[a]—cont'd

| | DNA virus | RNA virus | | | | | | | | | | | |
| | Begomo-virus | Tobamovirus | | | Potyvirus | | Polero-virus | Potex-virus | Phyto-reo-virus | Tenui-virus | Cucumovirus | | |
miRNA and targets[b]	ToLC-NDV	TMV	ShMV	ORMV	TuMV	PVY	CLR DV	PVX	RDV	RSV	CMV	TAV	References[c]
miR319	ns	ns	ns	ns	ns	ns	ns	ns	ns	ns	Sly↑	Sly↑	[11,12]
miR319*	ns	ns	ns	ns	ns	ns	ns	ns	ns	ns	Sly↑	ns	
TCP4	nd	nd	nd	nd	nd	nd	nd	nd	nd	nd	*Sly↑*	*Sly↑*	
miR396	ns	ns	ns	Ath↑	ns	ns	ns	ns	ns	ns	ns	ns	[16]
GRF	nd	nd	nd	*Ath↑*	nd	nd	nd	nd	nd	nd	nd	nd	
miR398	Sly↑	ns	ns	ns	Nbe↑	Nbe↑	ns	Nbe↑	ns	ns	ns	ns	[7,14]
CSD	*Sly↓*	nd	nd	nd	*Nbe↑*	*Nbe↑*	nd	*Nbe↑*	nd	nd	nd	nd	
miR472	ns	ns	ns	Ath↑	ns	ns	ns	ns	ns	ns	ns	ns	[16]
LRR	nd	nd	nd	*Ath↑*	nd	nd	nd	nd	nd	nd	nd	nd	

miR1425	ns	ns	ns	ns	ns	ns	ns	ns	Osa-	ns	[8]
miR1425*	ns	ns	ns	ns	ns	ns	ns	ns	Osaî	ns	ns
SAM	nd	nd	nd	nd	nd	nd	nd	nd	**Osaî**	nd	nd
miR1885	ns	ns	ns	ns	Braî	ns	ns	ns	ns	ns	[4]
LRR	nd	nd	nd	nd	**Braî**	nd	nd	nd	nd	nd	nd

aNote: see Table 2.1.
bmiRNA targets are in **bold italic**.
cReferences. See footnote in Table 2.1.
nt: not determined. ns: not shown.
ShMV: Sunn-hemp mosaic virus.
ACO1: Argonaute 1. AP2: Apetala 2-like transcription factor. ARF: Auxin response factor. ATHB-8: HD-ZIP III homeobox 8 transcription factor. CBF: CBF transcription factor. CSD: Copper superoxide dismutase. DCL1: RNAse III-like DICER-like I endonuclease. GRF: Growth-regulating factor. HD-ZIP: HD-ZIP transcription factor. LRR: TIR-NBS-LRR class R gene. MYB: MYB transcription factor. NAC1: NAC domain transcription factor. NAM: NAC domain transcription factor. PPR: Pentatricopeptide. SCL: Scarecrow-like transcription factor. SPL: Squamosa promoter binding protein-like transcription factor. SAM: (S-adenosyl-L-Met)-dependent carboxyl methyltransferase. TCP4: TCP transcription factor.

TABLE 2.3 Function of miRNA-targets responsive to virus infection in plants[a]

miRNA	Function	Target genes	Target protein class	References[b]
miR156/157	Leaf development Phase change and flowering	SPL	Squamosa promoter binding protein-like transcription factor	[1,7,16]
miR159	Hormone response Seed germination Leaf development	MYB	MYB-like transcription factor	[10,11,14]
miR160	Signaling Floral organ identity	ARF	Auxin responsive factor	[8,10]
miR161/400	RNA editing Mitochondrion biogenesis Plant growth	PPR	Pentatricopeptide repeat containing protein	[16]
miR162	miRNA biogenesis	DCL1	RNAse III-like Dicer-like I endonuclease	[5,11,12,14]
miR164	Plant morphogenesis Auxin signaling Root development Biotic and abiotic stress response	NAM NAC1	NAC domain transcription factor NAC domain transcription factor	[11,12,14]
miR165/166	Meristem formation Vascular development	ATHB-8 HD-ZIP	HD-ZIP homeobox 8 transcription factor HD-ZIP transcription factor	[1,8,11,12]
miR167	Signaling transduction	ARF8	Auxin responsive factor	[8,10]
miR168	miRNA function	AGO1	Argonaute protein	[8,10,11,12,14,16]
miR169	Abiotic stress response	CBF	CBF transcription factor	[14]

TABLE 2.3 Function of miRNA-targets responsive to virus infection in plants[a]—cont'd

miRNA	Function	Target genes	Target protein class	References[b]
miR171	Developmental patterning	SCL6	Scarecrow-like transcription factor	[1,7,8,10,14]
miR172	Floral development and phase change	AP2	Apetala 2-like transcription factor	[8,14]
miR319	Leaf development and cell division	TCP4	TCP transcription factor	[11,12]
miR396	Leaf development	GRF	Growth-regulating factor	[16]
miR398	Abiotic stress	CDS	Copper superoxide dismutases	[7,14]
miR472	Disease resistance	LRR	TIR-NBS-LRR class disease-resistant protein	[16]
miR1425	Floral scent production	SAM	(S-adenosyl-L-Met)-dependent carboxyl methyltransferase	[8]
miR1885	Disease resistance	LRR	TIR-NBS-LRR class disease-resistant protein	[4]

[a]See Tables 2.1 and 2.2.
[b]**References**. See footnote in Table 2.1.

in ORMV-infected *Arabidopsis*, CMV- or TAV-infected tomato, and PVY- or PVX-infected *N. benthamiana* (Cillo et al 2009, Feng et al 2009, 2012, Hu et al 2011, Pacheco et al 2012). These TF targets included squamosa promoter binding protein-like TF (*SPL*), growth-regulating factor (*GRF*), *ARF16/17*, *NAC1*, *HD-ZIP*, *TCP4*, *MYB* and *SCL6-IV*. For the non-TF targets, *CSD*, pentatricopeptide (*PPR*), 1-4-beta glucanase (*GLU*) and TIR-NBS-LRR class R gene (*LRR*), elevated expression levels of both the mRNA targets and the corresponding miRNAs were observed in PVY- or PVX-infected *N. benthamiana*, ORMV-infected *Arabidopsis* and DNA begomovirus ToLCNDV-infected tomato (Naqvi et al 2010, Hu et al 2011, Pacheco et al 2012).

Virus-, strain- or tissue-specific induced mRNA targets

Bazzini et al (2011) suggested that alterations of miRNA and target mRNA levels might be delayed in plants infected with a less severe virus. Such an impact was easily observed at the early stage of viral infection. For example, TMV and ShMV differ markedly in the symptoms they produced on tobacco, severe or mild, respectively (Bazzini et al 2011). TMV is referred to as a severe virus, and ShMV a mild one. A statistically significant increase in the *ATHB-8* mRNA level was detected at 6 dpi in TMV (severe virus)-infected tobacco plants compared to the mock-inoculated ones, and negatively correlated with the significantly decreased miR166 level. However, there is no statistically significant alteration in the levels of both *ATHB-8* mRNA and miR166, in ShMV (mild virus) infected tobacco at 6 dpi. The statistically significant increase in the *ATHB-8* mRNA level, together with the significant decrease in the miR166 level, was detected only at 11 dpi in ShMV-infected plants, and to a lesser extent compared to the TMV infection. In another study, *AGO1*, *HID-ZIP* and *ARF8* were increased in agreement with up-regulation of miR168 and down-regulation of miR166 and miR167, respectively, in RSV-infected rice (Du et al 2011). However, none of these genes showed significant changes in expression levels, as did their cognitive miRNAs in RDV-infected plants (Du et al 2011).

On the other hand, Cillo et al (2009) demonstrated that severe or mild CMV strains were able to interfere differentially with the accumulation of the mRNA targets in infected tomato. For example, *TCP4* and Phantastica (*PHAN*) were up-regulated only by the severe CMV-Fny strain but not by the mild one; and in comparison to the mild CMV-LS strain, the severe CMV-Fny strain induced higher levels of *AGO1* and *HD-ZIP* expression.

Some miRNA target genes were expressed in a tissue-specific manner upon virus infection. Transcripts of *CBF*, *DCL1*, *AP2*, *GLU* and *CSD1* were accumulated to almost a similar extent in flowers of the healthy and ToLCND-infected tomato plants, while these target genes were up- or down-regulated in the infected leaves (Naqvi et al 2010). *SCL6* and *NAM* was markedly reduced in flowers but enhanced in leaves of infected plants, whereas *LA*, *CSD2* and *UAE1* were down-regulated in both flowers and leaves (Naqvi et al 2010).

miRNA* targets

Several miRNA* targets were detected in virus-infected plants. The expression of Os11g15060 (S-adenosyl-L-Met-dependent carboxyl methyltransferase, *SAM*, a target of miR1425*), Os11g38140 and Os02g49240 (potential targets of miR160*) and Os03g38170 (potential target of miR171*) was decreased in RSV-infected rice, and correlated well with the increased accumulation of the corresponding miRNA* (Du et al 2011). Moreover, the miRNA*-regulated targets showed virus specificity. Although the expression of Os11g15060 was down-regulated in both RSV and RDV infected rice plants, the cleavage product and cleavage sites of the target were identified only in RSV-infected plants, but not in the RDV-infected ones (Du et al 2011). The expression of Os11g38140, Os02g49240 and Os03g38170 decreased in RSV infected rice compared with that in RDV infection.

miRNA targets in synergistic infection

Synergistic infection may induce the alteration of miRNA target expression to a greater extent than does a single infection. A greater accumulation of *SCL6-IV* mRNA (a target of miR171) and *CSD* (a target of miR398) was detected in double PVX-PVY-infected *N. benthamiana* plants as compared with plants infected with PVX or PVY alone (Pacheco et al 2012). Expression level of *SPL9* mRNA (a target of miR156) in PVX-PVY-infected *N. benthamiana* was higher than that in PVX infection, but showed no difference compared to that in PVY infection.

miRNA AND SYMPTOMS

Some studies suggested that severity of symptoms, induced by either DNA or RNA viruses, is correlated with miRNA accumulation (Bazzini et al 2007, Naqvi et al 2010, Lang et al 2011a, Amin et al 2011). Thus, these miRNAs might serve as potential biomarkers for virus infection.

miR159/319 and miR172 were specifically up-regulated in tomato leaves by infection with the DNA begomovirus ToLCNDV infection, which might be linked to leaf curl disease (Naqvi et al 2010). *MYB* factors, the miR159 targets, are well-established in determining leaf structure, and the observed leaf deformation in ToLCNDV-infected tomato plants could be due to altered levels of miR159. *TCP4*, the miR319 target, is a suppressor of growth in *Arabidopsis* (Nag et al 2009). The accumulation of miR319 in ToLCNDV-infected leaves will down-regulate the expression of tomato *TCP4* homologs. This would lead to uncontrolled cell growth reflected in leaf deformation (Naqvi et al 2010). In *N. benthamiana*, the DNA begomoviruses, ACMV, CbLCuV, CLCuMV and TYLCV, generally induced the accumulation of a common set of miRNAs and thus led to the decreased translation of the corresponding target genes that involved in plant development (Amin et al 2011). However, the extent of the enhanced levels of individual miRNAs differed for distinct begomoviruses, reflecting differences in severity of symptoms. Lang et al (2011a) found that miR169 in tobacco accumulated to a greater abundance in the plants infected with PVX, causing severe symptoms compared to CMV causing mild symptoms. In tobacco, TMV and *Tomato mosaic virus* (ToMV) inducing the most severe symptoms altered miRNA accumulation to a greater extent than *Tobacco etch virus* (TEV) and PVY inducing mild symptoms (Bazzini et al 2007).

INTERACTION BETWEEN miRNAs AND THEIR TARGETS AS WELL AS VIRAL SILENCING SUPPRESSORS

The default state of plant miRNAs is to bring their targets under translational as well as RNA stability control, with the two layers of regulation not necessarily coinciding spatially or temporally (Voinnet 2009). Voinnet (2009) proposed that miRNA regulatory circuits in plants may include, but may not limited to, spatial restriction, temporal regulation, mutual exclusion of the miRNA and target gene, as well as the dampening of target gene expression. Most miRNAs do not function independently but rather are involved in overlapping regulatory stress-miRNA networks (Khraiwesh et al 2012). The level of those conserved miRNAs appears to be regulated during stress, and target genes appear to be stress-regulated as well (Khraiwesh et al 2012).

In plant–virus interactions, virus infection caused Argonaute quenching and global changes in Dicer homeostasis (Azevedo et al 2010). Virus-induced accumulation of miR168, a counter-defense action of the invading virus, seems to be involved in repression of AGO1 protein accumulation (Várallyay et al 2010). Different mechanisms might act at the early and late stage of infection to alter the expression of miRNAs and their mRNA targets in the virus-infected plants (Bazzini et al 2011). Interference of unrelated viral silencing suppressors with miRNA-directed processes seems to be a general feature in virus–plant interactions (Kasschau et al 2003, Chapman et al 2004, Dunoyer et al 2004, Zamore 2004, Goto et al 2007, Várallyay et al 2010, Ahn et al 2011).

Argonaute quenching and global changes in Dicer homeostasis caused by viral infection

Plant DCL1 and AGO1, the two key enzymes in the miRNA biogenesis pathway, undergo sophisticated homeostatic regulations through the action of miR162 and miR168, respectively (Xie et al 2003, Vaucheret et al 2004, Voinnet 2009). Spatial or temporal changes in expression of these two miRNAs may influence the global levels of mature miRNA production (Voinnet 2009).

Recent findings demonstrated the *Arabidopsis* Argonaute quenching and global changes in Dicer homeostasis caused by the *Turnip crinkle virus* (TCV)-encoded glycine/tryptophane (GW)-containing capsid protein, P38 (Azevedo et al 2010). The authors proposed that, during initial phases of the infection, the TCV dsRNA is mainly processed by DCL4 into 21-nt siRNA, which incorporate into AGO1 to effect primary antiviral silencing. AGO1 also mediates the activity of cellular miRNAs, including that of miR162, which normally dampens DCL1 accumulation. In a second phase of the infection, P38 would bind to AGO1 through its GW motifs, resulting in a deficit of viral siRNA-loaded and cellular miRNA-loaded AGO1. In particular, reduced levels of miR162-loaded AGO1 would enhance DCL1 accumulation. The enhanced DCL1 accumulation would, in turn, promote a decrease in DCL4 and DCL3 levels by an unknown mechanism. Consequently, DCL2 would take over the antiviral function upon loading of its 22-nt siRNA products into AGO1. Finally, the continued action of P38 would result in the steady-state infection observed in TCV-infected *Arabidopsis*. The *Arabidopsis* DCL1-dependent miRNA pathway and the DCL4-/DCL2-dependent siRNA pathway have not been completely separated and share the common effector, AGO1 (Azevedo et al 2010).

Virus induction of miR168 is involved in repression of AGO1 protein accumulation

Induction of miR168 is ubiquitous in plant–virus interactions, and the increased miR168 is accompanied by *AGO1* mRNA induction. Várallyay et al (2010) suggested that the induction of miR168 in CymRSV-infected *N. benthamiana* is a counter-defense action of the invading virus. This counter-defense response is mediated by the viral silencing suppressor (p19), aiming to control the AGO1 protein, the main component of RISC. The induction of *AGO1* mRNA level is a part of the host defense reaction.

The transcriptional activation of the expression of the *MIR168a* precursor and its increased processing rate might be mainly responsible for the enhanced accumulation of miR168 in virus-infected plants (Várallyay et al 2010). In healthy plants, the miR168-mediated cleavage of *AGO1* mRNA has been described as the main mechanism for controlling AGO1 homeostasis (Rhoades et al 2002, Vaucheret et al 2004). However, the accumulation of full-length *AGO1* mRNA and the lack of significantly higher accumulation of potential 3' end cleavage products in virus-infected plants argue against the function of miRNA-mediated efficient cleavage of *AGO1* mRNAs (Várallyay et al 2010). Alternatively, AGO1 and mature miR168 might be associated with active polysomes, suggesting their involvement in translational repression (Lanet et al 2009). Várallyay et al (2010) demonstrated that miR168 is directly involved in the down-regulation of AGO1 protein. Elimination of the correct target site on the *AGO1* mRNA resulted in less effective inhibition of AGO1 protein accumulation, and elimination of miR168 from the infection process causes the loss of AGO1 protein down-regulation. In addition, the spatial overlap between the virus-infected zones and induced miR168 can ensure that cells accommodating replicating viruses contain controlled amounts of AGO1 protein, enabling efficient virus replication (Várallyay et al 2010). It is possible that the virus-infection-induced miR168 is sorted mainly into ZWILLE/PINHEAD/AGO10, which exerts its activity at the level of translational repression (Várallyay et al 2010).

On the other hand, using a CymRSV mutant (Cym19Stop) disabled in RNA silencing suppressor (p19) activity, Várallyay et al (2010) showed that the infected plant is able to induce the *AGO1* mRNA expression as a defense reaction, which results in enhanced accumulation of AGO1 protein. The elevated level of AGO1 protein could accommodate higher amounts of free virus siRNAs facilitating the efficient cleavage of viral RNAs.

A biphasic alteration of miRNAs and their targets' expression upon viral infection

Bazzini et al (2011) suggested that different mechanisms might act at the early and late stage of infection to alter the expression of miRNAs and their mRNA targets in the virus-infected plants. At an early stage of infection, when virus was not yet detected systemically, certain miRNAs were down-regulated along with up-regulation of the mRNA targets. The late stage of infection includes enhanced levels of both miRNAs and the targeted mRNAs, as well as viral protein accumulation.

A transcriptional component may be included in early stages of infection, as implied by the coordinated reduction of pre- and mature miR166 levels in TMV infected tobacco (Bazzini et al 2011). A previous study demonstrated that ORMV infection elevated transcriptional activity of the miR164a promoter in *Arabidopsis* (Bazzini et al 2009). The basal defenses, as well as the systemic signaling of viral associated molecular patterns (VAMPs) triggering innate immunity, may play a role in the initiation of this early phase of miRNA alteration (Bazzini et al 2011). In turn, this early phase may be orchestrating the antiviral defense (Ruiz-Ferrer & Voinnet 2009, Bazzini et al 2011). Moreover, changes in the precursors of the hormone metabolites, jasmonic acid (JA) and salicylic acid (SA), that are infection-specific hubs on the response network, may directly or indirectly trigger the alteration in systemic miRNA accumulation observed in the absence of virus (Bazzini et al 2011). Indeed, SA- and/or JA-responsive elements were detected in several promoters of the miRNAs that compose the cluster altered at the beginning of the infection stage (Liu et al 2008, Bazzini et al 2009).

In the later stage of infection, the viral proteins may inhibit miRNA activity, for example, by sequestering small RNAs (sRNAs) through the post-transcriptional gene silencing (PTGS) suppression activity of the p122 subunit of TMV replicase or TMV-Cg 126K replication protein (Bazzini et al 2007, Csorba et al 2007, Kurihara et al 2007, Vogler et al 2008, Hu et al 2011).

Species-, tissue-, virus-, strain-specific alteration of miRNAs and their targets' expression upon viral infection

Plant *MIR* genes are independent transcription units and thus have their own promoters (Xie et al 2005). In addition to biotic or abiotic stress responsive elements commonly detected in *MIR* promoters (Megraw et al 2006), tissue-specific or even cell-specific regulatory elements are likely to exist (Parizotto et al 2004, Válóczi et al 2006, Kawashima et al 2009). The transcriptional control might explain to some extent the observed tissue-specific alteration of miRNAs and their targets expression. Moreover, organ-specific competition between DCL1 and DCL3 for miRNA processing might constitute a broad regulatory mechanism controlling the production of active miRNAs in specific tissue (Voinnet 2009).

Distinct RISCs might simultaneously operate in virus-infected cells (Ruiz-Ferrer & Voinnet 2009). This may explain, to some extent, the virus- or strain-specific alteration of miRNAs and their targets. Selective sRNA loading into specific AGOs seem strongly (albeit not entirely) influenced by their 5' terminal nucleotide (Mi et al 2008, Montgomery et al 2008, Takeda et al 2008). Thus, miRNAs with predominant 5'-Us are frequently loaded onto AGO1, whereas most AGO2- and AGO5-assocoated sRNAs have 5'-G and 5'-U termini. Ruiz-Ferrer & Voinnet (2009) assume that similar rules apply to viral small RNAs (vsRNAs). The vsRNA allocation to plant AGOs might vary extensively from one virus to another, or between virus strains, owing to differences in vsRNA populations and 5'-nucleotide polymorphisms. Some plant AGOs might strongly inhibit specific virus subsets, but not others.

Additionally, Du et al (2011) demonstrated that alteration of rice miRNAs and miRNA*s by RSV, but not by RDV, indicating distinct virus–host interactions. RSV and RDV infections differentially modified the expression of rice RNA silencing pathway genes. During RDV infection, with the exception of RDR1 being enhanced, there were no significant changes in the expression of RNA silencing pathway genes. In contrast, during RSV infection, the expression levels of DCL3a and DCL3b were down-regulated, whereas the expression of DCL2 was enhanced. Three of the four AGO1 homologs, AGO1a, AGO1b and AGO1c, were up-regulated in RSV-infected plants. Whether the altered expression patterns of the DCLs, RDRs and AGOs are responsible for the altered miRNA/siRNA biogenesis/accumulation in the RSV- and RDV-infected rice plants remains to be determined (Du et al 2011).

Previous studies indicate that the 2b protein derived from the severe strain CMV-Fny, but not from the mild strains CMV-LS or CMV-Q, is able to block AGO1 activity and impair proper miRNA-guided mRNA cleavage in *Arabidopsis* (Chapman et al 2004, Zhang et al 2006b, Lewsey et al 2007). Although *CMV-Q2b* transgenic *Arabidopsis* lines expressed the *Q2b* transcript at levels as high as *CMV-FNY2b* transgenic lines, the Q2b protein was barely detectable (Chapman et al 2004). Therefore, the absence of developmental defects in *CMV-Q2b* lines may be explained by the low accumulation levels of intact CMV-Q2b protein (Chapman et al 2004, Zhang et al 2006b).

The regulation of miRNA expression appears to vary between plant species, such as those that are adapted to suboptimal conditions and domesticated species (Sunkar et al 2012). Khraiwesh et al (2012) also suggested that some of the stress regulation of miRNAs and their targets, which were observed for only a single species, might not be applicable to other ones. In another study, the difference in tissue specificity for conserved miRNAs between *Brassica* and *Arabidopsis* suggests that even the older miRNAs might possess their own precise regulation processes through the tissue-dependent miRNA biogenesis and target selection in different plant species (He et al 2008). Furthermore, although some miRNA-target modules are conserved within or beyond angiosperms, the biologic functions of the regulatory modules may vary in different species (Chen 2010).

miRNA*s

In RSV-infected rice plants, the special accumulation of miRNA*s, but not miRNAs, may be explained by an RSV-induced enhancement of the activities of certain RISCs that especially associate with miRNA*s or by interference with loading of some miRNAs onto RISCs (Du et al 2011). Du et al (2011) demonstrated that, among the miRNA*s accumulated during RSV infection, there was an 'A' bias in the 19th nucleotide from the 5′ terminus. The authors proposed that the 'A' bias may direct certain miRNA* sequences onto AGO2 or AGO18. It might be also the potential influence of AGO1, as the 'A' bias in the 19th nucleotide from the 5′ terminus of miRNA* corresponds to the 5′ terminal 'U' of miRNA (Du et al 2011). In ORMV-infected *Arabidopsis*, the strong enrichment of miRNA* sequences primarily concerns those initiating with a 5′-G, indicative of specific small RNA (sRNA) associated effector complexes formed upon virus infection (Hu et al 2011).

The alternative possibility might be stabilization of the duplex miRNAs. Schnettler et al (2010) demonstrated stabilization of a miR171c/miR171c* complex by several tospoviruses in infected *N. benthamiana*, along with elevated accumulation of miR171c*. This stabilization appears to be due to the activity of the tospoviral non-structural proteins (NSs), acting to suppress the antiviral RNA silencing (Schnettler et al 2010). In *Zucchini yellow mosaic virus* (ZYMV)-infected squash plants, the accumulation of miR159*, miR166* and miR168* was always higher than that in uninfected plants, suggesting that viral infection prevents their degradation (Shiboleth et al 2007).

In addition, there may be secondary effects, such as transcriptional enhancement due to loss of feedback inhibition and up-regulation of *DCL1* and *AGO1* gene expression through loss of miR162 and miR168 control (Mlotshwa et al 2005, Zhang et al 2006b).

Viral silencing suppressors

In addition to inhibition of the antiviral silencing, interference of unrelated viral silencing suppressors with miRNA-directed processes seems to be a common phenomenon in virus–plant interaction. However, different suppressors may function by distinct mechanisms. Three strong suppressors, P1/HC-Pro (potyvirus P1 serine proteinase/helper-component proteinase), p21 (*Beet yellows virus*, BYV, p21 protein) and p19 (*Tomato bushy stunt virus*, TBSV, p19 protein), were suggested to inhibit the turnover of miRNA* species (Chapman et al 2004).

Only p21 and p19, but not P1/HC-Pro, could be detected in a complex with miRNA/miRNA* *in vivo*. Sequestration of miRNA/miRNA* duplexes may occur in the cytoplasm after processing by DCL1 in the nucleus and subsequent transport to the cytoplasm (Zamore 2004). However, P1/HC-Pro clearly inhibited miRNA* turnover, suggesting that miRNA/miRNA* unwinding and RISC assembly was suppressed. A TuMV-encoded P1/HC-Pro interfered with the activity of *Arabidopsis* miR171, which directs cleavage of several mRNAs coding for Scarecrow-like TFs, by inhibiting its nucleolytic function (Kasschau et al 2003). The TBSV encoded p19 exhibited host-dependent effects on RNA silencing. Although p19 induced the accumulation of both miR168 and its target *AGO1* mRNA, but suppressed AGO1 translation via up-regulation of miR168 in *Arabidopsis* and *N. benthamiana* (Dunoyer et al 2004, Várallyay et al 2010), it does not affect the miRNA pathway in potato (Ahn et al 2011). In contrast to p19 of tombusvirus that can effectively bind miRNAs, the CMV 2b (CMV 2b protein) suppressor only weakly bound to a miRNA (miR171) duplex, but it interferes with the PTGS pathway by directly binding siRNAs or long double-stranded RNA (dsRNA) (Goto et al 2007).

Other virus-encoded proteins, such as the TMV movement protein and coat protein interactions, also alter accumulation of tobacco miRNAs (Bazzini et al 2007).

miRNAs enriched during virus infection may have no strong effects on the levels of their mRNA targets

In some cases, the enriched sRNA, including miRNA, levels in virus-infected plants may have no strong effects on the levels of their mRNA targets (Hu et al 2011). This does not necessarily indicate that these increased miRNAs are inactive. The targets of these miRNAs may be regulated by feedback mechanisms at the level of transcription, or may be controlled by established RISCs that are stable. Thus, such targets are resistant against virus-induced changes in miRNA levels. Alternatively, the miRNA enrichment may occur predominantly in specific tissues in which the majority of the corresponding targets is not expressed (Hu et al 2011). The enriched miRNAs may also function in miRNA-guided translational repression (Brodersen et al 2008, Brodersen & Voinnet 2009, Várallyay et al 2010), but not in target transcript cleavage. In addition, plant miRNAs can also mediate DNA methylation (Chellappan et al 2010, Wu et al 2010). Hence, the enriched miRNAs may also act at the transcriptional level.

THE ARMS RACE BETWEEN PLANTS AND PLANT VIRUSES

Plant endogenous small RNAs, including miRNAs, siRNAs and long siRNAs, play an important role in plant immunity (Jin 2008, Voinnet 2008, Padmanabhan et al 2009). Plant immunity systems include pathogen-associated molecular pattern (PAMP)-triggered immunity (PTI) and effector-triggered immunity (ETI)

(Chisholm et al 2006, Jones & Dangl 2006). The recognition of PAMP or a microbe-associated molecular pattern (MAMP) triggers small RNA pathways. Transcriptional and/or post-transriptional regulation of the host gene expression by the small RNAs triggers PTI and basal defenses. Pathogens, however, have evolved countermeasures, such as delivering effector proteins into the plant cell to suppress host PTI (Navarro et al 2008). The defense and counter-defense between host and pathogen never ends. Plants, in turn, have acquired resistance (R) proteins to recognize these pathogen effectors and trigger the ETI, a more robust and specific response.

Plant viruses are both the trigger and the target of RNA silencing (Chen 2010). Almost all plant viruses encode RNA silencing suppressors, and some suppressors especially affect the miRNA pathways (Kasschau et al 2003, Chapman et al 2004, Chen et al 2004, Dunoyer et al 2004, Zhang et al 2006b). This reinforces the concept that RNA silencing evolved as an antiviral defense mechanism, and highlights the arms race between plants and plant viruses (Chen 2010). The symptoms caused by viral infection might be in part the result of inhibitory effects on host miRNAs by viral RNA silencing suppressors (Kasschu et al 2003, Silhavy & Burgyán 2004).

Navarro et al (2006) provided the first example of miRNA that plays a role in plant PTI by negatively regulating the auxin signaling pathway. Plants activate defenses after perceiving PAMPs, such as bacterial flagellin. In *Arabidopsis*, perception of flagellin increases resistance to the bacterium *Pseudomonas syringae*. *Arabidopsis* miR393 was induced by a bacterial flagellin-derived peptide flg22 and down-regulated auxin receptors transport inhibitor response 1 (TIR1), auxin signaling F-box protein 2 (AFB2) and AFB3. Repression of auxin signaling restricts the growth of *P. syringae*, implicating auxin in disease susceptibility and miRNA-mediated suppression of auxin signaling in resistance. Moreover, the up-regulation of miR160 and miR393 in *Soybean mosaic virus* (SMV)-infected soybean might suggest that the defense responses triggered miRNA-mediated suppression of auxin signaling pathways (Yin et al 2013).

Chen et al (2004) proposed that miRNAs could play a pathogenic role in the induction of viral disease. A virulence factor, named p69, encoded by *Turnip yellow mosaic virus* (TYMV) suppresses the siRNA pathway upstream of dsRNA but promotes the miRNA pathway in *Arabidopsis* (Chen et al 2004). In the *P69* transgenic P69c plants, the abundance of *DCL1* mRNA, as well as miR156, miR157, miR158, miR162, miR164, miR167 and miR171, increased. Cleavage of the *SCL6-IV* mRNA by miR171, and of the *SPL3* mRNA by miR156, occurred in P69c plants. Consequently, the late flowering phenotype of the P69 plants may be, in part, attributable to the p69-stimulated miR165 knockdown of *SPL3* mRNA. It is known the constitutive over-expression of *SPL3* results in early flowering (Cardon et al 1997). Chen et al (2004) proposed that enhanced miRNA silencing is a consequence of a negative feedback regulation of the siRNA pathway triggered by p69 suppression.

Li et al (2012) proposed that *R*-gene miRNAs could involve in fine-tuning *R*-gene function during plant–virus coevolution. The authors demonstrated a conserved role for miRNAs in Toll/Interleukin-1 receptor-nucleotide binding site-leucine-rich repeat (TIR-NBS-LRR) immune receptor gene regulation and virus resistance in *Solanaceae*. The tobacco miR6019 and miR6020 guide cleavage of transcripts of the TIR-NBS-LRR immune receptor *N* from tobacco, which confers resistance to TMV. Coexpression of *N* with miR6019 and miR6020 resulted in attenuation of *N*-mediated resistance to TMV. The miRNA-mediated attenuation of *R*-gene expression may have coevolved with multicopy *R*-gene loci, and might be one of several mechanisms that contribute to limiting potential fitness costs associated with their evolutionary trajectories. This process would facilitate the continuing amplification and diversification of *R* genes. Viruses have evolved effectors that can suppress miRNA and siRNA pathways (Li & Ding 2006). In addition to contributing to pathogen spread, the temporal shutdown of miRNA (and siRNA) function by pathogen effectors might also enhance *R* gene function by blocking the formation or activity of the *R*-gene targeting miRNAs and, thus, provide some balance to resistance and susceptibility during host–microbe interactions.

In SMV-infected soybean, an increased expression level of miR1510 would lead to the down-regulation of its target *TIR-NBS-LRR* disease-resistance gene (Yin et al 2013). The change in expression levels of *R* gene could cause soybean plants to be susceptible to SMV. The new *Brassica* miR1885, targeting *TIR-NBS-LRR* class gene for cleavage, was specifically induced by TuMV but not by CMV or TMV (He et al 2008). Whether miR1885 is directly involved in the pathogenesis of TuMV, and whether miR1885 and its target *R* genes regulate host basal defense response, remain to be determined (He et al 2008, Padmanabhan et al 2009).

CONCLUSION

This chapter describes the experimental confirmation of the altered levels of host-encoded miRNAs and their mRNA targets in response to virus infection in a susceptible plant. A common set of miRNAs, mainly those related to plant development or described as responsive to different biotic and abiotic stresses, was summarized. miR168 was up-regulated by virus infection in a plant- and virus-independent manner. However, other miRNAs, as well as their targets, presented virus-, strain-, plant- and tissue-specificity. Virus infection may cause quenching of Argonaute and global changes in Dicer homeostasis. In virus-infected plants, the levels of miRNAs and their mRNA targets were altered antagonistically. In some cases, a parallel increase in the levels of target mRNAs and the corresponding miRNA species was observed. Moreover, some miRNAs and their targets exhibited a biphasic expression pattern. In plant–virus interaction, viral silencing suppressors affect the miRNA pathway by distinct mechanisms.

Abbreviations

ACMV: *African cassava mosaic virus.*
AGO1: Argonaute 1.
AP2: Apetala 2-like transcription factor.
ARF: Auxin response factor.
ATHB-8: HD-ZIP III homeobox 8 transcription factor.
BYV: *Beet yellows virus.*
CBF: CBF transcription factor.
CbLCuV: *Cabbage leaf curl virus.*
CLCuMV: *Cotton leaf curl Multan virus.*
CLRDV: *Cotton leafroll dwarf virus.*
CMV: *Cucumber mosaic virus.*
CMV 2b: *Cucumber mosaic virus* 2b protein.
CSD: Copper superoxide dismutase.
CymRSV: *Cymbidium ringspot virus.*
DCL1: RNAse III-like DICER-like I endonuclease.
dpi: Days post-inoculation.
dsRNA: Double-stranded RNA.
EBV: *Epstein–Barr virus.*
GLU: Glucanase.
GRF: Growth-regulating factor.
HASTY: The *Arabidopsis* ortholog of the nuclear export receptor Exportin 5.
HC-Pro: Helper-component proteinase.
HD-ZIP: HD-ZIP transcription factor.
JA: Jasmonic acid.
LA: Lanceolate.
LRR: TIR-NBS-LRR class R gene.
miRNA: microRNA.
miRNA*: antisense miRNA.
MNSV: *Melon necrotic spot carmovirus.*
mRNA: messenger RNA.
MYB: MYB transcription factor.
NAC1: NAC domain transcription factor.
NAM: NAC domain transcription factor.
NSs: non-structural proteins.
nt: Nucleotide.
ORMV: *Oilseed rape mosaic virus.*
P1: P1 serine proteinase.
p19: *Tomato bushy stunt virus* p19 protein.

p21: *Beet yellows virus* p21 protein.

PHAN: MYB domain transcription factor Phantastica.

PMMoV: *Pepper mild mottle tobamovirus.*

PPR: Pentatricopeptide.

PPV: *Plum pox virus.*

pre-miRNA: miRNA precursor.

pri-miRNA: primary miRNA.

PTGS: Post-transcriptional gene silencing.

PVX: *Potato virus X.*

PVY: *Potato virus Y.*

RDV: *Rice dwarf virus.*

RISC: RNA-induced silencing complex.

RSV: *Rice stripe virus.*

SA: Salicylic acid.

SAM: (S-adenosyl-L-Met)-dependent carboxyl methyltransferase.

SCL: Scarecrow-like transcription factor.

ShMV: *Sunn-hemp mosaic virus.*

siRNA: Small interfering RNA.

SPL: Squamosa promoter binding protein-like transcription factor.

sRNA: Small RNA.

TAV: *Tomato aspermy virus.*

TBSV: *Tomato bushy stunt virus.*

TCP: TEOSINTE BRANCHED/CYCLOIDEA/PCF transcription factor.

TCV: *Turnip crinkle virus.*

TEV: *Tobacco etch virus.*

TF: Transcription factor.

TMV: *Tobacco mosaic virus.*

ToLCNDV: *Tomato leaf curl New Delhi virus.*

ToMV: *Tomato mosaic virus.*

TRV: *Tobacco rattle virus.*

TuMV: *Turnip mosaic virus.*

TYLCV: *Tomato yellow leaf curl virus.*

UAE: Ubiquitin-activating enzyme.

VAMP: Viral associated molecular pattern.

vsRNA: Viral small RNA.

ZYMV: *Zucchini yellow mosaic virus.*

ACKNOWLEDGMENTS

The authors are indebted to Dr Baohong Zhang (Department of Biology, East Carolina University, Greenville, North Carolina, USA) for critical reading of the manuscript and for helpful suggestions on the text. The authors are grateful to the editor's valuable comments on improving the manuscript. The present research was supported by the National Science Centre of Poland (Grant 3044/B/P01/2010/39). The funder had no role in study design, data collection and analysis, decision to publish, or preparation of the manuscript.

REFERENCES

Ahn, J.W., Lee, J.S., Davarpanah, S.J., Jeon, J.H., Park, Y.I., Liu, J.R., Jeong, W.J., 2011. Host-dependent suppression of RNA silencing mediated by the viral suppressor p19 in potato. Planta 234, 1065–1072.

Akbergenov, R., Si-Ammour, A., Blevins, T., Amin, I., Kutter, C., Vanderschuren, H., Zhang, P., Gruissem, W., Meins Jr., F., Hohn, T., Pooggin, M.M., 2006. Molecular characterization of geminivirus-derived small RNAs in different plant species. Nucleic Acids Research 34, 462–471.

Amin, I., Patil, B.L., Briddon, R.W., Mansoor, S., Fauquet, C.M., 2011. A common set of developmental miRNAs are upregulated in *Nicotiana benthamiana* by diverse begomoviruses. Virology Journal 8, 143.

Azevedo, J., Garcia, D., Pontier, D., Ohnesorge, S., Yu, A., Garcia, S., Braun, L., Bergdoll, M., Hakimi, M.A., Lagrange, T., Voinnet, O., 2010. Argonaute quenching and global changes in Dicer homeostasis caused by a pathogen-encoded GW repeat protein. Genes & Development 24, 904–915.

Bao, N., Lye, K.W., Barton, M.K., 2004. MicroRNA binding sites in *Arabidopsis* class III HD-ZIP mRNAs are required for methylation of the template chromosome. Developmental Cell 7, 653–662.

Bartel, D.P., 2004. MicroRNAs: genomics, biogenesis, mechanism, and function. Cell 116, 281–297.

Baumberger, N., Baulcombe, D.C., 2005. *Arabidopsis* ARGONAUTE1 is an RNA Slicer that selectively recruits microRNAs and short interfering RNAs. Proceedings of the National Academy of Sciences of the United States of America 102, 11928–11933.

Bazzini, A.A., Hopp, H.E., Beachy, R.N., Asurmendi, S., 2007. Infection and coaccumulation of *tobacco mosaic virus* proteins alter microRNA levels, correlating with symptom and plant development. Proceedings of the National Academy of Sciences of the United States of America 29, 12157–12162.

Bazzini, A.A., Almasia, N.I., Manacorda, C.A., Mongelli, V.C., Conti, G., Maroniche, G.A., Rodriguez, M.C., Distéfano, A.J., Hopp, H.E., del Vas, M., Asurmendi, S., 2009. Virus infection elevates transcriptional activity of miR164a promoter in plants. BMC Plant Biology 9, 152.

Bazzini, A.A., Manacorda, C.A., Tohge, T., Conti, G., Rodriguez, M.C., Nunes-Nesi, A., Villanueva, S., Fernie, A.R., Carrari, F., Asurmendi, S., 2011. Metabolic and miRNA profiling of TMV infected plants reveals biphasic temporal changes. PLoS One 6, e28466.

Blevins, T., Rajeswaran, R., Shivaprasad, P.V., Beknazariants, D., Si-Ammour, A., Park, H.S., Vazquez, F., Robertson, D., Meins Jr., F., Hohn, T., Pooggin, M.M., 2006. Four plant Dicers mediate viral small RNA biogenesis and DNA virus induced silencing. Nucleic Acids Research 34, 6233–6246.

Bollman, K.M., Aukerman, M.J., Park, M.Y., Hunter, C., Berardini, T.Z., Poethig, R.S., 2003. HASTY, the *Arabidopsis* ortholog of exportin 5/MSN5, regulates phase change and morphogenesis. Development 130, 1493–1504.

Brodersen, P., Voinnet, O., 2009. Revisiting the principles of microRNA target recognition and mode of action. Nature Reviews Molecular Cell Biology 10, 141–148.

Brodersen, P., Sakvarelidze-Achard, L., Bruun-Rasmussen, M., Dunoyer, P., Yamamoto, Y.Y., Sieburth, L., Voinnet, O., 2008. Widespread translational inhibition by plant miRNAs and siRNAs. Science 320, 1185–1190.

Cardon, G.H., Höhmann, S., Nettesheim, K., Saedler, H., Huijser, P., 1997. Functional analysis of the *Arabidopsis thaliana* SBP-box gene *SPL3*: a novel gene involved in the floral transition. Plant Journal 12, 367–377.

Carrington, J.C., Ambros, V., 2003. Role of microRNAs in plant and animal development. Science 301, 336–338.

Chapman, E.J., Prokhnevsky, A.I., Gopinath, K., Dolja, V.V., Carrington, J.C., 2004. Viral RNA silencing suppressors inhibit the microRNA pathway at an intermediate step. Genes & Development 18, 1179–1186.

Chellappan, P., Xia, J., Zhou, X., Gao, S., Zhang, X., Coutino, G., Vazquez, F., Zhang, W., Jin, H., 2010. siRNAs from miRNA sites mediate DNA methylation of target genes. Nucleic Acids Research 38, 6883–6894.

Chen, X., 2010. Small RNAs—secrets and surprise of the genome. Plant Journal 61, 941–958.

Chen, J., Li, W.X., Xie, D., Peng, J.R., Ding, S.W., 2004. Viral virulence protein suppresses RNA silencing-mediated defense but upregulates the role of microRNA in host gene expression. Plant Cell 16, 1302–1313.

Chisholm, S.T., Coaker, G., Day, B., Staskawicz, B.J., 2006. Host–microbe interactions: shaping the evolution of the plant immune response. Cell 124, 803–814.

Cillo, F., Mascia, T., Pasciuto, M.M., Gallitelli, D., 2009. Differential effects of mild and severe *Cucumber mosaic virus* strains in the perturbation of microRNA-regulated gene expression in tomato map to the 3′ sequence of RNA 2. Molecular Plant–Microbe Interactions 22, 1239–1249.

Csorba, T., Bovi, A., Dalmay, T., Burgyán, J., 2007. The p122 subunit of Tobacco Mosaic Virus replicase is a potent silencing suppressor and compromises both small interfering RNA- and microRNA-mediated pathways. Journal of Virology 81, 11768–11780.

Diaz-Pendon, J.A., Li, F., Li, W.X., Ding, S.W., 2007. Suppression of antiviral silencing by *cucumber mosaic virus* 2b protein in *Arabidopsis* is associated with drastically reduced accumulation of three classes of viral small interfering RNAs. Plant Cell 19, 2053–2063.

Donaire, L., Wang, Y., Gonzalez-Ibeas, D., Mayer, K.F., Aranda, M.A., Llave, C., 2009. Deep-sequencing of plant viral small RNAs reveals effective and widespread targeting of viral genomes. Virology 392, 203–214.

Du, P., Wu, J., Zhang, J., Zhao, S., Zheng, H., Gao, G., Wei, L., Li, Y., 2011. Viral infection induces expression of novel phased microRNAs from conserved cellular microRNA precursors. PLoS Pathogens 7, e1002176.

Dunoyer, P., Lecellier, C.H., Parizotto, E.A., Himber, C., Voinnet, O., 2004. Probing the microRNA and small interfering RNA pathways with virus-encoded suppressors of RNA silencing. Plant Cell 16, 1235–1250.

Fahlgren, N., Howell, M.D., Kasschau, K.D., Chapman, E.J., Sullivan, C.M., Cumbie, J.S., Givan, S.A., Law, T.F., Grant, S.R., Dangl, J.L., Carrington, J.C., 2007. High-throughput sequencing of *Arabidopsis* microRNAs: evidence for frequent birth and death of MIRNA genes. PLoS One 2, e219.

Feng, J., Wang, K., Liu, X., Chen, S., Chen, J., 2009. The quantification of tomato microRNAs response to viral infection by stem-loop real-time RT-PCR. Gene 437, 14–21.

Feng, J., Lai, L., Lin, R., Jin, C., Chen, J., 2012. Differential effects of *Cucumber mosaic virus* satellite RNAs in the perturbation of microRNA-regulated gene expression in tomato. Molecular Biology Reports 39, 775–784.

Goto, K., Kobori, T., Kosaka, Y., Natsuaki, T., Masuta, C., 2007. Characterization of silencing suppressor 2b of *cucumber mosaic virus* based on examination of its small RNA-binding abilities. Plant & Cell Physiology 48, 1050–1060.

Griffiths-Jones, S., Saini, H.K., van Dongen, S., Enright, A.J., 2008. miRBase: tools for microRNA genomics. Nucleic Acids Research 36, D154–D158.

He, X.F., Fang, Y.Y., Feng, L., Guo, H.S., 2008. Characterization of conserved and novel microRNAs and their targets, including a TuMV-induced TIR-NBS-LRR class R gene-derived novel miRNA in *Brassica*. FEBS Letters 582, 2445–2452.

Hu, Q., Hollunder, J., Niehl, A., Kørner, C.J., Gereige, D., Windels, D., Arnold, A., Kuiper, M., Vazquez, F., Pooggin, M., Heinlein, M., 2011. Specific impact of tobamovirus infection on the *Arabidopsis* small RNA profile. PLoS One 6, e19549.

Hull, R., 2009. Overview of plant viruses. In: Hull (Ed.), Comparative plant virology, second ed. Elsevier, Oxford, Chapter 2.

Jin, H., 2008. Endogenous small RNAs and antibacterial immunity in plants. FEBS Letters 582, 2679–2684.

Jones, J.D., Dangl, J.L., 2006. The plant immune system. Nature 444, 323–329.

Jones-Rhoades, M.W., Bartel, D.P., Bartel, B., 2006. MicroRNAS and their regulatory roles in plants. Annual Review of Plant Biology 57, 19–53.

Kasschau, K.D., Xie, Z., Allen, E., Llave, C., Chapman, E.J., Krizan, K.A., Carrington, J.C., 2003. P1/HC-Pro, a viral suppressor of RNA silencing, interferes with *Arabidopsis* development and miRNA function. Developmental Cell 4, 205–217.

Kawashima, C.G., Yoshimoto, N., Maruyama-Nakashita, A., Tsuchiya, Y.N., Saito, K., Takahashi, H., Dalmay, T., 2009. Sulphur starvation induces the expression of microRNA-395 and one of its target genes but in different cell types. Plant Journal 57, 313–321.

Khraiwesh, B., Zhu, J.K., Zhu, J., 2012. Role of miRNAs and siRNAs in biotic and abiotic stress responses of plants. Biochimica et Biophysica Acta 1819, 137–148.

Kurihara, Y., Inaba, N., Kutsuna, N., Takeda, A., Tagami, Y., Watanabe, Y., 2007. Binding of tobamovirus replication protein with small RNA duplexes. Journal of General Virology 88, 2347–2352.

Lanet, E., Delannoy, E., Sormani, R., Floris, M., Brodersen, P., Crete, P., Voinnet, O., Robaglia, C., 2009. Biochemical evidence for translational repression by *Arabidopsis* microRNAs. Plant Cell 21, 1762–1768.

Lang, Q., Jin, C., Lai, L., Feng, J., Chen, S., Chen, J., 2011a. Tobacco microRNAs prediction and their expression infected with *Cucumber mosaic virus* and *Potato virus X*. Molecular Biology Reports 38, 1523–1531.

Lang, Q.L., Zhou, X.C., Zhang, X.L., Drabek, R., Zuo, Z.X., Ren, Y.L., Li, T.B., Chen, J.S., Gao, X.L., 2011b. Microarray-based identification of tomato microRNAs and time course analysis of their response to *Cucumber mosaic virus* infection. Journal of Zhejiang University-SCIENCE B 12, 116–125.

Leister, D., 2004. Tandem and segmental gene duplication and recombination in the evolution of plant disease resistance gene. Trends in Genetics 20, 116–122.

Lewsey, M., Robertson, F.C., Canto, T., Palukaitis, P., Carr, J.P., 2007. Selective targeting of miRNA-regulated plant development by a viral counter-silencing protein. Plant Journal 50, 240–252.

Li, F., Ding, S.W., 2006. Virus counterdefense: Diverse strategies for evading the RNA silencing immunity. Annual Review of Microbiology 60, 503–531.

Li, F., Pignatta, D., Bendix, C., Brunkard, J.O., Cohn, M.M., Tung, J., Sun, H., Kumar, P., Baker, B., 2012. MicroRNA regulation of plant innate immune receptors. Proceedings of the National Academy of Sciences of the United States of America 109, 1790–1795.

Liu, H.H., Tian, X., Li, Y.J., Wu, C.A., Zheng, C.C., 2008. Microarray-based analysis of stress-regulated microRNAs in *Arabidopsis thaliana*. RNA 14, 836–843.

Llave, C., Kasschau, K.D., Rector, M.A., Carrington, J.C., 2002. Endogenous and silencing-associated small RNAs in plants. The Plant Cell 14, 1605–1619.

Lu, Y.D., Gan, Q.H., Chi, X.Y., Qin, S., 2008. Roles of microRNA in plant defense and virus offense interaction. Plant Cell Reports 27, 1571–1579.

Mallory, A.C., Reinhart, B.J., Bartel, D., Vance, V.B., Bowman, L.H., 2002. A viral suppressor of RNA silencing differentially regulates the accumulation of short interfering RNAs and micro-RNAs in tobacco. Proceedings of the National Academy of Sciences of the United States of America 99, 15228–15233.

Megraw, M., Baev, V., Rusinov, V., Jensen, S.T., Kalantidis, K., Hatzigeorgiou, A.G., 2006. MicroRNA promoter element discovery in *Arabidopsis*. RNA 12, 1612–1619.

Meyers, B.C., Kozik, A., Griego, A., Kuang, H., Michelmore, R.W., 2003. Genome-wide analysis of NBS-LRR-encoding genes in *Arabidopsis*. Plant Cell 15, 809–834, Erratum in: Plant Cell, 15, 1683.

Mi, S., Cai, T., Hu, Y., Chen, Y., Hodges, E., Ni, F., Wu, L., Li, S., Zhou, H., Long, C., Chen, S., Hannon, G.J., Qi, Y., 2008. Sorting of small RNAs into *Arabidopsis* argonaute complexes is directed by the 5′ terminal nucleotide. Cell 133, 116–127.

miRBase, 2012. http://www.mirbase.org/cgi-bin/browse.pl. June 2013.

Mlotshwa, S., Schauer, S.E., Smith, T.H., Mallory, A.C., Herr Jr., J.M., Roth, B., Merchant, D.S., Ray, A., Bowman, L.H., Vance, V.B., 2005. Ectopic DICER-LIKE1 expression in P1/HC-Pro *Arabidopsis* rescues phenotypic anomalies but not defects in microRNA and silencing pathways. Plant Cell 17, 2873–2885.

Montgomery, T.A., Howell, M.D., Cuperus, J.T., Li, D., Hansen, J.E., Alexander, A.L., Chapman, E.J., Fahlgren, N., Allen, E., Carrington, J.C., 2008. Specificity of ARGONAUTE7-miR390 interaction and dual functionality in TAS3 trans-acting siRNA formation. Cell 133, 128–141.

Moxon, S., Jing, R., Szittya, G., Schwach, F., Rusholme Pilcher, R.L., Moulton, V., Dalmay, T., 2008. Deep sequencing of tomato short RNAs identifies microRNAs targeting genes involved in fruit ripening. Genome Research, 18, 1602–1609.

Nag, A., King, S., Jack, T., 2009. miR319a targeting of TCP4 is critical for petal growth and development in *Arabidopsis*. Proceedings of the National Academy of Sciences of the United States of America 106, 22534–22539.

Naqvi, A.R., Haq, Q.M., Mukherjee, S.K., 2010. MicroRNA profiling of *tomato leaf curl New Delhi virus* (tolcndv) infected tomato leaves indicates that deregulation of mir159/319 and mir172 might be linked with leaf curl disease. Virology Journal 7, 281.

Navarro, L., Dunoyer, P., Jay, F., Arnold, B., Dharmasiri, N., Estelle, M., Voinnet, O., Jones, J.D., 2006. A plant miRNA contributes to antibacterial resistance by repressing auxin signaling. Science 312, 436–439.

Navarro, L., Jay, F., Nomura, K., He, S.Y., Voinnet, O., 2008. Suppression of the microRNA pathway by bacterial effector proteins. Science 321, 964–967.

Pacheco, R., García-Marcos, A., Barajas, D., Martiáñez, J., Tenllado, F., 2012. PVX-potyvirus synergistic infections differentially alter microRNA accumulation in *Nicotiana benthamiana*. Virus Research 165, 231–235.

Padmanabhan, C., Zhang, X., Jin, H., 2009. Host small RNAs are big contributors to plant innate immunity. Current Opinion in Plant Biology 12, 465–472.

Parizotto, E.A., Dunoyer, P., Rahm, N., Himber, C., Voinnet, O., 2004. *In vivo* investigation of the transcription, processing, endonucleolytic activity, and functional relevance of the spatial distribution of a plant miRNA. Genes & Development 18, 2237–2242.

Park, W., Li, J., Song, R., Messing, J., Chen, X., 2002. CARPEL FACTORY, a Dicer homolog, and HEN1, a novel protein, act in microRNA metabolism in *Arabidopsis thaliana*. Current Biology 12, 1484–1495.

Pasquinelli, A.E., Ruvkun, G., 2002. Control of developmental timing by microRNAs and their targets. Annual Review of Cell and Developmental Biology 18, 495–513.

Pérez-Quintero, A.L., Neme, R., Zapata, A., López, C., 2010. Plant microRNAs and their role in defense against viruses: a bioinformatics approach. BMC Plant Biology 10, 138.

Pfeffer, S., Zavolan, M., Grässer, F.A., Chien, M., Russo, J.J., Ju, J., John, B., Enright, A.J., Marks, D., Sander, C., Tuschl, T., 2004. Identification of virus-encoded microRNAs. Science 304, 734–736.

Reinhart, B.J., Weinstein, E.G., Rhoades, M.W., Bartel, B., Bartel, D.P., 2002. MicroRNAs in plants. Genes & Development 16, 1616–1626.

Rhoades, M.W., Reinhart, B.J., Lim, L.P., Burge, C.B., Bartel, B., Bartel, D.P., 2002. Prediction of plant microRNA targets. Cell 110, 513–520.

Richly, E., Kurth, J., Leister, D., 2002. Mode of amplification and reorganization of resistance genes during recent *Arabidopsis thaliana* evolution. Molecular Biology and Evolution 19, 76–84.

Ruiz-Ferrer, V., Voinnet, O., 2009. Roles of plant small RNAs in biotic stress responses. Annual Review of Plant Biology 60, 485–510.

Sanan-Mishra, N., Kumar, V., Sopory, S.K., Mukherjee, S.K., 2009. Cloning and validation of novel miRNA from basmati rice indicates cross talk between abiotic and biotic stresses. Molecular Genetics and Genomics 282, 463–474.

Sarnow, P., Jopling, C.L., Norman, K.L., Schütz, S., Wehner, K.A., 2006. MicroRNAs: expression, avoidance and subversion by vertebrate viruses. Nature Reviews Microbiology 4, 651–659.

Schauer, S.E., Jacobsen, S.E., Meinke, D.W., Ray, A., 2002. DICER-LIKE1: blind men and elephants in *Arabidopsis* development. Trends in Plant Science 7, 487–491.

Schnettler, E., Hemmes, H., Huismann, R., Goldbach, R., Prins, M., Kormelink, R., 2010. Diverging affinity of tospovirus RNA silencing suppressor proteins, NSs, for various RNA duplex molecules. Journal of Virology 84, 11542–11554.

Shiboleth, Y.M., Haronsky, E., Leibman, D., Arazi, T., Wassenegger, M., Whitham, S.A., Gaba, V., Gal-On, A., 2007. The conserved FRNK box in HC-Pro, a plant viral suppressor of gene silencing, is required for small RNA binding and mediates symptom development. Journal of Virology 81, 13135–13148.

Shukla, L.I., Chinnusamy, V., Sunkar, R., 2008. The role of microRNAs and other endogenous small RNAs in plant stress responses. Biochimica et Biophysica Acta 1779, 743–748.

Silhavy, D., Burgyán, J., 2004. Effects and side-effects of viral RNA silencing suppressors on short RNAs. Trends in Plant Science 9, 76–83.

Silva, T.F., Romanel, E.A., Andrade, R.R., Farinelli, L., Østerås, M., Deluen, C., Corrêa, R.L., Schrago, C.E., Vaslin, M.F., 2011. Profile of small interfering RNAs from cotton plants infected with the polerovirus *Cotton leafroll dwarf virus*. BMC Molecular Biology 12, 40.

Sunkar, R., 2010. MicroRNAs with macro-effects on plant stress responses. Seminars in Cell & Developmental Biology 21, 805–811.

Sunkar, R., Li, Y.F., Jagadeeswaran, G., 2012. Functions of microRNAs in plant stress responses. Trends in Plant Science 17, 196–203.

Szittya, G., Moxon, S., Pantaleo, V., Toth, G., Rusholme Pilcher, R.L., Moulton, V., Burgyan, J., Dalmay, T., 2010. Structural and functional analysis of viral siRNAs. PLoS Pathogens 6, e1000838. doi:10.1371/journal.ppat.1000838.

Tagami, Y., Inaba, N., Kutsuna, N., Kurihara, Y., Watanabe, Y., 2007. Specific enrichment of miRNAs in *Arabidopsis thaliana* infected with *Tobacco mosaic virus*. DNA Research 14, 227–233.

Takeda, A., Iwasaki, S., Watanabe, T., Utsumi, M., Watanabe, Y., 2008. The mechanism selecting the guide strand from small RNA duplexes is different among argonaute proteins. Plant Cell Physiology 49, 493–500.

Trindade, I., Capitão, C., Dalmay, T., Fevereiro, M.P., Santos, D.M., 2010. miR398 and miR408 are up-regulated in response to water deficit in *Medicago truncatula*. Planta 231, 705–716.

Válóczi, A., Várallyay, E., Kauppinen, S., Burgyán, J., Havelda, Z., 2006. Spatio-temporal accumulation of microRNAs is highly coordinated in developing plant tissues. Plant Journal 47, 140–151.

Várallyay, E., Válóczi, A., Agyi, A., Burgyán, J., Havelda, Z., 2010. Plant virus-mediated induction of miR168 is associated with repression of ARGONAUTE1 accumulation. EMBO Journal 29, 3507–3519.

Vaucheret, H., Béclin, C., Fagard, M., 2001. Post-transcriptional gene silencing in plants. Journal of Cell Science 114, 3083–3091.

Vaucheret, H., Vazquez, F., Crété, P., Bartel, D.P., 2004. The action of ARGONAUTE1 in the miRNA pathway and its regulation by the miRNA pathway are crucial for plant development. Genes & Development 18, 1187–1197.

Vogler, H., Kwon, M.O., Dang, V., Sambade, A., Fasler, M., Ashby, J., Heinlein, M., 2008. *Tobacco mosaic virus* movement protein enhances the spread of RNA silencing. PLoS Pathogens 4, e1000038.

Voinnet, O., 2008. Post-transcriptional RNA silencing in plant-microbe interactions: a touch of robustness and versatility. Current Opinion in Plant Biology 11, 464–470.

Voinnet, O., 2009. Origin, biogenesis, and activity of plant microRNAs. Cell 136, 669–687.

Voinnet, O., Pinto, Y.M., Baulcombe, D.C., 1999. Suppression of gene silencing: a general strategy used by diverse DNA and RNA viruses of plants. Proccedings of the National Academy of Sciences of the United States of America 96, 14147–14152.

Wang, X.B., Wu, Q., Ito, T., Cillo, F., Li, W.X., Chen, X., Yu, J.L., Ding, S.W., 2010. RNAi-mediated viral immunity requires amplification of virus-derived siRNAs in *Arabidopsis thaliana*. Proceedings of the National Academy of Sciences of the United States of America 107, 484–489.

Waterhouse, P.M., Wang, M.B., Lough, T., 2001. Gene silencing as an adaptive defence against viruses. Nature 411, 834–842.

Wu, L., Zhou, H., Zhang, Q., Zhang, J., Ni, F., Liu, C., Qi, Y., 2010. DNA methylation mediated by a microRNA pathway. Molecular Cell 38, 465–475.

Xie, Z., Kasschau, K.D., Carrington, J.C., 2003. Negative feedback regulation of Dicer-Like1 in *Arabidopsis* by microRNA-guided mRNA degradation. Current Biology 13, 784–789.

Xie, Z., Allen, E., Fahlgren, N., Calamar, A., Givan, S.A., Carrington, J.C., 2005. Expression of *Arabidopsis* MIRNA genes. Plant Physiology 138, 2145–2154.

Yin, X., Wang, J., Cheng, H., Wang, X., Yu, D., 2013. Detection and evolutionary analysis of soybean miRNAs responsive to soybean mosaic virus. Planta 237, 1213–1225.

Yu, B., Yang, Z., Li, J., Minakhina, S., Yang, M., Padgett, R.W., Steward, R., Chen, X., 2005. Methylation as a crucial step in plant microRNA biogenesis. Science 307, 932–935.

Zamore, P.D., 2004. Plant RNAi: How a viral silencing suppressor inactivates siRNA. Current Biology 14, R198–200.

Zhang, B., Pan, X., Cobb, G.P., Anderson, T.A., 2006a. Plant microRNA: a small regulatory molecule with big impact. Developmental Biology 289, 3–16.

Zhang, X., Yuan, Y.R., Pei, Y., Lin, S.S., Tuschl, T., Patel, D.J., Chua, N.H., 2006b. *Cucumber mosaic virus*-encoded 2b suppressor inhibits *Arabidopsis* Argonaute1 cleavage activity to counter plant defense. Genes & Development 20, 3255–3268.

Zhang, W., Gao, S., Zhou, X., Xia, J., Chellappan, P., Zhou, X., Zhang, X., Jin, H., 2010. Multiple distinct small RNAs originate from the same microRNA precursors. Genome Biology 11, R81.

Chapter 3

Host–virus interactions in banana-infecting viruses

R. Selvarajan and V. Balasubramanian

National Research Centre for Banana (ICAR), Tiruchirappalli, India

INTRODUCTION

Bananas and plantains are the world's most important food crops and constitute an important staple nutrition source for millions of people in tropical and subtropical countries. Biotic and abiotic stresses are major threats in achieving the targeted production and productivity. Among biotic stresses, viral diseases are considered to be a significant barrier for achieving higher productivity. Four viral diseases, viz., banana bunchy top disease (BBTD) caused by *Banana bunchy top virus* (BBTV), bract mosaic disease (BBrMD) caused by *Banana bract mosaic virus* (BBrMV), banana streak disease (BSD) caused by different species of *Banana streak virus* (BSV), and banana mosaic or infectious chlorosis caused by *Cucumber mosaic virus* (CMV), occur in most of the banana-growing regions. In the lower Pulney hills of Tamil Nadu, India, a famous, uniquely flavored elite dessert banana cv. Virupakshi (Pome group, AAB) has been almost destroyed by the BBTD and the area under this banana has been reduced from 18,000 ha to 2000 ha (Kesavamoorthy 1980). A loss of about Rs 40 million annually has been reported in Kerala state alone due to this disease (Mehta et al 1964). The survey made during May 2009, in lower Pulney hills, Dindigul district of Tamil Nadu, recorded a 14–72% incidence of BBTD in Hill banana (Selvarajan et al 2011). Recent remerengences of bunchy top disease during 2007–2010 in Kodur, Andhra Pradesh and Jalagon, Maharashtra, to the extent of causing loss of 50 million US$ per annum, has drawn attention of the policy makers and plant pathologist in India. Banana streak virus disease (BSD) has become a major threat to banana production, international exchange, tissue culture and breeding programs. BSV is widely distributed in all banana-producing countries (Hull et al 2000). A yield loss of 49–48% has been recorded in cv. Poovan (Mysore, AAB) due to BSV (Thangavelu et al 2000). Estimated yield losses of between 7 and 90% attributed to the disease (Lockhart et al 1998, Davis et al 2000, Daniells et al 2001, Harper et al 2004, Agindotan et al 2006).

BBrMV was first reported in 1979 in the Philippines at Davao on the island of Mindanao (Thomas & Magnaye 1996) and is widespread throughout the Philippines, India, Sri Lanka, Vietnam and Western Samoa, Costa Rica, Uganda, Ghana, Zanzibar and South Africa (Rodoni et al 1999). Around 40% reduction of bunch weight of *kokkan* diseased plant over healthy was recorded in Kerala. The yield loss caused by BBrMD in Kerala was 52 and 70% in cvs. Nendran and Robusta (Cherian et al 2002). Selvarajan & Jeyabaskaran (2006) reported that the average yield reduction in cv. Nendran due to BBrMV was 30%; the yield loss varied from farm to farm having different soil fertility. Cucumber mosaic disease caused by CMV has attained a serious status in most of the banana-growing states of India. In India, it was reported in Maharastra in 1943 by Kamat & Patel (1951). Rao (1980) reported the incidence of banana mosaic disease in cultivars Robusta, Dwarf Cavendish and Rasthali from 2 to 7% in parts of Andhra Pradesh and Tamil Nadu. An outbreak of banana mosaic was recorded in Poovan in Tamil Nadu during 1986 and 1988 (Kathirvel et al 1986, Mohan & Laksmanan 1988). A serious outbreak of CMV in tissue culture plants of Grand Nain banana was recorded in Jalgaon, Maharastra, during 2008–2010.

Plant–virus interactions are extremely complex and have been studied in depth for more than half a century. As a consequence, the mechanisms linked to viral accumulation inside host cells, movement of virus within the plant as well as the plant defense mechanisms, have been partially elucidated (Hull 2009). Viruses need living tissue for their multiplication and thus do not normally cause the death of the host, although there are exceptions. A large body of evidence has recently shown that to accomplish their life cycle, plant viruses need to confront plant defense mechanisms and to hijack the functions of different host factors. As a consequence, viral components must interact and/or interfere with host components that, in turn, in some instances, would cause an alteration in the plant physiology resulting in the development of symptoms. Indeed, recent discoveries have evidenced that plant development is affected by plant–virus interactions, which interfere with a broad range of cellular processes, such as hormonal regulation, cell-cycle control and endogenous transport of macromolecules etc. Nevertheless, the identities of all host factors involved in the viral cycle are still unknown. In this chapter we have highlighted the present state of knowledge about plant virus interactions in banana.

VIRUS–HOST INTERACTION AT THE PLANT LEVEL

There are several factors involved in unequal distribution of virus concentrations between the initially infected leaf and the rest of the plant. Many of the changes in host plant metabolism are probably secondary consequences of virus infection and non-essential for virus replication. A single gene change in the host or a single mutation in the virus may change an almost symptomless infection into a severe disease (Hull 2009). Metabolic changes occur when a virus interacts with the host for its multiplication and these changes are expected to alter the quality and/or quantity of leaf exudates, which, in turn, will be reflected on the phyllosphere microflora of such plants. The leaf exudates of healthy banana plants contained higher concentrations of glucose and sucrose than those of the BBTV-infected plants (Balakrishnan Nair 1969) and the total microbial population on the leaf surface of healthy plants was found to be lower than that of the infected plants. Balakrishnan Nair & Wilson (1975) studied the microbial population in the phyllosphere of BBTV-infected banana plants in comparison with healthy plant in cv. Nendran. The results indicated that bacterial, actinomycete and total microbial populations on the leaves of BBTV-infected plants were higher than on the leaves of healthy plants. The populations of fungi on the middle and bottom leaves of BBTV-infected plants were higher than on the healthy plants, and the age of plants did not appear to influence the population of phyllosphere microflora. Virus infection has various effects on the growth of plants. The stunting of growth could be due to a change in the activity of growth hormones, a reduction in the availability of the products of carbon fixation by a direct effect on chloroplast structure, or translocation of fixed carbon and a reduction in uptake of nutrients. Magee (1927) reported that the first symptom of the disease was the appearance of dark-green streaks on the undersurface of the leaf. As the disease progresses, infected leaves become progressively stunted and malformed and have a more upright bearing than usual, eventually resulting in a 'bunchy' display. BBTV-affected plants show intermittent dark green dots, dash, streaks of variable length like 'Morse code' patterns on the leaf sheath, midrib, leaf veins and petioles of infected plants. Leaves produced are progressively shorter, brittle in texture, narrow, and give the appearance of bunchyness at the top (Fig. 3.1, A, B and C), hence the name (Magee 1927). In late infection of BBTV, or in the case of latency, the plant can throw a 'bunch' but the fingers never develop to maturity. Fruits of infected plants are malformed. In the case of Grand Naine, the BBTV-affected plants throw a bunch with an extremely long or very short peduncle. Sometimes, affected Grand Naine banana fingers appear like a non-Cavendish type. Marginal chlorosis and yellowing of whole leaf lamina, resembling iron deficiency, are common symptoms noticed in BBTV infection. Vein flecking symptoms in the lamina are also noticed. When infection occurs very late in the season the plant would show dark green streaks at the tip of the bracts of the male flower bud. Sometimes the bract's tip of male bud is converted into a green and leafy structure (Thomas et al 1994).

Zhang et al (1997) studied the content of three endogenous hormones, viz., gibberellic acid (GAs), isopentenyladenine group (iPAs) and abscisic acid (ABA), and virus movement in banana plants infected by BBTV, by enzyme-linked immunosorbent assay (ELISA). They reported that the content of GAs in inoculated plants at different infecting periods was lower than that in healthy plants, in spite of increasing slightly during infection. The iPAs content of infected plants decreased significantly after the 14th day of BBTV inoculation and was low-level during infection. ABA was greatly induced and accumulated in BBTV-infected banana plants. The ABA content of infected plants tested on the 35th day of

FIGURE 3.1　Banana bunchy top. (A) BBTV-infected Hill banana exhibiting marginal leaf chlorosis, narrow leaves with short petiole leading to bunchyness in the daughter suckers. (B) Bracts of male bud showing greenish leafy-like structure instead of purple colored normal bud. (C) Malformed choked banana bunch emerging from a BBTV infected cultivar Grand Naine banana; the banana fingers appear like a non-Cavendish type of banana. (D) Banana black aphid, *Pentalonia nigronervosa*, feeding on BBTV infected leaves.

BBTV inoculation was 3.34 times more than that of the control. The concentration of BBTV contents in inoculated leaves and in top leaves—determined by indirect ELISA—indicated that BBTV particles replicated greatly both in inoculated leaves and top leaves after the 21th day of inoculation and symptom appeared in top leaves after the 35th day of inoculation and these results indicated that the symptom of BBTD may be closely related to the unbalance of endogenous hormones of infected plants, but indirectly related to BBTV movement within the plant. Hook et al (2008) studied the effect of BBTV infection on growth and morphology of banana plants; their results revealed that BBTV-infected plants showed significant reduction in petiole size, plant canopy and height, leaf area, pseudostem girth and chlorophyll content, compared with healthy control plants.

There have been few studies on the distribution and movement of BBTV within the banana plant. Raj et al (1970) observed that BBTV had an incubation period of 5–15 days at the point of inoculation before migrating downward. The virus was observed to move to the lower regions of the plant at a very rapid rate, possibly in a few hours or less, following the incubation period based on symptom development rather than on detection of the virus. Wu & Su (1992) reported that BBTV was detected, using ELISA, in all young leaves of banana plants with symptoms, but the virus was either absent or present in low concentrations in the symptomless older leaves of the infected plants. This correlated with the virus being first detectable in the fourth leaf unfurling, usually about 27 days after inoculation (Thomas & Dietzgen 1991). Wu & Su (1992) also reported that BBTV was not detected in the corm (rhizome) and was either absent or present in low concentrations in the roots and the sheaths of the older symptomless leaves. Hafner et al (1995) reported that BBTV replicated for a short period at the site of inoculation and subsequently moved down the pseudostem to the basal meristematic region and ultimately into the roots and newly formed leaves; relatively high levels of BBTV DNA were found in the roots of infected plants. This study indicated that the virus has the ability to move into these leaves but may not have replicated or accumulated to significant levels. The appearance of multimeric forms of BBTV suggested that the virus may have replicated via a rolling-circle mechanism. BBTV has been found to express visual symptoms only 23 to 25 days after inoculation, but the virus could be detected early from young roots or cortex tissue even before expression of symptoms (Nancy 2003).

BSV-infected plants initially develop small dots with a golden yellow discoloration; later, the dots extend to form long streaks. The chlorotic streaks (Fig. 3.3, A and B) become necrotic, giving a blackish appearance to the lamina. Necrotic streaks are also observed on the midrib, petiole and pseudostem. Bunch choking, abortion of bunch and seediness in fingers are observed in infected plants. Sometimes, diseased plants are stunted and the fruit becomes distorted, with a thinner peel, and the bunches are small in size. On some occasions, the heart leaf rots and eventually the plant dies. Necrosis of cigar leaf and death of the entire plant has been recorded in plantain hybrids in Nigeria (Harper & Hull

1998). Leaf stripping symptoms are observed in infected cvs. Poovan, Grand Nain and Robusta. In India, bunches from infected plants bear a female phase followed by a short male phase again; a female phase has been observed in BSV infected in cv. Poovan. Emergence of bunch by breaking open the side of the pseudostem has frequently been observed in cv. Poovan. Fruit malformation and seediness in cv. Poovan are associated with the BSV infection. Pseudostem splitting and peel splitting are also observed with BSV infection (Lockhart & Jones 2000a). Depending on the infecting BSV species, highly susceptible banana cultivars can develop more severe symptoms, such as pseudostem splitting and necrosis, eventually leading to the death of infected plants (Lockhart & Jones 2000a, Daniells et al 2001).

Symptom incidence alone may not truly reflect the extent of virus invasion in plants because of the possibility of variation in invasion, replication or cell-to-cell spread of a virus among infected genotypes (Cooper & Jones 1983, Pataky et al 1990). This, together with an erratic distribution of BSV symptoms and antigens over individual leaves, as well as between different leaves of the same plant (Lockhart 1986, Dahal et al 1998), may explain why some genotypes suffer lower yield loss than others. In such cases, additional characteristics, such as symptom severity, virus titre and yield, have been used to determine and differentiate between 'tolerance' and 'resistance' (Pataky et al 1990, Kerns & Pataky 1997). Virus occurrence and symptom expression as well as the relative concentration of BSV antigens, fluctuated greatly between seasons during the cropping cycle, being high during the rainy season and low or negligible during the hot dry season. The natural incidence of plants with symptoms and BSV-infected plants varied between genotypes (Dahal et al 2000).

The symptoms produced by banana bract mosaic virus are very distinct. BBrMD is characterized by the presence of spindle-shaped pinkish to reddish streaks on bracts (Fig. 3.2, A and B), pseudostem, midrib and peduncle—and also observed on the fingers (Rodoni et al 1997, Thomas et al 1997, Rodoni et al 1999, Singh et al 2000). The characteristic mosaic symptoms on the flower bracts give the disease its common name. Necrotic streaks on fingers, leaf, pseudostem and mid rib are also recorded due to BBrMD. In Nendran, the leaf orientation changes in such a way that it gives the appearance of the 'Travelers palm' plant (Balakrishnan et al 1996, Singh et al 1996). Bunches with an unusually very long or very short peduncle, chocking of bunches, raised corky growth on the peduncle, are also observed (Selvarajan & Jeyabaskaran 2006). Infected plants bear bunches with an extended female phase after a short male phase has been observed. In Robusta, fingers of infected plants stop to develop and give the appearance of a 'pencil', explaining the local name 'pencil kai' (pencil-sized fruit). When the pseudostem was cut horizontally, the necrosis was found to be deep seated. Young suckers also exhibit reddish spindle-shaped streaks on the pseudostem. Plants with BBrMD showed a significant reduction in height, girth, leaf area, finger weight and girth over healthy plants. Though this disease bears the name of bract mosaic, it was reported that this disease

FIGURE 3.2   Bract mosaic. (A) Reddish discoloration of pseudostem, an initial symptom of BBrMD. (B) Dark reddish to purple colored spindle-shaped streak mosaic symptoms on bracts of a male bud. (C) *Aphis gossypii*, a vector of BBrMV. (D) Electron micrograph of BBrMV, flexuous rod-shaped virus particles measuring 750 nm in length.

could be diagnosed only when symptoms appears on bract; in Thiruvanantha-puram, Kerala, India, farmers have named the disease *Pola roga*, which means 'disease of pseudostem' in cultivar Nendran (Thangavelu et al 2000). Singh (2003) reported that the bract of the infected plants exhibited spindle-shaped discontinuous dark red streaks. Dark red to purple mosaic streaks were also observed on the pseudostem after the removal of the leaf sheath. The emerging suckers were deeply pigmented red to purple, and foliar symptoms appeared as chlorotic streaks parallel to veins and petioles. Dhanya et al (2006) reported that the production of polyphenol oxidase in BBrMV-infected plants was greater than in healthy plants.

Banana mosaic, or infectious chlorosis, is characterized by a range of symptoms—including diffused foliar mosaic (Fig. 3.4, A, B and C), severe chlorosis, chlorotic streaking or flecking, stripes, line patterns, ring spots, leaf curling, distortion, rosette appearance of leaf arrangement to stunting of the plant—and this disease is a serious threat to banana cultivation (Niblett et al 1994). Necrosis of emerging cigar leaves leads to varying degrees of necrosis in the unfurled leaf lamina. The internal tissue of the pseudostem also becomes necrotic. Affected plants throw small bunches with malformed fingers. Uneven ripening has been associated with the virus. The critical features of this disease are the variability and biologic properties of the pathogen. Most strains of the virus are so-called 'common' strains, which do not produce severe symptoms or cause significant crop damage (Lockhart & Jones 2000b). Mosaic symptoms are most pronounced during cool weather but do not persist; in contrast, the severe heart rot strains of CMV (Magee 1940a, Bouhida & Lockhart 1990) can cause damaging symptoms, which include chlorosis, cigar leaf necrosis, internal pseudostem necrosis and plant death.

THE VIRUS–VECTOR RELATIONSHIP

Viruses can enter the plant cell only passively through wounds caused by physical injuries due to environmental factors or by vectors. BBTV is transmitted by banana black aphid, *Pentalonia nigronervosa*, in a persistent manner (Fig. 3.1D) (Magee 1940b, Wu & Su 1990, Hu et al 1996). No other aphids are known to transmit BBTV. Magee (1940a) first provided evidence for the circulative, persistent nature of transmission by the aphid. It was demonstrated that the nymphs could retain the virus through moults and that the virus was retained by viruliferous aphids during daily transfers to fresh plants for as long as 13 days after removal from diseased plants. The nymphal stage of the aphid is more efficient in transmission. The aphids are usually found clustered around the unfurled heart-leaf and the sheathing leaf base of petioles which are ideal locations for feeding and protection. The aphids feed at the base of the plant petioles near the pseudostem on which the banana aphid feeds (Magee 1927, Robson et al 2007) and readily acquires the virus (Magee 1927, Anhalt & Almeida 2008). They are found on the base of the pseudostem and on very young suckers. The aphids flourish throughout the year in hills, but are more numerous during the rainy season. Both winged and wingless individuals occur in a normal aphid colony. Banana aphids produce large quantities of 'honey-dew' which attracts ants. The presence of ants is a good indication of the presence of aphids on a banana plant. An aphid acquires the virus after at least 4 hours of feeding on an infected plant and it can retain the virus through its adult life for a period of 15–20 days. BBTV is transmitted for at least 15–20 days post-acquisition and there is a detectable latent period in the vector estimated to be 20–28 hours (Magee 1940b, Hu et al 1996, Anhalt & Almeida 2008).

Magee's experiments indicated that an average of about 25 days' incubation is necessary for the development of banana bunchy top symptoms (Magee 1927). We conducted a transmission experiment with aphids in tissue culture plants; only 29.4% of the plants inoculated with aphids expressed the typical symptoms of BBTV. The transmission efficiency of many vector-borne plant viruses has been shown to be affected by temperature. Temperature has also been shown to have an affect on BBTD symptom development, spread, transmission efficiency and on vector biology. Wu & Su (1990) compared BBTV acquisition efficiency at 16, 20 and 27°C using groups of aphids for transmission experiments; they demonstrated that temperature affects efficiency, with no transmission at 16°C and maximum efficiency at 27°C. BBTD symptoms have been estimated to take 3 weeks to 4 months after inoculation to be observed; most of this variability is suggested to be due to temperature—epidemics can potentially increase in size and spread faster with temperature increments. Anhalt & Almeida (2008) reported that adult aphids transmitted the virus more efficiently at 25 and 30°C than at 20°C, but temperature had no affect on transmission efficiency by nymphs. Adult aphids transmitted BBTV more efficiently than third instar nymphs at all temperatures tested. By decoupling the relationship between temperature and aphid BBTV acquisition or inoculation, we determined that temperature affected inoculation events more strongly than acquisition. Longer plant access periods increased viral acquisition and inoculation efficiencies in a range of 60 minutes to 24 hours. Both BBTV acquisition and inoculation efficiencies peaked after 18 hours of the plant access period (Nancy 2003). It has been found that the optimum temperature range for acquisition of virus by the vector was 25°C to 27°C. Minimum inoculation feeding of 3 hours was required to successfully acquire the BBTV by the vector. Menon & Christudas (1967) reported the life history and population dynamics of *Pentalonia nigronervosa* in Kerala, India, and stated that climatic conditions existing during the summer months and rainy weather were unfavorable for the banana aphid in Kerala. Samraj et al (1970) reported that a minimum time of 5 days and a maximum of 10–15 days is required for the downward movement of the virus after inoculation with the aphid; this might change, depending on the vigor of the plant. Furthermore, aphids were not confined to the site of inoculation. Hafner et al (1995) reported that BBTV does not replicate within its aphid vector. To study the BBTV latency, BBTV was transferred through viruliferous aphids (*Pentalonia nigronervosa*) to virus-free tissue culture plants of the cultivar Grand Nain. Out of 24 plants, 7 plants did not express symptoms for up to 524 days, but they were PCR-positive and the virus remained latent (Selvarajan, unpublished).

Bressan & Watanabe (2011) stated that BBTV antigens specifically localized in the anterior midgut (AMG) and specific cells forming the principal salivary glands (PSGs) within its aphid vector; this was determined by an immunofluorescence assay. Watanabe & Bressan (2013) examined the translocation, compartmentalization and retention of BBTV in the aphid vector *Pentalonia nigronervosa*. Their results indicated that BBTV translocates rapidly through the aphid vector; it is internalized in the anterior midgut, in which it accumulates, and is retained at concentrations higher than either those in the haemolymph or in the principal salivary glands.

BSV has been shown to be experimentally transmitted by *Planococcus citri* and *Saccharicoccus sacchari*, both of which colonize banana (Lockhart et al 1992). BSV was detected in field-captured *Dysmicoccus brevipes*; this shows that *D. brevipes* might act as a vector (Kubiriba et al 2001). *Ferrisia virgata* (striped mealy bug) (Fig. 3.3C) was found to transmit BSV from banana to banana (Selvarajan et al 2006). Meyer et al (2008) reported transmission of activated episomal *Banana streak OL virus* (BSOLV) to banana cv. Williams by three mealy bug species, viz., *Dysmicoccus brevipes*, *Planococcus citri*, and *P. ficus*. BBrMV is non-persistently transmitted through several aphid species, viz., *Pentanlonia nigronervosa*, *Rhopalosiphum maidis*, *Aphis gossypii* (Fig. 3.2C) (Magnaye & Espino 1990, Munez 1992), and in addition the cowpea aphid *Aphis craccivora* has also been reported to transmit the disease (Selvarajan et al 2006). The aphids *Aphis gossypii*, *A. craccivora* (Fig. 3.4C), *Rhopalosiphum maidis*, *R. purnifolia*, *Myzus persicae* and *Macrosiphum pisi* have been reported to carry and spread CMV (Magee 1940b, Capoor & Varma 1968, Mali & Rajagore 1980, Rao 1980).

VIRUS–HOST INTERACTION AT THE MOLECULAR LEVEL

BBTV belongs to the genus *Babuvirus* of the family *Nanoviridae*. The size of the isometric virion is 18–20 nm in diameter. BBTV is a multi-component virus which consists of at least six particles. Each particle is packed with a distinct circular single-stranded DNA (cssDNA). Each genomic component is approximately 1 kb in size (BBTV DNA-R, -S, -M, -C, -N and -U) (Harding et al 1993, Burns et al 1995, Vetten et al 2005). A single protein of 19.6 kDa associated with the virions is assumed to be the coat protein (Harding et al 1991, Burns et al 1995, Wanitchakorn et al 1997). BBTV-DNA R encodes a master replication initiation protein (Rep). The functions of DNA-U is unknown, while DNA-M, -C and -N encode the movement protein, cell-cycle link (clink) protein, and the nuclear shuttle protein, respectively (Wanitchakorn et al 2000).

BSV is a pararetrovirus belonging to the genus *Badnavirus* in the family *Caulimoviridae*. The virions are non-enveloped, bacilliform in shape (Fig. 3.3D), on average 130–150 nm × 30 nm in size, and they contain a circular dsDNA as the genome (~ 7–8 kbp) (Lockhart & Olszewski 1993, Singh et al 2000) which is replicated by reverse transcription. Their genomes contain three open reading frames (ORF) on one strand. ORF I and II potentially encode two small proteins of unknown function of 20.8 and 14.5 kDa, respectively. ORF III

FIGURE 3.3 Streak disease. (A) Infected leaf showing golden yellowish chlorotic streaks and necrotic streaks originating from the midrib in cultivar Poovan (syn: Mysore). (B) Severe streak symptoms on the whole leaf lamina due to BSMYV infection in cv. Grand Nain. (C) *Ferrisia virgata*, a striped mealy bug vector of BSV. (D) Electron micrograph of BSMYV; the virions are bacilliform.

encodes a polyprotein of 208 kDa consisting of a putative cell-to-cell movement protein, coat protein, an aspartic protease, and viral replicase, which has reverse transcriptase (RT) and ribonuclease H (RNase H) functions (Harper & Hull 1998). This polyprotein is thought to be post-translationally cleaved into functional units by the aspartic protease. BSVs have a wide serologic

FIGURE 3.4 Banana mosaic. (A) Typical chlorotic mosaic symptom on lamina of a CMV-infected banana plant. (B) Severe mosaic on a field-grown banana plant of cv. Grand Nain. (C) Cigar leaf rotting symptoms due to infection of a severe strain of CMV. (D) *Aphis craccivora*, vector of CMV on a young cowpea plant seedling.

and molecular variability. Indeed, BSV is considered to be the generic name of several distinct virus species, with similar biologic properties, all infecting banana. Studies of partial sequences of BSV collected in Uganda and Mauritius (Harper et al 2004, Geering et al 2005a, Jaufeerally-Fakim et al 2006) revealed great genetic diversity, with up to 30% nucleotide divergence among BSV isolates infecting the same *Musa* host plant. Now, ten BSV isolates have been classified as independent species of the genus *Badnavirus*. They are: *Banana streak OL virus* (BSOLV), *Banana streak GF virus* (BSGFV), *Banana streak MY virus* (BSMYV), *Banana streak VN virus* (BSVNV), *Banana streak IM virus* (BSIMV), *Banana streak CA virus* (BSCAV), *Banana streak UA virus* (BSUAV), *Banana streak UI virus* (BSUIV), *Banana streak UL virus* (BSULV), and *Banana streak UM virus* (BSUMV) (James et al 2011). Integrated badnavirus sequences, termed endogenous pararetroviruses (EPRVs), are known to occur within the banana genome (Harper & Hull 1998; Harper et al 1999, Ndowora et al 1999, Geering et al 2001, 2005a, b). Although some EPRV sequences show homology to the genomes of recognized or tentative BSVs, including BSMYV, BSGFV, BSOLV and BSIMV, many others have no known episomal counterpart (Geering et al 2005a). Two types of integrated BSV sequence are known to occur in banana. The first type contains the majority of banana EPRVs and comprises incomplete virus genomes which are incapable of causing infections. The second type of integrated sequence, known as endogenous activatable BSVs (eaBSVs), consist of the entire genome of characterized episomal BSVs which exist as multiple non-contiguous regions of the virus DNA combined with host-genomic sequences. The origin of these field outbreaks was correlated with the presence of infectious endogenous BSV (eBSV) sequences present within the *M. balbisiana* genome (Harper et al 1999, Ndowora et al 1999, Dallot et al 2001, Lheureux et al 2003, Côte et al 2010). Different abiotic stresses have been identified that trigger the production of viruses from infectious eBSV: temperature differences, water stress, *in vitro* culture, and interspecific crosses (Dahal et al 1998, 2000, Dallot et al 2001, Lheureux et al 2003, Côte et al 2010). Indeed, extensive studies on micro propagation procedures have clearly shown that the proliferation stage was a major determinant in the spontaneous appearance of BSV viral particles in interspecific *Musa* cultivars, regardless of the nature of the hybrids used (Dallot et al 2001, Côte et al 2010). This indicates that such activation is a general phenomenon. Nevertheless, little is known about the exact mechanisms underlying the expression of functional viral genomes from eBSV. While incomplete integrants have been found in both the A- and B-genomes derived from the wild progenitors of domesticated banana, *Musa acuminata* (A) and *M. balbisiana* (B), respectively, the eaBSVs have been detected only in the B-genome of various banana accessions (Geering et al 2001, 2005b, Gayral et al 2008). Selvarajan et al (2004) reported the integration of banana streak virus genome in *Musa* germplasm accessions having one or more balbisiana (B) genome as their constituent. BBrMV belongs to the family *Potyviridae* and the genus *Potyvirus*. BBrMV has flexuous filamentous (Fig. 3.2D) particles (660–760 × 12 nm) with a single-stranded positive-sense RNA genome consisting of 9711 nucleotides, excluding the 3′ terminal poly (A) tail (Ha et al 2008, Balasubramanian & Selvarajan 2012). The genome consists of a single large open reading frame (ORF) of 9378 nucleotides. CMV belongs to the genus *Cucucmovirus* and the family Bromoviridae (Roossinck et al 1999). The virus particles are isometric in shape; they measure 29 nm in diameter, and each particle is composed of 180 subunits. CMV is a multi-component virus; its genome consists of single-stranded tripartite positive-sense RNAs (RNA 1, RNA 2 and RNA 3) and an additional sub-genomic RNA (RNA 4) derived from RNA 3 (Hubili & Francki 1974). RNA 1 and RNA 2 encode the 1a and 2a proteins, involved in virus replication (Hayes & Buck 1990), while RNA 3a encodes the 3a movement protein (MP) (Suzuki et al 1991) and 3b expressed from RNA 4 coat protein (CP) (Hubili & Francki 1974, Schwinghamer & Symons 1975) with 5′ cap structures and 3′ conserved regions.

In recent years, several studies have indicated that specific viral proteins function as silencing suppressors to overcome the challenge posed by the RNAi/post-transcriptional gene silencing (PTGS) machinery evolutionarily conserved among all plants. Recent evidence suggests that RNA silencing has a more general role in the regulation of gene expression in addition to its role in host defense against viral infection (Voinnet et al 2003). Helper component-proteinase (HC-Pro) of potyviruses, which was one of the first suppressors identified, interferes with RNA silencing at a step upstream of the production of siRNA (Anandalakshmi et al 1998, Brigneti et al 1998, Kasschau et al 2003). On the other hand, the 2b protein encoded by CMV is able to prevent the spread of RNA silencing signals by blocking their translocation (Brigneti et al 1998, Guo & Ding 2002). BBTV coat protein and putative movement protein have been shown to function as silencing suppressors in a study involving infection of *Potato virus X* on *Nicotiana benthamiana* (Niu et al 2009). They did not observe any such silencing suppression activity in inoculated transformed plants probably because the viral DNA could not replicate in the absence of Rep protein and hence no coat protein and movement protein were synthesized in those plants. The sequence-specific RNA degradation pathway, directed by small interfering RNAs (siRNAs), restricts the accumulation and spread of exogenous virus invaders (Mlotshwa et al 2008). To overcome this strategy, most plant viruses have evolved suppressor proteins to counteract host RNA silencing (Chapman et al 2004). Similar to siRNA-directed RNA degradation, miRNA metabolism can also be altered by the activity of viral silencing suppressors through attacking common elements of the two pathways (Bazzini et al 2007). Also, plants can generate miRNAs during viral infection; these are involved in the regulation of the virus defense process in plants or in targeting some key genes of virus development (Lu et al 2008). We identified 18 conserved miRNAs and detected 25 potential targeted genes in banana. Real-time PCR assays were performed to profile the expression levels of three miRNAs, viz., miR156, 159 and 166, after infection by *Banana streak Mysore virus*. The symptom severity was correlated with the miRNA accumulation, and increased expression of all three miRNAs during virus infection. This study will pave the way to an understanding of the plant–pathogen interactions and host defense signalling pathways (Mary Sheeba et al 2013).

TRANSGENIC PLANT AND VIRUS RESISTANCE

Genetic engineering has been successfully employed to incorporate virus resistance into existing desirable plant cultivars (Collinge et al 2010, Simón-Mateo & García 2011). Transgenic banana, with resistance against BBTV, has been attempted in Australia, Hawaii and India using a pathogen-derived resistance approach. Coat protein gene, full-length and truncated rep gene, RNAi vector using rep gene of BBTV have been applied. We have developed a few putative transgenic lines of Hill banana for BBTV resistance with a cp-gene-mediated approach and an RNAi approach using replicase (Selvarajan, unpublished). Borth et al (2011) have succeeded in generating banana plants resistant to BBTV using the viral Rep gene. They used a variety of constructs based on the BBTV Rep gene sequence and its regulatory regions to achieve this resistance. These included constructs based on a mutated Rep sequence, expression of an antisense strand of Rep, and an inverted repeat of the Rep sequence. The total number of BBTV-resistant banana plants, as a percentage of the total number of transgenic plants tested in BBTD bioassays in this study, did not exceed 13% for any of the four constructs used. Transgenic 'Cavendish' with potential BBrMV resistance has already been generated at Queensland University of Technology. These transgenic lines were transformed with the coat protein coding region of a Philippines isolate under the control of the maize polyubiquitin promoter using microprojectile bombardment (Dale & Harding 2003).

Post-transcriptional gene silencing (PTGS) using intron-hairpin-RNA (ihpRNA) is widely used to knock down the expression of a gene at the mRNA level in a variety of plants (López-Gomollón & Dalmay 2010). By analogy, the viral mRNAs of both RNA and DNA viruses should also be degraded through PTGS, and hence PTGS should be equally effective against both RNA and DNA viruses. Further, PTGS directed against vital viral proteins does not involve the production of any new proteins in the transgenic plants as once the dsRNA is synthesized, the host plant's machinery recognizes it as an aberrant RNA and cleaves any cognate mRNA formed later in that cell using the siRNAs and the related RNAi machinery.

Shekhawat et al (2012) explored the concept of using ihpRNA transcripts corresponding to viral master replication initiation protein (Rep) to generate BBTV-resistant transgenic banana plants. This study indicated that the use of an intron between the two complementary domains of Rep-derived sequences makes for efficient siRNA synthesis, and they have obtained 100% resistance against BBTV infection in transgenic plants. Elayabalan et al (2012) generated hill banana resistant to BBTV using an RNAi-BBTV Rep-mediated approach. The transformed plants were symptomless, and the replication of challenged BBTV was almost completely suppressed. This approach was shown to be effective in the management of BBTV in hill banana. There are no reported successful attempts to generate *Badnavirus* or *Caulimovirus* resistance. Banana is a difficult model in which to develop a strategy, and the hypervariability of episomal BSV would seem to make broad resistance difficult.

CONCLUSION

Banana viruses cause dreadful diseases in tropical and sub-tropical conditions. In the past two decades, with the advent of molecular techniques, the genomes of the banana viruses have been elucidated and their genetic diversity has been studied extensively. However, the molecular interactions of viral proteins with the host metabolites and cellular proteins have not been studied in depth.

Of late, the movement of virus within the vector and the host has been studied using confocal microscopy with respect to BBTV. The influence of environment on disease development and the spread of infection have been studied for BBTD in order to forecast epidemics; however, data are lacking for all the viruses infecting banana. In the case of banana streak virus, the role of EPRVs in infection has been proved to some extent in B genome-containing cultivars. The significance of dead eBSV sequences in the host genome of wide range of hybrids and wild bananas is not understood. Virus infection and its influence on plant miRNAs need critical research efforts to understand the reasons for the different symptoms induced by banana viruses. In future, the proteomics and transcriptomic approaches have to be used to learn more about virus–vector–host interactions with respect to banana viruses.

REFERENCES

Agindotan, B.O., Winter, S., Lesemann, D., Uwaifo, A., Mignouna, J., Hughes, J., Thottappilly, G., 2006. Diversity of banana streak-inducing viruses in Nigeria and Ghana: Twice as many sources detected by immunoelectron microscopy (IEM) than by TAS-ELISA or IC-PCR. African Journal of Biotechnology 5 (12), 1194–1203.

Anandalakshmi, R.G., Pruss, J., Ge, X., Marathe, R., Mallory, A.C., Smith, T.H., Vance, V.B., 1998. A viral suppressor of gene silencing in plants. Proceedings of the National Academy of Sciences of the United States of America 95, 13079–13084.

Anhalt, M.D., Almeida, R.P.P., 2008. Effect of temperature, vector life stage, and plant access period on transmission of Banana bunchy top virus to banana. Phytopathology 98, 743–748.

Balakrishnan, S., Gokulapalan, C., Paul, S., 1996. A widespread banana malady in Kerala, India. Infomusa 5 (1), 28–29.

Balakrishnan Nair, P.K., 1969. Effect of bunchytop virus infection on the chemical constituents, phyllosphere microfiora and the incidence of Cordana leaf spot in banana. M.Sc. Ag. thesis. Kerala University, India.

Balakrishnan Nair, P.K., Wilson, I., 1975. Phyllosphere microflora of banana plants in relation to bunchy top virus infection. Sydowia 28 (1/6), 162–165.

Balasubramanian., V., Selvarajan., R., 2012. Complete genome sequence of a banana bract mosaic virus isolate infecting the French plantain cv. Nendran in India. Archives of Virology 157, 397–400.

Bazzini, A.A., Hopp, H.E., Beachy, R.N., Asurmendi, S., 2007. Infection and co-accumulation of tobacco mosaic virus proteins alter microRNA levels, correlating with symptom and plant development. Proceedings of the National Academy of Sciences of the United States of America 104 (29), 12157–12162.

Bressan, A., Watanabe, S., 2011. Immunofluorescence localisation of *Banana bunchy top virus* (family Nanoviridae) within the aphid vector, *Pentalonia nigronervosa*, suggests a virus tropism distinct from aphid-transmitted luteoviruses. Virus Research 155 (2), 520–525.

Brigneti, G., Voinnet, O., Li, W.X., Ji, L.H., Ding, S.W., Baulcombe, D.C., 1998. Viral pathogenicity determinants are suppressors of transgene silencing in *Nicotiana benthamiana*. EMBO Journal 17, 6739–6746.

Borth, W., Perez, E., Cheah, K., Chen, Y., Xie, W.S., Gaskill, D., Khalil, S., Sether, D., Melzer, M., Wang, M., Manshardt, R., Gonsalves, D., Hu, J.S., 2011. Transgenic banana plants resistant to banana bunchy top virus infection. Acta Horticulture 897, 449–457.

Bouhida, M., Lockhart, B.E.L., 1990. Increase in importance of cucumber mosaic virus infection in greenhouse grown bananas in Morocco. Phytopathology 80, 981.

Burns, T.M., Harding, R.M., Dale, J.L., 1995. The genome organization of banana bunchy top virus: analysis of six ssDNA components. Journal of General Virology 76, 1471–1482.

Capoor, S.P., Varma, P.M., 1968. Banana mosaic in the Deccan. Indian Phytopath. Soc. Bull. 4, 11–14.

Chapman, E.J., Prokhnevsky, A.I., Gopinath, K., Dolja, V.V., Carrington, J.C., 2004. Viral RNA silencing suppressors inhibit the microRNA pathway at an intermediate step. Genes & Development 18, 1179–1186.

Cherian, A.K., Rema Menon, Suma, A., Shakunthala Nair, & Sudheesh, M.V. (2002). Impact of Banana bract mosaic diseases on the yield of commercial banana varieties of Kerala. Global Conference on Banana and Plantain, Bangalore, 28–31 October, 2002. Abstract p. 155.

Collinge, D.B., Jorgensen, H.J., Lund, O.S., Lyngkjaer, M.F., 2010. Engineering pathogen resistance in crop plants: current trends and future prospects. Annual Review of Phytopathology 48, 269–291.

Cooper, J.I., Jones, A.T., 1983. Responses of plant viruses: proposals for the use of term. Phytopathology, 73,127–128.

Côte, F.X., Galzi, S., Folliot, M., Lamagnère, Y., Teycheney, P.Y., Iskra-Caruana, M.L., 2010. Micropropagation by tissue culture triggers differential expression of infectious endogenous *Banana streak virus* sequences (eBSV) present in the B genome of natural and synthetic interspecific banana plantains. Molecular Plant Pathology 11 (1), 137–144.

Dahal, G., Pasberg-Gauhl, C., Gauhl, F., Thottappilly, G., Hughes Jd', A., 1998. Studies on a Nigerian isolate of banana streak badnavirus. II. Effect of intraplant variation on virus accumulation and reliability of diagnosis by ELISA. Annals of Applied Biology 132, 263–275.

Dahal, G., Ortiz, R., Tenkouano, A., Hughes, Jd'A., Thottappilly, G., Vuylsteke, D., Lockhart, B.E.L., 2000. Relationship between natural occurrence of banana streak badnavirus and symptom expression, relative concentration of viral antigen, and yield characteristics of some micropropagated Musa spp. Plant Pathology 49, 68–79.

Dale, J., Harding, R., 2003. Strategies for the generation of virus resistant bananas. In: Atkinson, H., Dale, J., Harding, R. (Eds.), Genetic transformation strategies to address the major constraints to banana and plantain production in Africa. INIBAP, Montpellier (France), p. 130.

Dallot, S., Acuna, P., Rivera, C., Ramirez, P., Cote, F., Lockhart, B.E.L., Caruana, M.L., 2001. Evidence that the proliferation stage of micropropagation procedure is determinant in the expression of *Banana streak virus* integrated into the genome of the FHIA 21 hybrid (Musa AAAB). Archives of Virology 146, 2179–2190.

Daniells, J.W., Geering, A.D.W., Bryde, N.J., Thomas, J.E., 2001. The effect of *Banana streak virus* on the growth and yield of dessert bananas in tropical Australia. Annals of Applied Biology 139, 51–60.

Davis, R.I., Geering, A.D.W., Thomas, J.E., Gunua, T.G., Rahamma, S., 2000. First records of *Banana streak virus* on the island of New Guinea. Australasian Plant Pathology 29, 281.

Dhanya, M.K., Rajagopalan, B., Umamaheswaran, K., Ayisha, R., 2006. Isozyme variation in banana (*Musa* sp.) in response to bract mosaic virus infection. Indian Journal of Crop Science 1 (1–2), 140–141.

Elayabalan, S., Kalaiponmani, K., Subramanian, S., Selvarajan, R., Radhs, P., Ramalatha, M., Kumar, K.K., Balasubramanian, P., 2012. Development of *Agrobacterium*-mediated transformation of highly valued hill banana cultivar Virupakshi (AAB) for resistance to BBTV disease. World Journal of Microbiology and Biotechnology, DOI:10.1007/s11274-012-1214-z.

Gayral, P., Noa-Carrazana, J.C., Lescot, M., Lheureux, F., Lockhart, B.E.L., Matsumoto, T., Piffanelli, P., Iskra-Caruana, M.L., 2008. A single *Banana streak virus* integration event in the banana genome as the origin of infectious endogenous pararetrovirus. Journal of Virology 82 (13), 6697–6710.

Geering, A.D.W., Olszewski, N.E., Dahal, G., Thomas, J.E., Lockhart, B.E.L., 2001. Analysis of the distribution and structure of integrated *Banana streak virus* DNA in a range of *Musa* cultivars. Molecular Plant Pathology 2, 207–213.

Geering, A.D.W., Olszewski, N.E., Harper, G., Lockhart, B.E.L., Hull, R., Thomas, J.E., 2005a. Banana contains a diverse array of endogenous badnaviruses. Journal of General Virology 86, 511–520.

Geering, A.D.W., Pooggin, M.M., Olszewski, N.E., Lockhart, B.E.L., Thomas, J.E., 2005b. Characterization of *Banana streak Mysore virus* and evidence that its DNA is integrated in the B genome of cultivated *Musa*. Archives of Virology 150, 787–796.

Guo, H.S., Ding, S.W., 2002. A viral protein inhibits the long range signaling activity of the gene silencing signal. EMBO Journal 21, 398–407.

Ha, C., Coombs, S., Revill, P.A., Harding, R.M., Vu, M., Dale, J.L., 2008. Design and application of two novel degenerate primer pairs for the detection and complete genomic characterization of potyviruses. Archives of Virology 153, 25–36.

Hafner, G.J., Harding, R.M., Dale, J.L., 1995. Movement and transmission of banana bunchy top virus DNA component one in bananas. Journal of General Virology 76, 2279–2285.

Harding, R.M., Burns, T.M., Dale, J.L., 1991. Virus-like particles associated with banana bunchy top disease contain small single-stranded DNA. Journal of General Virology 72, 225–230.

Harding, R.M., Burns, T.M., Hafner, G., Dietzgen, R.G., Dale, J.L., 1993. Nucleotide sequence of one component of the banana bunchy top virus genome contains the putative replicase gene. Journal of General Virology 74, 323–328.

Harper, G., Hull, R., 1998. Cloning and sequence analysis of *Banana streak virus* DNA. Virus Genes 17, 271–278.

Harper, G., Hart, D., Moult, S., Hull, R., 2004. Banana streak virus is very diverse in Uganda. Virus Research 100, 51–56.

Hooks, C.R.R., Wright, M.G., Kabasawal, D.S., Manandhar, R., Almeida, R.P.P., 2008. Effect of banana bunchy top virus infection on morphology and growth characteristics of banana. Annals of Applied Biology 153, 1–9.

Hu, J.S., Wang, M., Sether, D., Xie, W., Leonhardt, K.W., 1996. Use of polymerase chain reaction (PCR) to study transmission of banana bunchy top virus by the banana aphid (*Pentalonia nigronervosa*). Annals of Applied Biology 128, 55–64.

Hubili, N., Francki, R.I.B., 1974. Comparative studies on tomato aspermy and *cucumber mosaic viruses*. III. Further studies on relationship and construction of a virus from part of the two viral genomes. Virology 61, 443–449.

Hull, R., 2009. Comparative plant virology, second ed. Academic Press, Oxford.

Hull, R., Harper, G., Lockhart, B.E.L., 2000. Viral sequences integrated into plant genomes. Trends in Plant Science 5, 362–365.

James, A., Geijskes, R.J., Dale, J.L., Harding, R.M., 2011. Molecular characterisation of six badnavirus species associated with leaf streak disease of banana in East Africa. Annals of Applied Biology 158, 346–353.

Jaufeerally-Fakim, Y., Khorugdharry, A., Harper, G., 2006. Genetic variants of *Banana streak virus* in Mauritius. Virus Research 115, 91–98.

Kamat, M.N., Patel, K., 1951. Notes on two important plant diseases in Bombay. Stat. Bull. Pl. Prot., New Delhi, 3, 16.

Kathirvel, A.K., Sathiamoorthy, S., Baskaran, T.L., Letchoumanae, S., 1986. Outbreak of banana mosaic in Trichy District. TNAU News Letter 15, 3.

Kasschau, K.D., Xie, Z.X., Allen, E., Llave, C., Chapman, E.J., Krizan, K.A., Carrington, J.C., 2003. P1/HC-Pro, a viral suppressor of RNA silencing, interferes with *Arabidopsis* development and miRNA function. Dev Cell 4, 205–217.

Kerns, M.R., Pataky, J.K., 1997. Reactions of sweet corn hybrids with resistance to maize dwarf mosaic virus. Plant Disease 81, 460–464.

Kesavamoorthy, R.C., 1980. Radical changes in ecosystem in the Pulney hills. In: Muthukrishnan, C.R., Abdul Chaser, J.B.M. (Eds.), Proceedings of the 13th National Seminar on Banana Production Technology, TNAU, Coimbatore, pp. 23–28.

Kubiriba, J., Legg, J.P., Tushemereirwe, W., Adipala, E., 2001. Vector transmission of *Banana streak virus* in the screen house in Uganda. Annals of Applied Biology 139, 37–43.

Lheureux, F., Carreel, F., Jenny, C., Lockhart, B.E.L., Iskra-Caruana, M.L., 2003. Identification of genetic markers linked to banana streak disease expression in inter-specific *Musa* hybrids. Theoretical and Applied Genetics 106, 594–598.

Lockhart, B.E.L., 1986. Purification and serology of a bacilliform virus associated with banana streak disease. Phytopathology 76, 995–999.

Lockhart, B.E.L., Jones, D.R., 2000a. Banana streak. In: Jones, D.R. (Ed.), Diseases of banana, abacá and enset. CABI, Wallingford, UK, pp. 263–274.

Lockhart, B.E.L., Jones, D.R., 2000b. Banana mosaic. In: Jones, D.R. (Ed.), Diseases of banana, abacá and enset. CABI, Wallingford, UK, pp. 256–263.

Lockhart, B.E.L., Olszewski, N.E., 1993. Serological and genomic heterogeneity of banana streak badnavirus: implications for virus detection in *Musa* germplasm. In: Ganry, J. (Ed.), Breeding banana and plantain for resistance to diseases and pests. CIRAD, Montpellier, France, in collaboration with INIBAP (International Network for the Improvement of Banana and Plantain), pp. 105–113.

Lockhart, B.E.L., Autrey, L.J.C., Comstock, J.C., 1992. Partial purification and serology of *Sugarcane mild mosaic virus*, a mealybug-transmitted clorterolike virus. Phytopathology 82, 691–695.

Lockhart B.E.L., Ndowora T.C., Olszewski N.E., & Dahal, G. (1998). Studies on integration of banana streak *Badnavirus* sequences in *Musa*: identification of episomally-expressible badnaviral integrants in *Musa* genotypes. In: Frison, E.A., Sharrock, S.E. (Eds.), *Banana streak virus:* a unique virus–*Musa* interaction? Proceedings of a workshop of the PROMUSA Virology working group held in Montpellier, France, 19–21 January 1998.

López-Gomollón, S., Dalmay, T., 2010. Recent patents in RNA silencing in plants: constructs, methods and applications in plant biotechnology. Recent Patents on DNA and Gene Sequences 4, 155–166.

Lu, Y.D., Gan, Q.H., Chi, X.Y., Qin, S., 2008. Roles of miRNA in plant defense and virus offense interaction. Plant Cell Reports 27 (10), 1571–1579.

Magee, C.J.P., 1927. Investigation on the bunchy top disease of banana. Council for Scientific and Industrial Research, Melbourne, Australia, p. 86.

Magee, C.J.P., 1940. Transmission of infectious chlorosis or heart rot of banana and its relationship to cucumber mosaic. Journal of Australian Institute of Agriculture 6, 44–47.

Magee, C.J.P., 1940. Transmission studies on the Banana bunchy top virus. Journal of the Australian Institute of Agricultural Science 6, 109–110.

Magnaye, L.V., Espino, R.R.C., 1990. Note: Banana bract mosaic, a new disease of banana I. Symptomatology. Philippine Agriculturist 73, 55–59.

Mali, V.R., Rajagore, S.B., 1980. A *cucumber mosaic virus* of banana in India. Phytpathologische Zeitschrift 98, 127–136.

Mary Sheeba, M., Selvarajan, R., Mustaffa, M.M., 2013. Prediction and identification of microRNA from banana infected with *Banana streak Mysore virus* (BSMYV). Madras Agriculture Journal, (in press).

Menon, M.R., Christudas, S.P., 1967. Studies on the population of aphid *Pentalonia nigronervosa* Coq. on banana plants in Kerala. Agriculture Research Journal of Kerala 5, 84–86.

Metha, P.R., Joshi, N.C., Rao, M.H., Renjhen, P.L., 1964. Bunchy top-serious disease of banana in India. Science and Culture 30, 259–263.

Meyer, J.B., Kasdorf, G.G.F., Nel, L.H., Pietersen, G., 2008. Transmission of activated episomal Banana streak OL (badna) virus (BSOLV) to cv. Williams banana (Musa sp.) by three mealybug species. Plant Diseases 92, 1158–1163.

Mlotshwa, S., Pruss, G.J., Vance, V., 2008. Small RNAs in viral infection and host defense. Trends in Plant Science 13 (7), 375–382.

Mohan, S., Lakshmanan, P., 1988. Outbreak of *cucumber mosaic virus* on *Musa* sp. in Tamil Nadu, India. Phytoparasitica 16, 281–282.

Munez, A.R., 1992. Symptomatology, transmission and purification of banana bract mosaic virus (BBMV) in 'giant cavendish' banana. Faculty of Graduate School, Los Baños, University of Philippines 1–57.

Nancy, J., 2003. Movement of *banana bunchy top virus* and its relationship with its aphid vector (*Pentalonia nigronervosa*) MSc thesis. Bharathidasan University, India.

Ndowora, T., Dahal, G., LaFleur, D., Harper, G., Hull, R., Olszewski, N., Lockhart, B.E.L., 1999. Evidence that badnavirus infection in *Musa* can originate from integrated sequences. Virology 255, 214–220.

Niblett, C.L., Pappu, S.S., Bird, J., Lastra, R., 1994. Infectious chlorosis, mosaic and heart rot. Compendium of tropical fruit disease. APS Press, St Paul, MN, USA, pp. 18–19.

Niu, S., Wang, B., Guo, X., Yu, J., Wang, X., Xu, K., Zhai, Y., Wang, J., Liu, Z., 2009. Identification of two RNA silencing suppressors from banana bunchy top virus. Archives of Virology 154, 1775–1783.

Pataky, J.K., Murphy, J.F., D'Arcy, C.J., 1990. Resistance to maize dwarf mosaic virus, severity of symptoms, titer of virus, and yield of sweet corn. Plant Disease 74, 359–364.

Raj, J.S., Ramanatha Menon, M., Christudas, P., 1970. The movement of banana bunchy top virus in the plant. Agricultural Research Journal of Kerala 8, 106–109.

Rao, D.G., 1980. Studies on a new strain of banana mosaic virus in South India. In: Muthukrishnan, C.R., AbdulKhader, J.B.M. (Eds.), Proceedings of the National Seminar on Banana Production Technology, Tamilnadu Agriculturual University, Coimbatore, India, pp. 155–159.

Robson, J.D., Wright, M.G., Almeida, R.P.P., 2007. Biology of *Pentalonia nigronervosa* (Hemiptera, Aphididae) on banana using different rearing methods. Environmental Entomology 36, 46–52.

Rodoni, B.C., Ahlawat, Y.S., Varma, A., Dale, J.L., Harding, R.M., 1997. Identification and characterization of *Banana bract mosaic virus* in India. Plant Diseases 81, 669–672.

Rodoni, B.C., Dale, J.L., Harding, R.M., 1999. Characterization and expression of the coat protein-coding region of *Banana bract mosaic potyvirus*, development of diagnostic assays and detection of the virus in banana plants from five countries in Southeast Asia. Archives of Virology 144, 1725–1737.

Roossinck, M.J., Bujarski, J., Ding, S., Hajimorad., W.R., Hanad, K., Scott, S., Tousignant, M., 1999. Family *Bromoviridae*. Virus taxonomy. Eighth Report of the International Committee on Taxonomy of Viruses. Academic Press, San Diego, CA, USA, pp. 923–935.

Samraj, J., Menon, M.R., Christudas, S.P., 1970. The movement of banana bunchy top virus in plant. Agricultural Research Journal of Kerala 8, 106–108.

Schwinghamer, M.W., Symons, R.H., 1975. Fractionation of *cucumber mosaic virus* RNA and its translation in a wheat embryo cell-free system. Virology 63, 252–262.

Selvarajan, R., Jeyabaskaran, K.J., 2006. Effect of *Banana bract mosaic virus* (BBrMV) on growth and yield of cultivar Nendran (Plantain, AAB). Indian Phytopathology 59 (4), 496–500.

Selvarajan, R., Balasubramanian, V., Dayakar, S., Uma, S., Ahlawat, Y.S., Sathiamoorthy, S., 2004. Integration of *Banana streak virus* genome in *Musa* germplasm with B genome. Abstracts of the 4th International Symposium on Molecular and Cellular Biology of Bananas. Penang, Malaysia, 6–9 July, 2004, p. 102.

Selvarajan, R., Balasubramanian, V., Sathiamoorthy, S., 2006. Vector transmission of banana bract mosaic and *Banana streak viruses* in India. Abstracts of XVI Annual Convention and International Symposium on Management of Vector-borne Viruses. Hyderabad, ICRISAT 7–10th February, 2006, p. 110.

Selvarajan, R., Mary Sheeba, M., Balasubramanian, V., Rajmohan, R., Lakshmi Dhevi, N., Sasireka, T., 2011. Molecular characterization of geographically different *Banana bunchy top virus* (BBTV) isolates in India. Indian Journal of Virology 21 (20), 110–116.

Shekhawat, U.K.S., Ganapathi, T.R., Hadapad, A.B., 2012. Transgenic banana plants expressing siRNAs targeted against viral replication initiation gene display high-level resistance to *Banana bunchy top virus* infection. Journal of General Virology 93, 1804–1813.

Simón-Mateo, C., García, J.A., 2011. Antiviral strategies in plants based on 613 RNA silencing. Biochim Biophys Acta 1809, 722–731.

Singh, S.J., 2003. Viral diseases of banana, first ed. Kalyani, Ludhiana, India.

Singh, S.J., Selvarajan, R., Singh, H.P., 2000. Detection of bract mosaic virus (kokkan disease) by electron microscopy and serology. In: Singh, H.P., Chadha, K.L. (Eds.), Banana-improvement, production and utilization. Proceedings of the Conference on Challenges for Banana Production and Utilization in 21st Century, AIPUB, NRCB, Trichy, India, pp. 381–383.

Suzuki, M., Kuwata, S., Kataoka, J., Masuta, C., Nitta, N., Takanami, Y., 1991. Functional analysis of deletion mutation of *Cucumber mosaic virus* RNA3 using as *in vitro* transcriptions system. Virology 183, 108–113.

Thangavelu, R., Selvarajan, R., Singh, H.P., 2000. Status of *Banana streak virus* and *Banana bract mosaic virus* diseases in India. In: Singh, H.P., Chadha, K.L. (Eds.), Banana: improvement, production and utilization. Proceedings of the Conference on Challenges for Banana Production and Utilization in 21st Century, AIPUB, NRCB, Trichy, India, pp. 364–376.

Thomas, J.E., Dietzgen, R.G., 1991. Purification, characterization and serological detection of virus-like particles associated with *banana bunchy top virus* in Australia. Journal of General Virology 72, 217–224.

Thomas, J.E., Magnaye, L.V., 1996. Banana bract mosaic disease. *Musa* Disease Fact Sheet, No. 1. INIBAP, Montpellier, France.

Thomas, J.E., Iskra-Caruana, M.L., Jones, D.R., 1994. Banana bunchy top disease. *Musa* Disease Fact Sheet No. 4. INIBAP, Montpellier, France, p. 2.

Thomas, J.E., Geering, A.D.W., Gambley, C.F., Kessling, A.F., White, M., 1997. Purification, properties and diagnosis of *Banana bract mosaic potyvirus* and its distinction from abaca mosaic potyvirus. Phytopathology 87, 698–705.

Vetten, H.J., Chu, P.W., Dale, J.L., Harding, R.M., Hu, J., Katul, L., Kojima, M., Randles, J.W., Sano, Y., Thomas, J.E., 2005. Virus taxonomy. In: Fauquet, M.A. (Ed.), Eighth Report of the International Committee on Taxonomy of Viruses, Academic Press, San Diego, pp. 343–352.

Voinnet, O., Rivas, S., Mestre, P., Baulcombe, D.C., 2003. An enhanced transient expression system in plants based on suppression of gene silencing by the p19 protein of tomato bushystunt virus. Plant Journal 33, 949–956.

Wanitchakorn, R., Harding, R.M., Dale, J.L., 1997. *Banana bunchy top virus* DNA-3 encodes the viral coat protein. Archives of Virology 142, 1673–1680.

Wanitchakorn, R., Hafner, G.J., Harding, R.M., Dale, J.L., 2000. Functional analysis of proteins encoded by *Banana bunchy top virus* DNA-4 to -6. Journal of General Virology 81, 299–306.

Watanabe, S., Bressan, A., 2013. Tropism, compartmentalization and retention of banana bunchy top virus (Nanoviridae) in the aphid vector *Pentalonia nigronervosa*. Journal of General Virology 94, 209–219.

Wu, R.Y., Su, H.J., 1992. Detection of banana bunchy top virus in diseased and symptomless banana plants with monoclonal antibody. Tropical Agriculture (Trinidad) 69, 397–399.

Zhang, H., Zhu, X., Liu, H., 1997. Effect of banana bunchy top virus (BBTV) on endogenous hormone of banana plant. Acta Phytopathologica Sinica 27, 79–83.

Chapter 4

Recent advances on interactions between the whitefly *Bemisia tabaci* and begomoviruses, with emphasis on *Tomato yellow leaf curl virus*

Pakkianathan Britto Cathrin and Murad Ghanim

Department of Entomology, Agricultural Research Organization, The Volcani Center, Bet Dagan, Israel

INTRODUCTION

Tomato yellow leaf curl virus (TYLCV) (Begomovirus, Geminiviridae) is the causative agent of tomato yellow leaf curl disease (TYLCD) in tropical and subtropical regions, resulting in crop losses of up to 100%. TYLCV is also capable of infecting more than 30 other plant species spanning 12 plant families, including cultivated vegetables, ornamentals, weeds and wild plant species (Czosnek & Ghanim 2011). The virus is transmitted by the silverleaf whitefly *Bemisia tabaci* biotype B. Symptoms of TYLCD vary, depending on the growth stage at the time of initial infection, environmental conditions, and cultivars. In tomatoes, symptoms include severe stunting, marked reduction in leaf size, upward cupping, chlorosis of leaf margins, mottling, flower abscission and significant yield reduction. Symptoms in common bean include leaf thickening, leaf crumpling, upward curling of leaves, abnormal lateral shoot proliferation, deformation and reduction in the number of pods. The severity of the viral epidemic correlates with the proportion of the whitefly population that acts as a vector for TYLCV (Czosnek & Ghanim 2011). Subsequent application of insecticides against *B. tabaci* populations in the field and greenhouses is the most commonly used strategy to manage TYLCD. In the late 1970s, TYLCV-resistant tomato cultivars were introduced to control TYLCV fatality (Pico et al 1999). These cultivars consisted of introgressing resistant traits from wild tomato species. On the other hand, research is ongoing to understand the interactions between TYLCV, tomato and its only vector, *B. tabaci*.

Research on virus–plant interactions has included studies aimed at understanding virus movement, symptom development, replication, and the plant's response to the virus, whereas studies on virus–vector interaction were aimed at understanding the mechanisms of acquisition, retention and transmission of TYLCV by *B. tabaci* (Skaljac & Ghanim 2010). *B. tabaci* has numerous biotypes, and it transmits a large number of viruses that infect many important agricultural plants, causing a major economic impact (Brown & Czosnek 2002). *B. tabaci* biotypes are morphologically indistinguishable (Gill 1990, Rosell et al 1997); however, they vary considerably in their ability to transmit geminiviruses (Bedford et al 1994, Czosnek & Ghanim 2011), their ability to utilize different host plant (Brown & Bird 1995), and their rate of development (Wang & Tsai 1996). In addition to endogenous species, new invasive and better adapted biotypes—such as the well-known B and Q ones—ave invaded crop systems and increased the level of damage (Czosnek & Brown 2009). Generally, TYLCV is effectively transmitted by both the B and Q biotypes (Jiang et al 2004); however, in Israel, it was shown that the B biotype is a much better vector than the Q biotype for TYLCV (Gottlieb et al 2010). TYLCV is known today to occur in several continents around the globe, including Asia, Africa, Europe, and North America (Czosnek & Latterot 1997). The only vector, in all countries, is *B. tabaci*. Almost 50 years of research on TYLCV epidemics have provided a firm understanding of TYLCV diversity and its interactions with the vector. This chapter summarizes the major findings on these interactions.

TOMATO YELLOW LEAF CURL VIRUS CAUSING WORLDWIDE EPIDEMICS

During the 1960s, a new plant disease, reported in the Jordan valley in Israel, caused severe damage to a newly introduced tomato cultivar. This disease was later called tomato yellow leaf curl disease (TYLCD); it is caused by TYLCV and was found to be vectored by *B. tabaci* populations (Cohen & Nitzany 1966). Although symptoms of TYLCD on plants were observed as early as the 1930s, outbreaks of the diseases were not observed until *B. tabaci* populations greatly increased. Today, TYLCV is known to occur in several continents around the world, including Asia, Middle and Far East Africa, Europe, Caribbean Islands and North America (Czosnek & Latterot 1997). Further geographic investigation reported that TYLCV has been found in Japan (Kato et al 1998), Mexico (Ascencio-Ibañez et al 1999), and the United States of America (Momol et al 1999). There have been almost 40 years of research on TYLCV epidemics, and the virus–vector and virus–plant interactions, aimed at developing better means of controlling this disease.

Molecular characteristics of TYLCV

The development of molecular tools has enabled significant knowledge to be gained on geminiviruses, their genetic arrangements, and their role in causing

TYLCD. Geminiviruses are circular plant DNA viruses characterized by a $22 \text{ nm} \times 38 \text{ nm}$ germinate particle comprised of two joined, incomplete icosahedra encapsulating a genome of single-stranded DNA of about 2700 nucleotides (Goodmann 1977, Harrison et al 1977, Francki et al 1980, Zhang et al 2001). Geminiviruses transmitted by whiteflies are assigned to the genus *Begomovirus* (van Regenmortel et al 2000). Generally, begomoviruses possess two genomic components designated DNA-A and DNA-B (bipartite); however, the TYLCV species has only a single DNA-A-like genome component (monopartite: ~2.8 kb). The TYLCV genome has six partially overlapping open reading frames (ORFs) bi-directionally organized in two transcriptional units that are separated by an intergenic region (IR) of approximately 300 nucleotides (Rybicki et al 2000). V1, which encodes the coat protein (CP) responsible for encapsidation of the genome and involved in virus movement and vector recognition, and V2, which encodes a suppressor of gene silencing to overcome the plant defense system (Zrachya et al 2007), are the two ORFs present on the virion sense strand. The complementary virus strand has four ORFs: C1 which encodes a replication-associated protein (Rep) and essential for replication, C2 a transcription activator protein (TrAP) involved in the activation of transcription from the coat protein promoter, C3 a replication enhancer protein (REn) interacting with the C1 protein and enhancing viral DNA accumulation, and C4 (embedded within C1). Protein products encoded by the V2 'pre-coat' and the C4 ORFs have been implicated in symptom expression and virus movement. The non-coding IR region located upstream of the V2 and C1 ORFs contains key elements (stem–loop structure) for the replication and transcription of the viral genome (Jupin et al 1994, Wartig et al 1997, Noris et al 1998). The first complete sequences of TYLCV isolates were reported in 1991 for isolates from Sardinia (TYLCV-Sar) and Israel (TYLCV-Is) (Kheyr-Pour et al 1991, Navot et al 1991). Much genetic information for TYLCV isolates (either complete or partial sequences) from worldwide strains can be retrieved from public repositories.

TYLCV—a worldwide threat to tomato production

Knowledge regarding the TYLCV infection cycle in the plant is essential for developing efficient and novel control strategies or eradication. However, reports on the natural spread of TYLCV based on large-scale surveys are scarce. In general, available data indicate that TYLCV is widespread in weed hosts but it usually does not cause symptoms in those plants. In Spain and Italy, for example, before the virulent TYLCV-Sar had caused severe epidemics on tomato in the late 1980s and early 1990s, it was known only from annual weed species such as *Datura stramonium*, *Solanum nigrum*, and *Euphorbia* sp. (Bosco et al 1993, Davino et al 1994, Sanchez-Campos et al 1999, Moriones 2000, Moriones & Castillo 2000). Similarly, in Israel, infections with TYLCV-Is have been reported since 1931 (Cohen & Antignus 1994), and only some weed species, such as *Cynanchum acutum* and *Malva parviflora*, were found to be natural hosts of the virus (Cohen et al 1988). In Israel, *C. acutum* was shown to be a potent source of inoculum for the infection of tomato crops by marking *B. tabaci* adults feeding naturally on *C. acutum* with fluorescent dust (Cohen et al 1988). Marked individuals were trapped on sticky yellow traps located up to 7 km away in the main tomato production area. Thus, *C. acutum* was suggested to be a source of primary spread of TYLCV to tomato in that region. Currently, isolates of at least 11 different *Begomovirus* species have been associated with TYLCD (Fauquet et al 2008). Moreover, multiple species can contribute to the same epidemic; for example, TYLCD epidemics in the Mediterranean basin involve strains of at least four virus species (Monci et al 2002, García-Andrés et al 2006, 2007, Davino et al 2009). In addition, TYLCV-Is and TYLCV-Mld have the broadest geographic ranges, stretching in the Old World from Japan in the east (Sugiyama et al 2008) to Spain in the west (Navas-Castillo et al 1999) and the Indian Ocean island of Reunion (Peterschmitt et al 1999) and Australia (Stonor et al 2003) in the south. Additionally, TYLCV-Is has apparently jumped at least twice between the Old and New Worlds (McGlashan et al 1994, Duffy & Holmes 2007) and is spreading into North and South America (Czosnek & Laterrot 1997). As the international trade in crop varieties is relatively widespread, it is perhaps not surprising that a virus like TYLCV-Is could attain such a global distribution. Nevertheless, among geminiviruses, the TYLCV-Is geographic range is unusually vast. This incessant global spread of TYLCV epitomizes a serious threat to tomato production in all temperate parts of the world. Recently, Lefeuvre et al (2010) applied phylogeographic inference and recombination analyses with available TYLCV sequences and reconstructed a history of TYLCV's diversification and movements throughout the world. They have accorded with the previous report that TYLCV most probably arose the first time somewhere in the Middle East between the 1930s and 1950s (Bird & Maramorosch 1978, Duffus 1986, Brown & Bird 1992, Brown et al 1995) and that its global spread only began in the 1980s after the emergence of two strains TYLCV-Mld and TYLCV-Is. Also, they have reported that Iran and surrounding regions form the current center of TYLCV diversity and the site where the most intensive ongoing TYLCV evolution is taking place. However, as this region is epidemiologically isolated, novel TYLCV variants are probably not going to be a direct global threat.

Knowledge about TYLCV–*B. tabaci* relationships and ecologic studies may provide novel information for understanding TYLCV epidemics. Studies in Italy (Bosco & Caciagli 1998) showed that the occurrence of *B. tabaci* outdoors is limited to warmer regions, in which TYLCV epidemics occur. These authors determined the climatic conditions that limit the geographic distribution of *B. tabaci*. Such information helps to forecast the ability of whiteflies to overwinter and their establishment in open-field conditions—and, therefore, the occurrence of TYLCV epidemics. An ecologic study conducted in Spain, where two biotypes of *B. tabaci* (B and Q) and two TYLCV species (TYLCV-Sar and TYLCV-Is) coexist, for investigating TYLCV epidemics, concluded that epidemics in tomato were originally caused by TYLCV-Sar (Noris et al 1994). However, after the appearance of TYLCV-Is (Navas-Castillo et al 1997),

TYLCV-Sar was completely displaced in some areas. Ecologic factors—such as the preferential transmission of TYLCV-Is by local biotypes of *B. tabaci* and the use of common bean in crop rotations that serves as a bridge crop for TYLCV-Is—were associated with main factors that contributed to the displacement of TYLCV-Sar by TYLCV-Is (Sanchez-Campos et al 1999). These results may have important practical implications for the control of TYLCV because the development of resistant cultivars is the best approach for controlling TYLCV and there are differences in the response of breeding lines to different TYLCV species, such as TYLCV-Sar and TYLCV-Is (Fargette et al 1996). One important aspect of TYLCV–*B. tabaci* relationships that can aid in understanding TYLCV epidemics is the recent demonstration that TYLCV-Is can be maintained in *B. tabaci* biotype B populations through copulation and transovarial transmission (Ghanim et al 1998, Ghanim & Czosnek 2000, Czosnek et al 2001, Ghanim et al 2001a). Therefore, *B. tabaci* may contribute to the reservoir for TYLCV epidemics, in addition to wild weed species as previously demonstrated. Due to the relevance of this finding for TYLCV epidemics and control, additional research should be conducted to determine whether similar relationships occur in other TYLCV–*B. tabaci* systems.

THE WHITEFLY *BEMISIA TABACI* — PEST AND VECTOR STATUS

Nomenclature and host range

The genus *Bemisia* comprises 37 species and originated in Asia (Mound & Halsey 1978). *B. tabaci* was first described by Gennadius on poinsettia plants in 1889 as *Aleyrodes tabaci* and it was described under numerous names before its morphologic variability was recognized. Five distinct groups of *B. tabaci* have now been identified by comparing their 16S ribosomal subunits. These are: (1) New World (US, Mexico, Puerto Rico), (2) Southeast Asia (Thailand, Malaysia), (3) Mediterranean basin (Southwest Europe, North Africa, Middle East), (4) Indian subcontinent (India, Pakistan, Nepal), (5) Equatorial Africa (Cameroon, Mozambique, Uganda, and Zambia).

First reports of a newly evolved biotype of *B. tabaci*, the B biotype, appeared in the mid-1980s (Brown et al 1995b). Commonly referred to as the silverleaf whitefly or poinsettia strain, the B biotype has been shown to be highly polyphagous and almost twice as fecund as previously recorded strains; it has also been documented as being a separate species, *B. argentifolii* (Bellows et al 1994). The B biotype is able to cause phytotoxic disorders in certain plant species, e.g. silverleafing in squashes (*Cucurbita* sp.), and this permits an irrefutable method of identification (Bedford et al 1992, 1994a). A distinctive non-specific esterase banding pattern is also helpful in the identification (Brown et al 1995a), but it is not infallible (Byrne et al 1995). Based on these markers, the B biotype was reported to spread rapidly (Costa et al 1993). Several 'biotypes' were described based on esterase morphotypes.

The Q biotype of *B. tabaci* was first described as native to the Mediterranean Basin in 1997 (Guirao et al 1997). A closely related Q was also described in Israel in 2005 (Horowitz et al 2003, 2005) and another Q was introduced into the United States in 2005, although it was still confined to greenhouses (Mckenzie et al 2009).

B. tabaci was known mainly as a pest of field crops in tropical and subtropical and temperate countries: cassava (*Manihot esculenta*), cotton (*Gossypium*), sweet potatoes (*Ipomoea batatas*), tobacco (*Nicotiana*) and tomatoes (*Lycopersicon esculentum*). Its host plant range within any particular region was small, yet *B. tabaci* had a composite range of around 300 plant species within 63 families (Mound & Halsey 1978).

With the evolution of the highly polyphagous B biotype, *B. tabaci* has now become a pest of glasshouse crops in many parts of the world, especially *Capsicum*, squashes (*Cucurbita pepo*), cucumbers (*Cucumis sativus*), *Hibiscus, Gerbera, Gloxinia,* lettuces (*Lactuca sativa*), poinsettia *(Euphorbia pulcherrima)* and tomatoes (*Lycopersicon esculentum*). *B. tabaci* moves readily from one host species to another and is estimated as having a host range of around 900 species (Asteraceae, Brassicaceae, Convolvulaceae, Cucurbitaceae, Euphorbiaceae, Fabaceae, Malvaceae, Solanaceae, etc.).

Life cycle

Eggs are laid usually in circular groups, on the underside of the leaves. They are anchored by a pedicel which is inserted into a fine slit made by the female ovipositor in the leaf tissues, and not into stomata, as in the case of many other aleyrodids. Eggs are whitish when first laid but gradually turn brown. Hatching occurs after 5–9 days at 30°C; this depends very much on host species, temperature and humidity (Sharaf et al 1985).

On hatching, the first instar, or 'crawler', is flat, oval and scale-like. This first instar is the only larval stage of this insect which is mobile. It moves from the egg site to a suitable feeding location on the lower surface of the leaf where its legs are lost in the ensuing molt and the larva becomes sessile. It does not therefore move again throughout the remaining nymphal stages. The first three nymphal stages last 2–4 days each (this could, however, vary with temperature). The fourth nymphal stage is called the 'puparium', and is about 0.7 mm long and lasts about 6 days; it is within the latter period of this stage that the metamorphosis to adult occurs. The adult emerges through a 'T'-shaped rupture in the skin of the puparium and spreads its wings for several minutes before beginning to powder itself with a waxy secretion from glands on the abdomen. Copulation begins 12–20 hours after emergence and takes place several times throughout the life of the adult. The life span of the female could extend for up to 60 days. The life of the male is generally much shorter, being between 9 and 17 days. Each female lays up to 160 eggs during her lifetime, although the B biotype has been shown to lay twice as many, and each group of eggs is laid in an arc around the female. Eleven to fifteen generations can occur within 1 year.

Significance and symptoms of *B. tabaci* infestations

B. tabaci has been known as a minor pest of cotton and other tropical or sub-tropical crops in the warmer parts of the world and, until about two decades ago, has been easily controlled by insecticides. In the southern states of the United States in 1991, however, it was estimated to have caused combined losses of 500 million US dollars to the winter vegetable crops (Perring et al 1993) through feeding damage and plant virus transmission. *B. tabaci* is also a serious pest in greenhouses worldwide. Depending on the level of infestation, the whitefly can cause leaf yellowing, and those leafs are later shed. The honeydew produced by the feeding of the nymphs covers the surface of the leaves and can cause a reduction in photosynthetic potential when colonized by molds. Honeydew can also disfigure flowers and, in the case of cotton, can cause problems in processing the lint. With heavy infestations, plant height, number of internodes and quality and quantity of yield can be affected (e.g. in cotton). The larvae of the B biotype of *B. tabaci* are unique in their ability to cause phytotoxic responses to many plant and crop species (Costa et al 1993). These include a severe silvering of squash leaves, white stems in pumpkin, white streaking in leafy brassica crops, uneven ripening of tomato fruits, reduced growth, yellowing and stem blanching in lettuce and kai choy (*Brassica campestris*) and yellow veining in carrots and *Lonicera* (Bedford et al 1994a,b).

The significance of *B. tabaci* as a virus vector

B. tabaci is the vector of over 100 plant viruses in the genera *Geminivirus*, *Closterovirus*, *Nepovirus*, *Carlavirus*, *Potyvirus* and a rod-shaped DNA virus (Markham et al 1994). The geminiviruses are by far the most important viruses agriculturally, causing yield losses to crops between 20 and 100% (Brown & Bird 1992). Geminiviruses cause a range of different symptoms which include yellow mosaics, yellow veining, leaf curling, stunting and vein thickening. Estimates indicated that a million ha of cotton is being decimated in Pakistan by the *Cotton leaf curl virus* (CoLCV) (Mansoor et al 1993), and important African subsistence crops such as cassava are affected by disastrous geminiviruses such as the *African cassava mosaic virus* (ACMV). Tomato crops throughout the world are particularly susceptible to many different geminiviruses, and in most cases they exhibit yellow leaf curl symptoms. Most of these epidemics in the Old World are attributed to *Tomato yellow leaf curl virus* (TYLCV) but may also be caused by other geminiviruses. TYLCV has also been recorded in the New World, but several others, exclusively American, tomato geminiviruses have now been described, e.g. *Tomato mottle virus* (EPPO/CABI, 1996).

The emergence of the B biotype of *B. tabaci*, with its ability to feed on many different host plants, has given whitefly-transmitted viruses the potential to infect new plant species. Two viruses have been shown to be no longer transmissible by *B. tabaci*—*Tobacco leaf curl virus* (TLCV) and *Abutilon mosaic virus* (AbMV)—possibly through many years of vegetative propagation of their ornamental host plants (Bedford et al 1994a). The major virus transmitted by *B. tabaci* is TYLCV that is causing major crop losses within the tomato industries of Spain, Italy, Israel and recently China and the United States. Newly identified *B. tabaci*-transmitted closteroviruses are reported to cause severe damage to cucumbers and melons in Spain and other Mediterranean countries (Berdiales et al 1999).

During this long-lasting virus–vector relationship, begomoviruses might have optimized the conformation of their capsid to fit the receptors that mediate their circulation in the insect host and to interact with insect proteins. It is interesting to note that the adaptation of the local vector to the local begomovirus is reflected in the parameters of acquisition and transmission. Transmission of a begomovirus by *B. tabaci* from the same geographic region is more efficient than in the case where the virus and the insect originated from two different regions (McGrath & Harrison 1995), suggesting an adaptation between the viruses and their vectors in the same geographic area. Circulation of begomoviruses inside their whitefly vectors may be one mechanism developed to avoid the invasion of insect tissues by harmful viruses. In the latter case, it is clear that these efforts have been only partially successful because many begomoviruses remain associated with the insect vector for many days following a short acquisition access period (AAP) (Polston et al 1990, Caciagli et al 1995, Rubinstein & Czosnek 1997), and some begomoviruses are able to invade the reproductive system (Ghanim et al 1998, Bosco et al 2004, Wang et al 2010) and affect vital parameters (Rubinstein & Czosnek 1997, Jiu et al 2007, Matsuura & Hoshino 2009).

PARAMETERS FOR ACQUISITION, TRANSMISSION AND RETENTION OF TYLCV BY *B. TABACI*

Young leaves and apices are the best target for whitefly-mediated inoculation (Ber et al 1990). In these tissues, the viral DNA replicates at the site of inoculation and is transported first to the roots, then to the shoot apex, and finally to the neighboring leaves and flowers. Inoculation of the oldest leaves and cotyledons is inefficient. During previous years, bioassays have been the only tool available for the study of the acquisition of TYLCV by whiteflies. Using whitefly-mediated transmission assays, it has been shown that *B. tabaci* can acquire enough viruses during an acquisition access period. With the development of ELISA and molecular techniques, it is now possible to detect viral molecules in individual insects (Czosnek et al 1988, Navot et al 1989, Polston et al 1990, Zeidan & Czosnek, 1991).

Acquisition

While feeding on a tomato plant, the stylet of *B. tabaci* follows a convoluted path and penetrates through the parenchyma cells (Fig. 4.1). The penetration through the parenchyma is mostly intracellular for reaching the phloem. Begomoviruses are acquired when the stylet penetrates the vascular tissues and reaches the phloem (Pollard 1955). Based on biologic tests, the parameters of acquisition, retention, and transmission of a begomovirus were first defined for TYLCV-Is (Cohen & Harpaz 1964, Cohen & Nitzany 1966; reviewed by Czosnek & Ghanim 2011). The reported minimum acquisition access period (AAP) and inoculation access period (IAP) of Middle Eastern TYLCV isolates varied from 15 to 60 minutes and from 15 to 30 minutes, respectively (Cohen & Nitzany 1966, Ioannou 1985, Mansour & Al-Musa 1992, Mehta et al 1994). Similar values were reported for TYLCV-Sar from Italy (Caciagli et al 1995) and *Tomato leaf curl Bangalore virus* (ToLCBV-In) from India (Butter & Rataul 1977, Reddy & Yaraguntaiah 1981, Muniyappa et al 2000). The development of molecular tools has allowed a refinement of these studies. TYLCV was readily detected by Southern blot hybridization in DNA extracted from a single whitefly of the B biotype. The hybridization signals indicated that insects that had access to the same tissues for the same period of time could acquire variable amounts of viral DNA (Zeidan & Czosnek 1991). The polymerase chain reaction (PCR) has allowed the detection of TYLCV DNA in a single insect in

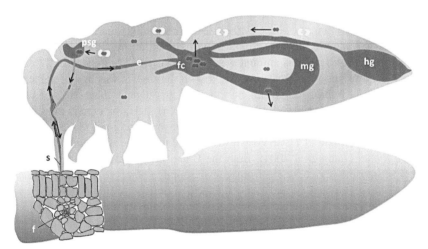

FIGURE 4.1   The circulative transmission pathway for begomoviruses (red particles) transmitted by whiteflies. These viruses are restricted to the plant phloem (f), which facilitates dispersal by sap-sucking insects. The filter chamber (fc) in the midgut (mg) is the first site of internalization into the vector, which occurs via endocytosis. After circulation in the insect, viruses interact with the GroEL protein (yellow particles) in the hemolymph and cross the insect primary salivary glands (psg) via endocytosis; the viruses are ejected into a host plant with salivary secretions. hg = hindgut; e = esophagus; s = stylet.

amounts even below the threshold of infectivity (Navot et al 1992). TYLCV DNA could be amplified in 20% of the individuals collected 5 minutes after the acquisition access period (AAP) and in all insects after a further 5 minutes (Navot et al 1992, Atzmon et al 1998, Ghanim et al 2001a). Analysis of the electronic waveforms produced during insect feeding indicated that, following a short probing period, the minimum phloem contact threshold period was 1.8 minutes for successful inoculation of TYLCV (Jiang et al 2000).

Transmission

A single insect is able to infect a tomato plant with TYLCV following a 24 hour AAP, and the efficiency of transmission reaches 100% when 5–15 insects are used (Cohen & Nitzany 1966, Mansour & Al-Musa 1992, Mehta et al 1994). A similar number of insects are necessary to achieve 100% transmission of the New World bipartite geminivirus *Squash leaf curl virus* (SLCV) (Cohen et al 1983). In addition, gender and age also affect transmission ability (Czosnek et al 2001). Nearly all the females of 1–2 weeks from synchronized populations of adult *B. tabaci* were able to infect tomato plants by about 48 hours of the inoculation access period (IAP), after 48 hours AAP. In comparison, only around 20% of the males of the same age under the same conditions were able to infect plants. Infection capability decreased with respect to age. While 60% of the 3-week-old females infected plants, the males were totally unable to infect tomato plants. Only 20% of the 6-week-old females were able to infect tomato plants. Aging insects acquire fewer viruses compared to younger individuals (Rubinstein & Czosnek 1997). Seventeen days after emergence, the adult insects acquired less than half the virus acquired by 10-day-old insects. At the age of 24 days, this amount was only about 10%. At the age of 28 days and thereafter, the viral DNA associated with the insects was undetectable by Southern blot hybridization although the insects retained about 20% of their initial inoculation capacity. Transmission efficiency of Q biotype is not essentially different from that of the B biotype. Transmission of a TYLCSV isolate from Murcia, Spain (TYLCSV-ES) was studied using the B, Q and S biotypes of *B. tabaci* (Jiang et al 2004). Both the B and Q biotypes of *B. tabaci* were able to transmit TYLCSV-ES from infected tomato plants to *Solanum nigrum* and *Datura stramonium* and *vice versa*. No significant difference was found in transmission efficiency from infected tomato plants to weed plants between the B and Q biotypes. The S biotype could not survive on tomato long enough to acquire or transmit TYLCSV-ES. In these studies, the age and gender of the whiteflies was not taken into account.

Retention

After 48 hours AAP, begomoviruses are retained in their whitefly vector for several weeks and sometimes for the entire life of the insect. SLCV and TYLCV remain associated with *B. tabaci* during the entire life of the vector (Cohen et al 1989, Rubinstein & Czosnek 1997) while TYLCSV is undetectable after approximately 20 days (Caciagli & Bosco 1997). Investigation on viral transmission and retention suggest that the viral DNA remained associated with the insects for much longer than transmission ability. For instance, TYLCSV DNA was detectable up to 20 days after the end of the 48 hour AAP, whereas transmission could occur for up to only 8 days (Caciagli et al 1995). Detection of viral DNA (by Southern blot hybridization or PCR) and CP (by Western blot immunodetection or IC-PCR) suggests that these are not retained in *B. tabaci* for the same time periods. Following the end of the 48 hour AAP, TYLCV DNA was detected throughout the 5-week life span of the insect, while the amount of TYLCV CP steadily decreased until it was undetectable at day 12 (Rubinstein & Czosnek 1997). The disappearance of the virus CP was associated with a rapid decrease in the ability of the whitefly to produce infected host plants, as shown for TYLCV (Rubinstein & Czosnek 1997) and SLCV (Cohen et al 1983). Similarly, the difference in the retention of viral DNA and CP in *B. tabaci* was also observed with an Israeli isolate of the non-transmissible bipartite begomovirus AbMV (Morin et al 2000). Czosnek et al 2002 stated that, following a 4-day AAP on infected abutilon plants, the TYLCV-Is DNA persisted with *B. tabaci* for about 15 days, while the CP was detectable for up to only 7 days. Besides, TYLCV was retained for a much shorter time in the non-vector *T. vaporariorum* than in the *B. tabaci* vector (Czosnek et al 2002). The comparison of retention periods of TYLCV DNA and CP in the two insect species under the same experimental setup showed that TYLCV DNA was detectable in *B. tabaci* over the entire 7 days of the experiment while the CP was detectable during the first 4 days only. In contrast, TYLCV DNA was detected in *T. vaporariorum* only during the first 6 hours that followed the end of the AAP, and the CP for up to 4 hours. Thus, TYLCV vanished very quickly from *T. vaporariorum* once acquisition feeding has ceased, but nonetheless, the DNA appears to be retained longer than the CP even in the non-vector.

TYLCV circulation in B. tabaci

TYLCV is vectored only by *B. tabaci*, in a persistent circulative manner (Fig. 4.1) (Ghanim et al 2001a). Once ingested by whiteflies, begomoviruses are not immediately available for infection. They need to translocate in the insect's digestive tract, penetrate the gut membranes into the hemolymph and cross the epithelial cells of the whitefly's digestive tract which bridge between the gut lumen and the hemolymph. From there, begomoviral particles reach the salivary systems and finally enter the salivary duct from where they are egested with the saliva. Translocation of begomoviruses from the digestive tract to the hemolymph and from the hemolymph to the salivary gland is thought to be mediated by as yet unidentified receptors. The rate of translocation of TYLCV in circulative transmission has been studied by Ghanim et al (2001) by using PCR on dissected organs. The presence of TYLCV in the stylets, the head, the midgut, a hemolymph sample, and the salivary glands was assessed by TYLCV-specific primers. TYLCV DNA was first detected in the head of *B. tabaci* after a 10-minute AAP. The virus was present in the midgut after 40 minutes and was first detected in the hemolymph after 90 minutes. TYLCV was found in the salivary glands 5.5 hours after it was first detected in the hemolymph. Moreover, the authors found the signal of encapsidated virions by immunodetection. These results suggest that at least part of the virus moves as a virion.

TYLCV has been traced in *B. tabaci* using antibodies raised against the CP. The virus was localized in the stylets, associated mainly with the food canal all along the lumen. Similarly, TYLCV was immunolocalized to the proximal part of the descending midgut, the filter chamber and the distal part of the descending midgut, and in the primary salivary glands (Brown & Czosnek 2002, Czosnek et al 2002). The localization patterns of the TYLCSV were similar to those of TYLCV. TYLCSV has been detected in the midgut, the microvilli, and in the cytoplasm of the primary salivary gland cells (Medina et al 2006, Ghanim & Medina 2007, Caciagli et al 2009). Although viral DNA fragments have been amplified from ovary tissue of whiteflies that acquired TYLCV (Ghanim et al 1998) and TYLCSV (Bosco et al 2004), no specific labeling of the TYLCSV CP in ovaries was detected (Caciagli et al 2009).

Velocity of TYLCV translocation in B. tabaci

Once ingested, begomoviruses are not immediately infective. The time it takes from the beginning of the AAP to the moment the whitefly efficiently transmits the virus to plants is called the latent period. It may vary due to the experimental conditions or to changes in virus and/or vector with time. For example, the latent period of TYLCV was reported to be 21 hours in the early 1960s (Cohen & Nitzany 1966) while later on it was found to be 8 hours (Ghanim et al 2001). The velocity of translocation of TYLCV DNA and CP was determined using whitefly stylets, head, midgut, hemolymph and salivary glands dissected from a single insect as substrate for PCR and immunocapture-PCR (Ghanim et al 2001). TYLCV was detected in the head of whiteflies as early as 10 minutes after the beginning of the AAP and in the midgut approximately after 40 minutes. TYLCV crossed the midgut and reached the hemolymph 30 minutes after it was first detected in the midgut, 90 minutes after the beginning of the AAP. TYLCV was detected in the salivary glands approximately 5.5 hours after it was first detected in the haemolymph, 7 hours after the beginning of the AAP. Whiteflies were able to infect tomato plants 1 hour after the virus was first detected in the salivary system, indicating that the threshold amount of virions necessary to obtain an efficient infection is low. Translocation timing of TYLCV DNA and CP overlapped, suggesting that the TYLCV moves as virions. The velocity of SLCV translocation in *B. tabaci* was similar (Rosell et al 1999).

TRANSOVARIAL TRANSMISSION OF TYLCV BY *B. TABACI*

Studies on the survival of the virus between the growing seasons of tomato gave rise to the following observations. (1) The virus is not transmitted through the seeds of infected plants. (2) There is no obvious alternative host to tomato that has been shown to be the likely reservoir of the virus between seasons, except for some weeds (Bosco et al 1993, Davino et al 1994, Sanchez-Campos et al 1999, Moriones 2000, Moriones & Castillo 2000). (3) Infection of tomatoes starts almost immediately after planting, even when the insect population is not at its peak (Cohen et al 1988). It was thus postulated that the whitefly serves as a source for the virus, which is passed between generations through the egg.

Ghanim et al (1998) tested the possibility that TYLCV is transmitted between generations. Their study has proven that TYLCV is transmitted to the progeny of a single viruliferous insect for at least two successive generations through the egg. Moreover, the progeny of viruliferous insects was able to infect tomato test plants. Dissection and analysis of the reproductive system of viruliferous whiteflies showed that both the ovaries and the maturing eggs contained TYLCV DNA (Ghanim et al 1998). The closely related TYLCSV was also found to be transmitted transovarially to the first generation progeny. TYLCSV was detected in eggs and nymphs as well as in adults of the first generation progeny (Bosco et al 2004). However, in contrast to TYLCV, the adult progeny of viruliferous insects were unable to infect tomato plants. It is interesting to note that the same scientists found that TYLCV was detected neither in instars nor in adult progeny of viruliferous females. These divergent results may be due to intrinsic differences in the highly inbred insect colonies raised in the laboratory and used in these experiments. The way in which TYLCV (Ghanim et al 1998) and TYLCSV (Bosco et al 2004) enter the whitefly reproductive system is unknown. It is possible that during the maturation of eggs in the ovaries, geminiviral particles penetrate the egg, together with the endosymbionts, via an aperture in the membrane (Costa et al 1995). Invading TYLCV may affect the development of some of the eggs, causing a decrease in fertility (Rubinstein & Czosnek 1997, Jiu et al 2007, Liu et al 2009). The vertical transmission of TYLCV and *Tomato yellow leaf curl china virus* (TYLCCNV) by the B and Q biotypes of *B. tabaci* was studied using virus isolates and whitefly colonies established in China (Wang et al 2009). Virus DNA was detected in eggs and nymphs but not in the adults of the first generation progeny, except in the combination of TYLCV and Q biotype whitefly, where only about 3% of the adults contained the virus DNA. The offspring adults produced by viruliferous females did not transmit the viruses to test plants. These results differed from those reported previously (Ghanim et al 1998, Bosco et al 2004).

SEX-MEDIATED TRANSMISSION OF TYLCV BY *B. TABACI*

TYLCV can be transmitted between *B. tabaci* B biotype males and females in a sex-dependent manner, in the absence of any other source of the virus (Ghanim &

Czosnek 2000). TYLCV was transmitted from viruliferous males to non-viruliferous females and from viruliferous females to non-viruliferous males, but not between insects of the same sex. Transmission took place when insects were caged in groups or in couples, in a feeding chamber or on TYLCV non-host cotton plants. Both viruliferous male and female whiteflies can transmit TYLCV to their counterparts; there was no significant difference in the efficiency of viral transmission between the two sexes. Both viral DNA and CP were detected in the recipient whiteflies, strongly indicating that the insects acquired encapsidated virions. The recipient insects were able to efficiently inoculate tomato test plants. These plants contained the virus genomic DNA, its replicative form and the virus CP, and showed the symptoms of a systemic infection. Therefore, whiteflies acquired the virus from sexual partners and had all the infectious properties characteristic of TYLCV virions ingested from infected tomato plants. Insect-to-insect transmission increased the number of whiteflies able to infect tomato test plants. In addition, TYLCV was observed in the hemolymph of whiteflies that had acquired the virus from sexual partners, indicating that the virus follows, at least in part, the circular path inherent to acquisition from plants. TYLCV reached the hemolymph more than 4 hours after the whiteflies had been caged with viruliferous insects of the other sex; in comparison, the virus was found in the hemolymph of insects caged with infected tomato plants after 1.5 hours (Ghanim et al 2001).

The fact that virus is found in the hemolymph of recipient males and females more than 4 hours after the start of sexual contact, and the progeny of these females also contain virus, points to several possible modes of transfer. Later studies by Ghanim et al (2001, 2007) revealed that the hemolymph plays an essential role in the transmission of TYLCV among *B. tabaci* individuals of opposite gender. TYLCV was first detected in the hemolymph of the recipient insects about 1.5 hours after caging, but was detected neither in the midgut nor in the head at this time. From there, TYLCV followed the pathway associated with acquisition from infected plants and did not cross the gut membranes back into the digestive system (Ghanim et al 2001, 2007). Hence TYLCV passes from one insect to another by exchange of fluids accompanying intercourse, and reaches the open blood circulative system of the sexual partners. Mating was obligatory in order for TYLCV to pass from one insect to another. Transmission of TYLCV in a gender-related manner was not exclusive to the *B. tabaci* B biotype, but was also shared with the Q biotype, indicating that this biologic feature might be widely shared among whiteflies (Ghanim et al 2007). The bipartite begomoviruses SLCV and *Watermelon chlorotic stunt viru* (WmCSV) were shown also to be transmitted horizontally among whiteflies of the B biotype with an efficacy similar to that of TYLCV.

The horizontal transmission of TYLCV and TYLCCNV by the B and Q biotypes of *B. tabaci* was studied (Wang et al 2009). Both TYLCV DNA and TYLCCNV DNA were shown to be transmitted horizontally by each of the two biotypes of the whitefly, but frequency of transmission was usually low. The overall percentage of horizontal transmission for either TYLCCNV or TYLCV in each of the two whitefly biotypes was below 5%. Neither virus species nor whitefly biotypes had a significant effect on the frequency of transmission. Caging together *B. tabaci* and *Trialeurodes vaporariorum*, two whitefly species that do not mate, confirmed that mating is obligatory for TYLCV transmission. The virus ingested by *B. tabaci* was not detected in *T. vaporariorum*, and the virus ingested by *T. vaporariorum* was not found in *B. tabaci*. It has to be noted that while TYLCV is found in the hemolymph of *B. tabaci* after feeding on infected tomato plants, the virus is ingested by *T. vaporariorum*, but it is unable to cross the gut/hemolymph barrier (Czosnek et al 2002), probably because the latter insect does not possess the begomoviral receptors that allow viruses to cross the gut wall. Interestingly, TYLCV was not transmitted when individuals of the B biotype where caged with individuals of the Q biotype (Ghanim et al 2007), indicating that B and Q biotypes do not mate (Pascual & Callejas 2004).

MOLECULAR INTERACTIONS BETWEEN TYLCV AND *B. TABACI*

Generally, during begomovirus transmission in the vector, the capsid protein is exposed to the whitefly primary organs and it is hypothesized to interact with insect receptors and chaperons (Morin et al 2000). TYLCV is acquired as a virion from the plant phloem and passes along the food canal in the stylet, with other substances from the phloem, and then reaches the esophagus of *B. tabaci*. The esophagus is a chitin-lined tissue that does not allow food/virion penetration to the hemolymph (Ghanim et al 2001b). The tissue through which virions can cross to the hemolymph is a modification of the digestive system called the filter chamber (Ghanim et al 2001b). The filter chamber is a complicated structure that combines tissue from the midgut, hindgut, and the caeca. Membranes from these organs interdigitate to form this complicated structure that ensures direct absorption of 'pure' useful substances for the insect into the hemolymph, while more 'complicated' food is pushed into the descending midgut by the muscular caeca. It is hypothesized that the majority of TYLCV virions are absorbed from the filter chamber into the hemolymph (Fig. 4.1), while a minority of the virions circulate into the descending then the ascending midguts, and cross the mid-gut epithelial cells to the hemolymph (Ghanim & Medina 2007; reviewed by Skaljac & Ghanim 2010). Microscopic studies have shown extensive location of TYLCV virions in the filter chamber area; their concentration decreases toward the descending and the ascending midguts (Ghanim et al 2009, and Fig. 4.1). Unlike aphids and luteoviruses, TYLCV virions cross the epithelial cells in the midgut, and not the hindgut, and the specificity resides in this area of the digestive system (Czosnek et al 2002). In the hemolymph, TYLCV virions interact with a 63 KDa GroEL protein produced by the endosymbiotic bacteria of *B. tabaci*, which protects the virions from proteolysis by the insect's immune system (Morin et al 1999, 2000, Gottlieb et al 2010). Virions cross the first barrier of the digestive system into the hemolymph within 1 hour (Ghanim et al 2001a). A second recognition barrier is thought to reside on the apical membrane of the primary salivary gland of *B. tabaci* (Brown & Czosnek 2002), unlike the aphid–luteovirus system, in which recognition resides in the accessory salivary glands (Gildow & Rochow 1980, Gildow & Gray 1993). Ohnesorge and Bejarano (2009) reported that a 16 kDa small heat shock protein, belonging to HSP20-a crystalline family, is bound with TYLCSV CP. The TYLCSV CP interaction domain with BtHSP16 was located within the conserved region of the N-terminal part of TYLCSV CP (amino acids 47–66), overlapping almost completely with the nuclear localization signal described for the CP of TYLCV (Kunik et al 1998). The region necessary for transmission of TYLCSV by *B. tabaci* (amino acids 129–152) is not directly involved in the specific interaction between the CP and the BtHSP16. Not much is known about the molecular interactions between TYLCV and *B. tabaci*. Many studies have been aimed at addressing the replication of TYLCV in *B. tabaci*, but it is believed that TYLCV and geminiviruses do not replicate in their insect vectors. One study reported accumulation of TYLCV transcripts in *B. tabaci* after acquisition from infected plants (Sinisterra et al 2005).

A genomic project was launched in 2002 and has sequenced more than 20,000 expressed sequence tags (ESTs) from adult whiteflies, as well as other developmental stages including nymphs, eggs, and viruliferous adults with TYLCV and ToMoV (Leshkowitz et al 2006). This large-scale sequencing of ESTs from *B. tabaci* led to a better understanding of the genetic makeup of the whitefly relative to that of other insect models. It was estimated that the genome of the whitefly is about five times bigger than the genome of *Drosophila melanogaster* (Brown et al 2005). Following this sequencing, a spotted DNA microarray containing 6000 unique ESTs from the whitefly was developed and used to study the resistance capability of the whitefly to insecticides (Ghanim & Kontsedalov 2007), its immune response to parasitoids (wasp *Eretmocerus mundus*) (Mahadav et al 2008), and the response to heat stress conditions in the B and the Q biotypes (Mahadav et al 2009). Efforts are still underway to sequence more ESTs from the whitefly. Recent studies using an advanced version of this microarray, which was prepared based on Agilent's technology, are aimed at studying the response of *B. tabaci* to feeding on plants modified with the contents of defense materials, its response to modified contents of nicotine in tobacco plants, and its response to the presence/absence of symbiotic bacteria. A recent study has shown that RNA interference (RNAi), an effective mechanism for silencing mRNA in many organisms, including insects, also occurs in *B. tabaci* (Ghanim et al 2007).

Although many of the described studies are still underway, the path to considering *B. tabaci* as an organism with a fully sequenced genome with rich genomic resources is still long. Recently, Su et al 2012 sequenced the transcriptome of the primary salivary glands (an organ with only 13–20 cells) of the Mediterranean species of the *B. tabaci* complex using an effective cDNA amplification method in combination with short-read sequencing. They have obtained 13,615 unigenes including 3159 sequences. The number of unigenes obtained from the salivary glands of the whitefly is at least four-fold greater than that obtained from the salivary gland genes of other plant-sucking insects. The functions of the primary glands were analysed by a sequence similarity search, and by comparisons with the whole transcriptome of the whitefly. The results showed that the genes related to metabolism and transport were significantly enriched in the primary salivary glands. Moreover, these authors have reported that a number of highly expressed genes in the salivary glands might be involved in secretory protein processing, secretion and virus transmission. These analyses provide a valuable resource for future investigations of the functions of salivary gland-specific genes and biologic processes during whitefly–plant interactions.

CONCLUDING REMARKS

The diversity among arthropod vectors, and the viruses they transmit, is expanding their economic importance worldwide. In particular, begomoviruses vectored by *B. tabaci* are causing the most devastating viral diseases in agricultural crops worldwide. While new and diverse pest control strategies are adopted for controlling whiteflies, they continue to pose great economic impact (Brown et al 1995, Frohlich et al 1999). Differences in host plant preference, host range, fecundity, dispersal behavior, vector competency, phytotoxic feeding effects, endosymbiont composition, invasiveness, and insecticide resistance, are all among the factors that directly influence the ability of *B. tabaci* to become a worldwide top-rated pest. Research on TYLCV–plant and TYLCV–*B. tabaci* interaction has resulted in hundreds of research papers devoted to an understanding the biologic, molecular and cellular events underlying these interactions.

Whitefly genomics research is expected to open important avenues into the discovery of novel strategies for whitefly and whitefly-transmitted virus management based on an improved understanding of molecular, cellular, and biologic processes. The genome sequence of *B. tabaci* will synergize projects underway to develop and sequence *B. tabaci*-expressed sequence tags (ESTs) or cDNA libraries for functional genomics and proteomics analysis. The benefits are far reaching and include identification of genes that combat abiotic and biotic stresses (that often lead to invasiveness and insecticide resistance), and an understanding of the basis of whitefly–virus specificity. Collectively, efforts in genomics, proteomics, and functional genomics will initiate further local, regional, national and international partnerships to expand present and future efforts aimed at determining the *B. tabaci* genome and proceed to undertake functional genomics aspects that are of high interest among a broad user community.

REFERENCES

Ascencio-Ibáñez, J.T., Díaz-Plaza, R., Méndez-Lozano, J., Monsalve-Fonnegra, Z.I., Argüello-Astorga, G.R., Rivera-Bustamante, R.F., 1999. First report of tomato yellow leaf curl geminivirus in Yucatán, México. Plant Disease 83, 1178.

Atzmon, G., van Hoss, H., Czosnek, H., 1998. PCR-amplification of tomato yellow leaf curl virus (TYLCV) from squashes of plants and insect vectors: application to the study of TYLCV acquisition and transmission. European Journal of Plant Pathology 104, 189–194.

Bedford, I. D., Briddon, R. W., Markham, P. G., Brown, J. K., Rosell, R.C., 1992. *Bemisia tabaci*—biotype characterisation and the threat of this whitefly species to agriculture. Proceedings of the 1992 British Crop Protection Conference—Pests and Diseases 3, 1235–1240.

Bedford, I.D., Briddon, R.W., Brown, J.K., Rosell, R.C., Markham, P.G., 1994a. Geminivirus transmission and biological characterisation of *Bemisia tabaci* (Gennadius) biotypes from different geographic regions. Annals of Applied Biology 125, 311–325.

Bedford, I. D., Pinner, M., Liu, S., Markham, P. G., 1994b. *Bemisia tabaci*—potential infestation, phytotoxicity and virus transmission within European agriculture. Proceedings of the 1994 British Crop Protection Conference—Pests and Diseases 2, 911–916.

Bellows, T.S., Perring, T.M., Gill, R.J., Headrick, D.H., 1994. Description of a species of *Bemisia* (Homoptera: *Aleyrodidae*). Annals of the Entomological Society of America 87, 195–206.

Ber, R., Navot, N., Zamir, D., Antignus, Y., Cohen, S., Czosnek, H.G., 1990. Infection of tomato by the tomato yellow leaf curl virus: susceptibility to infection, symptom development and accumulation of viral DNA. Archives of Virology 112, 169–180.

Berdiales, B., Bernal, J.J., Sáez, E., Woudt, B., Beitia, F., Rodríguez-Cerezo, E., 1999. Occurrence of cucurbit yellow stunting disorder virus (CYSDV) and beet pseudo-yellows virus in cucurbit crops in Spain and transmission of CYSDV by two biotypes of *Bemisia tabaci*. European Journal of Plant Pathology 105, 211–215.

Bird, J., Maramorosch, K., 1978. Viruses and virus diseases associated with whiteflies. Advances in Virus Research 22, 55–110.

Bosco, D., Caciagli, P., 1998. Bionomics and ecology of *Bemisia tabaci* (Sternorrhyncha: Aleyrodidae) in Italy. European Journal of Entomology 95, 519–527.

Bosco, D., Caciagli, P., Noris, E., 1993. Indagini epidemiologiche sul virus dell'accartocciamento fogliare giallo del pomodoro (TYLCV) in Italia. Informatore Fitopatologico 11, 33–36.

Bosco, D., Mason, G., Accotto, G.P., 2004. TYLCSV DNA, but not infectivity, can be transovarially inherited by the progeny of the whitefly vector *Bemisia tabaci* (Gennadius). Virology 323, 276–283.

Brown, J.K., Bird, J., 1992. Whitefly-transmitted geminiviruses and associated disorders in the Americas and the Caribbean Basin. Plant Disease 76, 220–225.

Brown, J.K., Bird, J., 1995. Variability within the *Bemisia tabaci* species complex and its relation to new epidemics caused by Geminivirus. CEIBA 36, 73–80.

Brown, J.K., Coats, S.A., Bedford, I.D., Markham, P.G., Bird, J., Frohlich, D.R., 1995a. Characterization and distribution of esterase electromorphs in the whitefly, *Bemisia tabaci* (Genn.) (Homoptera: *Aleyrodidae*). Biochemical Genetics 33, 205–214.

Brown, J.K., Frohlich, D.R., Rosell, R.C., 1995b. The sweetpotato or silverleaf whiteflies. Biotypes of *Bemisia tabaci* or a species complex. Annual Review of Entomology 40, 511–534.

Brown, J.K., Lambert, G.M., Ghanim, M., Czosnek, H., Galbraith, D.W., 2005. Nuclear DNA content of the whitefly *Bemisia tabaci* (Genn.) (Aleyrodidae: Homoptera/Hemiptera) estimated by flow cytometry. Bulletin of Entomological Research 95, 309–312.

Brown, J.K., Czosnek, H., 2002. Whitefly transmission of plant viruses. In: Plumb, R.T. (Ed.), Advances in botanical research, Academic press, New York, pp. 65–100.

Butter, N.S., Rataul, H.S., 1977. The virus–vector relationship of the *Tomato leafcurl virus* (TLCV) and its vector, *Bemisia tabaci* Gennadius (Hemiptera: Aleyrodidae). Phytoparasitica 5, 173–186.

Byrne, F.J., Bedford, I.D., Devonshire, A.L., Markham, P.G., 1995. Esterase variation and squash silverleaf induction in 'B' biotype *Bemisia tabaci* (Homoptera; Aleyrodidae). Bulletin of Entomological Research 85, 175–179.

Caciagli, P., Bosco, D., 1997. Quantitation over time of tomato yellow leaf curl geminivirus DNA in its whitefly vector. Phytopathology 87, 610–613.

Caciagli, P., Bosco, D., Al-Bitar, L., 1995. Relationships of the Sardinian isolate of tomato yellow leaf curl geminivirus with its whitefly vector *Bemisia tabaci* Gen. European Journal of Plant Pathology 101, 163–170.

Caciagli, P., Medina, V., Marian, D., Vecchiati, M., Masenga, V., Mason, G., Falcioni, T., Noris, E., 2009. Virion stability is important for the circulative transmission of Tomato yellow leaf curl sardinia virus by *Bemisia tabaci*, but virion access to salivary glands does not guarantee transmissibility. Journal of Virology 83, 5784–5795.

Cohen, S., Antignus, Y., 1994. Tomato yellow leaf curl virus (TYLCV), a whitefly-borne geminivirus of tomatoes. Advances in Disease Vector Research 10, 259–288.

Cohen, S., Harpaz, I., 1964. Periodic, rather than continual acquisition of a new tomato virus by its vector, the tobacco whitefly (*Bemisia tabaci* Gennadius). Entomologia experimentalis et Applicata 7, 155–166.

Cohen, S., Nitzany, F.E., 1966. Transmission and host range of the tomato yellow leaf curl virus. Phytopathology 56, 1127–1131.

Cohen, S., Duffus, J.E., Larsen, R.C., Liu, H.Y., Flock, R.A., 1983. Purification, serology, and vector relationships of Squash leaf curl virus, a whitefly-transmitted geminivirus. Phytopathology 73, 1669–1673.

Cohen, S., Kern, J., Harpaz, I., Ben-Joseph, R., 1988. Epidemiological studies of the tomato yellow leaf curl virus (TYLCV) in the Jordan Valley, Israel. Phytoparasitica 16, 259–270.

Cohen, S., Duffus, J.E., Liu, H.Y., 1989. Acquisition, interference, and retention of cucurbit leaf curl viruses in whiteflies. Phytopathology 79, 109–113.

Costa, H.S., Ullman, D.E., Johnson, M.W., Tabashnik, B.E., 1993. Squash silverleaf symptoms induced by immature, but not adult, *Bemisia tabaci*. Phytopathology 83, 763–766.

Costa, H.S., Westcot, D.M., Ullman, D.E., Rosell, R.C., Brown, J.K., Johnson, M.W., 1995. Morphological variation in *Bemisia* endosymbionts. Protoplasma 189, 194–202.

Czosnek, H., Brown, J., 2009. The whitefly genome—white paper: proposal to sequence multiple genomes of *Bemisia tabaci* (Gennadius). In: Stansly, P.A. (Ed.), Bemisia: bionomics and management, Springer, Dordrecht, pp. 503–532.

Czosnek, H., Ghanim, M., 2011. *Bemisia tabaci*–Tomato yellow leaf curl virus interaction causing worldwide epidemics. In: Thompson, W.M.O. (Ed.), The whitefly, *Bemisia tabaci* (Homoptera: *Aleyrodidae*) interaction with Geminivirus-infected host plants, Springer, Dordrecht, pp. 51–61.

Czosnek, H., Latterot, H., 1997. A world-wide survey of tomato yellow leaf curl viruses. Archives of Virology 142, 1391–1406.

Czosnek, H.G., Ber, R., Antignus, Y., Cohen, S., Navot, N., Zamir, D., 1988. Isolation of the tomato yellow leaf curl virus—a geminivirus. Phytopathology 78, 508–512.

Czosnek, H.G., Ghanim, M., Rubinstein, G., Morin, S., Fridman, V., Zeidan, M., 2001. Whiteflies: Vectors, and victims (?), of geminiviruses. In: Maramorosch, K. (Ed.), Advances in Virus Research, Academic Press, New York, pp. 291–322.

Czosnek, H., Ghanim, M., Ghanim, M., 2002. Circulative pathway of begomoviruses in the whitefly vector *Bemisia tabaci*—insights from studies with Tomato yellow leaf curl virus. Annals of Applied Biology 140, 215–231.

Davino, M., D'Urso, F., Areddia, R., Carbone, M., Mauromicale, G., 1994. Investigations on the epidemiology of tomato yellow leaf curl virus (TYLCV) in Sicily. Petria 4, 151–160.

Davino, S., Napoli, C., Dellacroce, C., Miozzi, L., Noris, E., Davino, M., Acctto, G.P., 2009. Two new natural begomovirus recombinants associated with the tomato yellow leaf curl disease co-exist with parental viruses in tomato epidemics in Italy. Virus Research 143, 15–23.

Duffus, J.E., 1986. Whitefly transmission of plant viruses. In: Harris, K. (Ed.), Current topics in vector research, vol. 4. Springer-Verlag, New York, pp. 73–91.

Duffy, S., Holmes, E.C., 2007. Multiple introductions of the Old World begomovirus *Tomato yellow leaf curl virus* into the New World. Applied and Environmental Microbiology 73, 7114–7117.

EPPO/CABI, 1996. Bean golden mosaic bigeminivirus; Lettuce infectious yellows closterovirus; Squash leaf curl bigeminivirus; Tomato mottle bigeminivirus; Tomato yellow leaf curl bigeminivirus. In: Smith, I.M. (Ed.), Quarantine pests for Europe, CAB International, Wallingford, UK.

Fargette, D., Leslie, M., Harrison, B.D., 1996. Serological studies on the accumulation and localisation of three tomato leaf curl geminiviruses in resistant and susceptible *Lycopersicon* species and tomato cultivars. Annals of Applied Biology 128, 317–328.

Fauquet, C.M., Briddon, R.W., Brown, J.K., Moriones, E., Stanley, J., Zerbini, M., Zhou, X., 2008. Geminivirus strain demarcation and nomenclature. Archives of Virology 153, 783–821.

Francki, R.I.B., Hatta, T., Boccardo, G., Randles, J.W., 1980. The composition of chlorotis striate mosaic virus, a geminivirus. Virology 101, 233–241.

Frohlich, D., Torres-Jerez, I., Bedford, I.D., Markham, P.G., Brown, J.K., 1999. A phylogeographic analysis of the *Bemisia tabaci* species complex based on mitochondrial DNA markers. Molecular Ecology 8, 1593–1602.

García-Andrés, S., Monci, F., Navas-Castillo, J., Moriones, E., 2006. Begomovirus genetic diversity in the native plant reservoir *Solanum nigrum*: evidence for the presence of a new virus species of recombinant nature. Virology 350, 433–442.

García-Andrés, S., Accotto, G.P., Navas-Castillo, J., Moriones, E., 2007. Founder effect, plant host, and recombination shape the emergent population of begomoviruses that cause the tomato yellow leaf curl disease in the Mediterranean basin. Virology 359, 302–312.

Ghanim, M., Czosnek, H., 2000. Tomato yellow leaf curl geminivirus (TYLCV-Is) is transmitted among whiteflies (*Bemisia tabaci*) in a sex-related manner. Journal of Virology 74, 4738–4745.

Ghanim, M., Kontsedalov, S., 2007. Gene expression in pyriproxyfen resistant *Bemisia tabaci* Q biotype. Pest Management Science 63, 776–783.

Ghanim, M., Medina, V., 2007. Localization of tomato yellow leaf curl virus in its whitefly vector *Bemisia tabaci*. In: Czosnek, H. (Ed.), Tomato yellow leaf curl virus disease: management, molecular biology, breeding for resistance, Springer, Dordrecht, pp. 171–183.

Ghanim, M., Morin, S., Zeidan, M., Czosnek, H.G., 1998. Evidence for transovarial transmission of tomato yellow leaf curl virus by its vector, the whitefly *Bemisia tabaci*. Virology 240, 295–303.

Ghanim, M., Morin, S., Czosnek, H., 2001a. Rate of *Tomato yellow leaf curl virus* (TYLCV) translocation in the circulative transmission pathway of its vector, the whitefly *Bemisia tabaci*. Phytopathology 91, 188–196.

Ghanim, M., Rosell, R.C., Campbell, L.R., Czosnek, H., Brown, J.K., Ullman, D.E., 2001b. Microscopic analysis of the digestive, salivary and reproductive organs of *Bemisia tabaci* (Gennadius) (Hemiptera: Aleyrodidae) biotype B. Journal of Morphology 248, 22–40.

Ghanim, M., Sobol, I., Ghanim, M., Czosnek, H., 2007. Horizontal transmission of begomoviruses between *Bemisia tabaci* biotypes. Arthropod-Plant Interactions 1, 195–204.

Ghanim, M., Brumin, M., Popvski, S., 2009. A simple, rapid and inexpensive method for local-ization of *Tomato yellow leaf curl virus* and *Potato leafroll virus* in plant and insect vectors. Journal of Virological Methods 159, 311–314.

Gildow, F.E., Gray, S.M., 1993. The aphid salivary gland basal lamina as a selective barrier associ-ated with vector-specific transmission of barley yellow dwarf luteovirus. Phytopathology 83, 1293–1302.

Gildow, F.E., Rochow, W.F., 1980. Role of accessory salivary glands in aphid transmission of barely yellow dwarf virus. Virology 104, 97–108.

Gill, R. J. (1990). The morphology of whiteflies. In: Gerlinga, D. (Ed.), Whiteflies: their bionomics, pest status, and management. Intercept, UK, pp. 11–44.

Goodman, R.M., 1977. Single-stranded DNA genome in a whitefly-transmitted plant virus. Virology 83, 171–179.

Gottlieb, Y., Zchori-Fein, E., Mozes Daube, N., Kontsedalov, S., Skaljac, M., Brumin, M., Sobol, I., Czosnek, H., Vavre, F., Fleury, F., Ghanim, M., 2010. The transmission efficiency of *Tomato yellow leaf curl virus* by the whitefly *Bemisia tabaci* is correlated with the presence of a specific symbiotic bacteriumspecies. Journal of Virology 84, 9310–9317.

Guirao, P., Beitia, F., Cenis, J.L., 1997. Biotype determination of Spanish populations of *Bemisia tabaci* (Hemiptera: *Aleyrodidae*). Bulletin of Entomological Research 87, 587–593.

Harrison, B.D., Barker, H., Bock, K.R., Guthrie, E.J., Meredith, G., Atkinson, M., 1977. Plant viruses with circular single stranded DNA. Nature 270, 760–762.

Horowitz, A.R., Gorman, K., Ross, G., Denholm, I., 2003. Inheritance of pyriproxyfen resistance in the whitefly, *Bemisia tabaci* (Q biotype). Archives of Insect Biochemistry and Physiology 54, 177–186.

Horowitz, A.R., Kontsedalov, S., Khasdan, V., Ishaaya, I., 2005. Biotypes B and Q of *Bemisia tabaci* and their relevance to neonicotinoid and pyriproxyfen resistance. Archives of Insect Biochem-istry and Physiology 58, 216–225.

Ioannou, N., 1985. Yellow leaf curl and other diseases of tomato in Cyprus. Plant Pathology 345, 428–434.

Jiang, Y.X., De Blas, C., Barrios, L., Fereres, A., 2000. A correlation between whitefly (Homoptera: Aleyrodidae) feeding behaviour and transmission of Tomato yellow leaf curl virus. Annals of the Entomological Society of America 93, 573–579.

Jiang, Y.X., Dde Blas, C., Bedford, I.D., Nombela, G., Muñiz, M., 2004. Effect of *Bemisia tabaci* biotype in the transmission of *tomato yellow leaf curl sardinia virus* (TYLCSV-ES) between tomato and common weeds. Spanish Journal of Agricultural Research 2, 115–119.

Jiu, M., Zhou, X.P., Tong, L., Xu, J., Yang, X., Wan, F.H., Liu, S.S., 2007. Vector–virus mutualism accelerates population increase of an invasive whitefly. PLoS ONE 2, e182.

Jupin, I., De Kouchkovsky, F., Jouanneau, F., Gronenborn, B., 1994. Movement of *tomato yellow leaf curl geminivirus* (TYLCV): involvement of the protein encoded by ORF C4. Virology 204, 82–90.

Kato, K., Onuki, M., Fuji, S., Hanada, K., 1998. The first occurrence of tomato yellow leaf curl virus in tomato (*Lycopersicon esculentum* Mill.) in Japan. Annals of the Phytopathological Society of Japan 64, 552–559.

Kheyr-Pour, A., Bendahmane, M., Matzeit, N., Accotto, G.P., Crespi, S., Gronenborn, B., 1991. Tomato yellow leaf curl virus from Sardinia is a whitefly-transmitted geminivirus. Nucleic Acids Research 19, 6763–6769.

Kunik, T., Palanichelvam, K., Czosnek, H., Citovsky, V., Gafni, Y., 1998. Nuclear import of a gemi-nivirus capsid protein in plant and insect cells: implications for the viral nuclear entry. Plant Journal 13, 121–129.

Lefeuvre, P., Martin, D.P., Harkins, G., Lemey, P., Gray, A.J.A., Meredith, S., Lakay, F., Monjane, A., Lett, J.M., Varsani, A., Heydarnejad, J., 2010. The spread of tomato yellow leaf curl virus from the Middle East to the world. PLOS Pathogens 6 (10), e1001164.

Leshkowitz, D., Gazit, S., Reuveni, E., Ghanim, M., Czosnek, H., McKenzie, C., Shatters, R.G., BrownJ, K., 2006. Whitefly (*Bemisia tabaci*) genome project: analysis of sequenced clones from egg, instar, and adult (viruliferous and non-viruliferous) cDNA libraries. BMC Genomics 7, 79.

Liu, J., Zhao, H., Jiang, K., Zhou, X.-P., Liu, S.S., 2009. Differential indirect effects of two plant viruses on an invasive and an indigenous whitefly vector: implications for competitive displacement. Annals of Applied Biology 155, 39–448.

Mahadav, A., Gerling, D., Gottlieb, Y., Czosnek, H., Ghanim, M., 2008. Parasitization by the wasp *Eretmocerus mundus* induces transcription of genes related to immune response and symbiotic bacteria proliferation in the whitefly *Bemisia tabaci*. BMC Genomics 9, 342.

Mahadav, A., Kontsedalov, S., Czosnek, H., Ghanim, M., 2009. Thermotolerance and gene expression following heat stress in the whitefly *Bemisia tabaci* B and Q biotypes. Insect Biochemistry Molecular Biology 39, 668–676.

Mansour, A., Al-Musa, A., 1992. Tomato yellow leaf curl virus: host range and virus-vector relationships. Plant Pathology 41, 122–125.

Mansoor, S., Bedford, I., Pinner, M., Stanley, J., Markham, P., 1993. A whitefly-transmitted geminivirus associated with cotton leaf curl disease in Pakistan. Pakistan Journal of Botany 25, 105–107.

Markham, P.G., Bedford, I.D., Liu, S., Pinner, M.S., 1994. The transmission of geminiviruses by *Bemisia tabaci*. Pesticide Science 42, 123–128.

Matsuura, S., Hoshino, S., 2009. Effect of tomato yellow leaf curl disease on reproduction of *Bemisia tabaci* Q biotype (Hemiptera: Aleyrodidae) on tomato plants. Applied Entomology and Zoology 44, 143–148.

McGlashan, D., Polston, J., Bois, D., 1994. Tomato Yellow Leaf Curl Geminivirus in Jamaica. Plant Disease 78, 1219–1219.

McGrath, P.F., Harrison, B.D., 1995. Transmission of tomato leaf curl geminivirus by *Bemisia tabaci* effects of virus isolate and vector biotype. Annals of Applied Biology 126, 307–316.

McKenzie, C.L., Hodges, G., Osborne, L.S., Byrne, F.J., Shatters, R.G.J., 2009. Distribution of *Bemisia tabaci* (Hemiptera: Aleyrodidae) biotypes in Florida—investigating the Q invasion. Journal of Economic Entomology 102, 670–676.

Medina, V., Pinner, M.S., Bedford, I.D., Achon, M.A., Gemeno, C., Markham, P.G., 2006. Immunolocalization of tomato yellow leaf curl Sardinia virus in natural host plants and its vector *Bemisia tabaci*. Journal of Plant Pathology 88, 299–308.

Mehta, P., Wyman, J.A., Nakhla, M.K., Maxwell, D.P., 1994. Transmission of tomato yellow leaf curl geminivirus by *Bemisia tabaci*. Journal of Economic Entolomogy 87, 1291–1297.

Momol, M.T., Simone, G.W., Dankers, W., Sprenkel, R.K., Olson, S.M., Mimol, E.A., Polston, J.E., Hiebert, E., 1999. First report of tomato yellow leaf curl virus in tomato in south Georgia. Plant Disease 83, 487.

Monci, F., Sánchez-Campos, S., Navas-Castillo, J., Moriones, E., 2002. A natural recombinant between the geminiviruses Tomato yellow leaf curl Sardinia virus and Tomato yellow leaf curl virus exhibits a novel pathogenic phenotype and is becoming prevalent in Spanish populations. Virology 303, 317–326.

Morin, S., Ghanim, M., Zeidan, M., Czosnek, H., Verbeek, M., van den Heuvel, J.F.J.M., 1999. A GroEL homologue from endosymbiotic bacteria of the whitefly *Bemisia tabaci* is implicated in the circulative transmission of tomato yellow leaf curl virus. Virology 30, 75–84.

Morin, S., Ghanim, M., Sobol, I., Czosnek, H., 2000. The GroEL protein of the whitefly *Bemisia tabaci* interacts with the coat protein of transmissible and non-transmissible begomoviruses in the yeast two-hybrid system. Virology 276, 404–416.

Moriones, E., 2000. TYLCV datasheet, EWSN, UK.

Moriones, E., Navas-Castillo, J., 2000. Tomato yello leaf curl virus, an emerging virus complex causing epidemics worldwide. Virus Research 71, 123–134.

Mound, L.A., Halsey, S.H., 1978. Whiteflies of the world, a systematic catalogue of the Aleyrodidae (Homoptera) with host plant and natural enemy data. British Museum (Natural History), London, UK.

Muniyappa, V., Venkatesh, H.M., Ramappa, H.K., Kulkarni, R.S., Zeidan, M., Tarba, C.-Y., Ghanim, M., Czosnek, H., 2000. Tomato leaf curl virus from Bangalore (ToLCV-Ban4): sequence comparison with Indian ToLCV isolates, detection in plants and insects, and vector relationships. Archives of Virology 145, 1583–1598.

Navas-Castillo, J., Sánchez-Campos, S., Díaz, J.A., Sáez-Alonso, E., Moriones, E., 1997. First report of tomato yellow leaf curl virus-Is in Spain: coexistence of two different geminiviruses in the same epidemic outbreak. Plant Disease 81, 1461.

Navas-Castillo, J., Sánchez-Campos, S., Diaz, J., Sáez-Alonso, E., Moriones, E., 1999. Tomato yellow leaf curl virus-Is causes a novel disease of common bean and severe epidemics in tomato in Spain. Plant Disease 83, 29–32.

Navot, N., Ber, R., Czosnek, H.G., 1989. Rapid detection of tomato yellow leaf curl virus in squashes of plants and insect vectors. Phytopathology 79, 562–568.

Navot, N., Pichersky, E., Zeidan, M., Zamir, D., Czosnek, H., 1991. Tomato yellow leaf curl virus: a whitefly-transmitted geminivirus with a single genomic component. Virology 185, 151–161.

Navot, N., Zeidan, M., Pichersky, E., Zamir, D., Czosnek, H., 1992. Use of polymerase chain reaction to amplify tomato yellow leaf curl virus DNA from infected plants and viruliferous whiteflies. Phytopathology 82, 1199–1202.

Noris, E., Hidalgo, E., Accotto, G.P., Moriones, E., 1994. High similarity among the tomato yellow leaf curl virus isolates from the West Mediterranean Basin: the nucleotide sequence of an infectious clone from Spain. Archives of Virology 135, 165–170.

Noris, E., Vaira, A.M., Caciagli, P., Masenga, V., Gronemborn, B., Accotto, G.P., 1998. Amino acids in the capsid protein of tomato yellow leaf curl virus that are crucial for systemic infection, particle formation, and insect transmission. Journal of Virology 72, 10050–10057.

Ohnesorge, S., Bejarano, E.R., 2009. Begomovirus coat protein interacts with a small heatshock protein of its transmission vector (*Bemisia tabaci*). Insect Molecular Biology 18, 693–703.

Pascual, S., Callejas, C., 2004. Intra- and interspecific competition between biotypes B and Q of *Bemisia tabaci* (Hemiptera: Aleyrodidae) from Spain. Bulletin of Entomological Research 94, 369–375.

Perring, T.M., Cooper, A.D., Rodriguez, R.J., Farrar, C.A., Bellows, T.S., 1993. Identification of a whitefly species by genomic and behavioural studies. Science 259, 74–77.

Peterschmitt, M., Granier, M., Mekdoud, R., Dalmon, A., Gambin, O., Vayssieres, J.F., Reynaud, B., 1999. First report of tomato yellow leaf curl virus in Reunion Island. Plant disease 83, 303–303.

Picó, B., Ferriol, M., Diez, M.J., Nuez, F., 1999. Developing tomato breeding lines resistant to tomato yellow leaf curl virus. Plant Breeding 118, 537–542.

Pollard, D.G., 1955. Feeding habits of the cotton whitefly. Annals of Applied Biology 43, 664–671.

Polston, J.E., Al-Musa, A., Perring, T.M., Dodds, J.A., 1990. Association of the nucleic acid of squash leaf curl geminivirus with the whitefly *Bemisia tabaci*. Phytopathology 80, 850–856.

Reddy, K.S., Yaraguntaiah, R.C., 1981. Virus–vector relationship in leaf curl disease of tomato. Indian Phytopathology 34, 310–313.

Rosell, R.C., Bedford, I.D., Frohlich, D.R., Gill, R.J., Brown, J.K., 1997. Analysis of morphological variation in distinct populations of *Bemisia tabaci* (Homoptera: Aleyrodidae). Annals of the Entomological Society of America 90, 575–589.

Rosell, R.C., Torres-Jerez, I., Brown, J.K., 1999. Tracing the geminivirus–whitefly transmission pathway by polymerase chain reaction in whitefly extracts, saliva, hemolymph, and honeydew. Phytopathology 89, 239–246.

Rubinstein, G., Czosnek, H.G., 1997. Long-term association of tomato yellow leaf curl virus (TYLCV) with its whitefly vector *Bemisia tabaci*: effect on the insect transmission capacity, longevity and fecundity. Journal of General Virology 78, 2683–2689.

Rybicki, E.P., Briddon, R.W., Brown, J.K., Fauquet, C.M., Maxwell, D.P., Stanley, J., Harrison, B.D., Markham, P.G., Bisaro, D.M., Robinson, D., 2000. Family Geminiviridae. In: Van Regenmortel, M.H.V., Fauquet, C.M., Bishop, D.H.L., Carstens, E., Estes, M., Lemon, S., Maniloff, J., Mayo, M.A., McGeoch, D., Pringle, C., Wickner, R. (Eds.), Virus taxonomy. Seventh Report of the International Committee on Taxonomy of Viruses, Academic Press, New York, pp. 285–297.

Sánchez-Campos, S., Navas-Castillo, J., Camero, R., Soria, C., Díaz, J.A., Moriones, E., 1999. Displacement of tomato yellow leaf curl virus (TYLCV)-Sr by TYLCV-Is in tomato epidemics in Spain. Phytopathology 89, 1038–1043.

Sharaf, N.S., Al-Musa, A.M., Batta, Y., 1985. Effect of different host plants on the population development of the sweetpotato whitefly (*Bemisia tabaci* Genn.; Homoptera: Aleyrodidae). Dirasat 12, 89–100.

Sinisterra, X.H., McKenzie, C.L., Hunter, W.B., Powell, C.A., Shatters, R.G., 2005. Differential transcriptional activity of plant-pathogenic begomoviruses in their whitefly vector (*Bemisia tabaci*, Gennadius: Hemiptera Aleyrodidae). Journal of General Virology 86, 1525–1532.

Skaljac, M., Ghanim, M., 2010. Tomato yellow leaf curl disease and plant–virus vector interactions. Israel Journal of Plant Sciences 58, 103–111.

Stonor, J., Hart, P., Gunther, M., DeBarro, P., Rezaian, M., 2003. Tomato leaf curl geminivirus in Australia: occurrence, detection, sequence diversity and host range. Plant Pathology 52, 379–388.

Su, Y.L., Li, J.M., Li, M., Luan, J.B., Ye, X.D., Wang, X.W., Liu, S.S., 2012. Transcriptomic analysis of the salivary glands of an invasive whitefly. PLoS ONE 7 (6), e39303.

Sugiyama, K., Matsuno, K., Doi, M., Tatara, A., Kato, M., Tagami, Y., 2008. TYLCV detection in *Bemisia tabaci* (Gennadius) (Hemiptera: Aleyrodidae) B and Q biotypes, and leaf curl symptom of tomato and other crops in winter greenhouses in Shizuoka Pref., Japan. Applied Entomology and Zoology 43, 593–598.

van Regenmortel, M.H.V., Fauquet, C.M., Bishop, D.H.L., Carstens, E.B., Estes, M.K., Lemon, S.M., Maniloff, J., Mayo, M.A., McGeoch, D.J., Pringle, C.R., 2000. Virus taxonomy classification and nomenclature of viruses. Seventh Report of the International Committee on Taxonomy of Viruses. Academic Press, San Diego, CA, p. 1167.

Wang, K., Tsai, J.H., 1996. Temperature effects on development and reproduction of silverleaf whitefly (Homoptera, Aleyrodidae). Annals of the Entomological Society of America 89, 375–384.

Wang, J., Zhao, H., Jian, L., Jiu, M., Qian, Y.-J., Liu, S.-S., 2009. Low frequency of horizontal and vertical transmission of two begomoviruses through whitefly *Bemisia tabaci* biotype B and Q. Annals of Applied Biology 157, 125–133.

Wang, X.W., Luan, J.B., Li, J.M., Bao, Y.Y., Zhang, C.X., Liu, S.S., 2010. De novo characterization of a whitefly transcriptome and analysis of its gene expression during development. BMC Genomics 11, 400.

Wartıg, L., Kheyr-Pour, A., Noris, E., De Kouchkovsky, F., Jouanneau, F., Gronemborn, B., Jupin, I., 1997. Genetic analysis of the monopartite tomato yellow leaf curl geminivirus: roles of V1, V2, and C2 ORFs in viral pathogenesis. Virology 228, 132–140.

Zeidan, M., Czosnek, H.G., 1991. Acquisition of tomato yellow leaf curl virus by the whitefly *Bemisia tabaci*. Journal of General Virology 72, 2607–2614.

Zhang, W., Olson, N.H., Baker, T.S., Faulkner, L., Agbandje-McKenna, M., Boulton, M.I., Davies, J.W., McKenna, R., 2001. Structure of the *Maize streak virus* geminate particle. Virology 279, 471–477.

Zrachya, A., Glick, E., Levy, Y., Arazi, T., Citovsky, V., Gafni, Y., 2007. Suppressor of RNA silencing encoded by *tomato yellow leaf curl virus*-Israel. Virology 358, 159–165.

Hosts and non-hosts in plant virology and the effects of plant viruses on host plants

András Takács
University of Pannonia, Georgikon Faculty of Agricultural Sciences, Institute for Plant Protection, Keszthely, Hungary

József Horváth
University of Pannonia, Georgikon Faculty of Agricultural Sciences, Institute for Plant Protection, Keszthely, Hungary, and Kaposvár University, Department of Botany and Plant Production, Kaposvár, Hungary

Richard Gáborjányi
University of Pannonia, Georgikon Faculty of Agricultural Sciences, Institute for Plant Protection, Keszthely, Hungary

Gabriella Kazinczi
Kaposvár University, Department of Botany and Plant Production, Kaposvár, Hungary

József Mikulás
Corvinus University of Budapest Institute of Viticulture and Oenology Research Station, Kecskemét, Hungary

TYPES OF HOST–VIRUS RELATIONSHIPS

Host–virus interactions differ widely in the mechanisms involved in the display of symptoms. Both compatible and incompatible host–virus relationships exist.

Compatible host–virus relationships

In compatible host–virus relationships, viruses infect the host cell, and depending on where virus symptoms appear, local and systemic hosts are distinguished. In some cases, local and systemic symptoms can appear at the same time. Local and systemic symptoms are called *external symptoms*. Environmental factors can greatly influence the type of virus symptoms. For example, at higher temperatures *Tobacco mosaic virus* (TMV) can translocate in *Nicotiana tabacum* 'Xanthi' plants, whereas at lower temperatures it creates only local chlorotic and necrotic lesions. The latest research proves that this phenomenon is associated with down-regulation of nicotinamide adenine dinucleotide phosphate-oxidase (NADPH oxidase) and superoxide, and stimulation of dehydroascorbate reductase (DHAR). On the other hand, enhanced glutathione (GSH) metabolism is correlated with sulfur-induced resistance in compatible tobacco–TMV relationships.

A special type of compatible host–virus relationship occurs when the virus particles in infected plants can move from cell to cell but where there are no visible external symptoms (e.g. *Alfalfa mosaic virus* [AMV] in the plant *Medicago sativa*). The presence of the virus can be detected by other diagnostic methods in the symptomless host.

Viruses can also cause microsymptoms (or *internal symptoms*) in an infected host cell. Some plant cells can react by forming special inclusion bodies in the cytoplasm of the infested host cell due to virus infection, which can be seen with a light microscope. Inclusion bodies can be characteristic of a certain class of viruses (e.g. pinwheel inclusion bodies are typical of infection by members of the *Potyvirus* genus).

Symptoms of viral infection can occur on all parts of the plant (root, stem, leaf, flower, crop) but occur most frequently on the leaves of the susceptible hosts. Table 5.1 lists the common external symptoms (in parentheses with their abbreviations), including host–virus examples. (See also Figs 5.1–5.3.)

A special type of host–virus relationship is the hypersensitive reaction (HR), when a virus infects a plant cell but the infected cell dies before the virus can translocate to other cells. This mechanism is a survival strategy for a plant to avoid spreading of virus particles or nucleic acids. Plants are susceptible to virus infection, but infection might not lead to symptoms. Plants are able to localize virus particles at the site of infection. The ability to produce an HR is a useful trait in breeding plants for virus resistance.

Non-compatible host–virus relationships

In the case of extreme resistance (immunity), a virus is not able to infect a host cell, so that symptoms do not occur and viruses cannot be detected.

The work of Kegler & Meyer (1987) showed that the resistance of plants to viruses may be either qualitative or quantitative. In the case of the former, a specific relationship exists between the plant's resistance genes and the genes of the virus. This qualitative resistance can be expressed as extreme resistance, hypersensitivity, or resistance to the spreading of viruses. In the case of quantitative resistance, no specific relationship exists between the plant's resistance genes and the genes of the virus. Types of quantitative resistance include resistance to virus replication and spreading, tolerance, field resistance, tolerance to plant disease, and real tolerance.

Fraser (1990) attempted to illustrate some of the processes in plant–virus relationships involving recognition events and the consequent responses of susceptibility or resistance. In a recent review, Király and colleagues (2007) summarized the different forms of plant resistance to different pathogens (including viruses) from the point of view of genetics and plant breeding (innate and aquired resistance).

TABLE 5.1 The most frequently observed virus symptoms in compatible host–virus relationships

Symptoms[a]	Host–virus relations[b]
Blistering (Bli)	*N.tabacum* 'Xanthi'–PVY
Color breaking (Cb)	*Tulipa* spp.–TBV
Chlorosis (Ch)	*Capsicum annuum*–AMV
Chlorotic lesions (Chl)	*Chenopodium amaranticolor*–AMV
Death of the plant (D)	*Solanum tuberosum*–PVY
Fern leaf (Fle)	*Vitis vinifera*–GFLV
Growth reduction (Gr)	*C. annuum*–TMV
Hypersensitive local necrotic lesions (HR)	*N. tabacum* 'Xanthi'–TMV
Latent (symptomless) (La)	*Medicago sativa*–AMV
Leaf deformation (Ldef)	*Lycopersicon esculentum*–ToMV
Leaf drop (Led)	*S. tuberosum*–PVY
Mosaic (M)	*Cucumis sativus*–CMV
Mottling (Mo)	*Cucumis melo*–WMV
Necrosis (N)	*S. tuberosum*–PVY
Necrotic lesions (Nl)	*N. glutinosa*–TMV
Chlorotic rings (ChRi)	*Prunus domestica*–PPV
Necrotic rings (NRi)	*S. tuberosum*–PVY[NTN]
Pattern (P)	*C. annuum*–TSWV
Top necrosis (Tn)	*S. tuberosum*–PVY
Vein banding (Vb)	*C. sativus*–ZYMV
Vein clearing (Vc)	*Brassica cretica* convar. *botrytis*–CaMV
Vein necrosis (Vn)	*C. anuum*–PVY
Vein netting (Vnt)	*Forsythia* spp.–ArMV

[a]*In the nomenclature for symptomatology, the symptoms are generally described in fractional form: the numerator of the abbreviation gives local symptoms, while the denominator indicates systemic symptoms, e.g. Nl/Ldef (local necrotic lesions/systemic leaf deformation) (see Baracsi 1999).*
[b]*TBV: Tulip breaking virus; PVY: Potato virus Y; PVY[NTN]: NTN strain of Potato virus Y; GFLV: Grapevine fanleaf virus; ToMV: Tomato mosaic virus; ZYMV: Zucchini yellow mosaic virus; CaMV: Cauliflower mosaic virus; ArMV: Arabis mosaic virus; WMV: Watermelon mosaic virus.*

FIGURE 5.1   Chlorotic (A) and necrotic (B) patterns on *Asclepias syriaca* leaves infected with *Tobacco mosaic virus* (TMV).

FIGURE 5.2   Local necrotic lesions on *Nicotiana glutinosa* leaves due to *Tobacco mosaic virus* (TMV) infection (left, healthy leaf; right, infected leaf).

THE ROLE OF HOST PLANTS IN VIRUS DIAGNOSIS

Early on in plant virus reseach, identification and detection of viruses were based on the presence and type of macro (external) symptoms. It is now believed that symptoms on plants may be unreliable because different viruses may cause similar or the same symptoms; the same virus may cause different symptoms, depending on the strain; complex viral infections or the presence/absence of a satellite RNA may greatly alter disease expression; different plants can manifest different symptoms

FIGURE 5.3   Systemic yellow mosaic on *Chenopodium album* plants after *Alfalfa mosaic virus* (AMV) infection.

as a result of infection by the same virus; and different environmental factors, especially temperature, may greatly influence the expression of disease symptoms.

On the other hand, certain symptoms can be characteristic of particular host–virus combinations, including the following examples: (1) flower break: *Tulipa* spp.–*Tulip breaking virus* (TBV); (2) fern-leaf: *V. vinifera–Grapevine fanleaf virus* (GFLV); enation: *Pisum sativum–Pea enation mosaic virus* (PeMV); vein necrosis: *N. tabacum–vein necrosis strain of *Potato virus Y* (PVY).

Based on the macro- and microsymptoms occurring on a test plant due to a viral infection, it is possible to estimate the presence or absence of some viruses. Although such biotesting methods are expensive, requiring much work and time, they are essential to virus diagnosis because pathologic characteristics of viruses can be examined only on their host plants. Therefore, even today, biotests are important diagnostic methods beside the other ones (e.g. ELISA, PCR techniques, etc.).

Some test plants are used to separate complex viral infections. Some plant species show varying symptoms of viral infections whereas others do not show symptoms at all. For example, a weed species, *Datura stramonium*, can be used to separate *Potato virus X* (PVX) and *Potato virus Y* (PVY). *D. stramonium* can host PVX while showing extreme resistance or immunity to PVY. Some other species can be used to separate two tobamoviruses, TMV and ToMV (see Table 5.2).

Other plants serve as propagative hosts for some viruses before virus isolation and purification.

Table 5.3 lists suggested species that can be used to identify some economically important individual viruses.

BIOLOGIC DECLINE OF PLANTS DUE TO VIRAL INFECTION

Viruses can seriously damage crops and weeds. In the case of crops, this damage is harmful, causing considerable yield losses and severe quantitative destruction. Earlier investigations were conducted on virus-infected crops. Little or no attention was paid to weed–virus interactions. Weeds can influence the crop yield directly by competing for nutrients, light and water, and they may play a considerable role in the epidemiology and overwintering of plant viruses. Weeds may be sources of primary infection in the spreading of plant diseases, and their seeds and vegetative reproduction organs (the latter in perennial weeds) may be reservoirs of obligate parasites such as viruses, by ensuring their overwintering. Furthermore, weeds as host plants of virus vectors may also be very important. In sustainable agricultural practice, considerable efforts are made to maintain biologic diversity; therefore, the purpose is not to kill all the weeds but rather to keep their levels under an economic threshold. Earlier investigatons from the perspectives of physiology and biochemistry were conducted on virus-infected crops rather than weeds. Although crop damage is harmful to plant production, the biologic decline of weeds due to viral infection may be an advantage for crop production. Viruses can considerably reduce growth, vegetative and generative biomass production, delay the generative development of weeds, reduce seed production, and may cause other changes in photosynthetic processes, nutrient uptake, drought tolerance, and germination characteristics. Although viruses cannot be used for biologic weed control because of their high genetic

TABLE 5.2 Separation of *Tobacco mosaic virus* (TMV) and *Tomato mosaic virus* (ToMV) (Horváth 1993)

Test plants	Viruses and symptoms[a]	
	TMV	ToMV
N. sylvestris	LS	L
C. amaranticolor	L	LS
C. quinoa	L	LS
Lycopersicon esculentum	S	S
Plantago major	L	LS

[a]L: local symptom; S: systemic symptom.

TABLE 5.3 Host and non-host plants of some important viruses (after Horváth 1993)

Viruses	Hosts	Non-hosts
Arabis mosaic virus (ArMV)	Chenopodium quinoa, Cucumis sativus	Capsicum annuum, Vicia faba
Beet necrotic yellow vein virus (BNYVV)	Beta macrocarpa, Nicotiana benthamiana, C. quinoa, Tetragonia expansa	C. sativus, Datura stramonium, N. glutinosa, P. vulgaris, V. faba
Cucumber mosaic virus (CMV)	C. quinoa, N. benthamiana, N. clevelandii; N. megalosiphon, Vigna unguiculata	Beta vulgaris
Plum pox virus (PPV)	C. foetidum, N. clevelandi	C. sativus cv. Lange Gele Tros
Potato virus Y (PVY)	C. amaranticolor, N. tabacum 'Xanthi', C. quinoa, Lycium spp., Physalis floridana, S. demissum x. S. tuberosum A6 hybrid	D. stramonium
Tobacco mosaic virus (TMV)	C. amaranticolor, N. glutinosa, N. sylvestris, N. tabacum 'Samsun'	Pisum sativum, Trifolium spp.
Tomato mosaic virus (ToMV)	C. amaranticolor, C. quinoa, D. stramonium, N, clevelandii, N. glutinosa	P. vulgaris cv. The Prince, Vigna sinensis cv. Blackeye
Tomato spotted wilt virus (TSWV)	N. clevelandii, Petunia hybrida, Tropaeolum majus	Tetragonia expansa

variability and poliphagous nature, viruses can indirectly contribute to weed control by reducing their competitive ability.

In experiments, three *Chenopodium* species (*C. album*, *C. murale*, and *C. quinoa*), two Solanaceous weeds (*D. stramonium* and *S. nigrum*), and two crops served as model species for studying biologic decline due to different viral infections (see Table 5.4).

The effect of viral infections on photosynthetic processes

Both *Cucumber mosaic virus* (CMV) and *Henbane mosaic virus* (HeMV) have been shown to reduce the photosynthetic pigment content of *D. stramonium* leaves. The most considerable (33%) reduction was detected in chlorophyll-*b* content, when plants were infected with CMV (see Fig. 5.4).

A later study focused on the effects of viral infections, herbicides, and plant extracts not only on photosynthetic pigment content, but also on fluorescence induction parameters (maximum quantum efficiency, actual quantum

TABLE 5.4 Model host–virus relationships for studying biologic decline due to viral infection

Hosts	Viruses[a]	Physiologic changes	References
D. stramonium	HeMV, CMV	Growth, photosynthetic pigment content	Kazinczi et al 1996
S. nigrum	TMV, ObPV, PepMV	Growth, nutrient content, seed production, photosynthetic pigment content, photosynthetic processes	Kazinczi et al 2001, 2002, Kazinczi et al 2006a,b,c
C. album, C. murale, C. quinoa	SoMV	Seed germination and seed viability, seed production	Kazinczi et al 1997, 2000
C. album, C. murale	SoMV	Phytochrome activity	Kazinczi et al 2000
S. nigrum	CMV	Drought resistance, photosynthetic pigment content, photosynthetic processes	Kazinczi et al 1998
N. tabacum 'Samsun'	ObPV	Photosynthetic pigment content, photosynthetic processes	Kazinczi et al 2006c
Capsicum annuum	ObPV	Photosynthetic pigment content, photosynthetic processes	Kazinczi et al 2006c

[a]CMV: *Cucumber mosaic virus*; HeMV: *Henbane mosaic virus*; ObPV: *Obuda pepper virus* (syn: ToMV-Ob); PepMV: *Pepino mosaic virus*; TMV: *Tobacco mosaic virus*.

efficiency, intristic energy utilization efficiency, ratio of open reaction centers, and non-photochemical quenching) and carbon assimilation. In these experiments, three systemic host relationships were used as models. In the *C. annuum–Obuda pepper virus* (ObPV) relationship, the activity of photosystem II (PSII) slightly increased fluorescence induction parameters. In ObPV-infected *N. tabacum*, the structural changes were similar to that of ObPV-infected *C. annuum*, but PSII efficiency did not significantly differ from that of the control. However, non-photochemical quenching (NPQ) increased because of the strongly decreasing CO_2 fixation activity. In the *S. nigrum–ObPV* relationship, the slightly increasing values of actual PSII quantum efficiency could be related to the likely elevated ratio of reaction center components (inreased chlorophyll *a/b* ratio) in the thylakoids. Application of fluazifop-p-buthyl herbicide as a post-emergence treatment prevented systemic viral infection; the virus-induced changes in photosynthesis are probably due to the inhibiting virus infection/replication process.

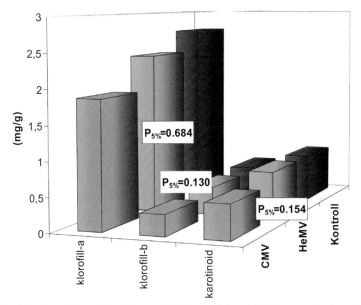

FIGURE 5.4   The effect of *Cucumber mosaic virus* (CMV) and *Henbane mosaic virus* (HeMV) infections on the chlorophyll and carotenoid content of *Datura stramonium.*

The effect of viral infections on plant biomass production

The shoot dry weight of test plants was reduced by 30–80%, depending on the host–virus relationships. (see Figs 5.5 and 5.6).

Sublethal water saturation deficit of healthy and virus-infected plants

In order to compare drought resistance of healthy and virus-infected *S. nigrum* plants, the sublethal water saturation deficit (WSD_{subl}) was determined. It was stated that even in healthy *Solanum* plants, the WSD_{subl} is low as compared to other summer annuals, such as *Digitaria sanguinalis, Ambrosia artemisiifolia,* and *Panicum dichotomiflorum.* This means that healthy *S. nigrum* leaves can lose only 36% of their maximum water content without suffering irreversible injuries. CMV infection reduced the WSD_{subl} by 6%, suggesting that viral infection caused disturbances in the water relationship and reduced drought tolerance (see Fig. 5.7, A and B).

The effect of viral infections on the germination characteristics of weeds

In germination bioassay studies, freshly harvested seeds of *C. album, C. murale, C. quinoa,* and *S. nigrum* were germinated under laboratory conditions at 25°C.

FIGURE 5.5   Uninfected (left) and *Tobacco mosaic virus* (TMV)-infected (right) *Solanum nigrum* plants.

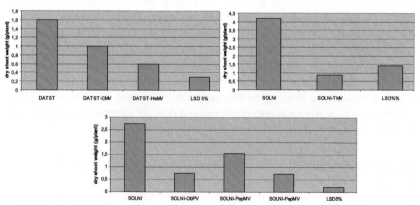

FIGURE 5.6   The effect of viral infections on the dry shoot weight of test plants (DATST: *Datura stramonium*; SOLNI: *Solanum nigrum*).

(A)

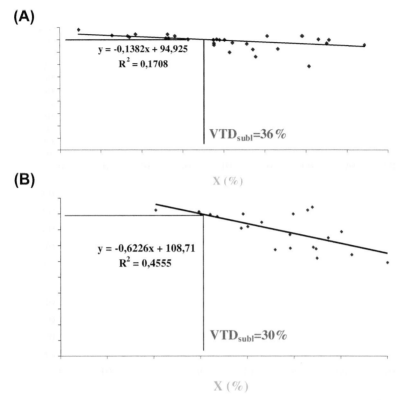

(B)

FIGURE 5.7   Sublethal water saturation deficit of healthy (A) and *Cucumber mosaic virus* (CMV)-infected (B) *Solanum nigrum* plants.

Chenopodium plants were infected with SoMV; *S. nigrum* plants were infected with ObPV, *Pepino mosaic virus* (PepMV), and ObPV–PepMV combinations, respectively. Seed viability was determined with TTC (2,3,5 triphenyltetrazolium chloride) methods, according to the internationally accepted seed testing standards.

A significant reduction in the germination rate was observed when *C. quinoa* and *C. murale* seeds were infected with SoMV, whereas germination of *C. album* was not influenced by viral infection. The seed viability of *C. album* was significantly reduced by SoMV infection.

Despite considerable reduction in the extent of seed germination, in the case of *S. nigrum* plants, seed viability was not significantly influenced by viral infection. This suggests that viral infection influenced seed dormancy characteristics rather than the viability of seeds (see Table 5.5).

Seeds derived from the virus-infected plants showed more intensive phytochrome activity as compared to those seeds derived from healthy plants. Phytochrome activity was highest after 3 minutes of red light after 3-day and 1-day imbibation periods with *C. album* and *C. murale*, respectively (see Fig. 5.8).

TABLE 5.5 Seed germination and seed viability of Chenopodium species and Solanum nigrum

	C. album			C. murale			C. quinoa		
	Control	SoMV	LSD5%	Control	SoMV	LSD5%	Control	SoMV	LSD5%
Germination (%)	63	62	7	50	40	6	100	96	3
Seed viability (%)	82	63	4	82	82	4	100	98	3

S. nigrum

	Control	ObPV	PepMV	ObPV +PepMV	LSD5%
Germination (%)	88	55	83	37	7
Seed viability (%)	79	86	84	88	13

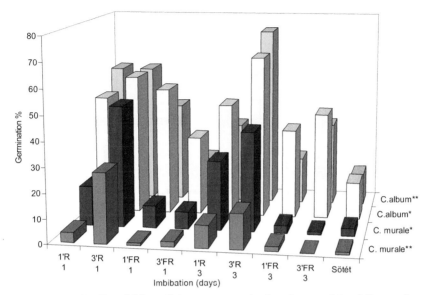

FIGURE 5.8   The effect of different light treatments on the phytochrome activity of *Chenopodium* seeds, derived from healthy and *Sowbane mosaic virus* (SoMV)-infected plants.

The effect of viral infections on the nutrient uptake of plants

In spite of the fact that many studies have dealt with the effect of nutrients on the development of pathogens and plant diseases, few data are available on the changes in nutrient content in host plants which occur as a result of infection by different pathogens. In our experiments there was no significant difference in the nitrogen (N), phosphorus (P), and calcium (Ca) content of TMV-infected *S. nigrum* plants, whereas the sodium (Na), magnesium (Mg), iron (Fe), manganese (Mn), zinc (Zn), and copper (Cu) content of leaves was significantly reduced compared to the healthy control. Enhanced concentration of potassium (K) was similarly observed after ObPV and PepMV infections (see Fig. 5.9). The physiologic cause of this unexpected observation is not yet clear.

The effect of viral infection on seed production in plants

ObPV+PepMV, ObPV, TMV, and PepMV caused 72%, 71%, 52% and 17% reductions in the seed production of *S. nigrum*, respectively (see Fig. 5.10). Generative development of *S. nigrum* plants was also delayed. This was well expressed in the higher proportions of the unripened berries within a plant.

VIRUS AND FUNGAL INTERACTIONS IN PLANTS

Studies on the interactions among plant pathogens and their hosts have mainly discussed the antagonistic or synergistic effects of different plant pathogens.

(A)

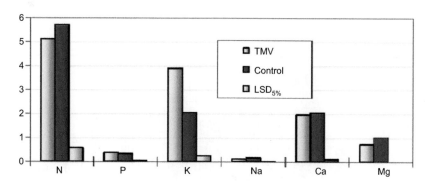

(B)

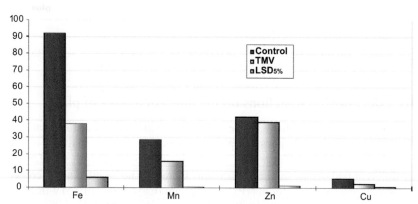

FIGURE 5.9 The effect of *Tobacco mosaic virus* (TMV) infection on the NPK (nitrogen, phosphorus, potassium), Na, Ca, Mg (%) (A) and Fe, Mn, Zn and Cu content (mg/kg) (B) of *Solanum nigrum*.

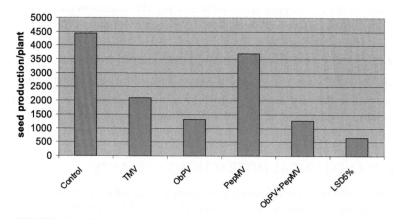

FIGURE 5.10 The effect of viral infections on the seed production of *Solanum nigrum*.

The effects of different mixed infections (i.e. virus + fungi) on host plants have rarely been taken into consideration. In our study, the effect of a fungus (*Macrophomina phaseolina*) and two viruses (CMV and PVY) on the development and nutrient content of pepper was studied. Viral infection alone did not significantly influence the development of pepper plants. *M. phaseolina* alone reduced only the fresh weight of pepper plants, suggesting the disturbance of root water uptake due to the fungal infection. In the case of several mixed infections, the injury effect increased. Fungal or viral single infections played an important role only in the early nutrient uptake of pepper plants. After flowering, no significant reduction in the nutrient content of pepper shoots was observed. In the case of mixed infection with viruses and *M. phaseolina*, the reduction in the nutrient content of pepper leaves was greater (see Table 5.6).

THE EFFECTS OF ARTIFICIAL AND NATURAL SUBSTANCES ON HOST–VIRUS RELATIONSHIPS

Virus particles create a very close biologic unit with the host cell. The synthesis of viral proteins occurs on the ribosomes of the infected plant cell. Therefore, chemical treatments for eliminating viruses *in vivo* cause not only the death of the virus, but also the death of the host plant cell. No *in vivo* viricides are known. However, some natural substances can inhibit virus replication and cell-to-cell movement and thus are able to reduce the virus concentration in infected plants. Some natural virus inhibitors, such as the proteins in *Phytolacca americana* (PAP), polysaccharides, glucoproteids, and phenol-like substances are naturally present in healthy plants. In other cases, these inhibitors are synthesized only after the stress effect, e.g. after virus infections (induced inhibitors). The latter ones can be divided into three groups: (1) phytoalexins, (2) pathogenesis-related (PR) proteins, and (3) chemically and physiologically heterogeneous inhibitors.

Plant allelochemicals, present in some plant species, can act as inhibitors to the development of susceptible recipient plants. In an experiment, the effects of some allelopathic weed species (*Abutilon theophrasti, Asclepias syriaca, Cirsium arvense*) on the development of plants and virus concentration in the virus-susceptible hosts were investigated simultaneously. Local (*C. quinoa, N. tabacum* 'Xanthi', and *N. glutinosa* to ObPV) and systemic host–virus relationships (*C. anuum, N. tabacum* 'Samsun', and *S. nigrum* to ObPV; *C. amaranticolor* to *Alfalfa mosaic* virus [AMV]) were used as models.

In the local host–virus relationship, the lesions were counted 5 days after the treatments. In the systemic host–virus relationships, the virus concentration was determined 5 weeks after the treatments by using DAS ELISA tests, where the virus concentration can be ascertained from the extinction values. No significant correlations were observed between the effect of plant extracts on the development of test plants and the effect of virus concentration. For example, *A. syriaca* and *A. theophrasti* shoot water extracts significantly reduced the fresh weight of the test plant (*N. tabacum* 'Samsun'), whereas significant reduction in ObPV concentration occurred due to *A. syriaca* root and *C. arvense* shoot water extracts (see Fig. 5.11).

TABLE 5.6 Effect of a mixed fungal and viral infection on the development and macronutrient content of pepper

Treatments[a]	Fresh weight (g/plant)		Dry weight (g/plant)		N (%)		P (%)		K (%)	
	Shoot	Root	Shoot	Root	V	G	V	G	V	G
Control	16.3	10	3.7	3.3	4.27	4.1	0.22	0.22	3.3	4.35
CMV-U/246	17.9	11.3	4.2	4.4	4.09	4.31	0.21	0.25	3.64	3.9
PVYNTN	16.8	8.8	3.6	3.4	3.74	4.28	0.19	0.24	3.38	3.85
M. phaseolina	14.1	7.4	3.2	3.3	3.51	4.05	0.17	0.21	3.16	4.21
CMV-U/246 + PVYNTN	14	10.5	3.3	3.8	3.64	4	0.17	0.21	3.5	3.92
CMV-U/246 + M. phaseolina	12.6	7.3	2.7	2.9	3.79	4.34	0.18	0.22	3.27	3.6
PVYNTN + M. phaseolina	14.3	8.5	3.3	2.7	4.37	4.67	0.23	0.26	2.99	3.42
CMV-U/246 + PVYNTN + M. phaseolina	12.4	7.2	2.6	2.5	4.16	4.26	0.19	0.24	3.14	3.66
LSD5%	1.78	1.57	0.51	0.56	0.28	0.44	0.19	0.24	3.66	3.14

[a]V: vegetative phenophase; G: generative phenophase.

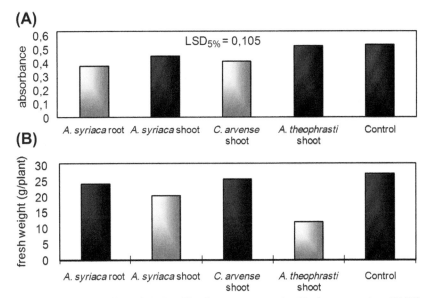

FIGURE 5.11   The effect of allelopathic plant extracts on the *Obuda pepper virus* (ObPV) concentration (A) and shoot fresh weight (B) of *Nicotiana tabacum* 'Samsun'.

The antiviral activity of several herbicides is well known, including triazine, carbamide, dinitroaniline, and auxin-type herbicides. Our experimental results regarding the favorable viral inhibitory side effect of some herbicides confirmed these previous results. It has been concluded that the inhibitory effect of the pendimethalin herbicide (a dinitroaniline that is used commonly for weed control in many crops) on plant viruses greatly depends on the host (species, cultivars), the viruses/virus strains, mode and dosage of herbicide application. In the *N. glutinosa–Obuda pepper virus* (ObPV) local host–virus relationship, the number of the local lesions was reduced by 55% by herbicide application. In systemic host–virus relationships, herbicide application delayed the appearance of systemic symptoms (e.g. *C. amaranticolor*–AMV) or significantly reduced the virus concentration (*N. tabacum* 'Samsun'–ObPV). A positive correlation between the herbicide dosage and the extent of virus inhibition was also observed. The viral inhibitory effect of other herbicides such as napropamide (commonly used for pepper weed control), and among ACCase inhibitors, fluasifop-p-butyl was also proven.

Our results call attention to the fact that certain herbicides may play an important role not only for weed control, but also for inhibiting the effects of economically important viruses, which occur in agroecosystems and cause diseases in both crops and weeds.

In addition to herbicides, other synthetic materials can act as virus inhibitors, including ribavirin and its analogues, nucleotide base analogues, cyclic substances, noncyclic substances, thiocarbamides, antibiotics, membrane lipid analogies, polysaccharides, and others.

REFERENCES

Baracsi, É., 1999. The role of biotest in plant virology. In: Horváth, J., Gáborjányi, R. (Eds.), Plant viruses and virological examination methods, Mezőgazda Kiadó, Budapest, pp. 139–167.

Fraser, R.S., 1990. The genetic of resistance to plant viruses. Annual Review of Phytopathology 28, 179–200.

Horváth, J., 1993. Host plants in diagnosis. In: Matthews, R.E.F. (Ed.), Diagnosis of plant virus diseases, CRC Press, Boca Raton FL, pp. 15–47.

Kegler, H., Meyer, U., 1987. Characterization and evaluation of quantitative virus resistance in plants. Archiv für Phytopathologie und Pflanzenschutz 23, 343–348.

Kazinczi, G., Horváth, J., Pogány, M., 1996. The effect of virus infection on the growth and photosynthetic pigment content of the virophilous Jimson weed (*Datura stramonium* L.). Acta Phytopathologica et Entomologica Hungarica 31, 175–179.

Kazinczi, G., Horváth, J., Hunyadi, K., Merkel, K., 1997. Effect of sowbane mosaic *Sobemovirus* (SoMV) on the germination biology of some *Chenopodium* species. Acta Phytopathologica et Entomologica Hungarica 32, 117–123.

Kazinczi, G., Horváth, J., Hunyadi, K., 1998. Sublethal water saturation deficit of the healthy and virus infected black nightshade (*Solanum nigrum* L.). Acta Phytopathologica et Entomologica Hungarica 33, 237–242.

Kazinczi, G., Horváth, J., Lukács, D., 2000. Germination characteristics of *Chenopodium* seeds derived from healthy and virus infected plants. Zeitschrift für Pflanzenkrankheiten und Pflanzenschutz Sonderheft 17, 63–67.

Kazinczi, G., Pribék, D., Takács, A., 2001. Biological decline of *Solanum nigrum* L. due to tobacco mosaic *tobamovirus* (TMV) infection. I. Growth and nutrient uptake. Acta Phytopathologica et Entomologica Hungarica 36, 9–14.

Kazinczi, G., Horváth, J., Takács, A.P., Pribék, D., 2002. Biological decline of *Solanum nigrum* L. due to tobacco mosaic *tobamovirus* (TMV) infection. II. Germination, seed transmission, seed viability and seed production. Acta Phytopathologica et Entomologica Hungarica 37, 329–333.

Kazinczi, G., Horváth, J., Takács, A., 2006a. On the biological decline of weeds due to virus infection. Acta Phytopathologica et Entomologica Hungarica 41, 213–221.

Kazinczi, G., Lukács, D., Takács, A., Horváth, J., Gáborjányi, R., Nádasy, M., Nádasy, E., 2006b. Biological decline of *Solanum nigrum* due to virus infection. Zeitschrift für Pflanzenkrankheiten und Pflanzenschutz 20, 781–786.

Kazinczi, G., Gáspár, L., Nyitrai, P., Gáborjányi, R., Sárvári, É., Takács, A., Horváth, J., 2006c. Herbicide-affected plant metabolism reduces virus propagation. Zeitschrift Naturforschung 61C, 692–698.

Király, L., Barna, B., Király, Z., 2007. Plant resistance to pathogen infection: forms and mechanisms of innate and acquired resistance. Plant Pathology 155, 385–396.

FURTHER READING

Almádi, L., 1976. Data to the water relation of Ambrosia elatior. Botanikai Közlemények 163, 199–204.

Bos, L., 1970. Symptoms of virus diseases in plants. Centre for Agricultural Publishing and Documentation, Wageningen, Netherlands.

Bos, L., 1999. Plant viruses, unique and intriguing pathogens. A texbook of plant virology. Backhuys, Leiden.

Bősze, Z., Kazinczi, G., Horváth, J., 1996. Reaction of unknown *Solanum stoloniferum* Schlechtd. et Bche and *Solanum demissum* Lindl. accessions to the tuber necrosis strain of potato Y *Potyvirus* (PVY^{NTN}). Acta Phytopathologica et Entomologica Hungarica 31, 169–174.

Christie, R.G., Edwardson, J.R., 1977. Light and electron microscopy of plant virus inclusions. Monograph Ser. 9. Florida Agr, Gainesville, USA Exp Stat.

Christie, R.G., Edwardson, J.R., 1986. Light microscopy techniques for detection of plant virus inclusions. Plant Disease 70, 273.

Clark, M.F., Adams, A.N., 1977. Characteristics of the microplate method of enzyme-linked immunosorbent assay for the detection of plant viruses. Journal of General Virology 34, 475–483.

Culver, J.N., Alwyn, G., Lindbeck, C., Dawson, W.O., 1991. Virus–host interactions: Induction of chlorotic and necrotic responses in plants by tobamoviruses. Annual Review of Phytopathology 29, 193–217.

Dijkstra, J., De Jager, C.P., 1998. Practical plant virology. Protocols and excercies. Springer, Berlin.

Edwardson, J.R., Christie, R.G., Purcifall, J.E., Peterson, M.A., 1993. Inclusions in diagnosing plant virus diseases. In: Matthews, R.E.F. (Ed.), Diagnosis of plant virus diseases, CRC Press, Boca Raton FL, USA.

Gáborjányi, R., Tóbiás, I., 1986. Inhibitors of virus infection inhibitors and virus replication in plants. Növénytermelés 35, 139–146.

Gera, A., Lawson, R.H., Hsu, H.T., 1995. Identification and assay. In: Loebenstein, G., Lawson, R.H., Brunt, A.A. (Eds.), Virus and virus-like diseases of bulb and flower crops, John Wiley, Chichester, pp. 165–180.

Gibbs, A., Harrison, B., 1976. Plant virology: the principles. Edward Arnold, London.

Goodmann, R.N., Király, Z., Wood, K., 1986. The biochemistry and physiology of plant disease. University of Missouri Press, Columbia, MO, USA.

Höller, K., Király, L., Künstler, A., Müller, M., Gullner, G., Fattinger, M., Zechmann, B., 2010. Enhanced glutathione metabolism is correlated with sulfur-induced resistance in *Tobacco mosaic virus*–infected genetically susceptible nicotiana tabacum plants. Molecular Plant–Microbe Interactions 23, 1448–1459.

Horváth, J., 1972. Hosts and non-hosts of plant viruses. Today and Tomorrow's Printers and Publishers, New Delhi.

Horváth, J., 1993. A list of proposed letter codes for hosts and non-hosts of plant viruses. Acta Phytopathologica et Entomologica Hungarica 28, 21–58.

Horváth, J., 1999. Plant virology. Keszthely, Hungary. University of Pannonia, Georgikon Faculty.

Horváth, J., Hunyadi, K., 1973. Studies on the effect of herbicides on virus multiplication. Acta Phytopathologica et Entomologica Hungarica 8, 347–350.

Horváth, J., Gáborjányi, R., 1999. Plant viruses and virological examination methods. Mezőgazda Kiadó, Budapest.

Horváth, J., Kazinczi, G., Takács, A.P., 2004. Sources of resistance against viruses in *Solanum* genus. Razprave IV. Razreda Sazu 14, 63–74.

Hunyadi, K., Béres, I., Kazinczi, G., 2011. Weeds, weed biology and weed control. Mezőgazda Kiadó, Budapest.

Kazinczi, G., 1994. Alternative hosts of viruses: weeds. DSc thesis, Hungary.

Kazinczi, G., Hunyadi, K., 1992. Water relations of some annual weeds. Zeitschrift für Pflanzenkrankheiten und Pflanzenschutz 12, 111–117.

Kazinczi, G., Horváth, J., Kadlicskó, S., 1998. The effect of *Macrophomina phaseolina* (Tassi) Goid. and two viruses on the nutrient content of pepper (*Capsicum annuum* L.) leaves. Acta Phytopathologica et Entomologica Hungarica 33, 305–311.

Kazinczi, G., Kadlicskó, S., Horváth, J., 1998. The effect of *Macrophomina phaseolina* (Tassi) Goid. and two viruses on pepper (*Capsicum annuum* L.). Acta Phytopathologica et Entomologica Hungarica 33, 61–68.

Kazinczi, G., Mikulás, J., Horváth, J., Torma, M., Hunyadi, K., 1999. Allelopathic effects of *Asclepias syriaca* roots on crops and weeds. Allelopathy Journal 6, 267–270.

Kazinczi, G., Béres, I., Narwal, S.S., 2001. Allelopathic plants. 1. Canada thistle [*Cirsium arvense* (L.) Scop]. Allelopathy Journal 8, 29–40.

Kazinczi, G., Béres, I., Narwal, S.S., 2001. Allelopathic plants. 3. Velvetleaf (*Abutilon theophrasti* Medic.). Allelopathy Journal 8, 179–188.

Kazinczi, G., Horváth, J., Lesemann, D.E., 2002. Perennial plants as new natural hosts of three viruses. Zeitschrift für Pflanzenkrankheiten und Pflanzenschutz 109, 301–310.

Kazinczi, G., Horváth, J., Béres, I., Takács, A.P., Lukács, D., 2002. The effect of pendimethalin (STOMP 330) on some host-virus relations. Zeitschrift für Pflanzenkrankheiten und Pflanzenschutz Sonderheft 18, 1093–1098.

Kazinczi, G., Horváth, J., & Takács, A.P. 2003. Interaction of viruses and herbicides on host plants. In: 6th Slovenian Conference on Plant Protection, Zrece (Slovenia), pp. 270–274.

Kazinczi, G., Horváth, J., Takács, A., Gáborjányi, R., Béres, I., 2004. Experimental and natural weed host–virus relations. Communications of Applied Biological Sciences 69, 53–60.

Kazinczi, G., Horváth, J., Takács, A.P., Béres, I., Gáborjányi, R., Nádasy, M., 2005. The role of allelopathy in host–virus relations. Cereal Research Communications 33, 105–108.

Kazinczi, G., Gáborjányi, R., Nádasy, E., Takács, A., Horváth, J., 2009. Plant–virus interactions. In: Narwal, S.S., Sampietro, D., Vattuone, M., Catalán, C., Politycka, B. (Eds.), Plant bioassays, Studium Press LLC, Houston TX, USA, pp. 207–234.

Király, L., Barna, B., Érsek, T., 1972. Hypersensitivity as a consequence, not the case of plant resistance to infection. Nature 239, 456–458.

Király, Z., Hafez, M.Y., Fodor, J., Király, Z., 2008. Suppression of tobacco mosaic virus-induced hypersensitive-type necrotization in tobacco at high temperature is associated with downregulation of NADPH oxidase and superoxidase and stimulation of dehydroascorbate reductase. Journal of General Virology 89, 799–808.

Mackenzie, D.R., Cole, H., Smith, C.B., Ercegovich, C., 1970. Effects of atrazine and maize dwarf mosaic virus infection on weight and macro and micro element constituents of maize seedlings in the greenhouse. Phytopathology 60, 272–279.

Magyar, L., Kazinczi, G., Keszthelyi, S., 2011. Drought tolerance of *Panicum dichotomiflorum*. Hungarian Weed Research and Technology 12, 41–47.

Matthews, R.E.F., 1977. Host plant responses to virus infection. In: Fraenkel-Conrat, H., Wagner, R.R. (Eds.), Comprehensive virology, Plenum Press, New York, pp. 297–361.

Moore, R.P., 1985. Handbook of tetrazolium testing. International Seed Testing Association, Zürich.

Rao, D.R., Raychaudhuri, S.P., Verma, V.S., 1994. Study on the effect of herbicides on the infectivity of cucumber mosaic virus. International Journal of Tropical Plant Disease 12, 177–185.

Reuveni, M., Agapov, V., Reuveni, R., 1995. Suppression of cucumber powdery mildew (*Sphaerotheca fuliginea*) by foliar sprays of phosphate and potassium salts. Plant Pathology 44, 31–39.

Schuster, G., 1972. Die Beeinflussung von Virussymptomen durch Herbizide und andere Pflanzenschutzmittel. Beitrage zur Tabakforschung 10, 14–21.

Schuster, G., 1988. Synthetic antiphytoviral substances. In: Kurstak, E., Maruysk, R.G., Murhy, F.A., van Regenmortel, M.H. (Eds.), Applied virology research, New vaccines and chemotherapy, vol. 1. Plenum, New York, pp. 265–283.

Takács, A., Kazinczi, G., Horváth, J., Pribék, D., 2000. Susceptibility of different *Solanum* species to PVX and TSWV. Mededelingen Van de Faculteit Landbouwwetenschappen Rijksuniversiteit, Gent 65 (2b), 593–595.

Vivanco, J.M., Queci, M., Salazar, L.F., 1999. Antiviral and antiviroid activity of MAP-containing extracts from *Mirabilis jalapa* roots. Plant Disease 83, 1116–1121.

Weinberger, P., Romero, M., Oliva, M., 1972. Ein metodischer Beitrag zur Bestimmung des subletalen (kritischen) Wassersattigungsdefizite. Flora 161, 555–561.

Chapter 6

Interference with insect transmission to control plant-pathogenic viruses

María Urizarna España and Juan José López-Moya
Centre for Research in Agricultural Genomics CRAG, CSIC-IRTA-UAB-UB, Campus UAB—Bellaterra, Barcelona, Spain

INTRODUCTION

Since the onset of civilization, agriculture has been an essential activity for securing the supply of food and other useful products for human beings—such as feed for farm animals and fibers. As plants are the basis of most sustainable agricultural processes, how to establish efficient strategies for protecting them against pests and pathogens has been a constant concern (Strange & Scott 2005). Nowadays, the increased demand for agricultural products and global changes are creating new challenges, considering the economic impact of production losses due to known pests or diseases and the emergence (or invasion) of new, well-adapted pathogens or pests.

Some studies have estimated that around 10% of the annual production of many crops on average can be lost (FAO 2000), with a substantial part being consumed by herbivores, mainly insect pests. In some cases, this rate of damage can be well above the expected productivity of the crops, for instance in terms of plant biomass allocated to reproduction (fruits and seeds): this clearly indicates the importance of dealing with, and controlling, insect pests. Furthermore, and besides the direct damage caused, some insects are responsible for spreading plant diseases, in particular those caused by viruses, which also can be quite detrimental to crop production. In this chapter we consider, in particular, this aspect of the problem by focusing on the transmission of viruses by insect vectors and on how ongoing research can provide innovative strategies to interfere with the process.

As indicated above, insects play an important role in the dissemination of most plant viruses (Hull 2001). Being obligate parasites, the maintenance and survival of viruses in nature depends on mechanisms that enable them to reach new hosts. Consequently, the transmission process is one of the most important steps in the biologic cycle of viruses, and the use of vector organisms is the most frequently adopted strategy to secure their dissemination. In addition to a few 'below-ground' organisms, such as chytrids, plasmodiophorids, and nematodes, the main vectors of plant viruses are arthropods, including mites and insects. Focusing on insects, plant viruses are spread from plant to plant by species belonging to several orders (Table 6.1), with hemipterans—in particular, species included in the suborder *Homoptera*, such as aphids, whiteflies, and leafhoppers— being the major vectors. Examples of viral taxons transmitted by the different categories of insect are included in Table 6.1.

As shown in Table 6.1, the transmission of members of a large number of virus groups is due to the action of insects of the families *Aphididae* and *Aleyrodidae*, aphids and whiteflies, respectively, as well as other insects such as leafhoppers, treehoppers, and planthoppers. All of these homopterans have in common mouthparts of a piercing-sucking type (Nault 1997), with stylets that are able to pierce the cell walls of plant tissues without causing major damage to the cell when accessing the phloem (Hewer et al 2011). Furthermore, the capacity for growth in colonies, particularly in the case of aphids, and the easy dispersion of these insects makes them collectively the most important groups of vectors of plant viruses.

The types of insect listed in Table 6.1 are all phytophagous, which means that they depend on plants for obtaining their food. To do so, they have to surmount a diverse arsenal of plant defenses, evolved in a classic arms race to stop the damage caused by their attacks. Adaptations by insects to overcome plant defenses might consist, for instance, of anatomic changes—such as the piercing-sucking mouthparts that allow them to reach, rapidly and efficiently, the vascular tissues of the host plant. The feeding behavior of the vectors might also affect their capacity for transmission of viruses, as the acquisition and inoculation steps occur during feeding. In the following sections we review the current knowledge of these aspects.

MODES OF TRANSMISSION

Mechanistically, successful transmission involves several steps. Step 1 is the acquisition of the virions from an infected plant. Next, the acquired virions need to be retained in the vector (Step 2), either at specific sites through the binding of virions to receptor-like elements in the digestive tract or circulating from different anatomic structures, mainly from the gut to the salivary glands. After this, delivery of virions from the retention sites is required (Step 3), in many cases following salivation. Finally, the virions have to be deposited in a susceptible cell of the host plant (Step 4), where they can again start an infectious cycle. The duration of these steps was used to propose the first classification of modes of virus transmission (Watson & Roberts 1939) using two main categories: nonpersistent and persistent. The length of the period during which the infectivity of the virus is retained in

TABLE 6.1 Main groups of insect vectors of plant viruses, with examples of representative viral taxa transmitted by them

Order	Suborder	Superfamily	Family	Type of insect (common name)	Examples of plant virus taxa[a]
Hemiptera	Homoptera	Aphidoidea	Aphididae	Aphids	Potyvirus, Macluravirus, Cucumovirus, Caulimovirus, Luteovirus, Polerovirus, Alfamovirus, Fabavirus, Closterovirus, Nanovirus, Sequivirus, Umbravirus…
		Aleyrodoidea	Aleyrodidae	Whiteflies	Begomovirus, Ipomovirus, Crinivirus, Carlavirus, Torradovirus…
		Membracoidea	Cicadellidae	Leafhoppers	Curtovirus, Mastrevirus, Phytoreovirus, Waikavirus
			Membracidae	Treehoppers	Topocuvirus
		Fulgoroidea	Delphacidae	Planthoppers	Fijivirus, Oryzavirus, Tenuivirus
Thysanoptera		Thripoidea	Thripidae	Thrips	Tospovirus
Coleoptera				Beetles	Bromovirus, Comovirus, Sobemovirus, Tymovirus

[a]Virus group names according to the database of the International Committee for Taxonomy of Viruses, ICTV, accessible at http://ictvonline.org/index.asp.

the vector serves to differentiate transmission into two modes. In the first case, the insect can transmit the virus to uninfected plants in a time frame ranging from seconds to few minutes after acquisition. Most importantly, the capacity to transmit the virus is lost rapidly. For *persistent transmission*, the acquisition of virus requires longer periods (around hours); subsequently, a retention period (from hours to days) is needed before the vector finally becomes viruliferous. In between these two modes of transmission, a third category called *semi-persistent transmission* is reserved for viruses with intermediate requirements in terms of acquisition and retention periods.

A further refinement of this classification is based on the route followed by the virus inside the insect vector (Harris 1977). In this case, non-circulative and circulative transmission are differentiated. In *non-circulative transmission*, the virus is temporarily associated and reversibly retained in the anterior tract of the digestive system of the vector (i.e., mouthparts or even the foregut). This type of transmission corresponds mainly to non- or semi-persistent viruses. When the acquisition period is longer, the efficiency of virus transmission might increase, in the case of semi-persistent viruses, or decrease rapidly, as often happens in nonpersistent viruses. This important difference seems to be associated with the stability of virion retention in the vector, which allows virions to be accumulated until the retention sites are saturated, increasing the chances of later transmission. Such a case has been observed with aphids and also other vectors, notably whiteflies and leafhoppers. On the other hand, aphids are apparently a unique type of vector that is able to transmit well-characterized nonpersistent viruses (Ng & Falk 2006), and retention seems to happen through a weak interaction, meaning that the virions can be rapidly lost when the feeding period is extended over a certain limit. Retention of the virus is limited in duration; this means that the virus can be almost immediately released and inoculated after acquisition, but the capacity to infect plants is retained only during short periods of time. Both acquisition and inoculation are thought to occur during feeding. With such a narrow time frame, it is indeed rather difficult to control the spread of these viruses using insecticide treatments because the window of opportunity passes quite rapidly.

Finally, *circulative transmission* requires the passage of virions through the insect. After feeding on an infected plant, the acquired virus must pass across the gut wall to reach the hemolymph, and eventually it could reach the salivary glands—getting access to the saliva that is ready for inoculation. The complete process might take days, and therefore a latency period occurs because, immediately after acquisition, the virus cannot be inoculated; the reason for this is that a rather long circulation inside the vector is required—during which the virus crosses cellular barriers within the insect. Two additional categories can be established in this case. When viral replication is restricted to the plant host, despite the movement of virions through the insect, the virus is considered to be *non-propagative*; however, in certain cases the virus can also replicate in the vector—in which case it is considered to be *propagative*. Interestingly, propagative viruses are parasites of both plants and insects, alternating between the two types of host to complete their cycles of dispersion.

MAIN GROUPS OF INSECT VECTORS OF PLANT VIRUSES

Because the feeding behavior of piercing-sucking insects could define their capacities for virus transmission (Stafford et al 2012), it is relevant to look at their characteristics before considering virus control strategies. In this section we review some important aspects and peculiarities of the main groups of insect vectors, including aphids, whiteflies, and leafhoppers.

Aphids

As shown in Table 6.1, aphids can spread a large number of plant viruses, such as potyviruses (a group of plant viruses comprising more than 150 definitive species), cucumoviruses, caulimoviruses, luteoviruses, poleroviruses, and many others. Using their characteristic piercing-sucking mouthparts, aphids can feed on plant sap. Aphid mouthparts have stylet morphology, adapted to access and feed on the contents of even individual plant cells. This stylet is composed of a pair of mandibular elements surrounding a pair of tightly interlocked maxillary elements that comprise two internal canals, one for injecting saliva and the other for sucking up plant fluids. Their searching behavior for plants, the range of available host plants, and their high reproduction rates contribute to the efficiency of aphids as carriers of viruses (Fereres & Moreno 2009). They are capable of producing an almost explosive growth in population thanks to their parthenogenetic system of reproduction, with adults giving birth daily to many individual nymphs (Fig. 6.1).

FIGURE 6.1 Aphids (*Myzus persicae*) feeding on a tobacco plant. An adult and several nymphs are shown. Photograph taken by M. Urizarna using an Olympus DP71.

Aphids can transmit viruses in all the previously described modes of transmission. In particular, apparently only aphids are capable of transmitting nonpersistent viruses. This might be related to their feeding behavior, which involves several short intracellular probings at the beginning of every feeding process—thought to be essential for acceptance of the plant as an adequate source of food. These short probings are likely to involve sucking up cellular contents to be tasted in more internal chemoreceptors, and acquisition of nonpersistent viruses is known to occur exactly during these probings. Furthermore, to facilitate penetration of the stylets, salivation is required, providing opportunities for inoculation of viruses previously acquired in other probes. The electrical penetration monitoring system (Tjallingii 1990) has served to provide conclusive experimental evidence for these processes (Martin et al 1997). Regarding retention sites for nonpersistent viruses, recent work with caulimoviruses has shown that the specific auxiliary factors required for transmission are retained near the distal part of the stylet, where the food canal and the salivary canal merge into a common duct (Uzest et al 2007). For this site, a particular anatomic structure, called the 'acrostyle,' has been recently described (Uzest et al 2010).

Other important elements for virus transmission are the barriers that circulative viruses are forced to cross inside the aphid (Gildow 1993). These include the cells lining the gut, either midgut or hindgut, from where viruses are internalized. Once in the hemolymph, an association with chaperon-like proteins of endosymbiont origin has been described (van den Heuvel et al 1994, Filichkin et al 1997), although it is unclear whether or not this association is essential for the process (Bouvaine et al 2011). Entry into the salivary gland normally occurs after the specific association of virions with the basal lamina, which constitutes another barrier for circulation (Gildow & Gray 1993).

Another important aspect in the relationship of aphids with viruses and host plants is the possibility that the presence of the virus might increase the attractiveness of the plant for aphid vectors. Indeed, symptoms such as yellowing could make the infected plant easier to spot by the aphids. Other changes associated with virus infections, such as the release of volatile compounds, might also serve to attract vectors. Moreover, these changes could also affect aphid feeding behavior to enhance the ability of acquisition and inoculation (Mauck et al 2010).

Whiteflies

Whiteflies can be considered the second most important type of vector due to their capacity to transmit many plant viruses, notably a large number of species of the genus *Begomovirus*, which currently comprises around 200 members. Other viruses transmitted by whiteflies include ipomoviruses of the family *Potyviridae* and criniviruses of the family *Closteroviridae* (Valverde et al 2004). As a direct pest of many crops (Fig. 6.2), whiteflies cause several important problems, although the major concern is associated with their role as virus vectors, especially in the case of emerging viral diseases (Navas-Castillo et al 2011).

FIGURE 6.2   Whitefly (*Bemisia tabaci*) feeding on a tobacco plant. Photograph taken by M. Urizarna using an Olympus DP71.

Whiteflies have piercing-sucking mouthparts similar to those of aphids, and they also feed by inserting their stylets into the plant to reach the phloem. However, some differences in the feeding behavior exist, notably in the frequency and duration of short probings, which might explain why most whitefly-transmitted viruses are classified as semi-persistently transmitted.

Leafhoppers

The first plant virus shown to be insect-transmitted was Rice dwarf virus, a phytoreovirus transmitted by a leafhopper (Fukushi 1934). Phytoreoviruses are circulative and propagative vectors, capable also of being transovarially transmitted to their descendants after prolonged periods of time without losing their capacity to infect plants (Honda et al 2007). Leafhopper vectors, in general, show a considerable degree of specificity for transmission of particular viruses. Leafhoppers are the vectors that transmit the largest number of propagative viruses of any vector group, in addition to their capacity to transmit semi-persistent viruses.

Unlike other vectors, leafhoppers can feed actively in the xylem and in the mesophyll, as well as in the phloem (Wayadande, 1990). Actually, their mouthparts, due to their larger size compared with those of other homopterans, can cause more damage to plants during feeding.

Other insect vectors

As shown in Table 6.1, additional insect vectors exist for certain plant viruses. The peculiar relationship of thrips and tospoviruses is worth mentioning here. In this case virus acquisition occurs during larval stages, while inoculation is restricted to adults (Whitfield et al 2005). A modification of vector behavior due to the virus presence has been recently described in this system (Stafford et al 2011). Also, the transmission of viruses by species of phytophagous beetles shows peculiarities derived from their chewing mechanism of feeding (Walters 1969, Mello et al 2010).

CONTROL OF INSECT VECTORS TO CONTROL VIRAL DISEASES

Once the importance of vector organisms in the dispersal of plant viruses was fully recognized, a logical follow up was the idea of controlling plant viruses through actions against their vectors. However, the effectiveness of insecticides was variable and depends on the mode of transmission (Perring et al 1999). More recently, there is a demand for safer and more socially acceptable systems of pest and pathogen control based mainly on Integrated Pest Management (IPM) concepts (Birch et al 2011), and this demand has been stimulated by serious environmental concerns concerning the massive use of pesticides in agriculture and the increasingly tight regulations.

An important component of any IPM program is the use of host plant resistance, which can affect the vector (Smith 2005), the virus (Maule et al 2007), or both. When analyzing resistance traits, it is critical to consider that the different terms used by plant breeders might have different meanings. For instance, a cultivar claimed to be resistant to a pest or a pathogen could possess only tolerance, meaning that the cultivar exhibits less damage than another one, despite the fact that the pest or the pathogen is present at similar levels. In the case of pests, the term *resistance* should be reserved for antixenosis or antibiosis. Antixenosis occurs when there is non-preference for the resistant plant compared to a susceptible one. When the life-history parameters (survival, development, fecundity) of the insect are affected, the term *antibiosis* can be applied. Obviously, both tolerances and true resistances are highly desirable in IPM programs in terms of ecology, economy, and environmental protection, but for virus control a simple tolerance against the vector might be useless because the transmission of the virus might be unaffected (for instance, in non-persistently transmitted viruses that require only short acquisition and inoculation periods).

Antibiosis and antixenosis reactions can derive from the enhancement in resistant plants of natural defensive traits against the damage caused by herbivores.

For instance, there is a wide range of direct physical and/or chemical defenses that plants might set in place to force out the insect feeding on them. Direct defenses include thorns, trichomes, or incorporation of silica, and also accumulation of components such as toxins or inhibitors of digestive enzymes. Furthermore, plants can produce indirect defenses, such as chemicals to attract predators/parasites of the herbivores (Hare 2011, Clavijo McCormick et al 2012).

Focusing on resistance genes specific for particular pests or pathogens, our current knowledge has improved recently thanks to the use of the powerful molecular biology tools (Kaloshian & Walling 2005, Westwood & Stevens 2010). In the case of vector transmission of viral diseases, two genes have been identified in plants, the *Mi* and *Vat* resistance genes, which correspond to the CC-NBS-LRR (coiled coil–nucleotide binding site–leucine rich repeat) subfamily of NBS-LRR resistance proteins. The *Mi-1.2* gene of tomato is known to confer resistance against root-knot nematodes (Milligan et al 1998), the potato aphid (Martinez de Ilarduya et al 2003), and some biotypes of *Bemisia tabaci* (Nombela et al 2003). Resistance occurs through activation of a programmed cell death response, following the interaction between *Mi-1.2* and elicitors from the pathogen side. Even more interesting is the *Vat* gene identified in melon, which confers resistance to the transmission process of viruses by *Aphis gossypii* (Silberstein et al 2003, Boissot et al 2010). It has been observed that, after an aphid lands on melon varieties carrying the *Vat* gene, there is an activation of typical plant responses such as phenol synthesis, callose deposition in the cell wall, and an oxidative burst (Villada et al 2009). However, and despite these two most interesting examples, the availability of other natural genes putatively targeting vectors in other plants species is not clear, and the process of identifying new sources of resistance traits could be a limiting factor in the use of this technology. As an alternative, genetic engineering methods can speed up the introgression of resistance traits (Collinge et al 2010), although there are important issues that should be considered, such as biosafety, societal opposition, and regulations.

An important aspect to consider when managing both natural and genetically engineered resistance is the possibility that the targeted organism could evolve to break the resistance. There are many examples of this, such as the case of Bt crops expressing insecticidal crystalline proteins derived from *Bacillus thuringensis*, with insects evolving to gain resistance against these toxins (Tabashnik et al 2003). This is not a surprise because adaptations in herbivores/pathogens and their host plants are potent driving forces in their respective coevolution. Therefore, an intelligent use of this technology is required in order to assure its sustainability.

INTERFERENCE WITH TRANSMISSION

Research in the fields of virus–plant, virus–vector, and plant–vector interactions could eventually lead to the design of novel strategies aiming to interfere with virus transmission. Filling the gaps in our current knowledge might serve to gain a better understanding of the transmission mechanisms and to identify key steps and/or specific elements essential for the process, which could turn out to be new targets for strategies aiming to block virus dissemination. In this section, we focus on recent work devoted to the search for specific virus receptors and available systems for interfering with gene expression in insects.

Identification of virus receptors

A good example of research that might be used to interfere with the transmission process is the characterization of vector elements acting as receptors for viruses. These receptors are likely implicated in transmission through their ability to confer specificity to the virus–vector interaction.

In the case of circulative viruses, it was soon recognized that vector specificity and tissue tropism might respond to the presence of some specific receptors or agents capable of interacting with viral proteins. The glycoproteins present at the surface of *Tomato spotted wilt virus* (TSWV) particles (Bandla et al 1998, German et al 1992) were found to interact with a 50 KDa protein identified in the thrips vector *Frankliniella occidentalis*, considered to be a candidate receptor essential for TSWV entry (Medeiros et al 2000). Another example can be found in luteoviruses and poleroviruses, which can be specifically transmitted in a circulative manner by a limited number of aphid species. The involvement in transmission of a minor readthrough form of the capsid protein (CP) present in virus particles has been demonstrated (Brault et al 1995). Several approaches have been used to identify putative receptors implicated in luteovirus transmission using aphid extracts (Miles 1972). Specifically, *Barley yellow dwarf virus* particles were used as a bait to identify aphid factors involved in virus recognition (Li et al 2001). Another protein proposed to be an important interactor for transmission of these viruses was GroEL, a chaperonin encoded by the *Buchnera* endosymbiotic bacteria of aphids (van den Heuvel et al 1994, Filichkin et al 1997). Mutants of *Beet western yellows virus* allowed mapping of the virion-GroEL binding to the readthrough domain of the CP, and it was suggested that this interaction might serve to protect virions during their circulation inside the aphid (van den Heuvel et al 1997). However, recent results question the availability of GroEL in the hemolymph, indicating that perhaps the *in vitro* observed interaction could not be essential for the transmission process (Bouvaine et al 2011).

Considering non-circulative plant viruses, an interesting case can be found in potyviruses (genus *Potyvirus*, family *Potyviridae*). The members of this genus are aphid-transmitted in a nonpersistent manner with the assistance of a viral encoded transmission factor (Govier & Kassanis 1974), the multifunctional helper-component protein (HCPro). HCPro has been proposed to act during transmission as a reversible connection (or a bridge) between the CP and hypothetical specific receptors in the aphid mouthparts (Raccah 2001). Following the bridge hypothesis and the recent identification of a particular anatomic structure at the tip of the aphid mouthparts, probably needed during transmission of caulimoviruses (Uzest et al 2010), research in this field is currently trying to identify the receptor(s) required for potyvirus acquisition, presumably located in or near this site in vector stylets. Previous work had shown the retention of *Tobacco etch virus* (TEV) virions on maxillary stylets in the presence of functionally active HCPro (Wang et al 1996). Because the major components of vector stylets are cuticular proteins, characterized by the presence of conserved R&R consensus domains (Dombrovsky et al 2007b), using the HCPro of *Zucchini yellow mosaic virus* (ZYMV) as a bait, a set of interactor cuticular proteins was identified (Dombrovsky et al 2007a). Later, the search was expanded to other proteins, finding a ribosomal protein S2 (RPS2) with a presumed dual function in the insect cuticle. The specific binding between RPS2 and the HCPro of TEV was confirmed after cloning and heterologous expression of the corresponding aphid gene (Fernandez-Calvino et al 2010). Experiments included, as a control, a mutant version of HCPro impaired in aphid transmission through alteration of a conserved motif likely involved in binding to aphid mouthparts (Blanc et al 1998). In these experiments, the mutated version of HCPro failed to interact specifically with RPS2, while retaining its capacity to bind the viral CP. However, the presence of RPS2 at or near the acrostyle has not yet been demonstrated, and probably other candidates must be explored before the receptor of potyvirus transmission is fully characterized (María Urizarna España, unpublished results).

RNA interference (RNAi) in insects

The recent discovery of the post-trancriptional gene silencing phenomenon in most eukaryotic organisms, including insects, has led to a large number of potential applications, ranging from its direct use in strategies for pest management (Huvenne & Smagghe 2010) to functional studies in insects (Belles 2010). The possibility of rapidly generating loss-of-function phenotypes using RNA interference (RNAi) in vectors could be a system for identifying elements essential for virus transmission. In short, upon introduction into the insect of a given amount of the triggering element, such as a double-stranded RNA (dsRNA) to be processed into small interfering RNAs (siRNAs), the endogenous silencing machinery will be activated, and the incorporation of siRNAs into the RNA-induced silencing complex (RISC) will serve to target specific sequences based on complementarity to the triggering element. In this way, specific silencing of certain genes could be achieved.

Despite the tremendous potential of this technology, it was soon recognized that not all species were equally susceptible to it, and these variations were specially clear in the case of insects, with some species being more sensitive than others to systemic RNAi. These differences are likely due to still unknown particularities during the amplification and spreading of the RNAi signals.

In practical terms, empirical approaches have to be adopted for each species, finding, for instance, the best system for delivering the triggering molecules, the most effective doses for the expected response, and the time frame of the effect. Among the delivery routes assayed we can mention microinjection of *in vitro* synthesized dsRNA into the insect hemocoel, which can provide transient knockdown of a given target gene (Dzitoyeva et al 2001). In some cases, a systemic effect throughout the insect body was demonstrated, showing knockdown of the targeted genes in a range of tissues. Interestingly, the microinjection system was very useful for the study of circulative viruses delivered into the body cavity (Tamborindeguy et al 2008). However, there are technical issues derived from the procedure: not all insects can survive the wound caused during the injection, and it is difficult to quantify the volume injected.

Another system for delivering the triggering element to a given organism is feeding on an artificial diet supplemented with *in vitro* generated dsRNAs. This system was demonstrated to work with the nematode *Caenorhabditis elegans* (Timmons et al 2001). A system for feeding aphids and other piercing-sucking insects through artificial membranes (Pirone 1964) can be used for these experiments. Providing dsRNA in an artificial diet confers silencing on targeted genes in different insects. In some cases oral delivery has failed to produce the expected effect, for instance in *Spodoptera litura* (Rajagopal et al 2002), suggesting that this system could not be suitable for all insect species.

In the case of phytophagous insects, oral delivery can be easily combined with the use of transgenic plants expressing siRNAs. This system can be denominated vegetal diet, and it has the potential to target insect genes specifically without causing alterations in the plant. The method has been tested in lepidopteran, coleopteran, and hemipteran species (Zha et al 2011). Interestingly, recent work shows that aphid genes are susceptible targets in this way (Pitino et al 2011). This system has the potential to be adopted in crop protection programs because it can be directed against specific pests, serving to knockdown transcripts in the insects feeding on the transgenic plants. In addition to plants stably transformed, transient expression systems can be used for the same purpose. A recent study shows, for instance, that a plant virus-based expression vector can be used to generate in the plant the dsRNA triggering elements (Kumar et al 2012). This is indeed an attractive switch to the idea of using RNAi to interfere with plant virus transmission: a viral expression vector, based on a plant virus, can be used to avoid the spreading of another unrelated plant pathogenic virus.

Figure 6.3 summarizes the available methods for RNAi in insects.

CONCLUSION

We have reviewed in this chapter our current knowledge on transmission of plant pathogenic viruses by insects and envisioned some future prospects for the control of both the vectors and the diseases they transmit, using strategies intended to interfere with the transmission process.

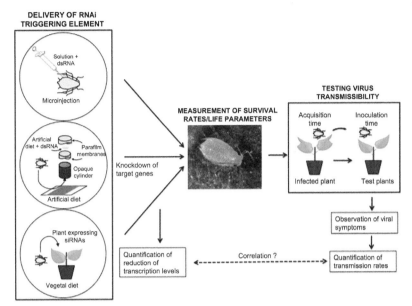

FIGURE 6.3   Experimental procedures for characterizing insect vector genes putatively involved in virus transmission. Aphids are shown as examples of insect vectors. On the left side of the scheme, delivery systems for inducing RNAi in insects are depicted: microinjection of dsRNA (above), an artificial feeding device with diet supplemented with dsRNA (center), or direct feeding on plant tissues expressing siRNAs (below). In the third case, the siRNAs can derive from stable transformation of plants, from transient expression mediated by agroinfiltration, or from infection with a plant virus-based expression vector. In the central part of the scheme, the effect of these treatments on the life parameters of the insect needs to be considered. Finally, on the right side, a plant-to-plant transmission test is shown, which could be used to quantify the effect on virus transmissibility of knocking down the specific target genes.

Some recent works show that strategies taking advantage of RNAi mechanisms can be adapted for crop protection against insect herbivores. However, there are still some gaps in the knowledge that should be filled before similar strategies could be applied to avoid virus dissemination by vector organisms in general. For instance, methods for suppressing gene expression in insects could eventually serve to target elements involved in viral transmission. In this case, there is a lack of sufficient knowledge in most cases regarding the identification of key molecules. In order to progress in this direction, new elements need to be unequivocally identified. The work done recently to characterize virus receptors in insect vectors is clearly aiming at this goal.

With a view to applying RNAi technology in the natural environment, there are some important points to take into consideration, including the use of a safe system for delivery of the triggering molecules (dsRNAs or siRNAs). Also, a detailed analysis of possible unwanted effects on untargeted genes, both in the vector species and in other off-target organisms, should be considered.

The availability of genome information of many insects, a trend that is likely to continue in the near future, will undoubtedly provide fuel for future developments. When whole-genome sequences are completed for most vectors, it could be possible to develop molecular markers for the identification of variants, with the potential to identify key elements required for virus transmission genetically. These tools would be critically important to fill some of the gaps in knowledge described above and to finally develop the novel approaches needed for virus control.

In the near future, further scientific developments are expected to contribute to more sustainable measures against plant diseases. In this context, plant viruses are still among the most challenging problems for numerous crops, and therefore, efforts to better understand and interfere with transmission, as described here, might result in decisive control of virus dissemination.

ACKNOWLEDGMENTS

Work in the authors' laboratory is supported by grant AGL2010-14949 from Ministerio de Economía y Competitividad (MINECO), Spain, and by funds from the Generalitat de Catalunya 2009SGR626. Maria Urizarna España is a recipient of an FPI fellowship from MINECO. We also thank Lluïsa Vilaplana and Ares Mingot for their contributions and assistance.

REFERENCES

Bandla, M.D., Campbell, L.R., Ullman, D.E., Sherwood, J.L., 1998. Interaction of Tomato spotted wilt tospovirus (TSWV) glycoproteins with a thrips midgut protein, a potential cellular receptor for TSWV. Phytopathology 88, 98–104.

Belles, X., 2010. Beyond *Drosophila*: RNAi in vivo and functional genomics in insects. Annual Review of Entomology 55, 111–128.

Birch, A.N.E., Begg, G.S., Squire, G.R., 2011. How agro-ecological research helps to address food security issues under new IPM and pesticide reduction policies for global crop production systems. Journal of Experimental Botany 62, 3251–3261.

Blanc, S., Ammar, E.D., Garcia-Lampasona, S., Dolja, V.V., Llave, C., Baker, J., Pirone, T.P., 1998. Mutations in the potyvirus helper component protein: effects on interactions with virions and aphid stylets. Journal of General Virology 79, 3119–3122.

Boissot, N., Thomas, S., Sauvion, N., Marchal, C., Pavis, C., Dogimont, C., 2010. Mapping and validation of QTLs for resistance to aphids and whiteflies in melon. Theoretical and Applied Genetics 121, 9–20.

Bouvaine, S., Boonham, N., Douglas, A.E., 2011. Interactions between a luteovirus and the GroEL chaperonin protein of the symbiotic bacterium *Buchnera aphidicola* of aphids. Journal of General Virology 92, 1467–1474.

Brault, V., van den Heuvel, J.F., Verbeek, M., Ziegler-Graff, V., Reutenauer, A., Herrbach, E., Garaud, J.C., Guilley, H., Richards, K., Jonard, G., 1995. Aphid transmission of beet western yellows luteovirus requires the minor capsid read-through protein P74. EMBO Journal 14, 650–659.

Clavijo McCormick, A., Unsicker, S.B., Gershenzon, J., 2012. The specificity of herbivore-induced plant volatiles in attracting herbivore enemies. Trends in Plant Sciences 17, 303–310.

Collinge, D.B., Jorgensen, H.J., Lund, O.S., Lyngkjaer, M.F., 2010. Engineering pathogen resistance in crop plants: current trends and future prospects. Annual Review of Phytopathology 48, 269–291.

Dombrovsky, A., Gollop, N., Chen, S., Chejanovsky, N., Raccah, B., 2007a. In vitro association between the helper component-proteinase of zucchini yellow mosaic virus and cuticle proteins of *Myzus persicae*. Journal of General Virology 88, 1602–1610.

Dombrovsky, A., Sobolev, I., Chejanovsky, N., Raccah, B., 2007b. Characterization of RR-1 and RR-2 cuticular proteins from *Myzus persicae*. Comparative Biochemistry and Physiology – Part B: Biochemistry & Molecular Biology 146, 256–264.

Dzitoyeva, S., Dimitrijevic, N., Manev, H., 2001. Intra-abdominal injection of double-stranded RNA into anesthetized adult *Drosophila* triggers RNA interference in the central nervous system. Molecular Psychiatry 6, 665–670.

FAO, 2000. The state of food insecurity in the world. SOFI, Rome.

Fereres, A., Moreno, A., 2009. Behavioural aspects influencing plant virus transmission by homopteran insects. Virus Research 141, 158–168.

Fernandez-Calvino, L., Goytia, E., Lopez-Abella, D., Giner, A., Urizarna, M., Vilaplana, L., Lopez-Moya, J.J., 2010. The helper-component protease transmission factor of tobacco etch potyvirus binds specifically to an aphid ribosomal protein homologous to the laminin receptor precursor. Journal of General Virology 91, 2862–2873.

Filichkin, S.A., Brumfield, S., Filichkin, T.P., Young, M.J., 1997. In vitro interactions of the aphid endosymbiotic SymL chaperonin with barley yellow dwarf virus. Journal of Virology 71, 569–577.

Fukushi, T., 1934. Studies on the dwarf disease of rice plant. Journal of the Faculty of Agricultural Hokkaido Imperial University 37, 41–164.

German, T.L., Ullman, D.E., Moyer, J.W., 1992. Tospoviruses: diagnosis, molecular biology, phylogeny, and vector relationships. Annual Review of Phytopathology 30, 315–348.

Gildow, F.E., 1993. Evidence for receptor-mediated endocytosis regulating luteovirus acquisition by aphids. Phytopathology 83, 270–278.

Gildow, F.E., Gray, S.M., 1993. The aphid salivary gland basal lamina as a selective barrier associated with vector-specific transmission of barley yellow dwarf luteoviruses. Phytopathology 83, 1293–1302.

Govier, D.A., Kassanis, B., 1974. Evidence that a component other than the virus particle is needed for aphid transmission of potato virus Y. Virology 57, 285–286.

Hare, J.D., 2011. Ecological role of volatiles produced by plants in response to damage by herbivorous insects. Annual Review of Entomology 56, 161–180.

Harris, K.F., 1977. An ingestion-egestion hypothesis of noncirculative virus transmission. In: Maramorosch, K.F.H.K. (Ed.), Aphids as virus vectors, Academic Press, New York.

Hewer, A., Becker, A., van Bel, A.J., 2011. An aphid's Odyssey – the cortical quest for the vascular bundle. Journal of Experimental Biology 214, 3868–3879.

Honda, K., Wei, T., Hagiwara, K., Higashi, T., Kimura, I., Akutsu, K., Omura, T., 2007. Retention of *Rice dwarf virus* by descendants of pairs of viruliferous vector insects after rearing for 6 years. Phytopathology 97, 712–716.

Hull, R., 2001. Matthews' plant virology, fourth ed. Academic Press, San Diego.

Huvenne, H., Smagghe, G., 2010. Mechanisms of dsRNA uptake in insects and potential of RNAi for pest control: a review. Journal of Insect Physiology 56, 227–235.

Kaloshian, I., Walling, L.L., 2005. Hemipterans as plant pathogens. Annual Review of Phytopathology 43, 491–521.

Kumar, P., Pandit, S.S., Baldwin, I.T., 2012. Tobacco rattle virus vector: a rapid and transient means of silencing manduca sexta genes by plant mediated RNA interference. PLoS One 7, e31347.

Li, C., Cox-Foster, D., Gray, S.M., Gildow, F., 2001. Vector specificity of barley yellow dwarf virus (BYDV) transmission: identification of potential cellular receptors binding BYDV-MAV in the aphid. *Sitobion avenae*. Virology 286, 125–133.

Martin, B., Collar, J.L., Tjallingii, W.F., Fereres, A., 1997. Intracellular ingestion and salivation by aphids may cause the acquisition and inoculation of non-persistently transmitted plant viruses. Journal of General Virology 78, 2701–2705.

Martinez de Ilarduya, O., Xie, Q., Kaloshian, I., 2003. Aphid-induced defence responses in Mi-1-mediated compatible and incompatible tomato interactions. Molecular Plant Microbe Interactactions 16, 699–708.

Mauck, K.E., De Moraes, C.M., Mescher, M.C., 2010. Deceptive chemical signals induced by a plant virus attract insect vectors to inferior hosts. Proceedings of the National Academy of Sciences of the USA 107, 3600–3605.

Maule, A.J., Caranta, C., Boulton, M.I., 2007. Sources of natural resistance to plant viruses: status and prospects. Molecular Plant Pathology 8, 223–231.

Medeiros, R.B., Ullman, D.E., Sherwood, J.L., German, T.L., 2000. Immunoprecipitation of a 50-kDa protein: a candidate receptor component for tomato spotted wilt tospovirus (Bunyaviridae) in its main vector, *Frankliniella occidentalis*. Virus Research 67, 109–118.

Mello, A.F., Clark, A.J., Perry, K.L., 2010. Capsid protein of cowpea chlorotic mottle virus is a determinant for vector transmission by a beetle. Journal of General Virology 91, 545–551.

Miles, P.W., 1972. The saliva of *Hemiptera*. Advances in Insect Physiology 9, 183–255.

Milligan, S.B., Bodeau, J., Yaghoobi, J., Kaloshian, I., Zabel, P., Williamson, V.M., 1998. The root knot nematode resistance gene Mi from tomato is a member of the leucine zipper, nucleotide binding, leucine-rich repeat family of plant genes. Plant Cell 10, 1307–1319.

Nault, L., 1997. Arthropod transmission of plant viruses: a new synthesis. Annals of the Entomological Society of America 90, 521–541.

Navas-Castillo, J., Fiallo-Olive, E., Sanchez-Campos, S., 2011. Emerging virus diseases transmitted by whiteflies. Annual Review of Phytopathology 49, 219–248.

Ng, J.C., Falk, B.W., 2006. Virus–vector interactions mediating non-persistent and semipersistent transmission of plant viruses. Annual Review of Phytopathology 44, 183–212.

Nombela, G., Williamson, V.M., Muniz, M., 2003. The root-knot nematode resistance gene Mi-1.2 of tomato is responsible for resistance against the whitefly *Bemisia tabaci*. Molecular Plant Microbe Interactions 16, 645–649.

Perring, T.M., Gruenhagen, N.M., Farrar, C.A., 1999. Management of plant viral diseases through chemical control of insect vectors. Annual Review of Entomology 44, 457–481.

Pirone, T.P., 1964. Aphid transmission of a purified stylet-borne virus acquired through a membrane. Virology 23, 107–108.

Pitino, M., Coleman, A.D., Maffei, M.E., Ridout, C.J., Hogenhout, S.A., 2011. Silencing of aphid genes by dsRNA feeding from plants. PLoS One 6, e25709.

Raccah, B., Huet, H., Blanc, S., 2001. Molecular basis of vector transmission: potyvirus. In: Harris, K.F., KF, Smith, O.P., Duffus, J.E. (Eds.), Virus–insect–plant interactions, Academic Press, New York, pp. 181–206.

Rajagopal, R., Sivakumar, S., Agrawal, N., Malhotra, P., Bhatnagar, R.K., 2002. Silencing of midgut aminopeptidase N of *Spodoptera litura* by double-stranded RNA establishes its role as *Bacillus thuringiensis* toxin receptor. Journal of Biological Chemistry 277, 46849–46851.

Silberstein, L., Kovalski, I., Brotman, Y., Perin, C., Dogimont, C., Pitrat, M., Klingler, J., Thompson, G., Portnoy, V., Katzir, N., Perl-Treves, R., 2003. Linkage map of *Cucumis melo* including phenotypic traits and sequence-characterized genes. Genome 46, 761–773.

Smith, C., 2005. Plant resistance to arthropods: molecular and conventional approaches. Springer, Dordrecht, The Netherlands.

Stafford, C.A., Walker, G.P., Ullman, D.E., 2011. Infection with a plant virus modifies vector feeding behavior. Proceedings of the National Academy of Sciences of the USA 108, 9350–9355.

Stafford, C.A., Walker, G.P., Ullman, D.E., 2012. Hitching a ride: vector feeding and virus transmission. Communicative & Integrative Biology 5, 43–49.

Strange, R.N., Scott, P.R., 2005. Plant disease: a threat to global food security. Annual Review of Phytopathology 43, 83–116.

Tabashnik, B.E., Carriere, Y., Dennehy, T.J., Morin, S., Sisterson, M.S., Roush, R.T., Shelton, A.M., Zhao, J.Z., 2003. Insect resistance to transgenic Bt crops: lessons from the laboratory and field. Journal of Economical Entomology 96, 1031–1038.

Tamborindeguy, C., Gray, S., Jander, G., 2008. Testing the physiological barriers to viral transmission in aphids using microinjection. Journal of Visualized Experiments 15, e700.

Timmons, L., Court, D.L., Fire, A., 2001. Ingestion of bacterially expressed dsRNAs can produce specific and potent genetic interference in *Caenorhabditis elegans*. Gene 263, 103–112.

Tjallingii, W.F., 1990. Continuous recording of stylet penetration activities by aphids. In: Eikenbary, C.R.D. (Ed.), Aphid–plant genotype interactions, Elsevier, Amsterdam, pp. 89–99.

Uzest, M., Gargani, D., Drucker, M., Hebrard, E., Garzo, E., Candresse, T., Fereres, A., Blanc, S., 2007. A protein key to plant virus transmission at the tip of the insect vector stylet. Proceedings of the National Academy of Sciences of the USA 104, 17959–17964.

Uzest, M., Gargani, D., Dombrovsky, A., Cazevieille, C., Cot, D., Blanc, S., 2010. The 'acrostyle': a newly described anatomical structure in aphid stylets. Arthropod Structure & Development 39, 221–229.

Valverde, R.A., Sim, J., Lotrakul, P., 2004. Whitefly transmission of sweet potato viruses. Virus Research 100, 123–128.

van den Heuvel, J.F., Verbeek, M., van der Wilk, F., 1994. Endosymbiotic bacteria associated with circulative transmission of potato leafroll virus by *Myzus persicae*. Journal of General Virology 75, 2559–2565.

van den Heuvel, J.F., Bruyere, A., Hogenhout, S.A., Ziegler-Graff, V., Brault, V., Verbeek, M., van der Wilk, F., Richards, K., 1997. The N-terminal region of the luteovirus readthrough domain determines virus binding to *Buchnera* GroEL and is essential for virus persistence in the aphid. Journal of Virology 71, 7258–7265.

Villada, E.S., Gonzalez, E.G., Lopez-Sese, A.I., Castiel, A.F., Gomez-Guillamon, M.L., 2009. Hypersensitive response to *Aphis gossypii* Glover in melon genotypes carrying the Vat gene. Journal of Experimental Botany 60, 3269–3277.

Walters, H.J., 1969. Beetle transmission of plant viruses. Advances in Virus Research 15, 339–363.

Wang, R.Y., Ammar, E.D., Thornbury, D.W., Lopez-Moya, J.J., Pirone, T.P., 1996. Loss of potyvirus transmissibility and helper-component activity correlate with non-retention of virions in aphid stylets. Journal of General Virology 77, 861–867.

Watson, M.A., Roberts, F.M., 1939. A comparative study of the transmission of Hyocyamus virus 3, potato virus Y and cucumber mosaic virus by the vector *Myzus persicae* (Sulz), *M. circumflexus* (Buckton) and *Macrosiphum geri* (Koch). Proceedings of the Royal Society of London B, 543–576.

Wayadande, A., 1990. Electronic monitoring of leafhoppers and planthoppers: feeding behavior and application in host-plant resistance studies. In: Ellsbury, E.M.B.E., Ullman, D.L. (Eds.), History, development, and application of AC electronic insect feeding monitors, Thomas Say Publications in Entomology, Lanham, pp. 86–105.

Westwood, J.H., Stevens, M., 2010. Resistance to aphid vectors of virus disease. Advances in Virus Research 76, 179–210.

Whitfield, A.E., Ullman, D.E., German, T.L., 2005. Tospovirus-thrips interactions. Annual Review of Phytopathology 43, 459–489.

Zha, W., Peng, X., Chen, R., Du, B., Zhu, L., He, G., 2011. Knockdown of midgut genes by dsRNA-transgenic plant-mediated RNA interference in the hemipteran insect Nilaparvata lugens. PLoS One 6, e20504.

Chapter 7

Transmission and host interaction of *Geminivirus* in weeds

Avinash Marwal and Anurag Kumar Sahu
Department of Science, Faculty of Arts, Science and Commerce, Mody Institute of Technology and Science, Lakshmangarh, Rajasthan, India

R.K. Gaur
Department of Science, Faculty of Arts, Science and Commerce, Mody Institute of Technology and Science, Lakshmangarh, Rajasthan, India

Avinash Marwal and Anurag Kumar Sahu contributed equally to this chapter.

INTRODUCTION

Recombination has played, and continues to play, a pivotal role in the evolution of geminiviruses and may be contributing to the emergence of new variants of *Geminivirus* because the high frequency of mixed infections provides an opportunity for the emergence of new viruses arising from recombination among strains and/or species (Harrison & Robinson 1999, Power 2000). In some cases, the recombinants exhibit a new pathogenic phenotype, which is often more virulent than that of the parents (Zhou et al 1997).

With the development of reliable computational tools for detecting recombination and an increasing number of genome sequences now available, many studies have reported evidence for recombination in a wide range of viral genera (Marwal et al 2011). Computational analysis suggests that such interspecific recombination events result in a remarkable diversity among viruses of the genus *Geminivirus* and could be a major cause of the emergence of new diseases caused by members of this genus (Padidam et al 1999).

Several weeds serve as the natural host of *Geminivirus*. Weeds show biologic competition with field crops, and in several cases this allows the transmission of *Geminivirus* from weeds to field crops and ornamental plants. This will increase the host range of *Geminivirus* and bring about losses of 80–100%. To combat this, RNAi technology (Marwal et al 2012a) provides a wonderful tool for checking on the devastating nature of those geminiviruses that infect weeds. Recently, inhibition of *Tomato yellow leaf curl virus*

(TYLCV) using whey proteins (whey protein is a mixture of globular proteins isolated from whey, the liquid material created as a by-product of cheese production) has been reported (Abdelbacki et al 2010). Moreover, structural bioinformatics is concerned with computational approaches to predict and analyze the spatial structure of proteins and nucleic acids by using different tools and techniques. Therefore, docking studies between α-lactalbumin, coat protein, and the Rep protein of *Begomovirus*, using modeling and docking (Hex 6.3) software, is a valuable approach (Prajapat et al 2011a). The results of these techniques were effectively applied for disease management and for the development of quarantine strategies for handling and transportation of infected plant samples.

INTERNATIONAL RECOGNITION OF *GEMINIVIRIDAE*

In recent years the family *Geminiviridae* has received a great deal of attention; it is one of the most important of the families of plant viruses that have been studied. Several reviews on the genus *Geminivirus* have covered different aspects of the biology of these viruses: epidemiology, serologic properties, and molecular biology (Polston & Anderson 1997, Hanley-Bowdoin et al 1999, 2000, Morales & Anderson 2001). Newly emerging strains of *Geminivirus* are causing severe disease epidemics in cotton, grain, legumes, tomato, and other staple food and cash crops in tropical and subtropical regions (Khan 2000, Boulton 2003). *Geminivirus* infection in the host plant drastically reduces photosynthesis, growth, fruit growth, and fruit quality, although the effects depend on the number of plants infected and on the age of the plant at the time of infection. In recent years, epidemics have affected from 30–100% of the crops. Symptoms now known to be associated with *Geminivirus* infection have been observed in plants grown in tropical and subtropical regions of the world since the mid-1800s (Wege et al 2000). These viruses cause a variety of symptoms, including vein yellowing, yellow mosaic, and leaf curl, and are spreading at an alarming pace due to a high rate of recombination (Briddon et al 2003, Mansoor et al 2003). Control of the spread of these viruses is difficult because they are transmitted through insect vectors such as whitefly (*Bemisia tabaci*) (Fauquet et al 2003). In addition, several weeds serve as alternative hosts for these viruses in the absence of the main crops.

Geminiviruses were recognized in 1978 by the International Committee on the Taxonomy of viruses (ICTV) on the basis of their unique virion morphology and their single-stranded DNA (ssDNA) genome (Matthews 1979, Goodman 1981). The family *Geminiviridae* is one of the largest families of plant viruses (Fauquet & Stanley 2005); its members have a circular ssDNA genome of approximately 2.7–5.2 kb (Harrison & Robinson 1999) encapsulated within twinned (geminate) icosahedral virions. The protein coat of members of the *Geminiviridae* consists of one type of protein molecule of about 28 kD molecular weight.

TRANSMISSION VECTORS OF THE FAMILY *GEMINIVIRIDAE*

Geminiviridae is the family that contains the greatest number of viruses (Fauquet et al 2005). General symptoms of diseases caused by *Geminivirus* are curling of leaves, yellowing of veins, yellow mosaic patterns, and dwarfing of leaves. Viruses belonging to the family *Geminiviridae* are plant viruses that are obligate intracellular parasites, having no machinery to replicate themselves (Stanley et al 2005). The family comprises four genera: *Begomovirus, Curtovirus, Mastrevirus*, and *Topocuvirus*.

Begomovirus is one the biggest genera of the family (Medina-Ramos et al 2008). It comprises about 200 species that are found worldwide. *Begomovirus* principally affects dicotyledonous plant species. These viruses are transmitted by whitefly (Markham et al 1994).

Whitefly (*Bemisia tabaci*) belongs to the family *Aleyrodidae* of the class Insecta (Fig. 7.1, A and B). This fly prevails more in the tropical and subtropical regions of the globe (Sidhu et al 2009). *Begomovirus* is the only genus of the *Geminiviidae* family to have a bipartite genome with virus genes resident on two different circular ssDNA molecules (i.e., DNA A and DNA B). Both DNA A and DNA B range from 2.7 to 3 kb (Hanley-Bowdoin et al 1999). DNA B is dependent on DNA A for its function. Moreover, in an up-to-date publication, an additional circular DNA α and β satellite has been reported, which also contributes to *Begomovirus* virulation (Briddon et al 2001) and nanovirus-like DNA satellite molecules (alphasatellites) (Briddon & Stanley 2006).

Curtovirus has seven known species: *Beet curly top Iran virus, Beet curly top virus, Beet mild curly top virus, Beet severe curly top virus, Horseradish curly top virus, Pepper curly top virus*, and *Spinach curly top virus* (Bolok Yazdi et al 2008). Curtoviruses are transmitted by the leafhopper (Fig. 7.1C) or by the treehopper (*Micrutalis malleifera*). They have a monopartite genome that comprises a circular single-stranded DNA molecule of up to 3 kb (Baliji et al 2004).

Mastrevirus is the second largest genus of this family. It comprises 14 species according to a recent ICTV publication (Brown et al 2012). Infection with *Mastrevirus* is mediated by the leafhopper, which is its chief host. The leafhopper belongs to the family *Cicadellidae*. 'Leafhopper' is a common name applied to any species of the family *Cicadellidae* (Ing-Ming et al 2000). Leafhoppers exist all over the world and do not have a pupae stage while turning into the adult from the nymph stage. *Mastrevirus* affects both dicot and monocot plant species. The genome is monopartite, up to 2.7 kb.

Topocuvirus is the smallest genus of the family *Geminiviridae* (Briddon & Markham 2001). It consists of only one species: *Tomato pseudo-curly top virus*. It infects dicotyledonous plants, mainly tomato. It has a monopartite, closed, circular, single-stranded DNA genome, about 2.8 kb long. The virus is transmitted by the treehopper (Hunter et al 1998).

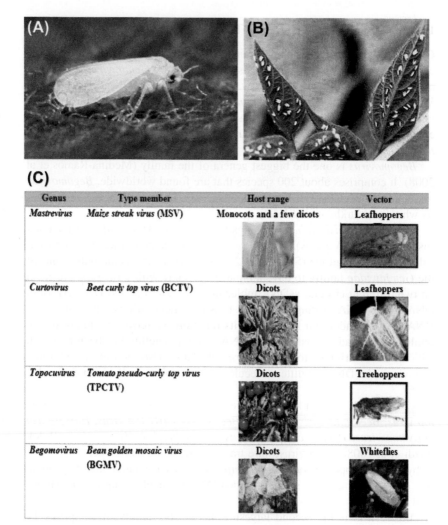

FIGURE 7.1 (A, B) Whitefly residing on the under surface of leaf. (C) Carrier vectors of *Geminivirus*.

GENOMIC ORGANIZATION IN DIFFERENT GENERA

Begomovirus DNA A has six open reading frames (ORFs): *AC1, AC2, AC3, AC4, AV1*, and *AV2*. *AC1* (*AL1*) encodes the replication initiation protein (Rep) (Saunders et al 2008). *AC2* (*AL2*) encodes a transcription-activator protein (TrAP). *AC3* (*AL3*) encodes replication enhancer proteins (Tiendrebeogo et al 2008). *AC4* determines the expression of symptoms (Fig. 7.2 [A]). *AV1* (*AR1*) encodes the coat protein (CP). *AV2* encodes the movement protein; it is also called the 'precoat' ORF. DNA B (Fig. 7.2 [B]) has two ORFs. *BV1* (*BR1*) encodes the nuclear shuttle protein (NSP). *BC1* (*BL1*) encodes the movement

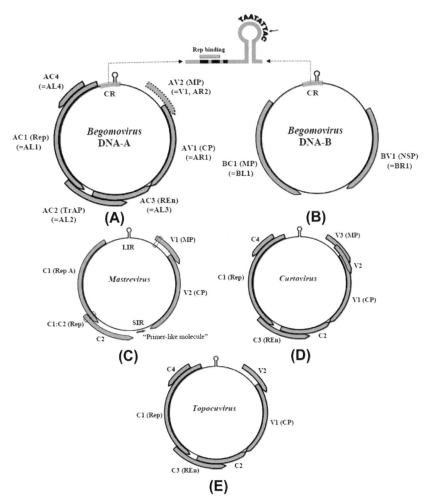

FIGURE 7.2 Genomic organization of different genera of the family *Geminiviridae*. Maps of the type members of each genus are shown: [A, B] *Begomovirus* (*Bean golden mosaic virus*; BGMV); [C] *Mastrevirus* (*Maize strek virus*, MSV); [D] *Curtovirus* (*Beet curly top virus*, BCTV); and [E] *Topocuvirus* (*Tomato pseudo-curly top virus*, TPCTV).

protein, involved in cell-to-cell transfer (Rojas et al 2005). *AV1, AV2*, and *BV1* are the plus (+) virion sense strand, whereas *AC1, AC2 AC3, AC4*, and *BC1* represent the negative (−) complementary sense strand (Yadava et al 2010).

Curtovirus has seven ORFs (Park et al 1999). Three of the ORFs, *V1, V2*, and *V3* are in the positive (virion) sense, while the other four—*C1, C2, C3*, and *C4*—are in the negative (complementary) sense. *V1* encodes the coat protein, *V2* encodes an ss/dsDNA regulator, and *V3* encodes a movement protein (Fig. 7.2 [D]). *C1* encodes a replication protein, whereas *C2* has no known function. *C3* encodes a protein similar to the replication enhancer protein of *Begomovirus*. The *C4* product is an initiator of cell division and is also involved in the determination of symptoms (Park et al 2003).

There are four ORFs in the *Mastrevirus* genome (Boulton 2002). Two of them are on the virus (+) sense (Fig. 7.2 [C]) and the other two are on the complementary (–) sense (Heyraud-Nitschke et al 1995). *V1* is the larger plus (+) sense ORF, which encodes the coat protein (CP) (Gutierrez et al 2004). *V2* encodes a cell-to-cell movement protein (Nahid et al 2008). A unique feature of the mastreviruses is the expression of two Rep proteins. First is the full-length Rep, which is translated from a spliced transcript of the *C1* and *C2* ORFs. The second is RepA, which is translated from a *C1* transcript (Xie et al 1999).

The *Topocuvirus* genome (shown in Fig. 7.2 [E]) has six ORFs. *V1* and *V2* encode a coat protein and a movement protein, respectively (Varma et al 2003). Replicase A and replicase B are, respectively, encoded by *C1* and *C2*. *C3* encodes a replication-enhancer protein with a function similar to that of the *Begomovirus* replication-enhancer protein. Cell division is initiated by the *C4*-encoded product. *V1* and *V2* are in the plus sense and *C1*, *C2*, and *C3* are in the negative sense region (Govindappa et al 2011).

HOST INTERACTION OF *GEMINIVIRUS* IN WEEDS

Datura inoxia showing typical symptoms of *Begomovirus* infection (i.e., leaf curling, decreased leaf size and growth stunting) was observed in the fields of Rajasthan, India (Fig. 7.3A). The PCR product was partially sequenced (JN000702), and a 550-nucleotide sequence was used to construct a phylogenetic tree using the neighbor-joining method in DNASIS Pro ver.2.6 (Marwal et al 2012b). Sequence analysis of the virus under study showed 82–83% nucleotide sequence identity with the corresponding region of *Chilli leaf curl virus* (FM210476), *Tomato leaf curl virus* (AJ810360), and *Croton yellow vein virus* (FN543112).

The uncharacterized beta-satellite of *Begomovirus* associated with *Calotropis procera* (Fig. 7.3B) was characterized by using molecular and *in silico* tools and techniques. Attempts to identify the presence of a DNA-β in the infected *C. procera* samples, using rolling circle amplification (RCA) followed

FIGURE 7.3 (A) *Datura inoxia* showing leaf curl symptoms in Lakshmangarh, Rajasthan, India, in 2011. (B) *Calotropis procera* showing yellow mosaic symptoms. (C) *Begomovirus* symptoms of yellowing of leaf was observed in *Verbesina encelioides*.

by restriction digestion, produced a ca. 1.4 kb product, corresponding to that expected for a full-length amplicon from a beta-satellite, which was sequenced (accession number HQ631430). During BLASTp, analysis of a second reading frame of HQ631430 (HQ631430/2-f) against Protein Databank revealed 35% identity with tryptophanyl-tRNA synthetase of *Giardia lamblia* (3FOC). A Ramachandran plot of HQ631430/2-f.pdb had only 57.1% residues in the most favored region, while 3FOC.pdb had 94.2% residues in the most favored region; therefore, only the template 3FOC.pdb model could be placed in a good-quality category. The protein-binding function was predicted for HQ631430/2-f as an important functional site of the model with a 0.29 confidence level through 3d2GO. The *Croton yellow vein mosaic* beta-satellite (GU111995 CroYVMB) serves as a major parent, and the *Croton yellow vein mosaic* beta-satellite-Panipat 8 (HM143908 PaLCuVM) as a minor parent for HQ631430. This may be the first report of recombination in the *Croton yellow vein mosaic* beta-satellite (Prajapat et al 2012).

Natural infection of *Tomato leaf curl virus* was observed in the plants *Parthenium hystrophorus* and *Sonchus asper* in Uttar Pradesh, India. *B. tabaci* could transmit the virus from these symptomless infected weeds to tomato plants, which exhibited leaf curling and twisting symptoms after 15–25 days of transmission (Ansari & Tewari 2005). Yellow vein mosaic disease (YVMD) of Kenaf (*Hibiscus cannabinus*) was observed in northern India and caused significant reduction in plant height and crop yield. Southern hybridization with a *Begomovirus*-specific probe, and PCR amplification with DNA β and coat protein primer, confirmed the association of *Begomovirus* with the disease (Ghosh et al 2007). Full-length DNA-A of a *Begomovirus* infecting mesta (*Hibiscus cannabinus*) was cloned and sequenced. The component nucleotide sequence of the DNA-A molecule shared the highest sequence identity (83.5%) with an Indian *Begomovirus* causing cotton leaf curl disease. Thus it was considered as a novel *Begomovirus* species as *Mesta yellow vein mosaic* (Chatterjee & Ghosh 2007a).

Bean golden mosaic Begomovirus (BGMV) was reported to cause bright yellow mosaic symptoms in *Macroptilium lathyroides*, a weed common in Puerto Rico (Idris et al 1999). For *Tomato yellow leaf curl virus* (TYLCV), weeds act as a reservoir or a 'transmission bridge' between cropping and non-cropping seasons (Salati et al 2002). The report of begomoviruses in *Lythrum hyssopifolia* from China suggests that *L. hyssopifolia* was an adaptive host for begomoviruses (Guo & Zhou 2005). A DNA β component was identified associated with *Mesta yellow vein mosaic* (YVM) disease (Chatterjee & Ghosh 2007b). Six beta-satellite isolates were characterized; they were associated with YVM disease in a mesta crop from three different geographic locations in India. Another incidence of YVM disease of mesta in northern India was recorded, and the cause, *Begomovirus*, was identified as a distinct monopartite species associated with a DNA β satellite. This *Begomovirus* species shared a low sequence identity with the *Mesta yellow vein mosaic virus* (Das et al 2008a,b).

The uncharacterized alpha-satellite of *Begomovirus* associated with a common weed, *Verbesina encelioides* (Fig. 7.3C), was characterized by using molecular and *in silico* tools and techniques. *Verbesina encelioides* leaf curl alpha-satellite (HQ631431) shows 87% nucleotide sequence identity with *Sida yellow vein disease*-associated DNA1 (FN806782). Translated 3-frame of HQ631431 (HQ631431/3-f.pdb) and its homologous 2HWT.pdb had, respectively, 67% and 77.2% residues in the most favorable region of its Ramachandran plot, so neither of these two models can be said to have good quality. The 3d2GO server showed hydrolase activity as a possible function for HQ631431/3-f.pdb, with a 0.58% confidence level. *In silico* prediction of results can be used to confirm *Begomovirus* not only in host weeds but also in other crops (Prajapat et al 2011c).

Honeysuckle yellow vein virus had been isolated from honeysuckle (*Lonicera japonica*), a perennial weed, with yellow vein symptoms, suggesting infection by a *Begomovirus* (Ogawa et al 2008). A *Begomovirus* was detected in the weed *Mimosa invisa*, collected from vegetable- and fruit-growing areas of the Malaysia Peninsula (Mahmoudieh et al 2008). It was found that a *Begomovirus* disease complex was associated with yellow vein disease of a common weed, *Digera arvensis*. The presence of multiple and recombinant beta-satellites in *D. arvensis* indicates that weeds can be important sources of multiple *Begomovirus* components that affect crop plants (Mubin et al 2009). Infection by new bipartite *Begomovirus* in two common weeds, *Malvastrum americanum* and *Sida spinosa*, was reported from Jamaica (Graham et al 2007). Weeds of the genus *Sida*, collected in Brazil, had harbored several geminiviruses persistently over decades of vegetative propagation. They serve as cradles for new geminiviruses originating from pseudo-recombination or molecular recombination, as had been exemplified by *Sida micrantha mosaic-associated viruses* (SimMV) (Jovel et al 2007). *Alternanthera yellow vein virus* (AlYVV) in *Sonchus arvensis* was associated with satellites shown previously to be associated with other begomoviruses in Pakistan. The monopartite begomoviruses may associate with distinct satellites that were prevalent in the region (Mubin et al 2010).

Begomoviruses were associated with *Rhynchosia minima* yellow mosaic disease and identified through molecular characterization. Sequence comparison shows a maximum identity of 84% with an isolate of *Velvet bean severe mosaic virus* (India (Lucknow) 2009) [Accession no. FN543425] (Jyothsna et al 2011). Plants of the common weed *Jatropha gossypifolia* with yellow mosaic symptoms typical of *Geminivirus* infection were often found growing among crops in Jamaica. These crops are known to be hosts to several begomoviruses in Jamaica (Roye et al 1999). That was the first report of a *Begomovirus* associated with *J. gossypifolia* in Jamaica and had been tentatively named *Jatropha mosaic virus* (JMV) (Roye et al 2005). Leaf curl in *Zinnia elegans* was observed in a subtemperate region in northern India, and the causal organism was identified as *Ageratum enation virus* (AEV). They show the association of nanovirus-like satellite DNA1 along with DNA-A (Kumar et al 2010).

Macroptilium lathyroides is a widely distributed weed in Cuba that is infected with a new *Begomovirus* species, *Macroptilium yellow mosaic virus* (MaYMV)

(Ramos et al 2002). *Calopogonium golden mosaic virus* (CalGMV) was isolated from a weed, *Calopogonium* sp., collected in 1991 near Quepos, Costa Rica (Diaz et al 2002). Shibuya et al (2007) reported the PepYLCIDV infection in *Ageratum conyzoides* plants that were affected with yellow vein disease in Indonesia. The recent discovery that monopartite begomoviruses on *Ageratum* and cotton essentially require a DNA satellite called DNA β (Saunders et al 2000) is leading to the identification of several other hosts that have similar disease complexes. The yellow vein disease on *Croton bonplandianus* is associated with a monopartite *Begomovirus* and a distinct DNA β (Amin et al 2002). The complete sequences of a Begomovirus and an associated beta-satellite were identified from Croton bonplandianus that originated from Pakistan (Hussain et al 2011).

DNA β molecules associated with yellow vein mosaic disease of *Urena lobata, Croton bonplandianum, Sida acuminate*, and *S. rhombifolia* indicated the widespread variation of geminiviruses in different host plants (Chatterjee et al 2007). The *Geminivirus* isolated from the common weed *Ageratum* sp. in Sri Lanka also appears to be a distinct new *Geminivirus*, not related to the *Geminivirus* infecting *Ageratum* sp. in Singapore and Malaysia (AVRDC 2000). Samples of *Emilia sonchifolia* leaves showing conspicuous yellow veins were collected in the Chinese province of Fujian. Molecular characterization showed that a *Begomovirus* [*Emilia yellow vein virus* (EmYVV)] was associated with *E. sonchifolia* yellow vein disease (Yang et al 2008). The natural occurrence of *Sweet potato leaf curl virus* (SPLCV) was reported in *Ipomoea batatas* (*Convolvulaceae*) or *I. indica* (*Convolvulaceae*) in several countries, including Italy and China (Luan et al 2007). That was the first report of the natural occurrence of SPLCV in *I. purpurea*, a common weed species in China (Yang et al 2009).

RNAi TECHNOLOGY: A SPECTACULAR APPROACH AGAINST *GEMINIVIRUS*

RNAi is an evolutionarily conserved silencing pathway in which double-stranded RNA is broken down into small interfering RNA (siRNA) with the help of Dicer and the RNA-induced silencing complex (RISC) in a series of steps (Fig. 7.4). The two components of Dicer, dcr-2/r2d2, which are ATP-dependent, bind to siRNA and help it to load into the RISC by forming the RISC-loading complex (RLC). The RLC recruits associate components to form the effective machinery for gene silencing; it may remain bound to the complementary mRNA or may degrade the target. In this way, the activated RISC could potentially target multiple mRNAs and thus function catalytically (Somyaparna et al 2011).

RNAi-mediated virus resistance was first reported against *Potato virus Y* (PVY) in a transgenic tobacco plant (Waterhouse et al 1998). RNAi technology was used as an antiviral approach against human cell lines (Novina et al 2002), but it can also be used for developing resistance against plant viruses (Waterhouse et al 2001). RNAi technology, when used against a *Geminivirus* (*African cassava mosaic virus* (ACMV)), showed a 99% decrease in Rep transcripts and a 66% reduction in viral DNA. The siRNA was transiently

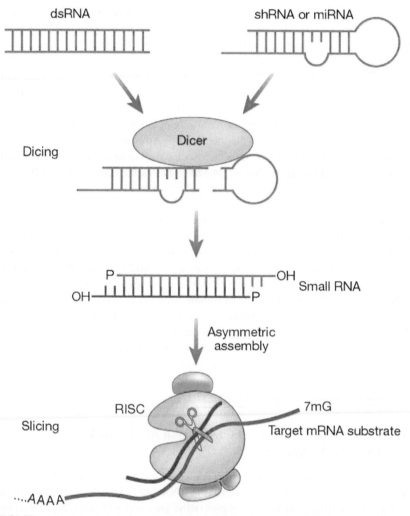

FIGURE 7.4 Mechanism of RNA silencing in different systems. Long dsRNA and miRNA precursors are processed to siRNA/miRNA duplexes by the RNase-III-like enzyme Dicer. These short dsRNAs are subsequently unwound and assembled into effector complexes, RISCs, which can direct RNA cleavage, mediate translational repression, or induce chromatin modification. *S. pombe, C. elegans*, and mammals carry only one Dicer gene. 7mG, 7-methylguanine; AAAA, poly-adenosine tail; Me, methyl group; P, 5′-phosphate. *Courtesy of Gregory J Hannon and John J Rossi, 2004*

transferred into the protoplast, making it effective against the replicase (Rep)-encoding sequence of the ACMV (Vanitharani et al 2003). Short interfering RNAs (siRNAs), the 21- to 23-nt double-stranded intermediates of this natural defense mechanism, are becoming powerful tools for reducing gene expression and countering viral infection in a variety of cells (Rougemaille et al 2012).

A novel single-stranded DNA of approximately 1350 nucleotides has been identified associated with an infected yellow vein mosaic croton; it was called DNA-β, and it requires a *Begomovirus* for replication, encapsidation, insect transmission, and movement in plants. It exhibits the properties of a satellite molecule, having an approximately 80-nucleotide conserved region, which has been suggested to be important in *trans*-replication of DNA-β by the *Begomoviruses* Rep, possibly containing cryptic Rep-binding sites. The resulting amplicons were cloned in the pCambia 1300. The resulting binary construct was introduced into *Agrobacterium tumefaciens* LBA4404 by electroporation with a Gene Pulser apparatus (Bio-Rad). Seeds of T1 lines were grown on MS, and 2-week-old seedlings were infected with the infectious clones of *Croton yellow vein mosaic* using the Bio Rad particle-delivery system. The relationship between transgene transcription and immunity was confirmed; it was also confirmed that the transgene protein does not mediate immunity (Marwal et al 2012c).

Transgenic resistance against begomoviruses has been achieved in a number of plants using a variety of strategies. The results presented above show that the RNAi approach has been investigated extensively; it is a powerful tool for biochemical studies and for developing transgenic plants. It provides a ray of hope for various challenging geminiviral diseases against weeds.

ANTIVIRAL AGENT: DOCKING FRAMEWORK

Molecular docking tools are used in structural molecular biology and structure-based drug discovery (Kartasasmita et al 2010). Docking predicts the preferred orientation of one molecule to a second when bound to each other to form a stable complex; it is used to predict the binding orientation of small-molecule drug or antiviral agents to their protein targets and is also used for protein–protein docking in order to predict the affinity and activity of the small molecule (Kitchen et al 2004).

A homology modeling study of the coat protein of *Mimosa yellow vein virus* (ADW83735) and docking between α-lactalbumin and the coat protein was carried out by using modeling and docking software (Hex 6.3). The ADW83735 model was validated by using the Procheck server for reliability that resulted in only 49.2% of residues present in core region (Fig. 7.5). In docking, the binding sites exhibit chemical specificity and the affinity that measure strength of the chemical bond for which root mean square (RMS) was calculated. The best docking confirmations have the lowest binding energy and greater number of conformation per cluster. Therefore on the basis of the RMS and energy values the best docking orientation was selected. The better RMS value of docking was −1.00.

The FASTA sequence of α-lactalbumin (accession ACI62509; source: *Bos taurus*) was mined from GenBank-NCBI, and with the help of the 3D-Jigsawn server its PDB file was designed. Automated comparative docking was done in between α-lactalbumin (ACI62509) and coat protein (ADW83735) by using program Hex 6.3 (Ritchie et al 2008). Hex is primarily a docking program for demonstrating the potential for performing fast 3D superpositions using the

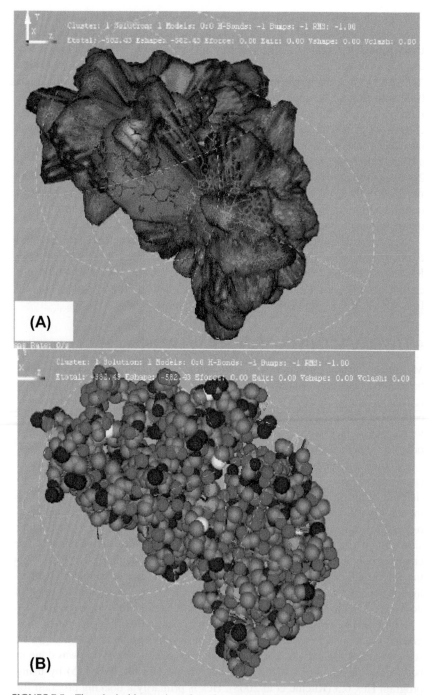

FIGURE 7.5 The spherical harmonic surfaces for receptor coat protein (ADW83735) and ligand α-lactalbumin (ACI62509). (A) Display surface of coat protein/α-lactalbumin complex. (B) Binding site model ADW83735 and ACI62509.

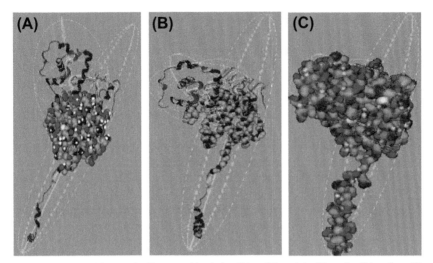

FIGURE 7.6   Docking results illustrate side chains (A), solid models (B), and solid surface (C) view forms of the Rep protein/α-lactalbumin complex.

SPF correlation approach. The PDB files of coat protein (ADW83735) and α-lactalbumin were uploaded as inputs into Hex for protein–protein docking. These are treated as a receptor and a ligand, respectively. Regularization is a procedure for fitting a protein model with the ideal covalent geometry of residues to the atomic positions of the target PDB structure (Ritchie 2008). Based on the energy minimization, the best pose of the docked complex was selected.

This study is very useful for the screening of inhibitors against *Mimosa yellow vein virus* proteins and can be further applied in the future design of antiviral agents against *Geminiviruses* (Prajapat et al 2011b).

Similarly, docking between α-lactalbumin and the Rep protein of *Begomovirus* by using modeling and docking (Hex 6.3) software was carried out (Fig. 7.6). This model shows that 35% identity is enough for the receptor-based design of antiviral agents. The closest homolog of the Rep protein was 1L2M|A, with the highest sequence identity of 81%, which was selected as a representative model using homology modeling software. All the input files and the constructed model were analyzed using the protein docking and spherical harmonic surfaces of the Hex. Structure refinement and energy minimization were performed with Hex itself. Hex sorts the generated orientations by docking energy and prints a summary of the 10,000 highest scoring (lowest energy) orientations (Prajapat et al 2011a). Docking can be used in several ways—for example, to study the mechanism of an enzymatic reaction, to identify possible binding modes for a ligand, and to screen a database.

CONCLUSION

The recognition that geminiviruses are capable of rapidly diverging through multiple mechanisms underscores the need for accurate molecularly based methods that permit detection and tracking of biologically significant variants in weeds. Molecular approaches must combine (i) a knowledge of biology and ecology and the ability to monitor both conserved sequences and specific sites most likely to undergo alteration with (ii) phylogenetic predictions to facilitate accurate identification and tracking of *Geminivirus* variants and to recognize new or resurgent viruses residing in weeds. Establishing databases of baseline sequences for extant viruses will permit future comparisons in establishing and interpreting disease patterns and associated trends for vector populations.

Weeds are potential sources of primary inoculums of viruses and they play an important role in the persistence and spread of viruses. In most places crops stay in the field for a particular season, while, during the off-season, a particular crop is not present. However, different weeds grow in or around these agricultural fields throughout the year. Most of these weeds demonstrated vein yellowing and yellow mosaic symptoms and were likely to be carriers of geminiviruses.

An expected consequence of this scenario would be recombination, which plays an important role in the evolution of new *Geminivirus* strains in India; these new strains are responsible for heavy losses of the new host variety. Thus, there is an urgent need to control *Geminivirus* infections. The use of computational and molecular techniques (e.g., RNAi) is a potential tool for reducing the prevalence of various *Geminivirus* diseases.

In this chapter we talked about the use of bioinformatics techniques for identification of possible ligand molecules that are effective against begomovirus replication. Moreover we highlighted the use of RNAi techniques for development of virus-resistant plants. Therefore, results of these techniques should be effectively applied to disease management, crop protection, and development of quarantine strategies at state and national levels in India.

ACKNOWLEDGMENTS

The authors would like to give a vote of thanks to the Department of Biotechnology (DBT project No. BT/PR13129/GBD/27/197/2009) and the Department of Science and Technology (DST project no. SR/FT/LS-042/2009), India, for their financial support.

REFERENCES

Abdelbacki, A.M., Taha, S.H., Sitohy, M.Z., Dawood, A.I.A., Hamid, M.M.A.E., Rezk, A.A., 2010. Inhibition of *Tomato yellow leaf curl virus* TYLCV using whey proteins. Virology Journal 7, 26–26.

Amin, I., Mansoor, S., Iram, S., Khan, M.A., Hussain, M., Zafar, Y., 2002. Association of a monopartite Begomovirus producing subgenomic DNA and a distinct DNA Beta on Croton bonplandianus showing yellow vein symptoms in Pakistan. Plant Disease 86, 444.

Ansari, N.A., Tewari, J.P., 2005. Hitherto unrecorded weed reservoirs of *Tomato leaf curl virus* (TLCV) in Eastern U.P. and its role in epidemiology of the disease. Journal of the Living World 12, 22–26.

Baliji, S., Black, M.C., French, R., Stenger, D.C., Sunter, G., 2004. Spinach curly top virus: a newly described *Curtovirus* species from Southwest Texas with incongruent gene phylogenies. Phytopathology 94 (7), 772–779.

Bolok Yazdi, H.R., Heydarnejad, J., Massumi, H., 2008. Genome characterization and genetic diversity of *Beet curly top Iran virus*: a *Geminivirus* with a novel nonanucleotide. Virus Genes 36 (3), 539–545.

Boulton, M.I., 2002. Functions and interactions of *Mastrevirus* gene products. Physiological and Molecular Plant Pathology 60, 243–255.

Boulton, M., 2003. Geminiviruses: major threats to world agriculture. Annals of Applied Biology 142, 143.

Briddon, R.W., Markham, P.G., 2001. Complementation of bipartite *Begomovirus* movement functions by *Topocuviruses* and *Curtoviruses*. Archives of Virology 146 (9), 1811–1819.

Briddon, R.W., Stanley, J., 2006. Sub-viral agents associated with plant single stranded DNA viruses. Virology 344 (1), 198–210.

Briddon, R.W., Bull, S.E., Amin, I., Idres, A.M., Mansoor, S., Bedford, I.D., Dhawan, P., Rishi, N., Siwatch, S.S., Abdel-Salam, A.M., Brown, J.K., Zafar, Y., Markham, P.G., 2003. Diversity of DNA β, a satellite molecule associated with some monopartite *Begomoviruses*. Virology 312, 106–121.

Brown, J.K., Fauquet, C.M., Briddon, R.W., et al., 2012. Family *Geminiviridae*. In: King, A.M.Q. (Ed.), Virus taxonomy: classification and nomenclature of viruses. Ninth Report of the International Committee on Taxonomy of Viruses, Academic Press, USA, pp. 351–373.

Chatterjee, A., Ghosh, S.K., 2007a. Association of a satellite DNA β molecule with Mesta yellow vein mosaic disease. Virus Genes 35, 835–844.

Chatterjee, A., Ghosh, S.K., 2007b. A new monopartite *Begomovirus* isolated from *Hibiscus cannabinus* L. in India. Archives of Virology 152, 2113–2118.

Chatterjee, A., Roy, A., Ghosh, S.K., 2007. First record of a *Begomovirus* associated with yellow vein mosaic disease of *Urena lobata* in India. Australasian Plant Disease Notes 2, 27–28.

Das, S., Ghosh, R., Paul, S., Roy, A., Ghosh, S.K., 2008a. Complete nucleotide sequence of a monopartite *Begomovirus* associated with yellow vein mosaic disease of mesta from north India. Archives of Virology 153, 1791–1796.

Das, S., Roy, A., Ghosh, R., Paul, S., Acharyya, S., Ghosh, S.K., 2008b. Sequence variability and phylogenetic relationship of betasatellite isolates associated with yellow vein mosaic disease of mesta in India. Virus Genes 37, 414–424.

Diaz, M., Maxwell, D.P., Karkashian, J.P., Ramírez, P., 2002. Calopogonium golden mosaic virus Identified in *Phaseolus vulgaris* from Western and Northern Regions of Costa Rica. Plant Disease 86, 188.

Fauquet, C.M., Stanley, J., 2005. Revising the way we conceive and name viruses below the species level: a review of *Geminivirus* taxonomy calls for the Nile and Mediterranean basins. Phytopathology 95, 549–555.

Fauquet, C.M., Bisaro, D.M., Briddon, R.W., Brown, J.K., Harrison, B.D., Rybicki, E.P., Stenger, D.C., Stanley, J., 2003. Revision of taxonomic criteria for species demarcation in the family *Geminiviridae*, and an updated list of begomovirus species. Archives of Virology 148, 405–421.

Fauquet, C.M., Mayo, M.A., Maniloff, J., Desselberger, U., Ball, L.A., 2005. Report of the International Committee on Taxonomy of Viruses. Academic Press, San Diego, USA p. 1259.

Ghosh, R., Paul, S., Roy, A., Mir, J.I., Ghosh, S.K., Srivastava, R.K., Yadav, U.S., 2007. Occurrence of *Begomovirus* associated with Yellow vein mosaic disease of kenaf (*Hibiscus cannabinus*) in northern India. Online: Plant Health Progress. http://dx.doi.org/10.1094/PHP-2007-0508-01-RS.

Goodman, R.M., 1981. Geminiviruses. Journal of General Virology 54, 9–21.

Govindappa, M.R., Shankergoud, I., Shankarappa, K.S., Wickramaarachchi, W.A.R.T., Reddy, B.A., Rangaswamy, K.T., 2011. Molecular detection and partial characterization of begomovirus associated with leaf curl disease of sunflower (*Helianthus annuus*) in Southern India. Plant Pathology Journal 10, 29–35.

Graham, A.P., Stewart, C.S., Roye, M.E., 2007. First report of a *Begomovirus* infecting two common weeds: *Malvastrum americanum* and *Sida spinosa* in Jamaica. Plant Pathology 56, 340.

Guo, X.J., Zhou, X.P., 2005. Molecular characterization of *Alternanthera yellow vein virus*: a new *Begomovirus* species infecting *Alternanthera philoxeroides*. Journal of Phytopathology 153, 694–696.

Gutierrez, C., Ramirez-Parra, E., Castellano, M.M., Sanz-Burgos, A.P., Luque, A., Missich, R., 2004. *Geminivirus* DNA replication and cell cycle interactions. Veterinary Microbiology 98, 111–119.

Hanley-Bowdoin, L., Settlage, S.B., Orozco, B.M., Nagar, S., Robertson, D., 1999. *Geminiviruses*: models for plant DNA replication, transcription, and cell cycle regulation. Critical Reviews in Plant Sciences 18, 71–106.

Hanley-Bowdoin, L., Settlage, S.B., Orozco, B.M., Nagar, S., Robertson, D., 2000. *Geminiviruses*: models for plant DNA replication, transcription, and cell cycle regulation. Critical Reviews in Biochemistry and Molecular Biology 35, 105–140.

Harrison, B.D., Robinson, D.J., 1999. Natural genomic and antigenic variation in whitefly-transmitted *Geminiviruses* (*Begomoviruses*). Annual Review of Phytopathology 37, 369–398.

Heyraud-Nitschke, F., Schumacher, S., Laufs, J., Schaefer, S., Schell, J., Gronenborn, B., 1995. Determination of the origin cleavage and joining domain of *Geminivirus* Rep proteins. Nucleic Acids Research 23, 910–916.

Hunter, W.B., Hierbert, E., Webb, S.E., Tsai, J.H., Polston, J.E., 1998. Location of *Geminiviruses* in the whitefly *Bemisia tabaci* (Homoptera: *Aleyrodidae*). Plant Disease 82, 1147–1151. http://dx.doi.org/10.1094/PDIS.1998.82.10.1147.

Hussain, K., Hussain, M., Mansoor, S., Briddon, R.W., 2011. Complete nucleotide sequence of a *Begomovirus* and associated betasatellite infecting croton (*Croton bonplandianus*) in Pakistan. Archives of Virology 156, 1101–1105.

Idris, A.M., Bird, J., Brown, J.K., 1999. First report of a bean-infecting *Begomovirus* from *Macroptilium lathyroides* in Puerto Rico that is distinct from *Bean golden mosaic virus*. Plant Disease 83, 1071–1071.

Ing-Ming, L., Davis, E.R., Dawn, E., Gundersen, R., 2000. Phytoplasma: phytopathogenic mollicutes. Annual Review of Microbiology 54, 221–255. http://dx.doi.org/10.1146/annurev.micro.54.1.221, PMID 11018129.

Jovel, J., Preiß, W., Jeske, H., 2007. Characterization of DNA intermediates of an arising Geminivirus. Virus Research 130, 63–70.

Jyothsna, P., Rawat, R., Malathi, V.G., 2011. Molecular characterization of a new *Begomovirus* infecting a leguminous weed *Rhynchosia minima* in India. Virus Genes 42, 407–414.

Kartasasmita, R.E., Herowati, R., Gusdinar, T., 2010. Docking study of quercetin derivatives on inducible nitric oxide synthase and prediction of their absorption and distribution properties. Journal of Applied Science 10, 3098–3104.

Khan, J.A., 2000. Detection of Tomato leaf curl Geminivirus in its vector *Bemisia tabaci*. Indian Journal of Experimental Biology 38, 512–515.

Kitchen, D.B., Decornez, H., Furr, J.R., Bajorath, J., 2004. Docking and scoring in virtual screening for drug discovery: methods and applications. Nature Reviews in Drug Discovery 3, 935–949.

Kumar, Y., Bhardwaj, P., Hallan, V., Zaidi, A.A., 2010. Detection and characterization of *Ageratum enation virus* and a nanovirus-like satellite DNA1 from zinnia causing leaf curl symptoms in India. Journal of General Plant Pathology 76, 395–398.

Luan, Y.S., Zhang, J., Liu, D.M., Li, W.L., 2007. Molecular characterization of *Sweet potato leaf curl virus* isolate from China (SPLCV-CN) and its phylogenetic relationship with other members of the *Geminiviridae*. Virus Genes 35, 379–385.

Mahmoudieh, K.M., Mohamad-Roff, M.N., Othman, R.Y., 2008. First report of a whitefly-transmitted *Geminivirus* infecting *Mimosa invisa* in Malaysia. Australasian Plant Disease Notes 3, 25–26.

Mansoor, S., Briddon, R.W., Zafar, Y., Stanley, J., 2003. *Geminivirus* disease complexes: an emerging threat. Trends in Plant Science 8, 128–134.

Markham, P.G., Bedford, S.L., Pinner, M.S., 1994. The transmission of *Geminivirus* by *Bemisia tabaci*. Pesticide Science 42, 123–128.

Marwal, A., Prajapat, R., Sahu, A., Gaur, R.K., 2011. *In silico* recombination analysis: a study for *Geminivirus* host mobility. Asian Journal of Biological Science 5 (1), 1–8.

Marwal, A., Prajapat, R., Sahu, A., Gaur, R.K., 2012a. Current status of *Geminivirus* in India: RNAi technology, a challenging cure. Asian Journal of Biological Science. http://dx.doi.org/10.3923/ajbs.2012.

Marwal, A., Sahu, A., Prajapat, R., Choudhary, D.K., Gaur, R.K., 2012b. First report of association of *Begomovirus* with the leaf curl disease of a common weed, *Datura inoxia*. Indian Journal of Virology 23 (1), pp. 83–84.

Marwal, A., Sahu, A.K., Prajapati, R., Choudhary, D.K., Gaur, R.K., 2012c. RNA silencing suppressor encoded by betasatellite DNA associated with Croton yellow vein mosaic virus. http://dx.doi.org/10.4172/scientificreports.153.

Matthews, R.E.F., 1979. Classification and nomenclature of viruses. Intervirology 12, 129–296.

Medina-Ramos, G., De La Torre-Almaraz, R., Bujanos-Muniz, R., Guevara-Gonzalez, R.G., Tierranegra-Garcia, N., et al., 2008. Co-transmission of *Pepper huastecoyellow vein virus* and *Pepper golden mosaic virus* in chili pepper by *Bemisia tabaci* (Genn.). Journal of Entomology 5, 176–184.

Morales, F.J., Anderson, P., 2001. The emergence and distribution of whitefly-transmitted *Geminiviruses* in Latin America. Archives of Virology 146, 415–441.

Mubin, M., Briddon, R.W., Mansoor, S., 2009. Diverse and recombinant DNA betasatellites are associated with a *Begomovirus* disease complex of *Digera arvensis*, a weed host. Virus Research 142, 208–212.

Mubin, M., Shahid, M.S., Tahir, M.N., Briddon, R.W., Mansoor, S., 2010. Characterization of *Begomovirus* components from a weed suggests that *Begomoviruses* may associate with multiple distinct DNA satellites. Virus Genes 40, 452–457.

Nahid, N., Amin, I., Mansoor, S., Rybicki, E.P., van der Walt, E., Briddon, R.W., 2008. Two dicot-infecting *Mastreviruses* (family *Geminiviridae*) occur in Pakistan. Archives of Virology 153, 1441–1451.

Novina, C.D., Murray, M.F., Dykxhoorn, D.M., Beresford, P.J., Riess, J., et al., 2002. siRNA-directed inhibition of HIV-1 infection. Nature Medicine 8, 681–686.

Ogawa, T., Sharma, P., Ikegami, M., 2008. First report of a strain of *Tobacco leaf curl Japan virus* associated with a satellite DNA in honeysuckle in Japan. Plant Pathology 57, 391.

Padidam, M., Sawyer, S., Fauquet, C.M., 1999. Possible emergence of new geminiviruses by frequent recombination. Virology 265, 218–225.

Park, E., Hwang, H., Lee, S., 1999. Molecular analysis of *Geminivirus* ORFs on symptom development. Plant Pathology Journal 15, 38–43.

Park, J., Hwang, H., Shim, H., Im, K., Auh, C., Lee, S., Davis, K., 2003. Altered cell shapes, hyperplasia, and secondary growth in Arabidopsis caused by Beet curly top *Geminivirus* infection. Molecular Cells 17, 117–124.

Polston, J.E., Anderson, P.K., 1997. The emergence of whitefly-transmitted geminiviruses in tomato in the western hemisphere. Plant Disease 81, 1358–1369.

Power, A.G., 2000. Insect transmission of plant viruses: a constraint on virus variability. Current Opinion in Plant Biology 3, 336–340.

Prajapat, R., Gaur, R.K., Marwal, A., 2011a. Homology modeling and docking studies between AC1 Rep protein of *Begomovirus* and whey α-lactalbumin. Asian Journal of Biological Sciences 4 (4), 352–361, 2011.

Prajapat, R., Marwal, A., Sahu, A., Gaur, R.K., 2011b. Phylogenetics and *in silico* docking studies between coat protein of *Mimosa yellow vein virus* and whey α-lactalbumin. American Journal of Biochemistry and Molecular Biology 1 (3), 265–274.

Prajapat, R., Marwal, A., Bajpai, V., Gaur, R.K., 2011c. Genomics and proteomoics characterization of Alphasatellite in weed associated with *Begomovirus*. International Journal of Plant Pathology 2 (1), 1–14.

Prajapat, R., Marwal, A., Sahu, A.K., Gaur, R.K., 2012. Molecular *in silico* structure and recombination analysis of betasatellite in *Calotropis procera* associated with *Begomovirus*. Archives of Phytopathology and Plant Protection 45 (16), 1980–1990.

Ramos, P.L., Fernández, A., 2002. Macroptilium yellow mosaic virus, a new *Begomovirus* infecting *Macroptilium lathyroides* in Cuba. Plant Disease 86, 1049.

Ritchie, D.W., 2008. Recent progress and future directions in protein-protein docking. Current Protein Peptide Science 9, 1–15.

Ritchie, D.W., Kozakov, D., Vajda, S., 2008. Accelerating protein–protein docking correlations using a six-dimensional analytic FFT generating function. Bioinformatics 24, 1865–1873.

Rougemaille, M., Braun, S., Coyle, S., Dumesic, P.A., Garcia, J.F., Isaac, R.S., Libri, D., Narlikar, G.J., Madhani, H.D., 2012. Ers1 links HP1 to RNAi. www.pnas.org/lookup/suppl/2012.

Roye, M.E., Wernecke, M.E., McLaughlin, W.A., Nakhla, M.K., Maxwell, D.P., 1999. *Tomato dwarf leaf curl virus*, a new bipartite *Geminivirus* associated with tomatoes and peppers in Jamaica and mixed infection with Tomato leaf curl virus. Plant Pathology 48, 370–378.

Roye, M., Collins, A., Maxwell, D.P., 2005. The first report of a *Begomovirus* associated with the common weed *Jatropha gossypifolia* in Jamaica. New Disease Reports 11, 46.

Salati, R., Nahkla, M.K., Rojas, M.R., Guzman, P., Jaquez, J., Maxwell, D., Gilbertson, R.L., 2002. Tomato *yellow leaf curl virus* in the Dominican Republic: characterization of an infectious clone, virus monitoring in whiteflies and identification of reservoir hosts. Pytopathology 92, 487–496.

Saunders, K., Bedford, I.D., Briddon, R.W., Markham, P.G., Wong, S.M., Stanley, J., 2000. A unique virus complex causes Ageratum yellow vein disease. Proceedings of the National Academy of Sciences of the USA 97, 6890–6895.

Saunders, K., Briddon, R.W., Stanley, J., 2008. Replication promiscuity of DNA-β satellites associated with monopartite *Begomoviruses*; deletion mutagenesis of the *Ageratum yellow vein virus* DNA-β satellite localizes sequences involved in replication. Journal of General Virology 89 (12), 3165–3172.

Shibuya, Y., Sakata, J., Sukamto, N., Kon, T., Sharma, P., Ikegami, M., 2007. First report of *Pepper yellow leaf curl Indonesia virus* in *Ageratum conyzoides* in Indonesia. Plant Disease 91, 1198.

Sidhu, J.S., Mann, R.S., Butter, N.S., 2009. Deleterious effects of *Cotton leaf curl virus* on longevity and fecundity of whitefly, *Bemisia tabaci* (Gennadius). Journal of Entomology 6, 62–66.

Stanley, J., Bisaro, D.M., Briddon, R.W., Brown, J.K., Fauquet, C.M., Harrison, B.D., Rybicki, E.P. and Stenger, D.C. 2005. Family *Geminiviridae*. In Virus taxonomy Eighth Report of the International Committee on Taxonomy of Viruses, pp. 301–326. Edited by C. M. Fauquet, M. A. Mayo, J. Maniloff, U. Desselberger, & L. A. Ball. London: Elsevier Academic Press.

Tiendrebeogo, F., Traore, V.S.E., Barro, N., Traore, A.S., Konate, G., Traore, O., 2008. Characterization of *Pepper yellow vein mali virus* in *Capsicum* sp. in Burkina Faso. Plant Pathol. Journal 7, 155–161.

Vanitharani, R., Chellappan, P., Fauquet, C.M., 2003. Short interfering RNA-mediated interference of gene expression and viral DNA accumulation in cultured plant cells. Proceedings of the National Academy of Sciences of the USA 100, 9632–9636.

Varma, A., Malathi, V.G., 2003. Emerging *geminivirus* problems: a serious threat to crop production. Annals of Applied Biology 142, 145–164.

Waterhouse, P.M., Graham, M.W., Wang, M.B., 1998. Virus resistance and gene silencing in plants can be induced by simultaneous expression of sense and antisense RNA. Proceedings of the National Academy of Sciences of the USA 95, 13959–13964.

Waterhouse, P.M., Wang, M.B., Lough, T., 2001. Gene silencing as an adaptive defense against viruses. Nature 411, 834–842.

Wege, C., Gotthardt, R.D., Frischmuth, T., Jeske, H., 2000. Fulfilling Koch's postulates for *Abutilon mosaic virus*. Archives of Virology 145, 2217–2225.

Xie, Q., Sanz-Burgos, A.P., Guo, H., Garcia, J.A., Gutierrez, C., 1999. GRAB proteins, novel members of the NAC domain family, isolated by their interaction with a *Geminivirus* protein. Plant Molecular Biology 39, 647–656.

Yadava, P., Suyal, G., Mukherjee, S.K., 2010. *Begomovirus* DNA replication and pathogenicity. Current Science 98, 360–368.

Yang, C.X., Cui, G.J., Zhang, J., Weng, X.F., Xie, L.H., Wu, Z.J., 2008. Molecular characterization of a distinct *Begomovirus* species isolated from *Emilia sonchifolia*. Journal of Plant Pathology 90, 475–478.

Yang, C.X., Wu, Z.J., Xie, L.H., 2009. First report of the occurrence of *sweet potato leaf curl virus* in tall morning glory (*Ipomoea purpurea*) in China. Plant Disease 93, 764.

Zhou, X., Liu, Y., Calvert, L., Munoz, C., Otim-Nape, G.W., Robinson, D.J., Harrison, B.D., 1997. Evidence that DNA-A of a *Geminivirus* associated with severe cassava mosaic disease in Uganda has arisen by interspecific recombination. Journal of General Virology 78, 2101–2111.

Chapter 8

Tombusvirus-induced multivesicular bodies: origin and role in virus–host interaction

L. Rubino and M. Russo
Istituto di Virologia Vegetale del CNR, UOS Bari, Bari, Italy

G.P. Martelli
Dipartimento di Scienze del Suolo, della Pianta e degli Alimenti, Università di Bari Aldo Moro, Bari, Italy, and Istituto di Virologia Vegetale del CNR, UOS Bari, Bari, Italy

INTRODUCTION

Cytopathologic structures—which were thought to be made up of membranous vesicles and endoplasmic reticulum strands aggregated in an ovoidal pseudo-organellar form—were first observed in thin-sectioned cells of *Datura stramonium* plants infected by *Tomato bushy stunt virus* (TBSV) (Russo & Martelli 1972), the type species of the genus *Tombusvirus* (Martelli & Russo 1995). Seemingly identical structures were shortly afterwards found in great numbers in mesophyll cells of *Chenopodium quinoa* infected by *Artichoke mottled crinkle virus* (AMCV), another tombusvirus (Martelli & Russo 1973). These membranous inclusions were the size of a large mitochondrion or a small plastid and consisted of three major components: (i) a discontinuous, irregularly thickened enveloping membrane with a contour broken by several breaches; (ii) a scanty granular matrix; and (iii) a great number of globose to ovoid membranous vesicles 80 to 150 nm in diameter, many of which contained a network of fine fibrils. These structures were denoted 'multivesicular bodies' (MVBs) (Martelli & Russo 1973). It is worth noting that tombusvirus-induced MVBs differ morphologically, structurally, and functionally from endosomes (i.e., normal constituents of mammalian and plant cells (Piper & Katzmann 2007, Otegui et al 2010)), which are also referred to as MVBs.

It soon became clear that MVBs were a hallmark of tombusvirus infections and a genus-specific feature (Martelli 1981). The problem of their genesis and function, however, persisted and was not solved by a very detailed ultrastructural investigation of *Cymbidium ringspot virus* (CymRSV)-induced MVBs, in which a diagrammatic representation of their structure was proposed (Martelli

& Russo 1981). In the course of these investigations, however, it was noticed that, occasionally, some of the MVBs had crystalline inclusions resembling catalase crystals, thus suggesting that they could have a peroxisomal origin. Cytochemical evidence that the MVB matrix contains catalase and glycolate oxidase confirmed that MVBs were indeed modified peroxisomes (Russo et al 1983). In the same study, it was also ascertained by differential RNase digestion that the fibrils within the MVB vesicles consisted of double-stranded RNA. This made attractive the hypothesis that MVBs were the sites of virus replication (Russo et al 1983).

In 1984, MVBs differing from those observed until that time were found in cells infected by *Carnation Italian ringspot virus* (CIRV), a further member of the genus *Tombusvirus*. Their outward aspect suggested a mitochondrial rather than a peroxisomal origin, a likelihood that proved to be true when these MVBs were shown to contain cytochrome oxidase. As with CymRSV, the vesicles of CIRV-induced MVBs contained dsRNA (Di Franco et al 1984).

Current knowledge indicates that tombusvirus-induced MVBs are invariably associated with tombusvirus infections, regardless of the viral species, the host, or the tissue examined (Russo et al 1987). However, they also occur in plants infected by viruses belonging to other taxonomic groups (e.g., carmoviruses (Russo & Martelli 1982) and closteroviruses (Kim et al 1989, Faoro 1997)), where they appear to have a mitochondrial origin and are likely to afford the same function, as specified further ahead for tombusviruses.

The role of MVBs in tombusvirus replication has been investigated over the years, and the results of these studies are summarized below.

GENERAL FEATURES OF TOMBUSVIRUSES

Tombusviruses are small (ca. 30 nm) isometric plant viruses with a monopartite, single-stranded, positive-sense RNA genome, classified in the genus *Tombusvirus*, family *Tombusviridae* (Rochon et al 2012). This family comprises eight genera, characterized by a genome encompassing a conserved RNA-dependent RNA polymerase lacking helicase and nucleotide-triphosphate binding motifs but containing the eight conserved motifs of RNA-dependent RNA polymerases of positive-stranded RNA [(+)RNA] viruses (Koonin 1991). A further common trait of the *Tombusviridae* is that the RNA polymerase is localized in the 5'-proximal open reading frame (ORF) and is expressed by ribosomal frameshifting or readthrough of a stop codon interrupting the frame at about one third of its length. Two RNA polymerase-related proteins are therefore synthesized, both essential for virus replication (Dalmay et al 1993). Because the *Tombusviridae* genome lacks the 5' cap and the 3' poly(A) tail, circularization is obtained by RNA–RNA interaction to promote translation. Internal genes are expressed via subgenomic RNAs generated by internal initiation or premature termination during minus strand synthesis (Rochon et al 2012).

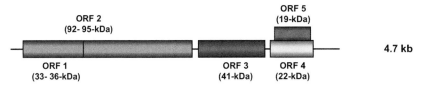

FIGURE 8.1   Diagrammatic representation of a tombusvirus RNA genome. Coding regions (ORFs) are shown as boxes and the sizes of the corresponding products are indicated. The 5' and 3' nontranslated regions and intergenic regions are shown by lines.

As shown in Figure 8.1, the tombusvirus genome consists of five ORFs, the first of which (ORF1) encodes a 33–36 kDa protein (p33/p36), whereas ORF2 encodes the 92–95 kDa polymerase (p92/p95) expressed by readthrough of the amber termination codon of ORF1. ORF3 encodes the 41 kDa coat protein (CP) and the two nested ORFs 4 and 5 encode the 22 kDa movement protein (p22) and the 19 kDa (p19) suppressor of RNA silencing, respectively (Russo et al 1994, White & Nagy 2004).

Defective interfering (DI) RNAs are often found in tombusvirus-infected cells. These are shortened forms of genomic RNA lacking all genes for replication, encapsidation, and movement but still able to be replicated *in trans* by the full-length genomic RNA (Russo et al 1994, White & Nagy 2004) or by mutants expressing only the replicase genes (Rubino et al 2004). DI RNAs replicate in plant cells (Kollar & Burgyan 1994, Rubino & Russo 1995) and yeast cells (Panavas & Nagy 2003, Pantaleo et al 2003) that express the viral polymerase only.

The replication strategy of tombusviruses conforms to that of (+)RNA viruses, a hallmark of which is the recruitment of cell membranes for the formation of replication complexes in which template RNA, viral, and host proteins gather, thus increasing replication efficiency, and are protected from the host defense response (Novoa et al 2005, Salonen et al 2005, den Boon & Ahlquist 2010). The assembly of the virus replication complex on intracellular membranes requires a sequence of coordinated steps in which viral replicase, viral RNA template, and host factors are involved. In the case of TBSV, the replicase proteins p33 and p92 are synthesized in the cytoplasm at an approximate ratio of 20:1, then p33 dimers and p33:p92 oligomers are formed through specific interacting domains (Rajendran & Nagy 2004, Panavas et al 2005). The interaction between a specific RNA-binding domain on p33/p92 (Rajendran & Nagy 2003) and a recognition element on the RNA (Pogany et al 2005) allows viral RNA binding. The complex is then targeted to the peroxisomal membrane where the complementary (−)RNA is first transcribed and used for several rounds of (+) RNA synthesis. When viral RNA synthesis comes to an end, the replication complex is disassembled (Nagy & Pogany 2006).

Tombusviruses replicate either in vesicles formed by the rearrangement of the single lining membrane of peroxisomes or in vesicles derived from the outer membrane of mitochondria (Russo et al 1987). In all cases, the localization

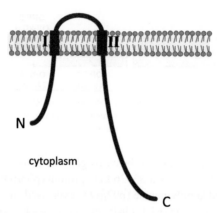

N

cytoplasm

C

FIGURE 8.2 Predicted model for the insertion of CymRSV p33 or CIRV p36 in the peroxisomal or mitochondrial outer membrane. The positions of transmembrane domains TMD 1 (I) and TMD 2 (II) are indicated. The N- and C-termini of the proteins are localized in the cytoplasmic portion of the MVB vesicles.

of the tombusviral replication complex is mediated exclusively by ORF1- and ORF2-encoded proteins, in particular by determinants located in a sequence of ca. 600 nucleotides (nts) at the 5' end of ORF1. This was conclusively demonstrated by exchanging different parts of the genome in full-length infectious clones of CymRSV (forming MVBs from peroxisomes) and CIRV (forming MVBs from mitochondria) (Burgyan et al 1996). The 5'-proximal 600 nts correspond to a sequence of ca. 200 amino acids (aa) containing two hydrophobic transmembrane domains (TMD 1 and TMD 2) separated by a hydrophilic loop. A model was predicted for the insertion of tombusvirus p33/p92 or p36/p95 in host membranes, according to which anchoring to peroxisomes or mitochondia is obtained by integration of TMD 1 and TMD 2 into the organelle's membrane, leaving the short connecting loop in the peroxisomal matrix or in the mitochondrial intermembrane space, while the N- and C-termini are localized in the interior of the vesicles that open to the cytoplasm (Rubino & Russo 1998) (Fig. 8.2). The viral replicase is stably anchored to cell membranes as shown by its resistance to extraction with carbonate, urea, or high salt (Rubino & Russo 1998, Rubino et al 2000, Navarro et al 2004).

PEROXISOME-DERIVED MVBs IN PLANT CELLS

Peroxisome-derived MVBs have been detected in plant cells infected with several tombusviruses (i.e. TBSV, AMCV, CymRSV, *Eggplant mottled crinkle virus* (EMCV), *Moroccan pepper virus* (MPV), *Neckar river virus* (NRV), *Petunia asteroid mosaic virus* (PAMV), *Pelargonium leafcurl virus* (PLCV) (Russo et al 1987), *Cucumber Bulgarian latent virus* (CBLV) (Kostova et al 2003), and *Cucumber necrosis virus* (CNV) (M. Russo, unpublished information)).

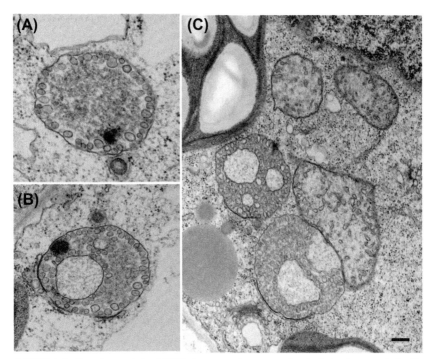

FIGURE 8.3   Peroxisome-derived MVBs in progressive stages of vesiculation in mesophyll cells of *N. benthamiana* infected by CymRSV. Bar: 0.2 μm.

Figure 8.3 shows the typical structure of peroxisomal MVBs in CymRSV-infected cells. Since the organelle's limiting single membrane undergoes progressive vesiculation, in the earlier stages of MVB formation only a few vesicles are visible (Fig. 8.3A). These appear as invaginations of the peroxisomal membrane, 80 to 150 nm in diameter, that contain a network of fine fibrils and open to the cytoplasm through a tiny neck. At later stages, the whole peroxisomal membrane is affected by vesiculation (Fig. 8.3B) and, in further advanced stages, the membrane proliferates extensively. Many more vesicles are thus formed, resulting in the production of membranous appendages that fold back on the peroxisomal body (Fig. 8.3C). In the early stages of alteration, peroxisomes may still be functional, as suggested by the presence of catalase and glycolate oxidase activity (Russo et al 1983). An interesting feature observed in peroxisomal MVBs is the topological association of endoplasmic reticulum (ER) strands with the peroxisomal membrane, which may be so closely appressed to look as if the two membranes were fused (Russo & Martelli 1972).

As mentioned, the fibrillar network of MVB vesicles is made up of double-stranded RNA molecules, interpreted as replicative forms of the viral genome (Russo et al 1983, Appiano et al 1986). The likelihood that these structures are viral replication sites was confirmed by the immunological detection of the replicase proteins p33/p92 within CymRSV-induced MVBs (Lupo et al 1994, Bleve-Zacheo et al 1997, Rubino & Russo 1998).

The relationship between the ER and peroxisomes was studied in detail by McCartney et al (2005), who, by expressing TBSV-encoded p33 in tobacco BY2 cells, identified three specific peroxisomal targeting signals (-$K_{11}K_{12}$-, -$K_{76}R_{77}R_{78}R_{80}$-, and –$R_{124}K_{129}K_{130}$-) and a single ER targeting signal (-K_5K_6-). These authors suggested that p33 is first targeted to the peroxisomal membrane, in which it is inserted via the transmembrane domains TMD 1 and TMD 2; then, along with p92 and template RNA, it forms the replication complex inside the vesicles. Part of the p33 molecules could mediate budding of the peroxisomal membrane and sorting to the ER. Vesicle formation from the ER elicited by p33 would produce additional membrane constituents, whose tranfer to peroxisomes contributes to MVB biogenesis and increases the membrane surface available to the replication complex.

Electron microscope observations of transgenic (Bleve-Zacheo et al 1997) or agroinfiltrated (L Rubino & M Russo, unpublished information) N. benthamiana plants expressing CymRSV p33/p92 showed membrane proliferation and an increased number of aggregated and misshapen peroxisomes. However, no typical MVBs were detected in either transgenic or transiently transformed cells, indicating that the simple expression of p33/p92 is not sufficient to produce MVBs in cells where no active virus replication takes place.

MITOCHONDRION-DERIVED MVBs IN PLANT CELLS

MVBs found in CIRV-infected cells originate from the proliferating activity of the outer mitochondrial envelope (Di Franco et al 1984). In the early stages of infection, just a few vesicles clearly opening to the cytoplasm through a neck are visible in the space between the outer and inner enveloping membrane (Fig. 8.4A). These vesicles have the same aspect and size (80–150 nm) of those of peroxisomal MVBs and contain dsRNAs. In later stages of infection the outer mitochondrial membrane becomes progressively more vesiculated (Fig. 8.4B) and finally completely disorganized (Fig. 8.4C).

As mentioned, the analysis of infectious hybrid CymRSV/CIRV clones showed that the determinants for the localization of the CIRV replication complex reside in a stretch of ca. 200 aa in the N-terminal region of the p36 protein encoded by the viral ORF1 (Burgyan et al 1996). Similarly to protein p33 of tombusviruses replicating in peroxisomal MVBs, CIRV p36 contains two hydrophobic transmembrane domains (TMD 1 and TMD 2) separated by a hydrophilic loop, which are thought to be part of a signal for targeting and anchoring this protein to the outer membrane of the mitochondrial envelope (Rubino & Russo 1998). To prove this point two different approaches were used: (i) MVBs were recovered from CIRV-infected leaf tissues with a cell fractionation procedure and found to contain p36 in protein extracts by Western blot analysis (Rubino et al 2001); (ii) p36 fused to the green fluorescent protein (GFP) was

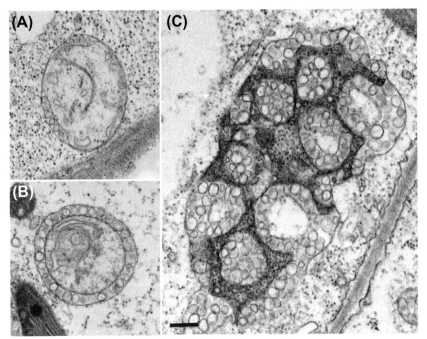

FIGURE 8.4   Mitochondria-derived MVBs in progressive stages of vesiculation in mesophyll cells of *N. benthamiana* infected by CIRV. A single highly vesiculated mitochondrion is shown in (C). Bar: 0.2 μm.

transiently expressed in *N. benthamiana* protoplasts and BY2 cells transfected with plasmids, or in *Agrobacterium*-infiltrated *N. benthamiana* leaves, and its mitochondrial localization was ascertained by fluorescence microscopy (Rubino et al 2001).

Electron microscopy of agroinfiltrated *N. benthamiana* leaves express-ing either p36 (Rubino et al 2001) or p36/p95 (L Rubino & M Russo, unpub-lished information) showed that mitochondria were misshapen and clumped, had fewer, larger, and irregularly shaped cristae, and an enlarged intermembrane space, suggestive of outer membrane overgrowth. The cytoplasmic side of the outer membrane was covered with electron-dense material, which was tenta-tively identified as accumulated p36. As for CymRSV, MVBs were not observed.

These observations pointed to the putative presence of a mitochondrial tar-geting signal (MTS; von Heijne 1986) in p36 that elicits overgrowth rather than vesiculation of the mitochondrial outer membrane. Initially, the p36 addressing pathway to mitochondria was hypothesized to rely upon a signal located in the N-terminal hydrophilic sequence of this protein (Rubino & Russo 1998). How-ever, further mutational analysis showed that the sequence (aa 1 to 98), contain-ing the putative MTS (aa 32 to 45), does not contribute to sorting and insertion of p36 in the mitochondrial membrane (Weber-Lotfi et al 2002). Furthermore, there is no apparent intervention of the surface-exposed protein import receptors (Weber-Lotfi et al 2002) typical of matrix-imported mitochondrial proteins (Herrmann & Neupert 2000). Since no MTS is present downstream of the second transmembrane domain, a 'stop-transfer' pathway (Nguyen et al 1988) can also be excluded. Experimental evidence proved more plausible the intervention of a 'signal-anchor mechanism' (Waizenegger et al 2003), constituted by the two hydrophobic domains, their flanking regions and the positively charged face of the amphipathic helix within the intervening loop sequence (Weber-Lotfi et al 2002, Hwang et al 2008), that interacted with the outer membrane translocase (TOM) complex (Hwang et al 2008).

Pelargonium necrotic spot virus (PNSV), a species in the *Tombusvirus* genus unrelated to CIRV, is the second member of the genus eliciting mitochondrial MVBs. PNSV ORF1 encodes a product of 36 kDa, with a high identity at the nt and aa level with CIRV p36. In PNSV-infected cells, mitochondria are swollen and form vesicular structures similar to CIRV-induced MVBs (Heinze et al 2004).

Interestingly, five viral isolates from plants or water samples identified as CIRV strains because of the serologic relationship with this virus had the ORF1-encoded protein with a size (33 kDa) and a composition similar to that of tombusviruses replicating in peroxisomal MVBs (Koenig et al 2009). In accordance, MVBs induced by all of these viral isolates were of peroxisomal origin, thus strengthening the notion of the specificity of tombusviral p33 in determining the type of organelle transformed into an MVB in infected cells. Incidentally, it was hypothesized that CIRV was generated by recombination events at the level of ORF1 between PNSV and one of these five strains (Koenig et al 2009).

TOMBUSVIRUS REPLICATION COMPLEX IN YEAST CELLS

Following the pioneering work of the P Ahlquist group (Janda & Ahlquist 1993, Ishikawa et al 1997), which used the yeast *Saccharomyces cerevisiae* as an alternative host for studying the replication of *Brome mosaic virus* (BMV), a similar system was developed for investigating the replication of tombusviruses (Rubino et al 2000, Panavas & Nagy 2003, Pantaleo et al 2003). In particular, proteins p33/p92 or p36/p95 and DI RNA were expressed in yeast cells transformed with appropriate plasmids. Immunofluorescence and electron microscopy showed that the peroxisomal and mitochondrial targeting and anchoring of tombusvirus replicase proteins was maintained in *S. cerevisiae* (Rubino et al 2000, Weber-Lotfi et al 2002, Navarro et al 2004, 2006). In particular, yeast cells expressing CIRV p36 or CymRSV p33 contained accumulations of misshapen mitochondria or peroxisomes, respectively, and proliferated cytoplasmic membranes surrounding the clumps of aggregated organelles (Rubino et al 2000, Navarro et al 2004, 2006).

Replication of CNV or CymRSV (Panavas & Nagy 2003, Navarro et al 2006) and of CIRV in yeasts (Pantaleo et al 2003) involves peroxisomes or

mitochondria, respectively, as in plant cells. However, cytopathic structures differing from typical MVBs were observed in yeast cells supporting DI RNA
replication. For instance, yeast cells in which DI RNA was replicating in the
presence of CymRSV polymerase contained massive aggregates of peroxisomes and accumulations of membranous elements, as in the cells expressing
the virus replicase alone (Navarro et al 2004, 2006) (Fig. 8.5).

Using the yeast system, the requirement of (+)RNA viruses for the assembly
of their replication complex was shown to be not restricted to a specific type

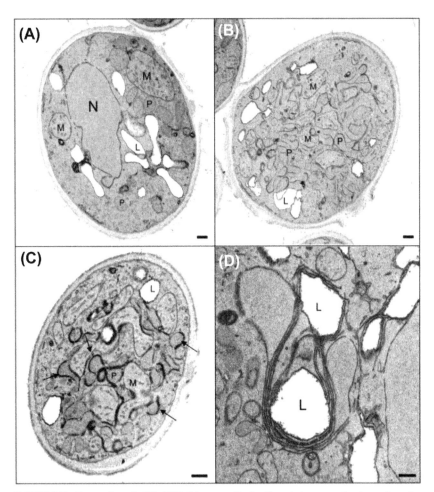

FIGURE 8.5   Expression of p33 in UTL-7A yeast cells significantly increases the size and number
of peroxisomes. (A) A cell transformed with empty vectors. (B–D) Cells expressing p33 together
with p92 and DI RNA transcripts displaying clusters of peroxisomes (B, C), many of which are partially encircled by electron-dense tubular structures (C, arrows), and accumulations of membranes
(D) free in the cytoplasm or encircling lipid droplets. P, peroxisomes; M, mitochondria; N, nucleus;
L, lipid droplets. Bar, 0.25 μm. *Reproduced with permission from Navarro et al 2006.*

of cell membrane. In fact, in a yeast strain defective for peroxisome biogenesis, tombusvirus replicase proteins and DI RNA template were redirected to ER membranes, where successful DI RNA replication occurred (Jonczyk et al 2007, Rubino et al 2007). ER membranes were induced to proliferate but were not rearranged in the form of MVBs (Rubino et al 2007).

The identification and characterization of host factors involved in virus replication is a major but still little explored field. The use of yeasts as model hosts for studying tombusvirus and other RNA virus replication may help to secure information in this specific area (Nagy & Pogany 2006, Nagy 2008). Several yeast mutant libraries are available that have already allowed the identification of more than 100 host genes involved in tombusvirus replication (Nagy et al 2012, Nawaz-ul-Rehman et al 2012). For instance, the chaperone HSP70 and the ESCRT proteins were shown to be needed for the correct assembly of the tombusvirus replication complex in cell membranes (peroxisomes or ER) (Serva & Nagy 2006, Barajas et al 2009, Wang et al 2009), whereas the Pex19p peroxisomal transport protein is involved in the specific recruitment of p33 to peroxisomal membranes (Jonczyk et al 2007, Rubino et al 2007). Finally, genes involved in the biosynthesis of phospholipids, sterols, and fatty acids directly affect tombusvirus replication by interfering with cell membrane synthesis (Sharma et al 2010, 2011).

REFERENCES

Appiano, A., D'Agostino, G., Bassi, M., Barbieri, N., Viale, G., Dell'Orto, P., 1986. Origin and function of *Tomato bushy stunt virus*-induced inclusion bodies. II. Quantitative electron microscope autoradiography and immunogold cytochemistry. Journal of Ultrastructure Research 97, 31–38.

Barajas, D., Jiang, Y., Nagy, P.D., 2009. A unique role for the host ESCRT proteins in replication of *Tomato bushy stunt virus*. PLoS Pathogens 5, e1000705.

Bleve-Zacheo, T., Rubino, L., Melillo, M.T., Russo, M., 1997. The 33K protein encoded by *Cymbidium ringspot virus* localizes to peroxisomes of infected cells and of uninfected transgenic plants. Journal of Plant Pathology 79, 197–202.

Burgyan, J., Rubino, L., Russo, M., 1996. The 5'-terminal region of a tombusvirus genome determines the origin of multivesicular bodies. Journal of General Virology 77, 1967–1974.

Dalmay, T., Rubino, L., Burgyan, J., Kollar, A., Russo, M., 1993. Functional analysis of *Cymbidium ringspot virus* genome. Virology 194, 697–704.

den Boon, J.A., Ahlquist, P., 2010. Organelle-like membrane compartmentalization of positive-strand virus replication factories. Annual Review of Microbiology 64, 241–256.

Di Franco, A., Russo, M., Martelli, G.P., 1984. Ultrastructure and origin of multivesicular bodies induced by *Carnation Italian ringspot virus*. Journal of General Virology 65, 1233–1237.

Faoro, F., 1997. Cytopathology of closteroviruses and trichoviruses infecting grapevines. In: Monette, P.L. (Ed.), Filamentous viruses of woody plants, Research Signpost, Trivandrum, pp. 29–47.

Heinze, C., Wobbe, V., Lesemann, D.E., Zhang, D.Y., Willingmann, P., Adam, G., 2004. Pelargonium necrotic spot virus: a new member of a genus *Tombusvirus*. Archives of Virology 149, 1527–1539.

Herrmann, J.M., Neupert, W., 2000. Protein transport into mitochondria. Current Opinion in Microbiology 3, 210–214.

Hwang, Y.T., McCartney, A.W., Gidda, S.K., Mullen, R.T., 2008. Localization of the *Carnation Italian ringspot virus* replication protein p36 to the mitochondrial outer membrane is mediated by an internal targeting signal and the TOM complex. BMC Cell Biology 9, 54–80.

Ishikawa, M., Janda, M., Kroll, M.A., Ahlquist, P., 1997. *In vivo* DNA expression of functional *Brome mosaic virus* RNA replicons in *Saccharomyces cerevisiae*. Journal of Virology 71, 7781–7790.

Janda, M., Ahlquist, P., 1993. RNA-dependent replication, transcription, and persistence of *Brome mosaic virus* RNA replicons in *S. cerevisiae*. Cell 72, 961–970.

Jonczyk, M., Pathak, K.B., Sharma, M., Nagy, P.D., 2007. Exploiting alternative subcellular localization for replication: tombusvirus replication switches to the endoplasmic reticulum in the absence of peroxisomes. Virology 362, 320–330.

Kim, K.S., Gonsalves, D., Teliz, D., Lee, K.W., 1989. Ultrastructure and mitochondrial vesiculation associated with closteroviruslike particles in leafroll-diseased grapevines. Phytopathology 79, 357–360.

Koenig, R., Lesemann, D.E., Pfeilstetter, E., 2009. New isolates of *Carnation Italian ringspot virus* differ from the original one by having replication-associated proteins with a typical tombusvirus-like N-terminus and by inducing peroxisome – rather than mitochondrion-derived multivesicular bodies. Archives of Virology 154, 1695–1698.

Kollar, A., Burgyan, J., 1994. Evidence that ORF 1 and 2 are the only virus-encoded replicase genes of *Cymbidium ringspot tombusvirus*. Virology 201, 169–172.

Koonin, E.V., 1991. The phylogeny of RNA-dependent RNA polymerases of positive-strand viruses. Journal of General Virology 72, 2197–2206.

Kostova, D., Lisa, V., Rubino, L., Marzachì, C., Roggero, P., Russo, M., 2003. Properties of *Cucumber Bulgarian latent virus*, a new species in the genus *Tombusvirus*. Journal of Plant Pathology 85, 27–33.

Lupo, R., Rubino, L., Russo, M., 1994. Immunodetection of the 33K/92K polymerase proteins in *Cymbidium ringspot virus*-infected and in transgenic plant tissue extracts. Archives of Virology 138, 135–142.

Martelli, G.P., 1981. Tombusviruses. In: Kurstak, E. (Ed.), Handbook of plant infection and comparative diagnosis, Elsevier/North Holland Biomedical Press, Amsterdam, pp. 61–90.

Martelli, G.P., Russo, M., 1973. Electron microscopy of *Artichoke mottled crinkle virus* in leaves of *Chenopodium quinoa* Willd. Journal of Ultrastructure Research 42, 93–107.

Martelli, G.P., Russo, M., 1981. The fine structure of *Cymbidium ringspot virus* in host tissues. I. Electron microscopy of systemic infections. Journal of Ultrastructure Research 77, 93–104.

Martelli, G.P., Russo, M., 1995. Family *Tombusviridae*. In: Murphy, F.A. (Ed.), Virus taxonomy. Sixth report of the International Committee on Taxonomy of Viruses, Springer Verlag, Vienna, pp. 392–397.

McCartney, A.W., Greenwood, J.S., Fabian, M.R., White, K.A., Mullen, R.T., 2005. Localization of the tomato bushy stunt virus replication protein p33 reveals a peroxisome-to-endoplasmic reticulum sorting pathway. Plant Cell 17, 3513–3531.

Nagy, P.D., 2008. Yeast as a model host to explore plant virus–host interactions. Annual Review of Phytopathology 46, 217–242.

Nagy, P.D., Pogany, J., 2006. Yeast as a model host to dissect functions of viral and host factors in tombusvirus replication. Virology 344, 211–220.

Nagy, P.D., Barajas, D., Pogany, J., 2012. Host factors with regulatory roles in tombusvirus replication. Current Opinion in Virology 2, 691–698.

Navarro, B., Rubino, L., Russo, M., 2004. Expression of *Cymbidium ringspot virus* 33-kilodalton protein in *Saccharomyces cerevisiae* and molecular dissection of the peroxisomal targeting signal. Journal of Virology 78, 4744–4752.

Navarro, B., Russo, M., Pantaleo, V., Rubino, L., 2006. Cytological analysis of *Saccharomyces cerevisiae* supporting *Cymbidium ringspot virus* defective interfering RNA replication. Journal of General Virology 87, 705–714.

Nawaz-ul-Rehman, M.S., Prasanth, K.R., Baker, J., Nagy, P.D., 2012. Yeast screens for host factors in positive-strand RNA virus replication based on a library of temperature-sensitive mutants. Methods. http:/dx.doi.org/10.1016/j.ymeth.2012.11.001.

Nguyen, M., Bell, A.W., Shore, G.C., 1988. Protein sorting between mitochondrial membranes specified by position of the stop-transfer domain. Journal of Cell Biology 106, 1499–1505.

Novoa, R.R., Calderita, G., Arranz, R., Fontana, J., Granzow, H., Risco, C., 2005. Virus factories: associations of cell organelles for viral replication and morphogenesis. Biology of the Cell 97, 147–172.

Otegui, M., Reyes, F.C., 2010. Endosomes in plants. Nature Education 3 (9), 23–28.

Panavas, T., Nagy, P.D., 2003. Yeast as a model host to study replication and recombination of defective interfering RNA of *Tomato bushy stunt virus*. Virology 314, 315–325.

Panavas, T., Serviene, E., Brasher, J., Nagy, P.D., 2005. Yeast genome-wide screen reveals dissimilar sets of host genes affecting replication of RNA viruses. Proceedings of the National Academy of Sciences of the USA 102, 7326–7331.

Pantaleo, V., Rubino, L., Russo, M., 2003. Replication of *Carnation Italian ringspot virus* defective interfering RNA in *Saccharomyces cerevisiae*. Journal of Virology 77, 2116–2123.

Piper, R.C., Katzmann, D.J., 2007. Biogenesis and function of multivesicular bodies. Annual Review of Cell and Developmental Biology 23, 519–547.

Pogany, J., White, K.A., Nagy, P.D., 2005. Specific binding of tombusvirus replication protein p33 to an internal replication elemement in the viral RNA is essential for replication. Journal of Virology 79, 4859–4869.

Rajendran, K.S., Nagy, P.D., 2003. Characterization of the RNA-binding domains in the replicase proteins of *Tomato bushy stunt virus*. Journal of Virology 77, 9244–9258.

Rajendran, K.S., Nagy, P.D., 2004. Interaction between the replicase proteins of *Tomato bushy stunt virus in vitro* and *in vivo*. Virology 326, 250–261.

Rochon, D., Lommel, S., Martelli, G.P., Rubino, L., Russo, M., 2012. *Tombusviridae*. In: King, A.M.Q. (Ed.), Virus taxonomy, Elsevier, Oxford, pp. 1111–1138.

Rubino, L., Russo, M., 1995. Characterization of resistance to *Cymbidium ringspot virus* in transgenic plants expressing a full-length viral replicase gene. Virology 212, 240–243.

Rubino, L., Russo, M., 1998. Membrane targeting sequences in tombusvirus infections. Virology 252, 431–437.

Rubino, L., Di Franco, A., Russo, M., 2000. Expression of a plant virus non-structural protein in *Saccharomyces cerevisiae* causes membrane proliferation and altered mitochondrial morphology. Journal of General Virology 81, 278–286.

Rubino, L., Weber-Lotfi, F., Dietrich, A., Stussi-Garaud, C., Russo, M., 2001. The open reading frame 1-encoded ('36K') protein of *Carnation Italian ringspot virus* localizes to mitochondria. Journal of General Virology 82, 29–34.

Rubino, L., Pantaleo, V., Navarro, B., Russo, M., 2004. Expression of tombusvirus open reading frames 1 and 2 is sufficient for the replication of defective interfering, but not satellite, RNA. Journal of General Virology 85, 3115–3122.

Rubino, L., Navarro, B., Russo, M., 2007. *Cymbidium ringspot virus* defective interfering RNA replication in yeast cells occurs on endoplasmic reticulum-derived membranes in the absence of peroxisomes. Journal of General Virology 88, 1634–1642.

Russo, M., Martelli, G.P., 1972. Ultrastructural observations on *Tomato bushy stunt virus* in plant cells. Virology 49, 122–129.

Russo, M., Martelli, G.P., 1982. Ultrastructure of *Turnip crinkle-* and *Saguaro cactus virus*-infected tissues. Virology 118, 109–116.

Russo, M., Di Franco, A., Martelli, G.P., 1983. The fine structure of *Cymbidium ringspot virus* infections in host tissues. III. Role of peroxisomes in the genesis of multivesicular bodies. Journal of Ultrastructure Research 82, 52–63.

Russo, M., Di Franco, A., Martelli, G.P., 1987. Cytopathology in the identification and classification of tombusviruses. Intervirology 28, 134–143.

Russo, M., Burgyan, J., Martelli, G.P., 1994. Molecular biology of *Tombusviridae*. Advances in Virus Research 44, 381–428.

Salonen, A., Ahola, T., Kaarlainen, L., 2005. Viral replication in association with cellular membranes. Current Topics in Microbiology and Immunology 285, 139–173.

Serva, S., Nagy, P.D., 2006. Proteomics analysis of the tombusvirus replicase: Hsp70 molecular chaperone is associated with the replicase and enhances viral RNA replication. Journal of Virology 80, 2162–2169.

Sharma, M., Sasvari, Z., Nagy, P.D., 2010. Inhibition of sterol biosynthesis reduces tombusvirus replication in yeast and plants. Journal of Virology 84, 2270–2281.

Sharma, M., Sasvari, Z., Nagy, P.D., 2011. Inhibition of phospholipid biosynthesis decreases the activity of tombusvirus replicase and alters the subcellular localization of replication proteins. Virology 415, 141–152.

von Heijne, G., 1986. Mitochondrial targeting sequences may form amphiphilic helices. EMBO Journal 7, 1335–1342.

Waizenegger, T., Stan, T., Neupert, W., Rapaport, D., 2003. Signal-anchor domains of proteins of the outer membrane of mitochondria. Journal of Biological Chemistry 278, 42064–42071.

Wang, R.Y., Stork, J., Nagy, P.D., 2009. A key role for heat shock protein 70 in the localization and insertion of tombusvirus replication proteins to intracellular membranes. Journal of Virology 83, 3276–3287.

Weber-Lotfi, F., Dietrich, A., Russo, M., Rubino, L., 2002. Mitochondrial targeting and membrane anchoring of a viral replicase in plant and yeast cells. Journal of Virology 76, 10485–10496.

White, K.A., Nagy, P.D., 2004. Advances in the molecular biology of tombusviruses: gene expression, replication and recombination. Progress in Nucleic Acid Research & Molecular Biology 78, 188–226.

Papaya ringspot virus-P: overcoming limitations of resistance breeding in Carica papaya L.

Sunil Kumar Sharma and Savarni Tripathi

Indian Agricultural Research Institute, Regional Station, Agricultural College Estate, Shivajinagar, Pune, India

INTRODUCTION

Papaya (*Carica papaya* L.) is one of the major tropical fruit crops, having a production of more than eleven million metric tons per annum. It is cultivated in all five continents, but the major share of its total production of 11.57 million metric tons (m MT) from 433,057 hectares in 2010 came from Asia, Central America, and Africa. India, Brazil, Nigeria, Indonesia, and Mexico are among the major papaya-producing countries (FAO 2012). Commercial cultivation of papaya is unable to achieve its full potential due to the widespread incidence of viral diseases. Among various viral diseases affecting papaya cultivation, *Papaya ringspot virus* strain papaya (PRSV-P) is the most devastating one in all major papaya-growing areas of the world (Fig. 9.1). The natural spread of PRSV-P is rapid; therefore, the virus may infect up to 100% of plants in a given area. The disease is so devastating that farmers have stopped growing papaya in severely affected areas. The use of transgenic cultivars, a successful strategy for managing the virus in Hawaii, has not, so far, been scaled up in other papaya-cultivating areas because of some technical reasons (virus sequence homology-dependent resistance) and environmental activism. Other approaches to managing PRSV-P have had only limited success. Therefore, the approach of introgression of virus resistance in papaya from highland papaya (species of *Vasconcellea*) by conventional breeding has become the only viable option in the present scenario. This chapter describes earlier research efforts, their progression, and the latest status of PRSV-P resistance breeding at the leading centers.

FIGURE 9.1   Healthy (A) and virus-infected (B) papaya plantations.

PAPAYA RINGSPOT VIRUS

The disease caused by *Papaya ringspot virus* was first described by Lindner et al (1945) in Hawaii (US) and was shown to be viral in nature by Jensen (1949). The name of the disease, ringspot, is taken from the occurrence of ring spots on the fruit of infected plants. Other symptoms produced by the infections are mosaic and chlorosis of the leaf lamina, water-soaked oily streaks on the petiole and upper part of the trunk, and a distortion of young leaves that sometimes results in shoe-string-like symptoms that resemble mite damage; infected plants lose vigor and become stunted (Fig. 9.2). Fruits from infected trees are of poor quality and generally have lower sugar concentrations. Plants subjected to early infection (before flowering) with the severe strain of the virus usually do not produce marketable fruits. The virus spreads mainly in the field via several species of aphid vector in a nonpersistent manner. It can be transmitted mechanically, and by grafting, but not by nematodes and seeds. Although papaya is the most important primary and secondary source for the spread of the virus, PRSV-P also infects plants of the *Cucurbitacae*. PRSV, a member of the genus *Potyvirus*, is further classified into two types: type P (PRSV-P), which infects cucurbits and papaya, and type W (PRSV-W), which infects cucurbits but not papaya (Purcifull et al 1984, Tripathi et al 2008, Gonsalves et al 2010). The biotypes P and W are serologically indistinguishable. The virions are non-enveloped, flexuous, and filamentous in shape; they measure $760\text{–}800 \times 12$ nm (Gonsalves & Ishii 1980). Virus particles contain 94.5% protein and 5.5% nucleic acid. The protein component consists of the virus coat protein (CP), which has a molecular weight (M_r) of 36,000 to 36,500. The density of the sedimenting component in purified PRSV-P preparations is 1.32 g cm^{-3} in CsCl. The genomic RNA of the virus consists of 10,326 nucleotides and has the typical array of genes found in potyviruses (Shukla et al 1994). The genome of PRSV-P consists of ssRNA with positive polarity and has the typical array of genes present in potyviruses (Yeh & Gonsalves 1985, Yeh et al 1992, Shukla et al 1994). The genome is monocistronic and is expressed via a large polypeptide of 381 kDa that is subsequently cleaved by the virus-encoded proteinases to yield functional proteins. Like other potyviruses, the functional proteins are produced—as proposed for PRSV-P and other potyviruses—by a

FIGURE 9.2   Symptoms of PRSV-P infection on papaya plant: (A) mosaic; (B) leaf reduction; (C) leaf deformation and blister formation; (D) shoe-string formation; (E) oily spots and streaks on petiole and stem; (F) reduction in fruit production; (G) ring spots on fruits.

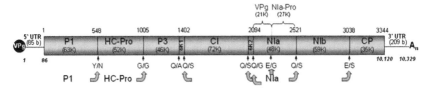

FIGURE 9.3   The PRSV genome. Numbers at the bottom indicate nucleic acid position. Numbers at the top indicate the amino acid position in the polyprotein. The molecular weights (M_r) of the individual viral proteins are shown below the viral protein name in parentheses. The 5′ and 3′ untranslated regions (UTRs) are marked along with their lengths (in bases, b) in parentheses. Amino acids flanking the cleavage sites of proteases P1, HC-Pro, and NIa (shown with block arrows) are indicated below black triangles marking the relative cleavage site position. The black circle labeled VPg represents the genome-linked protein (Tripathi et al 2008, p 270).

combination of cotranslational, post-translational, autoproteolytic, and transproteolytic processing by the three virus-encoded endoproteases P1, HC-Pro, and NIa (Yeh & Gonsalves 1985, Yeh et al 1992). The genetic organization of PRSV RNA is VPg-5′ leader –P1 (63K)-HC Pro-P3 (46K)-CI-P5 (6K)-NIa-NIb-CP-3′ noncoding region_poly(A) tract (Yeh et al 1992; Fig. 9.3). Phylogenetic studies showed that, within PRSV, the coat-protein gene sequences can diverge by as much as 14% at the nucleotide level and by 10% at the amino acid level (Jain et al 2004). The complete nucleotide sequence of the PRSV genome has been reported from several geographic isolates. However, the coat protein sequence of numerous strains has been analyzed by various laboratories (Tripathi et al 2006).

PRESENT STRATEGIES FOR PRSV-P MANAGEMENT

Because there is no prophylactic or therapeutic control measure for PRSV-P infection, the major emphasis is on minimizing yield losses by management of

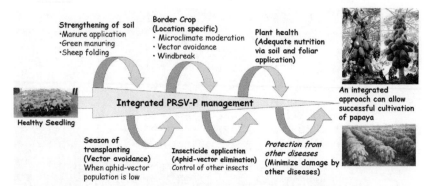

FIGURE 9.4 Factors contributing towards successful management of PRSV-P in papaya.

the disease. The current approach of disease management is to avoid infection at an early stage of plant growth, for yield losses are associated with the age of the plant at the time of infection and with the severity of infection. The disease is mainly transmitted from infected plants to healthy plants via aphid vectors; avoiding and/or reducing the population size of the vectors is the main strategy used for management. It has been established that simply adjusting one or the other factor of cultivation cannot in itself prevent PRSV-P infection or effectively reduce its further spread. However, adopting a strategic integrated management of cultural practices can help (e.g., using healthy (virus-free) seedlings of an appropriate cultivar(s), selecting the season of transplanting when the vector population is naturally low, planting a border crop around the papaya plantation, systematic roguing of infected plants, and controlling the population of aphid vectors in such a way that PRSV-P infection is avoided during the initial stage of plant growth (Fig. 9.4; Sharma et al 2010)). In addition to this strategy, the following approaches are also applied for disease management:

- *Shifting cultivation.* Cultivating papaya in an area for a short period, and then shifting to other areas, is a strategy most successfully used by commercial farmers in those parts of India where agricultural land is available for short-term lease of 1 to 2 years.
- *Isolation distance.* The required isolation distance of 400 m can be applied only in the sparsely cultivated areas.
- *Annual cultivation.* Owing to heavy infestation in the first year of papaya cultivation, the second or third years' crops are no longer economically viable. Farmers cultivate papaya as an annual crop in most parts of India (Sharma et al 2010). This strategy ensures that the damage caused by viral infection is not carried forward to the next season.
- *Season of transplanting.* In addition to the low aphid population, another factor affecting the season of papaya transplantation is the selling price of papaya fruits. It is usually higher during Ramadan (the fasting month for Muslims). Farmers go for papaya transplantation at a time when they can harvest

maximum fruits during the month of Ramadan (National Horticulture Board 2012). Because Ramadan occurs at a different time each year, the transplantation time of papaya is also adjusted accordingly. Therefore, the season of papaya transplanting should be chosen to achieve a balance between the aphid population and the market price of papaya fruits.

- *Raising papaya under a net house.* The population of aphid vectors can be reduced by cultivating papaya trees under protective netting. However, under net cultivation, fruits do not develop well and have a lower sugar content due to limited sunlight. When the nets are removed to allow better growth of the fruit, trees become infected. Moreover, net cultivation adds an additional cost to papaya cultivation (Gonsalves 1998).
- *Barrier crops.* Raising a border crop around a papaya plantation can reduce the entry of aphid vectors into the main plantation. Reducing the population of virus-carrying aphid vectors in a papaya plantation is likely to delay infection (Sharma et al 2010).
- *Application of insecticide.* Because PRSV-P is nonpersistently transmitted by aphid vectors, the application of insecticides has given no conclusive evidence of a reduction in PRSV-P. However, judicious use of botanicals and chemical insecticides can keep the aphid population under control and can keep plants free from other harmful insect pests (Sharma et al 2010).
- *Cross-protection.* When leaf extract of a plant infected with a mild strain of PRSV-P is applied to healthy papaya seedlings at the nursery stage, the seedlings develop immunity against a severe strain of the virus. When these seedlings are transplanted in the field and infected with a severe strain of PRSV-P, they show temporary resistance/tolerance against the disease. This technique—cross-protection—is not popular because of (i) its dependence on the availability of a mild strain homologous to the severe strain of the virus for each geographic area and (ii) the inconsistent results obtained (Gonsalves 1998).
- *Transgenic resistance.* Transgenic resistance is based on the concept of parasite-derived resistance (PDR) whereby the transformed plants, containing genes of a parasite, are protected against the detrimental effects of the same or related pathogens. In transgenic papaya, gene sequences of the coat protein of a mild strain of PRSV-P are transformed to the target papaya cultivars to make them resistant to a severe strain while maintaining other horticultural traits. Two varieties that use this technique (SunUp and UH Rainbow) were released for commercial cultivation in Hawaii. Later, a new transgenic hybrid, Laie Gold, between Rainbow F_2 and non-transgenic Kamiya, was developed (Gonsalves 1998). Transgenic resistance may not be effective against a different/heterologous strain of infecting PRSV-P. Moreover, a strong global environmental activism is creating an adverse political environment against the use of transgenic plants. Therefore, in the present scenario, the ideal approach for controlling PRSV-P is through introgression of resistance genes via conventional breeding.

BREEDING PRSV-P-RESISTANT PAPAYA CULTIVARS

In the absence of any viable alternative for the management of PRSV-P infection, there is renewed interest in the development of resistant cultivars. Although some papaya cultivars show a mild reaction to PRSV-P infection, there is no established source of PRSV-P resistance in *C. papaya* (Cook & Zettler 1970). Therefore, all attempts to develop PRSV-P resistance in *C. papaya* were centered on transferring resistance from highland papaya (species of *Vasconcellea*, Fig. 9.5). Both genera, *Carica* and *Vasconcellea*, belong to the family *Caricaceae*. The genus *Carica* has only one species, *papaya*, while the genus *Vasconcellea* has 21 species (Badillo 2000, 2001 cited in Van Droogenbroeck et al 2004, p 1477). Both *Carica* and *Vasconcellea* are diploid, and their species have 18 chromosomes (Storey 1976, Manshardt & Drew 1998). According to the earlier understanding of the family *Caricaceae*, all species of the genus *Vasconcellea* were part of the genus *Carica*. Therefore, in the literature prior to 2000, *Vasconcellea* species were referred to as *Carica* species and the crosses were described as 'interspecific' while they were, in fact, 'intergeneric.' Some species of the genus *Vasconcellea* show varying degree of PRSV-P tolerance/resistance/immunity in different parts of the world (Table 9.1). Variation in the PRSV-P reaction to *Vasconcellea* species is likely to be due to the genetic differences in the virus strain and/ or plant material, environmental conditions, and different methods of diagnosis. Out of these species, three *Vasconcellea* species, namely, *V. cauliflora, V. quercifolia,* and *V. cundinamarcensis* (*V. pubenscens*), via the bridge species *V. parviflora*, were used at different levels of the PRSV-P resistance breeding program. A time line of development of PRSV-P-resistant papaya by breeding is given in Figure 9.6.

Earlier attempts

Earlier attempts at resistance breeding (before 1998) were aimed at introgression of PRSV-P resistance genes from highland papaya (mainly *V. cauliflora*)

FIGURE 9.5   Hybrid (C) between *Carica papaya* (A) and *Vasconcellea cauliflora* (B).

TABLE 9.1 Reaction of various *Vasconcellea* species to PRSV-P infection

Species	Place	Reaction to PRSV-P	Reference
V. cauliflora	Florida	Susceptible	Conover 1962
	Venezuela	Resistant	Horovitz & Jiménez 1967 (cited in Gonsalves et al 2006, p 64)
	Mexico	Resistant	Alvizo & Rojkind 1987 (cited in Gonsalves et al 2006, p 64)
	Australia	Resistant	Magdalita et al 1998
	Venezuela	Susceptible	Gonzalez 2000 (cited in Gonsalves et al 2006, p 64)
V. quercifolia	Venezuela	Susceptible	Horovitz & Jiménez 1967
	Florida	Resistant	Conover 1964
	Hawaii	Resistant	Manshardt & Wenslaff 1989b
	Australia	Resistant	Drew et al 2006a
V. cauliflora, V. pubescens, V. quercifolia	Florida	Resistant	Conover 1964
V. parviflora, V. stipulata, V. goudotiana	Venezuela & Australia	Susceptible	Horovitz & Jiménez 1967
	Venezuela & Australia	Susceptible	Magdalita et al 1988
V. stipulata, V. pubescens, V. candicans V. × heilbornii nm pentagona	Venezuela	Resistant	Horovitz & Jiménez 1967
V. cundinamarcensis	Puerto Rico	Resistant	Adsuar 1971 (cited in Singh 1990, p 140)

(Horovitz & Jiménez 1958, cited in Ram 2005, p 9; Sawant 1958a,b, Padnis et al 1970, Moore & Litz 1984). PRSV-P resistance has been reported often in crosses between *C. papaya* and *V. cauliflora*; however, lack of vigor and the infertility of these hybrids has prevented further backcrossing (Horovitz & Jiménez 1967, Litz & Conover 1978, Manshardt & Wenslaff 1989a,b, Magdalita et al 1996). Many interspecific (intergeneric) and reciprocal combinations among various species of *C. papaya* and *V. species* were reported by Sawant (1958a,b). IIHR (1987) also reported success in creating interspecific hybridization between *C. papaya* and *V. cauliflora* and produced an F$_1$ hybrid. After

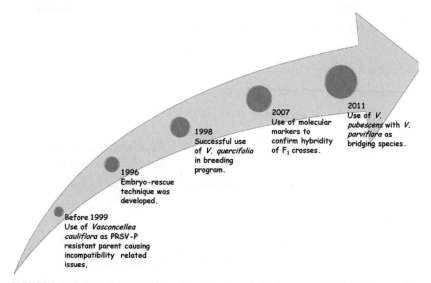

FIGURE 9.6   Milestones in PRSV-P resistance breeding in *Carica papaya* involving *Vasconcellea* species.

backcrossing the F_1 hybrid with *C. papaya*, a PRSV-P-tolerant line 21-19, with normal fruit quality, was developed, but they were unable to release any variety. Another report of successful production of PRSV-P-resistant viable F_1 and F_2 populations came from Khuspe et al (1980) who further reported a PRSV-P-resistant F_1 population segregated for the resistance in the F_2 population with a 3:1 ratio. There was no reliable method to confirm hybridity of F_1 at that time. They were also unable to take their work forward and release any PRSV-P-resistant papaya genotype. Field testing of subsequent generations of this hybrid in Australia showed susceptibility to the Australian strain of PRSV-P and morphologic similarity to the papaya genotype in fruit characteristics. However, some F_1 crosses between *C. papaya* and other *Vasconcellea* species had shown resistance to PRSV-P. All *V. pubescens* hybrids were resistant to PRSV-P when manually inoculated. Out of a large population of *C. papaya* × *V. quercifolia* hybrids 75% were resistant and 25% produced symptoms (Drew et al 1998).

Limitations of breeding between *C. papaya* and *Vasconcellea* species

Different *Vasconcellea* species have varying degrees of incompatibility with *C. papaya*. Considering their cross-compatibility, *Vasconcellea* species were arranged in the following three groups: (i) *C. monoica*, *C. cauliflora*, *C. microcarpa*, and *C. cundinamarcensis*; (ii) *Carica papaya*; and (iii) *C. goudotiana*. Crosses between group (i) and (ii) did not form mature seed, although such immature embryos can be cultured artificially. Crosses between groups (ii) and (iii) were never successful (Ram 2005). Despite sporadic success reported in hybridization between the wild species, incompatibility between *C. papaya* and *Vasconcellea* species has been a major limitation to the production of PRSV-P-resistant hybrids (Mekako & Nakasone 1975). Progressing beyond intergeneric F_1 hybrids has been very difficult, and the only success has resulted from backcrossing of *C. papaya* × *V. quercifolia* to *C. papaya* when papaya was the female parent but not when the intergeneric hybrid was the female parent. In Hawaii, female F_1 hybrids produced only unreduced gametes in backcrosses, yielding sesquidiploid plants that were sterile (Manshardt &Wenslaff 1989b, Manshardt & Drew 1998). Manshardt & Wenslaff (1989a, cited in Gonsalves et al 2006, p 64) also reported that 'hybrids between *C. papaya* and *V. cauliflora* lack vigor, rarely survive till flowering, and if they do, are infertile.' Many factors, such as variable chromosome numbers, the presence of univalents, lagging chromosomes at anaphase, and meiotic irregularities, were likely causes of nonfunctional gametes and infertility (Drew et al 2006a). Similarly, hybrids between *C. papaya* and other species of *Vasconcellea* (*V. stipulata*) were reported to have limited vigor and viability (Horovitz & Jiménez 1967). It is possible that mitochondrial DNA is important in obtaining fertility (Gonsalves et al 2006). Sawant (1958b, cited in Ram 2005, p 9), and Warmke et al (1954, cited in Singh 1990, p 14)—utilizing species such as *C. goudotiana*, *C. monoica*, *C. cundinamarcensis*, *C. cauliflora*, *C. grandis*, and *C. erythrocarpa*—reported various degrees of sterility and vigor. For example, fruits of *C. goudotiana* and *C. papaya* were dropped at 2 to 2.5 months with 90% set at the beginning. *C. goudotiana* and *C. monoica* pollen held its fruits 1–1.5 months with 15% set, and when crossed with *C. cauliflora* held the fruits 3–4 weeks with 5–6% fruits set (Warmke et al 1954, cited in Singh 1990, p 14). Varying degrees of fruit setting and retention were explained on the basis of the genetic proximity among various species by using molecular taxonomy (Aradhya et al 1999, Van Droogenbroeck et al 2004). Success with *V. quercifolia* is consistent with studies on genetic diversity, which showed that *C. papaya* was more closely related to *V. quercifolia* than other *Vasconcellea* species (Jobin-De´cor et al 1997, Drew et al 1998; Fig. 9.7).

Solving the problems of incompatibility

Climate plays a role in the success of the crossing program. Being a tropical crop, greater success was achieved in the tropical countries compared to limited success in the subtropical regions. For example, crosses were successful only in late spring or early summer in Southeast Queensland, Australia (Drew et al 2006b). In addition to the simple technique of using ten male flowers to pollinate one female *C. papaya* flower, the approaches followed to solve the problem of incompatibility between *C. papaya* and *Vasconcellea* species can be grouped in the following three categories:

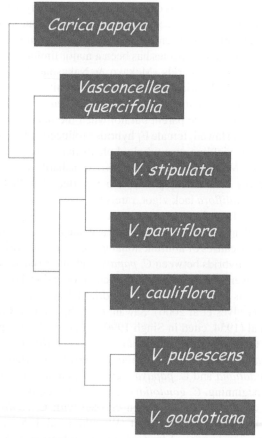

FIGURE 9.7 Genetic relationships among *Vasconcellea* species relevant to PRSV-P resistance breeding.

Use of sucrose solution

The earlier idea of breaking intergeneric crossing barriers by applying sucrose solution on stigma (Subramanyam & Iyer 1982, Iyer & Subramanyam 1984) was successfully exploited in India. In a breeding program of providing PRSV-P resistance to *C. papaya* cv. Surya by transferring a resistance gene from *V. cauliflora*, Dinesh et al (2007) got the maximum number of seeds to set (13.37%) when 5% sucrose solution was applied to the stigmatic surface of the flower. They demonstrated the role of sucrose in enhancing pollen germination and pollen tube growth. The developed progenies were confirmed to be true hybrids by using Inner Simple Sequence Repeat (ISSR) primers when amplifying DNA from hybrids and parents. Four primers, UBC 807 (5'-AGA GAG AGA GAG AGA GT-3'), 810 (5'-CAC ACA CAC ACA CAC AA-3'), 814 (5'-CTC TCT CTC TCT CTC TG-3'), and 861 (5'-ACC ACC ACC ACC ACC ACC-3') clearly amplified male specific bands, which were present in progenies but absent in the female. In a subsequent similar study in India using the same approach involving crosses between *C. papaya* (cvs. CO-7, Pusa Nanha and line CP 50) with *V. cauliflora*, Jayavalli et al (2011) used 5% sucrose solution and achieved successful crosses. Six hybrids (from CO-7V1 to CO-7V6) were free from PRSV-P symptoms. The hybridity of the F_1 population (three from CO-7, eight from Pusa Nanha, and seven from CP 50) was confirmed using ISSR primers by the amplification of DNA from progenies and their parents. ISSR primers UB-856 and UBC-807 and ISSR primer combinations UBC856-817, UBC 810-817, UBC 861-817, UBC 856-810, UBC 861-810, and UBC856-817 clearly amplified specific bands of the male parent, which were present in F_1 progenies and not in female parents. The primer UBC-856 (5'-ACA CAC ACA CAC ACA CYA-3') produced unique banding patterns, which were present in *Vasconcellea* and *C. papaya* cv. CO7 × *V. cauliflora* (CO-7V3) and *C. papaya* cv. Pusa Nanha × *V. cauliflora* (PNV9). In the case of the UBC-807 primer (5'-AGA GAG AGA GAG AGA GT-3'), one prominent band was observed only in the male parent and the hybrid, *C. papaya* Line CP 50 × *V. cauliflora* (CPV23), confirming hybridity of the crosses. However, there was no benefit in adding micronutrients to the sucrose solution in the success of the crosses.

Use of those Vasconcellea *species that are genetically* closer to C. papaya

Learning from past mistakes, researchers shifted to the use of more compatible *Vasconcellea* species rather than insisting on *V. cauliflora*. Species such as *V. quercifolia* became popular in the breeding program as they were compatible with *C. papaya*, and they carried PRSV-P resistance. Drew et al (1998) made crosses between *C. papaya* and *V. pubescens, V. quercifolia, V. parviflora*, and *V. goudotiana*, F_1 plants that were vigorous both in the greenhouses and in the field in Australia. While most F_1 hybrids were infertile, some male plants of hybrids formed between *C. papaya* and *V. quercifolia* and between *C. papaya* and *V. parviflora* produced some viable pollen. Earlier, some backcross generations were tried in Hawaii but were sterile (Manshardt & Zee 1994, cited in Gonsalves et al 2006, p 65). Infertility was not a problem subsequent to the BC_2 generations, which suggested that if preferential elimination of *V. quercifolia* DNA had occurred, it happened during production of the F_1 and BC_1 generations. Drew & O'Brien (2001) successfully used pollen from the *C. papaya* × *V. quercifolia* male plants to pollinate *C. papaya* flowers. One staminate plant of the backcross (BC_1) population demonstrated some PRSV-P resistance and produced fertile pollen. During field testing of these plants, one clone delayed symptoms for 12 months in the presence of inoculums and aphid vectors. Pollen from this plant was used to produce 200 BC_2 plants, of which 26 remained symptom-free after inoculation and progressively produced symptoms of varied severity over a

9-month period when planted in the field. Protocols to produce large numbers of vigorous intergeneric hybrids between *C. papaya* and *V. quercifolia* was reported in Australia (Drew et al 2006a), and there was success in crossing from *C. papaya* × *V. quercifolia* F_1 hybrids to *C. papaya* to produce BC_1 and BC_2 generations. These authors reported on further backcrossing to produce BC_1, BC_2, BC_3, and BC_4 generations to *C. papaya*, sib-cross populations, and the subsequent development of PRSV-P-resistant BC_3, SbC_3 and BC_4 plants. Vegas et al (2003, cited in Gonsalves et al 2006, p 64) obtained regeneration and evaluation of intergeneric hybrids between *C. papaya* and *V. cauliflora*. Alamery & Drew (2011) reported successful introgression of a PRSV-P resistance gene(s) into the papaya gene pool through intergeneric hybridization with *V. quercifolia* followed by backcrosses. In a recent report, Siar et al (2011) reported successful production of PRSV-P-resistant papaya by crossing *C. papaya* with *V. quercifolia*. One PRSV-P-resistant backcross 1 (BC_1) male plant was selected from 700 plantlets screened in a greenhouse and in the field under exposure to high inoculum and aphid vectors. 1465 seed-raised plants [137 BC_2, 546 SbC_2 (BC_2 sib-crosses), 147 BC_3, 379 SbC_3, and 256 BC_4] from this PRSV-P-resistant plant were grown under high disease pressure. Virus-resistant BC_3 and BC_4 plants were selected and confirmed by ELISA. They generally developed mild symptoms from 5 to 18 months in the field, but many of them showed the ability to produce symptom-free new growth. Some BC_3 and BC_4 plants were virus-free even after 18 months. Subsequently, all plants developed mild symptoms on foliage and fruits. However, they continued to grow vigorously and to produce good quality marketable fruit for 3 years—while all control plants developed severe symptoms after 3 months in the field.

Use of the embryo rescue technique and in vitro culture of hybrids

The study of reproductive barriers indicated that post-zygotic barriers, embryo abortion, and lack of endosperm development limited the success of the hybridization (Manshardt & Wenslaff 1989a). Therefore, efforts were targeted on the rescue of the embryo before it is aborted. Magdalita et al (1996) developed the protocol to rescue and germinate *C. papaya* × *V. cauliflora* immature embryos. Later, Drew et al (1998) developed procedures for hybridization of papaya with related PRSV-P-resistant *Vasconcellea* species (*V. cauliflora*, *V. quercifolia*, and *V. pubescens*). Multiple hybrid plants were produced from embryogenic cultures formed by germinated embryos on hormone-free agar-solidified medium. Because hybrids between *C. papaya* and many species of *Vasconcellea* lack vigor (and are generally infertile), the protocol has been adapted to produce hybrids between *C. papaya* and other PRSV-P-resistant species, *V. quercifolia*, and *V. pubescens*. Using this technique, intergeneric hybrid plants had been produced between *C. papaya* and *V. goudotiana* and *V. parviflora*.

RECENT DEVELOPMENTS

Use of PRSV-P-immune *V. pubescens* via a bridge species

The use of genetically compatible *V. quercifolia* in the breeding program had some limitations. The resistant hybrids showed mild virus symptoms. They were neither uniform in virus resistance nor stable, while the use of PRSV-P-immune *V. pubescens* as a parent had the serious limitation of being incompatible with *C. papaya*. The concept of a bridge species (*V. parviflora*, a compatible species with both *C. papaya* and *V. pubescens*) was used by Drew and his group in Australia to overcome the incompatibility between *C. papaya* and *V. pubescens*. Intergeneric hybrid populations of *C. papaya* × *V. pubescens* Lenne et Koch, *C. papaya* × *V. parviflora* A. DC., and interspecific populations of *V. pubescens* × *V. parviflora* were produced and evaluated for morphologic characteristics and disease resistance. F_2 and F_3 populations were produced from the *V. pubescens* × *V. parviflora* F_1; disease-resistant individuals with homozygous genotypes (RR) were selected with the aid of a codominant CAPS marker and backcrossed to *V. parviflora* or outcrossed to *C. papaya*. The disease-resistance gene from *V. pubescens* was successfully backcrossed into *V. parviflora* from F_2 hybrids, and in the F_1 hybrids between *C. papaya* and *V. parviflora*, 45% pollen fertility was obtained. The authors proposed *V. parviflora* as a potential bridging species between *C. papaya* and *V. pubescens* (O'Brien & Drew 2009).

Use of molecular markers and the nature of PRSV-P resistance

The papaya microsatellite loci published by de Oliveira et al (2010) provided the option of genetic markers for detailed studies of population genetic structure, hybridization among populations, and paternity, particularly those functioning as QTL (quantitative trait loci). These markers may be useful in encoding important agricultural traits such as disease resistance, yield, fruit type, and fruit size. Selection of PRSV-P-resistant hybrids with the help of molecular markers can reduce the time taken in the breeding program. Molecular markers have been developed for PRSV-P resistance in *V. pubescens* and are being developed for *V. quercifolia*. Out of 16 primer pairs tested on the segregating populations of *C. papaya* and *V. quercifolia* by Alamery & Drew (2011), no SSR markers were linked to the resistance gene. However, the marker SP16 confirmed the hybridity of F_1 hybrids and suggested some chromosomal elimination from *V. quercifolia*. Dillon et al (2005) converted the DNA marker OPK41R into an easily detectable codominant marker, which may be applied to the selection of homozygous PRSV-P-resistant hybrids of *V. cundinamarcensis*. The CAPS marker, Psilk4, was shown to correctly identify a resistant genotype with 99% accuracy when applied to the F_2 progeny of *V. parviflora* × *V. cundinamarcensis* segregating for PRSV-P resistance. This confirms the presence of a major gene controlling PRSV-P resistance in *V. cundinamarcensis*. The Psilk4 marker can assist in selection of PRSV-P-resistant progeny in the breeding program involving *V. cundinamarcensis* (Dillon et al 2006). These markers are being used to select PRSV-resistant *Vasconcellea* parents that are genetically closer to *C. papaya* and deciding genetic fidelity of the hybrids.

Earlier evidence suggested that PRSV resistance could be based on a single dominant gene (Micheletti 1962, cited in Gonsalves et al 2006, p 64). It is still not yet clear whether the resistance phenotype in *V. quercifolia* is controlled by a single dominant gene or by multiple genes. Based on a PRSV-P reaction of a large population of *C. papaya* × *V. quercifolia* hybrids, Drew et al (1998) indicated that PRSV-P resistance in *V. quercifolia* is not controlled by a single resistant gene. PRSV-P resistance in *V. cundinamarcensis* (*V. pubescens*) was reported to be controlled by a single dominant gene, or by a group of genes at a single locus. This gene was mapped by Alamery & Drew (2011) and assumed to be a kinase gene. Putative resistance gene(s) for PRSV-P resistance in *V. pubescens* are being sequenced and characterized. It is yet to be confirmed whether they confer PRSV-P resistance to *C. papaya* on transfer (Drew 2011). A homologous kinase gene and its mode of inheritance were studied in *V. quercifolia*. Segregating population was manually inoculated and screened for PRSV-P resistance/susceptibility. The presence of the resistance phenotype among the majority of hybrids indicated high stability of gene inheritance through successive backcrossed generations. The F_2 population of *V. parviflora* and *V. cundinamarcensis* showed Mendelian segregation for the expected ratio of 3:1 or 1:1, but when *Vasconcellea* species or their hybrids were outcrossed to *C. papaya*, inheritance patterns did not always follow Mendelian ratios, suggesting abnormal pairing of chromosomes or preferential elimination of the *Vasconcellea* genes (Drew at al 1998). Cytologic observations revealed variable chromosome numbers, the presence of univalents, and lagging chromosomes at anaphase; meiotic irregularities (Drew et al 2006a) would be expected to cause preferential elimination of *V. quercifolia* DNA. This was demonstrated by the ratio of susceptible to resistant plants, which ranged from 4:1 to 1:3. The resistance was confirmed against Philippines and Australian PRSV-P virus strains, which suggested that the resistance gene could be effective to confer the resistance to other virus strains in different countries.

CONCLUSION

The use of intergeneric hybridization seems to be a potent option for the development of PRSV-P-resistant *C. papaya* with acceptable horticultural traits. The initial steps of selection of papaya genotypes compatible with the wild species with better chances of producing viable hybrids have been done successfully. Analysis of RAPD profiles and comparison of DNA sequences from nuclear and mitochondrial genes show varying degree of proximity between *C. papaya* and various *Vasconcellea* species. *V. quercifolia* is closest and

V. cauliflora is the most distant to *C. papaya*. Efforts were redirected towards making crosses between *C. papaya* and compatible *Vasconcellea* species in the 2000s. In a study by Siar et al (2011), dissection of 114,839 seeds yielded 1011 embryos, of which 700 plants were established in a greenhouse. Of these, only one PRSV-P-resistant male BC1 was fertile and was used for further back-crossing. The above study emphasized the importance of producing large numbers of plants at each stage. This may be achieved by refining protocols for large-scale crossing, micro-propagation, and plantlet production. The essential condition of successful production of viable hybrids between *C. papaya* and *Vasconcellea* species has been reduced by refinement of the embryo rescue technique and *in vitro* clonal population. Easy evaluation of a large number of crosses has facilitated selection of a few male intergeneric hybrids with both PRSV-P resistance and some fertility required for backcrossing with *C. papaya*. Recent inclusion of PRSV-P-immune *V. pubescens* in the breeding program has raised the possibility of obtaining some hybrids with stable PRSV-P resistance and good fruit qualities. The recent discovery of Carvalho & Renner (2012) that *C. papaya* is closest to a clade of herbaceous or thin-stemmed species has added a new option for plant breeders who have so far tried to cross papaya only with woody highland papayas (*Vasconcellea*). However, concentrated efforts by a team with skills in breeding, embryo rescue, *in vitro* culture, acclimatization, cytology, and virology are required for the success of the breeding program.

REFERENCES

Adsuar, J., 1971. Resistance of *Carica candamarcensis* to the mosaic viruses affecting papaya (*C. papaya*) in Puerto Rico. Journal of Agriculture of the University of Puerto Rico 55 (9), 265–266.

Alamery, S., Drew, R.A., 2011. Studies on the genetics of PRSV-P resistance genes in intergeneric hybrids between *Carica papaya* and *Vasconcellea quercifolia*. International Society of Horticultural Sciences. (Abstract only.) Third International Symposium on Papaya, Chiang Mai, Thailand 19–22 December 2011, p. 16.

Alvizo, H.F., Rojkind, C., 1987. Resistencia al virus mancha anular del papaya en *Carica cauliflora*. Revista Mexicana de Fitopatología 5 (1), 61–62.

Aradhya, M.K., Manshardt, R.M., Zee, F., Morden, C.W., 1999. A phylogenetic analysis of the genus *Carica* L. (*Caricaceae*) based on restriction fragment length variation in a cpDNA intergenic spacer region. Genetic Resources and Crop Evolution 46 (6), 579–586.

Badillo, V.M., 2000. *Carica* L. vs. *Vasconcellea* St.-Hil. (*Caricaceae*) con la rehabilitacion de este ultimo. Ernstia 10 (1), 74–79.

Badillo, V.M., 2001. Nota correctiva *Vasconcellea* St. Hill. (Caricaceae) y no *Vasconcellea* (*Caricaceae*). Ernstia 11 (1), 75–76.

Carvalho, F.A., Renner, S.S., 2012. A dated phylogeny of the papaya family (*Caricaceae*) reveals the crop's closest relatives and the family's biogeographic history. Molecular Phylogenetics and Evaluation 65 (1), 46–53.

Conover, R.A., 1962. Virus diseases of papaya in Florida. (Abstract only.) Phytopathology 52 (1), 6.

Conover, R.A., 1964. Distortion ring spot, a severe virus disease of papaya in Florida. Proceedings of Florida State Horticulture Science 77, 440–444.

Cook, A.A., Zettler, F.W., 1970. Susceptibility of papaya cultivars to papaya ringspot and papaya mosaic viruses. Plant Disease Reporter 54 (4), 893–895.

de Oliveira, E.J., Amorim, V.B.O., Matos, E.L.S., Costa, J.L., Castellen, M.S., Pádua, J.G., Dantas, J.L.L., 2010. Polymorphism of microsatellite markers in papaya (*Carica papaya* L.). Plant Molecular Biology Reporter 28 (3), 519–530.

Dillon, S.K., Drew, R.A., Ramage, C., 2005. Development of a co-dominant SCAR marker linked to a putative PRSV-P resistance locus in 'wild papaya'. Acta Horticulturae 694, 101–104.

Dillon, S., Ramage, C., Ashmore, S., Drew, R.A., 2006. Development of a codominant CAPS marker linked to PRSV-P resistance in highland papaya. Theoretical and Applied Genetics 113 (6), 1159–1169.

Dinesh, M.R., Rekha, A., Ravishankar, K.V., Praveen, K.S., Santosh, L.C., 2007. Breaking the inter-generic crossing barrier in papaya using sucrose treatment. Scientia Horticulturae 114 (1), 33–36.

Drew, R.A., 2011. The use of non-transgenic technologies for the development of Papaya ringsopt resistance in *Carica papaya* (L.). International Society of Horticultural Sciences. (Abstract only.) Third International Symposium on Papaya, Chiang Mai, Thailand 19–22 December 2011, p. 8.

Drew, R.A., O'Brien, C.M., 2001. Progress with *Carica* interspecific hybridization to develop plants with resistance to PRSV-P. (Abstract only.) Second International Symposium on Bio-technology of Tropical and Subtropical Species, Taipei, Taiwan 5–9 November 2001, p. 22.

Drew, R.A., O'Brien, C.M., Magdalita, P.M., 1998. Development of interspecific *Carica* hybrids. Acta Horticulturae 461, 285–292.

Drew, R.A., Siar, S.V., O'Brien, C.M., Magdalita, P.M., Sajise, A.G.C., 2006a. Breeding for papaya ringspot virus resistance in *Carica papaya* L. via hybridisation with *Vasconcellea quercifolia*. Australian Journal of Experimental Agriculture 46 (3), 413–418.

Drew, R.A., Siar, S.V., O'Brien, C.M., Sajise, A.G.C., 2006b. Progress in backcrossing between *Carica papaya* × *Vasconcellea quercifolia* intergeneric hybrids and *C. papaya*. Australian Journal of Experimental Agriculture 46 (3), 419–424.

FAO (Food and Agricultural Organization of the United Nations), 2012. FAOSTAT. [Online] Available at: <htpp://www.fao.org/corp/statistics> [Accessed: 31.12. 2012].

Gonsalves, D., 1998. Control of papaya ringspot virus in papaya: a case study. Annual Review of Phytopathology 36, 415–437.

Gonsalves, D., Ishii, M., 1980. Purification and serology of papaya ringspot virus. Phytopathology 70 (11), 1028–1032.

Gonsalves, D., Vegas, A., Prasartsee, V., Drew, A., Suzuki, J.Y., Tripathi, S., 2006. Developing papaya to control papaya ringspot virus by transgenic resistance, intergeneric hybridization, and tolerance breeding. Plant Breeding Review 26, 35–78.

Gonsalves, D., Tripathi, S., Carr, J.B., Suzuki, J.Y., 2010. Papaya ringspot virus. [Online] Plant Health Instructor APS online: Available at: <http://www.apsnet.org/edcenter/intropp/lessons/viruses/Pages/PapayaRingspotvirus.aspx:> [Accessed: 31.12. 2012].

Gonzalez, A., 2000. Selection de cepas atenuadas y desarrollo de una tecnica de diagnostic precisa para el virus de la mancha anillada de la lechosa (*Carica papaya* L.). MSc thesis Universidad Central de Venezuela, Caracas, Venezuela.

Horovitz, S., Jiménez, H., 1958. Cruzabilidad entre especies de *Carica*. Agronomie Tropicale 7, 207–215.

Horovitz, S., Jiménez, H., 1967. Cruzamientos interespecíficos e intergenéricos en Caricáceas y sus implicaciones fitotécnicas. Agronomie Tropicale 17, 323–344.

IIHR, 1987. Interspecific hybridization in genus *Carica*. Annual Report 1987 Indian Institute of Horticultural Research, Bangalore.

Iyer, C.P.A., Subramanyam, M.D., 1984. Using of bridging species in interspecific hybridization in genus *Carica*. Current Science 53 (24), 1300–1301.

Jain, R.K., Sharma, J., Sivakumar, A.S., Sharma, P.K., Byadgi, A.S., Verma, A.K., Varma, A., 2004. Variability in the coat protein gene of *Papaya ringspot virus* isolates from multiple locations in India. Archives of Virology 149 (12), 2435–2442.

Jayavalli, R., Balamohan, T.N., Manivannan, M., Govindaraj, N., 2011. Breaking the intergeneric hybridization barrier in *Carica papaya* and *Vasconcellea cauliflora*. Scientia Horticulturae 130 (4), 787–794.

Jensen, D.D., 1949. Papaya virus diseases with special reference to papaya ringspot. Phytopathology 39 (3), 191–211.

Jobin-Décor, M.P., Graham, G.C., Henr, R.J., Drew, R.A., 1997. RAPD and isozyme analysis of genetic relationships between *Carica papaya* and wild relatives. Genetic Resources Crop Evolution 44 (5), 471–477.

Khuspe, S.S., Hendre, S.S., Mascarenhas, A.F., Jagannathan, V., Thobre, V., Joshi, A., 1980. Utilization of tissue culture to isolate interspecific hybrids in *Carica* L. In: Rao, P.M., Heble, M., Chadha, M. (Eds.), National symposium on plant tissue culture, genetic manipulations and somatic hybridization of plant cells. Bhabha Atomic Research Center, Bombay, India 27-29, pp. 198–205February 1980.

Lindner, R.C., Jensen, D.D., Ikeda, W., 1945. Ringspot: new papaya plunderer. Hawaii Farm and Home 8 (10), 10–14.

Litz, R.E., Conover, R.A., 1978. *In vitro* propagation of papaya. HortScience 13 (3), 241–242.

Magdalita, P.M., Villegas, V.N., Pimentel, R.B., Bayot, R.G., 1988. Reaction of papaya (*Carica papaya* L.) and related *Carica* species to ringspot virus. Philippine Journal of Crop Science 13 (3), 129–132.

Magdalita, P.M., Adkins, S.W., Godwin, I.D., Drew, R.A., 1996. An improved embryo-rescue protocol for *Carica* interspecific hybrids. Australian Journal of Botany 44 (3), 343–353.

Magdalita, P.M., Drew, R.A., Godwin, I.D., Adkins, S.W., 1998. An efficient interspecific hybridization protocol for *Carica papaya* L. × *C. cauliflora* Jacq. Australian Journal of Experimental Agriculture 38 (4), 523–530.

Manshardt, R.M., Drew, R.A., 1998. Biotechnology of papaya. Acta Horticulturae 461 (1), 65–73.

Manshardt, R.M., Wenslaff, T.F., 1989a. Interspecific hybridization of papaya with other *Carica* species. Journal of the American Society of Horticultural Science 114 (4), 689–694.

Manshardt, R.M., Wenslaff, T.F., 1989b. Zygotic polyembryony in interspecific hybrids of *Carica papaya* and *C. cauliflora*. Journal of the American Society of Horticultural Science 114 (4), 684–689.

Manshardt, R.M., Zee, F., 1994. Papaya germplasm and breeding in Hawaii. Fruit Varieties Journal 48 (3), 146–152.

Mekako, H.U., Nakasone, H.Y., 1975. Interspecific hybridization among six *Carica* species. Journal of the American Society of Horticultural Science 100 (3), 237–242.

Micheletti, D., 1962. Descripcion morfologica y cytological de una planta segregante de un cruce entre *Carica cauliflora* Jacq and *C. monoica* Desf. Agron Tropical (Maracay) 11 (2), 193–200.

Moore, G.A., Litz, R.E., 1984. Biochemical markers for *Carica papaya, C. cauliflora*, and plants from somatic embryos of their hybrid. Journal of the American Society of Horticultural Science 109 (2), 213–218.

National Horticulture Board, 2012. Indian Horticulture Database – 2011. National Horticulture Board, Ministry of Agricultural, Government of India, Gurgaon, p. 278. Available online at: <htpp://www.nhb.go.in> [Accessed: 31.12. 2012].

O'Brien, C.M., Drew, R.A., 2009. Potential for using *Vasconcellea parviflora* as a bridging species in intergeneric hybridisation between *V. pubescens* and *Carica papaya*. Australian Journal of Botany 57 (7), 592–601.

Padnis, N.A., Budrukkar, N.D., Kaulgud, S.N., 1970. Embryo culture technique in papaya (*C. papaya.* L.). Poona Agricultural College Magazine 60, 101–104.

Purcifull, D., Edwardson, J., Hiebert, J.E., Gonsalves, D., 1984. Papaya ringspot virus. CMI/AAB Descriptions of plant viruses, No. 292. (No. 84 revised, July 1984.) 8 pp CAB International, Wallingford, UK.

Ram, M., 2005. Papaya. Indian Council of Agricultural Research, New Delhi.

Sawant, A.C., 1958a. A study of the interspecific hybrids, *Carica monoica* × *C. cauliflora*. Proceedings of the American Society Horticultural Science 71, 330–333.

Sawant, A.C., 1958b. Crossing relationship in the genus *Carica*. Evolution 12, 263–266.

Sharma, S.K., Zote, K.K., Kadam, U.M., Tomar, S.P.S., Dhale, S.P.S., Sonawane, A.U., 2010. Integrated management of *Papaya ringspot virus*. Acta Horticulturae 851 (1), 473–480.

Shukla, D.D., Ward, C.W., Brunt, A.A. (Eds.), 1994. The *Potyviridae*, CAB International, Wallingford.

Siar, S.V., Beligan, G.A., Sajise, A.J.C., Villegas, V.N., Drew, R.A., 2011. *Papaya ringspot virus* resistance in *Carica papaya* via introgression from *Vasconcellea quercifolia*. Euphytica 181 (2), 159–168.

Singh, I.D., 1990. Papaya. Oxford and IBH Publishing Company Private, New Delhi.

Storey, W.B., 1976. Papaya, *Carica papaya*. In: Simmonds, N.W. (Ed.), Evaluation of crop plants. Longman, London, pp. 21–24.

Subramanyam, M.D., Iyer, C.P.A., 1982. Cytological studies in *Carica* spp and hybrids. Genetica Agraria 36 (1), 73–86.

Tripathi, S., Suzuki, J., Gonsalves, D., 2006. Development of genetically engineered resistant papaya for *Papaya ringspot virus* in a timely manner—a comprehensive and successful approach. In: Ronald, P. (Ed.), Plant–pathogen interactions: methods and protocols, vol. 354. Humana Press, Totowa, NJ, pp. 197–240.

Tripathi, S., Suzuki, J.Y., Ferreira, S.A., Gonsalves, D., 2008. *Papaya ringspot virus-P*: characteristics, pathogenicity, sequence variability and control. Molecular Plant Pathology 9 (3), 269–280.

Van Droogenbroeck, B., Kyndt, T., Maertens, I., Romeijn-Peeters, E., Scheldeman, X., Romero-Motochi, J.P., Van Damme, P., Goetghebeur, P., Gheysen, G., 2004. Phylogenetic analysis of the highland papayas (*Vasconcellea*) and allied genera (*Caricaceae*) using PCR-RFLP. Theoretical and Applied Genetics 108, 1473–1486.

Vegas, A., Trujillo, G., Sandrea, Y., Mata, J., 2003. Obtención, regeneración y evaluación de híbridos intergenéricos entre *Carica papaya y Vasconcellea cauliflora*. Interciencia 28 (12), 710–714.

Warmke, H.E., Cabanillas, E., Cruzado, U.J., 1954. An interspecific hybrid in genus *Carica*. Proceedings of the American Society of Horticultural Sciences 64, 284–288.

Yeh, S.D., Gonsalves, D., 1985. Translation of papaya ringspot virus RNA *in vitro*: detection of a possible polyprotein that is processed for capsid protein, cylindrical-inclusion protein, and amorphous-inclusion protein. Virology 143 (1), 260–271.

Yeh, S.D., Jan, F.J., Chiang, C.H., Doong, P.J., Chen, M.C., Chung, P.H., Bau, H.J., 1992. Complete nucleotide sequence and genetic organization of papaya ringspot virus RNA. Journal of General Virology 739 (10), 2531–2541.

Synergism in plant–virus interactions: a case study of CMV and PVY in mixed infection in tomato

Tiziana Mascia and Donato Gallitelli

Dipartimento di Scienze del Suolo, della Pianta e degli Alimenti, Università degli Studi di Bari Aldo Moro, Bari, Italy

INTRODUCTION

Cucumber mosaic virus (CMV) is recognized worldwide as a harmful pathogen for many horticultural, woody, and ornamental crops (Tomlinson 1987, García-Arenal & Palukaitis 2008). In the Mediterranean basin the economic importance of this virus correlates with recurrent outbreaks in melon (Alonso-Prados et al 1997, Luis-Arteaga et al 1998, Lecoq & Desbiez 2012), pepper (Moury & Verdin 2012), and tomato crops grown for canning (Gallitelli et al 1995, Varveri & Boutsika 1999, Aramburu et al 2007, Hanssen & Lapidot 2012), with incidences of 30% to nearly 100%. After the severe outbreak in Alsace (France) in 1972, causing tomato necrosis, other CMV epidemics have been reported with tomato plants displaying chlorosis, fruit internal necrosis, leaf curling, severe stunting, and plant death, in addition to the strong reduction of the leaflet blade known as filiformism or shoe-string. Most of the new syndromes were induced by the so-called 'Asian strains' (Gallitelli 2000) and by the presence of distinct variants of the CMV satellite RNA (CMV-satRNA), some of which co-determined necrosis and stunting (Gallitelli et al 1988, Jordá et al 1992). To date, a number of reports indicate that CMV populations are well established in Mediterranean areas (Aramburu et al 2007, Hanssen & Lapidot 2012), where they are often found in solanaceous, cucurbit, and legume crops in mixed infections with potyviruses (Gallitelli et al 1988, Hanssen & Lapidot 2012).

In most of such mixed infections, the viruses are known to interact synergistically (Palukaitis & García-Arenal 2003, Wege & Siegmund 2007), inducing exacerbation of disease symptoms, an increase in the virus titer, and accumulation at high levels also in nonhost plants and resistant varieties (Rochow 1972, Palukaitis & Kaplan 1997, Ryabov et al 2001, Sáenz et al 2002, Palukaitis & Garcia-Arenal 2003, Wege & Siegmund 2007, Martin & Elena 2009), complementation of movement defects and of insect-mediated transmission, and broadening of the range of host plants (Syller 2012, Takeshita et al 2012).

This chapter describes the interactions between CMV and PVY in tomato as a case study with features that are unique to this model and may be relevant to a better understanding of the biologic and molecular viral interactions in plants with mixed infection and of their outcomes from an eco-epidemiologic point of view.

VIRUS PROPERTIES

CMV is the type species of the genus *Cucumovirus* in the family *Bromoviridae* (Bujarski et al 2012). It has a positive-sense RNA genome split into three single-stranded segments that encode a total of five proteins. Like most other plant viruses with a divided genome, CMV genomic RNAs are packaged in separate icosahedral particles, all of which are required for infection. RNA1 encodes a single product of 110 kDa (1a protein) with methyltransferase and RNA helicase activity. RNA2 is bicistronic, encoding a large product of 98 kDa (2a protein) with hallmarks of viral polymerases and a small 12 kDa polypeptide (2b protein), which is encoded by a 3′-proximal open reading frame (ORF), partially overlapping the 98 kDa ORF, and is translated during plant infection from a subgenomic RNA called RNA 4A. The 2b protein is multifunctional, with a key role in pathogenesis because (i) it inhibits antiviral RNA silencing by binding short-interfering (si)RNAs; (ii) it perturbates micro (mi)RNA-regulated gene expression and DNA methylation by interacting with the host silencing proteins, AGO1 and 4, respectively (Ye et al 2004, Goto et al 2007, Lewsey et al 2007, 2009, Cillo et al 2009, Gonzales et al 2010); (iii) it gives the virus some protection from salicylic acid-induced antiviral defenses (Ji & Ding 2001), and (iv) it influences cell-to-cell and systemic virus movement (Ding et al 1995, Shi et al 2002, Soards et al 2002). RNA3 is also bicistronic, encoding the 3a protein in its 5′ proximal half and containing the coat protein (CP) ORF in its 3′ proximal half. The 3a protein, designated MP, is involved in virus movement while the CP cistron is translated via the subgenomic RNA4.

In addition to genomic RNAs, some strains of CMV encapsidate a satellite RNA (satRNA), which is a small RNA molecule that is completely dependent on the viral genome for its replication and spread but does not supply the helper virus with any essential function. CMV satRNAs range from 335 to 405 nucleotides in size, apparently do not contain any functional ORF, and occur in nature as several variants—which can attenuate or aggravate disease symptoms induced by the helper virus. This effect is particularly relevant in tomato plants in which the CMV infection phenotype may span from no symptoms to plant death. More than 100 satRNA variants have been described associated with 65 CMV strains (García-Arenal & Palukaitis 2008).

CMV strains can be divided into subgroups I and II on the basis of their sequence similarity and serologic relationships. Surveys of naturally infected crops suggest that strains of subgroup I are more frequent than those of subgroup II and sometimes they represent more than 80% of all isolates (Crescenzi et al 1993, Fraile et al 1997, Aramburu et al 2007). Strains in the same subgroup differ by only 2–3% of their sequence homology. Because sequence data show that a number of CMV strains within subgroup I and originating from Asia differ by 7–12% in sequence arrangement from other subgroup I strains, Palukaitis & Zaitlin (1997) proposed that subgroup I should be split by placing the 'Asian strains' in subgroup IB and the others in subgroup IA.

Potato virus Y (PVY) is the type species of the genus *Potyvirus*, one of the seven genera in the family *Potyviridae* (Adams et al 2012). The viral particles are non-enveloped flexuous rods with helical symmetry, 730–740 nm long and 11–12 nm in width, which encapsidate a 10-kb positive-sense single-stranded RNA with a 5′-covalently linked protein (VPg) and a 3′ poly (A) tail. The sequences of about 30 PVY isolates have been reported with homologies of 93–99% (Adams et al 2012). RNA of PVY is translated into a large polyprotein that is cleaved co- and post-translationally into functional proteins by three virus-encoded proteases. In the 5′→3′ direction, the PVY proteins are referred to as P1 (first protein), HC or HC–Pro (helper component protein), P3 (third protein), 6K1 (first 6 kDa protein), CI (cytoplasmic inclusion protein), 6K2 (second 6 kDa protein), NIa (small nuclear inclusion protein), NIb (large nuclear inclusion protein), and CP (coat protein).

The P1 protein has a serine protease domain at its C-terminal region, cleaving itself from the adjacent helper component protease (HC-Pro). It is the least conserved in sequence and the most variable in size and it plays a significant role in virus replication—probably due to the stimulation of the gene-silencing suppressor HC-Pro.

HC-Pro, as the 2b protein of CMV, plays an important role in the infectivity process as it is a determinant of pathogenicity, is involved in the aphid-mediated transmission of the virus, in the movement of the virus in the host plant, and in the suppression of RNA-dependent gene silencing by binding siRNAs (Lakatos et al 2006). The C-terminal part of the HC-pro protein acts as cysteine protease, cleaving itself from the precursor polyprotein.

The P3 protein is involved in virus replication in association with the proteins of the 'replication complex block' composed of the CI protein, VPg protease, and polymerase. The P3 protein seems also to be involved in host range and symptom development.

The CI protein forms peculiar cylindrical inclusions (denoted pinwheels) in infected cells, has an associated helicase activity, and is probably involved in viral cell-to-cell movement. CI is bordered by two small proteins, 6K1 and 6K2. The 6K1 protein has no known function, while the 6K2 is a transmembrane protein, probably needed to anchor the replication complex to the endoplasmic reticulum.

Protein NIa forms nuclear inclusions and has two domains: the VPg and the protease domain that cleaves all proteins at the C-terminal half of the precursor. The VPg product is involved in replication, translation, and plant susceptibility to potyviruses through the interaction with several isoforms of the host translation factor eIF4E (Whitham & Wang 2004).

The NIb protein is the putative RNA-dependent RNA polymerase, which also forms nuclear inclusions.

The coat protein (CP) is involved in cell-to-cell movement and vector transmission together with HC-Pro. Recent studies have shown the presence of an additional short ORF (PIPO = 'pretty interesting potyvirus ORF') embedded within the P3 cistron and expressed as a fusion product via a frameshift mechanism. This protein has been shown to be essential for virus intercellular movement (Adams et al 2012).

MOVING FROM FIELD CROPS INTO A MODEL SYSTEM

Mixed infections of PVY and CMV, often carrying a CMV-satRNA, were detected in commercial fields of tomato crops during the CMV outbreaks that occurred in Italy between the mid-1980s and the mid-1990s (Gallitelli et al 1988, Gallitelli 2000). A CMV strain of subgroup II, called CMV-PG, was responsible for the lethal necrosis disease, which was co-determined by a necrogenic variant of the satRNA, called PG-satRNA (Kaper et al 1990). A different symptomatology was induced by CMV subgroup IB strains, namely, CMV-Tfn and CMV-TTS, which were responsible for the tomato fruit internal necrosis and tomato top stunting diseases, respectively. These 'Asian strains' were new to Italy and were most probably introduced to the Mediterranean countries in the mid-1980s. Also, in these instances, two satRNA variants were detected in infected plants, but while the Tfn-satRNA was not responsible for the tomato fruit internal necrosis syndrome, the TTS-satRNA co-determined the stunting phenotype (Gallitelli 2000 and references therein). PVY was present in mixed infection with CMV-PG and CMV-Tfn, but its role in the phenotypes of crop disease was not demonstrated (Kaper et al 1990).

In spite of the economic relevance of such mixed infections, poor attention has been paid to the mechanisms they rely on. Therefore, a laboratory model has been set up to mimic field conditions and to study, in more detail, the interaction between CMV and PVY (Mascia et al 2010). Field condition were reproduced in tomato plants of the UC82 variety challenged with CMV-LS and CMV-Fny, which are strains of subgroups II and IA, respectively, and with the SON-41 strain of PVY. CMV-Fny is very aggressive in tomato while CMV-LS induces mild symptoms. The SON41 strain of PVY induces very mild mosaic symptoms on the first two systemically infected leaves within 2 weeks after inoculation, then the plant gradually recovers from this condition and no symptoms are visible in the new leaves up to 1 month after inoculation. For its limited pathogenicity in tomato, PVY-SON41 seemed a good candidate for evaluating any specific symptom change in tomato plants infected by the different inoculum combinations. Within 2 weeks after mechanical inoculation, tomato plants infected by CMV-Fny showed strong filimorphism and whole-plant growth reduction while those infected by CMV-LS did not develop stunting and filimorphism but only a minimal leaf blade reduction and deformation. In comparison to single infections, coinfection of tomato plants with PVY-SON41 and either CMV strain (LS or Fny) resulted in increased systemic symptoms and, as already observed in single infections, symptoms elicited by CMV-Fny strain in mixed infection with PVY-SON41 were more severe than those induced by CMV-LS—as in the latter case the plants were moderately stunted and developed leaf malformations but not a true leaf filimorphism.

THE LOSS-OF-FUNCTION APPROACH TO DISSECT THE SINGLE CELL, LOCAL, AND SYSTEMIC PATTERNS OF MIXED INFECTIONS

As new experimental evidence is provided, it becomes more and more evident that the plant's defense mechanism against viral infections—based on post-transcriptional gene silencing (PTGS)—and the viral counter defenses based on virus-encoded RNA silencing suppressors (VsRS) are involved in defining tissue invasion patterns during double infections of different hosts (Syller 2012). The 2b protein encoded by CMV and the helper component protease (HC-Pro) encoded by PVY are among the best-characterized VsRSs. The 2b protein uses complex activities to suppress RNA silencing—among which is inhibition of the systemic spread of the silencing signal into newly developing leaves—but it seems unable to suppress the RNA-silencing machinery already established in plant tissues prior to virus invasion. The potyviral HC-Pro is a double-stranded RNA-binding protein that interacts physically with siRNA duplexes and thereby prevents the assemblage of the RNA-induced silencing complex (RISC) (Lakatos et al 2006). However, PVY HC-Pro protein could not inhibit the activity of already-assembled RISC (Lakatos et al 2006), although, in transgenic plants, it could reverse a previously established silencing of the transgene.

Due to the different mode of action on the PTGS pathway, the two CMV and PVY VsRSs may or may not act synergistically when the two viruses are present in mixed infection in the same plant. A simple way to dissect this interaction could be a loss-of-function approach. CMV-FnyΔ2b is a mutant of the CMV-Fny strain with a deletion in the sequence encoding the 2b protein induced by site-directed mutagenesis (Ryabov et al 2001). Thus, this strain is unable to translate the 2b protein so that it can be fruitfully employed to better understand whether the RNA silencing pathway and VsRS proteins may be involved in the regulation of the biologic and molecular interactions between CMV and PVY in mixed infection in tomato.

In mixed-infected tomato protoplasts, the presence of PVY-SON41 enhanced the accumulation of CMV-Fny while that of CMV-FnyΔ2b remained at a very low level in either single or mixed infection. This demonstrates that, at single-cell level, functions absent in the CMV mutant could not be compensated for by PVY-SON41. However, this behavior seems to be host-dependent because, in tobacco protoplasts, the absence of the 2b protein delays but does not inhibit the replication of the CMV-FnyΔ2b mutant (Soards et al 2002). By contrast, CMV-Fny and CMV-FnyΔ2b have a differential effect on PVY-SON41 replication, which, compared with single infections, is reduced by half in mixed infection with CMV-Fny whereas it is doubled in the presence of the CMV-FnyΔ2b mutant. Thus, at single-cell level, the replication of PVY-SON41 is depressed by the 2b protein because protoplasts infected by PVY-SON41 and CMV-FnyΔ2b do not show such an inhibitory effect; rather, the deletion mutant favors the accumulation of the potyvirus. These results are in partial agreement with those obtained with cucumber protoplasts infected with CMV and the potyvirus *Zucchini yellow mosaic virus* (ZYMV) because the effect of coinfection with ZYMV resulted clearly in increased levels of CMV RNA, while the RNA accumulation level of ZYMV remained substantially unaffected (Wang et al 2004).

In locally infected tomato cotyledons, RNAs of both CMV strains accumulated essentially at the same level, while CMV-FnyΔ2b accumulated to a somewhat lesser extent and regardless of the presence or absence of the PVY-SON41. On the contrary, PVY-SON41 in mixed infection with either CMV strain accumulated to slightly higher levels than in single infection, while its accumulation was doubled in mixed infection with CMV-FnyΔ2b. These results are congruent with those obtained at the single-cell level and confirm an inhibitory effect of CMV 2b protein against PVY-SON41. Takeshita et al (2012) called this phenomenon 'local interference,' as it was also observed in *Nicotiana benthamiana* plants mixed-infected by CMV and *Turnip mosaic virus* (TuMV).

In systemically infected leaves, the presence of PVY-SON41 enhanced the accumulation of CMV-Fny and CMV-LS, and this picture correlated positively with the increase in severity of disease symptoms observed in the new vegetation (see above). The CMV-FnyΔ2b mutant was unable to move systemically, but its movement was restored in plants with mixed infection of PVY-SON41. At 1 month after inoculation the abundance of PVY-SON41 in mixed-infected plants was reduced by one-third compared with that of plants infected by the potyvirus alone, confirming the inhibitory effect of CMV. The overall pattern described for both CMV-Fny and CMV-LS remained unchanged at 60 days post-inoculation (dpi) because both CMV strains stayed upregulated and more uniformly distributed in plant tissues in the presence of the potyvirus. Surprisingly, at this time point, PVY-SON41 was upregulated, and better distributed as well, when in mixed infection with CMV-Fny—while in single infections it showed a tendency to decrease and to colonize tomato tissues erratically. In summary, while at 1 month after inoculation the positive effects of PVY-SON41 on CMV-Fny and, at the same time, the inhibitory effects of CMV-Fny on PVY-SON41 are consistent with the observations made in single cells, at very late stages of infection both the viruses seem to profit from double infections and appear to be more uniformly distributed in the tomato plant. This was true even for CMV-FnyΔ2b, which itself showed a poor accumulation, whereas it strongly aided the distribution of PVY-SON41 in the upper leaves. Similar behavior was also seen in the mixed infections of CMV and TuMV in *N. benthamiana* (Takeshita et al 2012) and was also consistent with the results of Zeng and associates (2007), who found that mixed infection of CMV and ZYMV in cucumber and bottle gourd stimulated early stage replication of CMV RNAs and delayed the decline of their accumulation levels.

PVY-SON41 COMPLEMENTS MOVEMENT DEFECTS OF THE CMV-FNYΔ2B MUTANT

In cotyledons of single-infected tomato plants, CMV-FnyΔ2b particles were abundant only in the mesophyll cells but remained confined mostly to the vascular bundle sheath, being unable to enter true-leaf phloem companion cells and immature sieve elements. In plants mixed-infected with PVY-SON41, the deletion mutant was detected in companion cells and immature sieve elements, suggesting a complementation by PVY-SON41 to enter these tissues. Interestingly, PVY-SON41 and CMV-FnyΔ2b were found consistently in the same true-leaf phloem companion cell of plants with mixed infection, whereas this was not the case with CMV-Fny expressing a fully functional 2b protein. Since an inhibitory effect of CMV-Fny 2b protein can be seen also in tomato cotyledons, this observation suggests a bias of CMV-Fny against ingress or replication of PVY-SON41 in the same cell of the vascular tissue. Because it has been proposed that RNA silencing may be hyperactivated in cells that control access to the phloem (Marathe et al 2000), this observation supports the hypothesis that there may be a correlation between the inability of CMV-FnyΔ2b to move systemically in tomato and RNA silencing. Indeed, the complementation for CMV-FnyΔ2b systemic movement is not necessary in tomato plants grown at 15°C, while it is necessary in plants grown at 22°C.

Recent studies have shown that RNA silencing is temperature-dependent, and that it is significantly enhanced at high temperatures (Szittya et al 2003, Chellappan et al 2005, Qu et al 2005) but inhibited at low temperatures. Thus the CMV-FnyΔ2b mutant could not move from the site of infection to the newly developed vegetation as a consequence of its inability to encode a 2b protein, which plays a key role in suppressing plant-driven RNA silencing. Evidence for a differential accumulation of small-interfering RNAs (siRNAs) in the distinct combinations of inocula substantiated this hypothesis, while the low abundance of PVY-SON41-specific siRNAs in mixed infection further confirmed the inhibitory effect of CMV-Fny against the potyvirus. In fact the accumulation of virus-specific siRNAs in infected cells is thought to be proportional to virus replication/accumulation. These results are in partial agreement with the model of Takeshita et al (2012) to study the CMV-TuMV interactions in *N. benthamiana*. In this model a major role of the 2b protein to unload CMV from vascular tissues and the spatial competition between CMV and TuMV under synergism, denoted 'local interference,' was confirmed, while RNA silencing seemed not to be involved in limiting the egression of CMV from vascular tissues and its systemic spread, a role that could be covered by a hitherto unknown mechanism by the 2b protein (Takeshita et al 2012).

CMV-satRNA MAKES A MORE COMPLEX PATTERN

Plants infected by CMV-Fny plus Tfn-satRNA in mixed infection with PVY-SON41 were substantially asymptomatic, as were plants infected by CMV-Fny plus Tfn-satRNA, while control plants infected by CMV-Fny, PVY-SON41, or by a mixture of the two exhibited the symptoms already described. Cillo and associates (2004, 2007) showed that when the CMV-Fny inoculum contains the ameliorative Tfn-satRNA variant, the genomic RNAs are reduced to barely detectable levels and the infection is symptomless. Similarly, tomato plants with a trilateral mixed infection of CMV-Fny, Tfn-satRNA, and PVY-SON41 showed a strong reduction on the accumulation of CMV-Fny while that of the potyvirus was increased and the infection was asymptomatic. The accumulation level of Tfn-satRNA remained substantially unchanged, regardless of the presence of the potyvirus. Although a direct effect of Tfn-satRNA on PVY-SON41 replication could be not ruled out, it seems very likely that in tomato plants with mixed infection, Tfn-satRNA indirectly aided the accumulation of PVY-SON41 by attenuating CMV antagonistic effects through its downregulation. Thus, Tfn-satRNA did perturb the interplay between CMV and PVY-SON41 in tomato, acting as a dominant factor in the interaction because it (i) mitigated the inhibition against PVY-SON41 in that the virus accumulated to high levels; (ii) abolished disease symptoms; and (iii) provided indirect evidence that exacerbation of disease symptoms in double-infected plants was very likely attributable to pathogenicity determinants encoded by CMV. These results do not seem congruent with those reported in other instances in which different hosts, CMV satellites, and potyviruses were used (Pruss et al 1997, Wang et al 2002), thus suggesting that the overall phenotype observed in this study is peculiar to tomato.

CONCLUSION

In this case study it has been shown that synergistic interactions may occur between two unrelated viruses in the same tomato plant with some features that are peculiar to this host. The most striking aspect of this interaction is probably the spatial distribution of the two viruses within the host plant, which seems to be the manifestation of antagonism rather than of synergism. Nonetheless, the final outcome of the interaction is undoubtedly of the synergistic type and is the most interesting and most important from both the virologic and agricultural points of view. From an ecologic and epidemiologic perspective, it is relevant that both viruses accumulated to an evidently increased titer and with a very good distribution up to 2 months after inoculation compared with single infections, thus being available for aphid transmission in any part of the plant where aphid populations feed. It is also relevant that in mixed infection in tomato, CMV-satRNA reduces the accumulation of its helper virus, causing an enhancement of the potyviral counterpart. This could cause a reduced probability of CMV being transmitted by aphids while it enhances the probabilities of transmission for PVY-SON41. A decrease in the efficiency of CMV transmission from CMV-satRNA-infected tomato plants has been documented previously (Escriu et al 2000). In a recent paper, Ziebell et al (2011) showed that tobacco plants systemically infected with CMV-FnyΔ2b exhibited strong resistance to the aphid *Myzus persicae*, indicated by increased numbers of dead aphids. In contrast, aphid survival and colony development was improved on CMV-infected plants compared to mock-inoculated controls. These results highlight a general role for viral silencing suppressor proteins in enhancing vector survival (Ziebell et al 2011). As a matter of speculation it could be hypothesized that in downregulating replication of CMV, the satRNA probably mitigates the positive effects of the 2b protein on aphid populations without abolishing them as in the case of the CMV-FnyΔ2b deletion mutant. This might favor a more efficient transmission of the PVY with relevant eco-epidemiologic consequences that warrant a specific investigation.

REFERENCES

Adams, M.J., Zerbini, F.M., French, R., Rabenstein, F., Stenger, D.C., Valkonen, J.P.T., et al., 2012. Family *Potyviridae*. In: King, A.M.Q. (Ed.), Virus taxonomy, Elsevier, Oxford, pp. 1069–1089.

Alonso-Prados, J.L., Fraile, A., Garcia-Arenal, F., 1997. Impact of cucumber mosaic virus and watermelon mosaic virus 2 infection on melon production in central Spain. Journal of Plant Pathology 79, 131–134.

Aramburu, J., Galipienso, L., Lopez, C., 2007. Reappearance of *Cucumber mosaic virus* isolates belonging to subgroup ib in tomato plants in north-eastern Spain. Journal of Phytopathology 155, 513–518.

Bujarski, J., Figlerowicz, M., Gallitelli, D., Roossinck, M.J., Scott, S.W., et al., 2012. Family *Bromoviridae*. In: King, A.M.Q. (Ed.), Virus taxonomy, Elsevier, Oxford, pp. 965–976.

Chellappan, P., Vanitharani, R., Ogbe, F., Fauquet, C.M., 2005. Effect of temperature on geminivirus-induced RNA silencing in plants. Plant Physiology 138, 1828–1841.

Cillo, F., Finetti-Sialer, M.M., Papanice, M.A., Gallitelli, D., 2004. Analysis of mechanisms involved in the *Cucumber mosaic virus* satellite RNA-mediated transgenic resistance in tomato plants. Molecular Plant-Microbe Interactions 17, 98–108.

Cillo, F., Pasciuto, M.M., De Giovanni, C., Finetti-Sialer, M.M., Ricciardi, L., Gallitelli, D., 2007. Response of tomato and its wild relatives in the genus *Solanum* to *Cucumber mosaic virus* and satellite RNA combinations. Journal of General Virology 88, 3166–3176.

Cillo, F., Mascia, T., Pasciuto, M.M., Gallitelli, D., 2009. Differential effects of mild and severe *Cucumber mosaic virus* strains in the perturbation of microRNA-regulated gene expression map to the 39 sequence of RNA 2. Molecular Plant-Microbe Interactions 22, 1239–1249.

Crescenzi, A., Barbarossa, L., Gallitelli, D., Martelli, G.P., 1993. Cucumber mosaic cucumovirus populations in Italy under natural epidemic conditions and after a satellite-mediated protection test. Plant Disease 77, 28–33.

Ding, S.W., Li, W.X., Symons, R.H., 1995. A novel naturally occurring hybrid gene encoded by a plant RNA virus facilitates long distance virus movement. EMBO Journal 14, 5762–5772.

Escriu, F., Perry, K.L., García-Arenal, F., 2000. Transmissibility of *Cucumber mosaic virus* by *Aphis gossypii* correlates with viral accumulation and is affected by the presence of its satellite RNA. Phytopathology 90, 1086–1072.

Fraile, A., Alonso-Prados, J.L., Aranda, M.A., Bernal, J.J., Malpica, J., García-Arenal, F., 1997. Genetic exchange by recombination or reassortment is infrequent in natural populations of a tripartite RNA plant virus. Journal of Virology 71, 934–940.

Gallitelli, D., 2000. The ecology of *Cucumber mosaic virus* and sustainable agriculture. Virus Research 71, 9–21.

Gallitelli, D., Di Franco, A., Vovlas, C., Kaper, J.M., 1988. Infezioni miste del virus del mosaico del cetriolo (CMV) e di potyvirus in colture ortive di Puglia e Basilicata. Informatore fitopatologico 12, 57–64.

Gallitelli, D., Martelli, G.P., Gebre-Selassie, K., Marchoux, G., 1995. Progress in the biological and molecular studies of some important viruses of *Solanaceae* in the Mediterranean. Acta Horticulturae 412, 503–514.

García-Arenal, F., Palukaitis, P., 2008. Cucumber mosaic virus. In: Mahy, B.W.J., Van Regenmortel, M.H.V. (Eds.), Desk encyclopedia of plant and fungal virology, Elsevier, Amsterdam, pp. 171–176.

González, I., Martínez, L., Rakitina, D.V., Lewsey, M.G., Atencio, F.A., Llave, C., Kalinina, N.O., Carr, J.P., Palukaitis, P., Canto, T., 2010. *Cucumber mosaic virus* 2b protein subcellular targets and interactions: their significance to RNA silencing suppressor activity. Molecular Plant-Microbe Interactions 23, 294–303.

Goto, K., Kobori, T., Kosaka, Y., Natsuaki, T., Masuta, C., 2007. Characterization of silencing suppressor 2b of cucumber mosaic virus based on examination of its small RNA-binding abilities. Plant Cell Physiology 48, 1050–1060.

Hanssen, I., Lapidot, M., 2012. Major tomato viruses in the Mediterranean basin. Advances in Virus Research 84, 31–66.

Ji, L.H., Ding, S.W., 2001. The suppressor of transgene RNA silencing encoded by cucumber mosaic virus interferes with salicylic acid-mediated virus resistance. Molecular Plant-Microbe Interactions 14, 715–724.

Jordá, C., Alfaro, A., Aranda, M.A., Moriones, E., García-Arenal, F., 1992. Epidemic of cucumber mosaic virus plus satellite RNA in tomatoes in Eastern Spain. Plant Disease 76, 363–366.

Kaper, J.M., Gallitelli, D., Tousignant, M., 1990. Identification of a 334-ribonucleotide viral satellite as principal aetiological agent in a tomato necrosis epidemic. Research in Virology 141, 81–95.

Lakatos, L., Csorba, T., Pantaleo, V., Chapman, E. J., Carrington, J. C., Liu, Y.-P, Dolja, V.V., Dolja, L F., López-Moya, J., Burgyàn, J., 2006. Small RNA binding is a common strategy to suppress RNA silencing by several viral suppressors. EMBO Journal 25, 2768–2780.

Lecoq, H., Desbiez, C., 2012. Viruses of cucurbit crops in the Mediterranean region: an ever-changing picture. Advances in Virus Research 84, 67–126.

Lewsey, M., Robertson, F.C., Canto, T., Palukaitis, P., Carr, J.P., 2007. Selective targeting of miRNA-regulated plant development by a viral counter-silencing protein. Plant Journal 50, 240–252.

Lewsey, M., Surette, M., Robertson, F.C., Ziebell, H., Choi, S.H., Ryu, K.H., Canto, T., Palukaitis, P., Payne, T., Walsh, J.A., Carr, J.P., 2009. The role of the *Cucumber mosaic virus* 2b protein in viral movement and symptom induction. Molecular Plant-Microbe Interactions 22, 642–654.

Luis-Arteaga, M., Alvarez, J.M., Alonso-Prados, J.L., Bernal, J.J., Garcia-Arenal, F., Laviña, A., Batlle, A., Moriones, E., 1998. Occurrence, distribution and relative incidence of mosaic viruses infecting field-grown melon in Spain. Plant Disease 82, 979–982.

Marathe, R., Anandalakshmi, R., Smith, T.H., Pruss, G.J., Vance, V.B., 2000. RNA viruses as inducers, suppressors and targets of post-transcriptional gene silencing. Plant Molecular Biology 43, 295–306.

Martin, S., Elena, S.F., 2009. Application of the game theory to the interaction between plant viruses during mixed infections. Journal of General Virology 90, 2815–2820.

Mascia, T., Cillo, F., Fanelli, V., Finetti-Sialer, M.M., De Stradis, A., Palukaitis, P., Gallitelli, D., 2010. Characterization of the interactions between *Cucumber mosaic virus* and *Potato virus Y* in mixed infections in tomato. Molecular Plant-Microbe Interactions 23, 1514–1524.

Moury, B., Verdin, E., 2012. Viruses of pepper crops in the Mediterranean basin: a remarkable stasis. Advances in Virus Research 84, 127–162.

Palukaitis, P., García-Arenal, F., 2003. Cucumoviruses. Advances in Virus Research 62, 241–323.

Palukaitis, P., Kaplan, I.B., 1997. Synergy of virus accumulation and pathology in transgenic plants expressing viral sequences. In: Tepfer, M., Balázs, E. (Eds.), Virus-resistant transgenic plants: potential ecological impact, Springer-Verlag, Berlin, pp. 77–84.

Palukaitis, P., Zaitlin, M., 1997. Replicase-mediated resistance to plant virus disease. Advances in Virus Research 48, 349–377.

Pruss, G., Ge, X., Shi, X.M., Carrington, J.C., Vance, B.V., 1997. Plant viral synergism: the potyviral genome encodes a broad range pathogenicity enhancer that trans-activates replication of heterologous viruses. Plant Cell 9, 859–868.

Qu, F., Ye, X.H., Hou, G.C., Sato, S., Clemente, T.E., Morris, T.J., 2005. RDR6 has a broad-spectrum but temperature-dependent antiviral defense role in *Nicotiana benthamiana*. Journal of Virology 79, 15209–15217.

Rochow, W.F., 1972. The role of mixed infections in the transmission of plant viruses by aphids. Annual Review of Phytopathology 10, 101–124.

Ryabov, E.V., Fraser, G., Mayo, M.A., Barker, H., Taliansky, M., 2001. *Umbravirus* gene expression helps *Potato leafroll virus* to invade mesophyll tissues and to be transmitted mechanically between plants. Virology 286, 363–372.

Sáenz, P., Salvador, B., Simón-Mateo, C., Kasschau, C.K., Carrington, J.C., García, J.A., 2002. Host-specific involvement of the HC protein in the long-distance movement of potyviruses. Journal of Virology 76, 1922–1931.

Shi, B.-J., Palukaitis, P., Symons, R.H., 2002. Differential virulence by strains of *Cucumber mosaic virus* is mediated by the 2b gene. Molecular Plant-Microbe Interactions 15, 947–955.

Soards, A.J., Murphy, A.M., Palukaitis, P., Carr, J.P., 2002. Virulence and differential local and systemic spread of *Cucumber mosaic virus* in tobacco are affected by the CMV 2b protein. Molecular Plant-Microbe Interactions 15, 647–653.

Syller, J., 2012. Facilitative and antagonistic interactions between plant viruses in mixed infections. Molecular Plant Pathology 13, 2014–2216.

Szittya, G., Silhavy, D., Molnar, A., Havelda, Z., Lovas, A., Lakatos, L., Banfalvi, Z., Burgyan, J., 2003. Low temperature inhibits RNA silencing-mediated defence by the control of siRNA generation. EMBO Journal 22, 633–640.

Takeshita, M., Koizumi, E., Noguchi, M., Sueda, K., Shimura, H., Ishikawa, N., Matsuura, H., Ohshima, K., Natsuaki, T., Kuwata, S., Furuya, N., Tsuchiya, K., Masuta, C., 2012. Infection dynamics in viral spread and interference under the synergism between *Cucumber mosaic virus* and *Turnip mosaic virus*. Molecular Plant-Microbe Interactions 25, 18–27.

Tomlinson, J.A., 1987. Epidemiology and control of virus diseases of vegetables. Annals of Applied Biology 110, 661–681.

Varveri, C., Boutsika, K., 1999. Characterization of cucumber mosaic cucumovirus isolates in Greece. Plant Pathology 48, 95–100.

Wang, Y., Gaba, V., Yang, J., Palukaitis, P., Gal-on, A., 2002. Characterization of synergy between *Cucumber mosaic virus* and potyviruses in cucurbit hosts. Phytopathology 92, 51–58.

Wang, Y., Lee, K.C., Gaba, V., Wong, S.M., Palukaitis, P., Gal-On, A., 2004. Breakage of resistance to *Cucumber mosaic virus* by co-infection with *Zucchini yellow mosaic virus*: enhancement of CMV accumulation independent of symptom expression. Archives of Virology 149, 379–396.

Wege, C., Siegmund, D., 2007. Synergism of a DNA and an RNA virus: enhanced tissue infiltration of the begomovirus *Abutilon mosaic virus* (AbMV) mediated by *Cucumber mosaic virus* (CMV). Virology 357, 10–28.

Whitham, S.A., Wang, Y., 2004. Roles for host factors in plant viral pathogenicity. Current Opinion in Plant Biology 7, 365–371.

Ye, J., Qua, J., Zhang, J.F., Geng, Y.F., Fang, R.X., 2004. A critical domain of the *Cucumber mosaic virus* 2b protein for RNA silencing suppressor activity. FEBS Letters 583, 101–106.

Zeng, R., Liao, Q., Feng, J., Li, D., Chen, J., 2007. Synergy between *Cucumber mosaic virus* and *Zucchini yellow mosaic virus* on *Cucurbitaceae* hosts tested by real-time reverse transcription-polymerase chain reaction. Acta Biochimica Biophysica Sinica (Shanghai) 39, 431–437.

Ziebell, H., Murphy, A.M., Groen, S.C., Tungadi, T., Westwood, J.H., Lewsey, M.J., Moulin, M., Kleczkowski, A., Smith, A.G., Stevens, M., Powell, G., Carr, J.P., 2011. *Cucumber mosaic virus* and its 2b RNA silencing suppressor modify plant-aphid interactions in tobacco. Scientific Reports 1, 187. DOI:10.1038/srep00187.

Chapter 11

Methods of diagnosis, stability, transmission, and host interaction of *Papaya lethal yellowing virus* in papaya

J. Albersio A. Lima and Aline Kelly Q. Nascimento
Federal University of Ceará, Laboratory of Plant Virology, Fortaleza, Brazil

Roberto C.A. Lima
BioClone, Eusébio—CE, Brazil

Verônica C. Oliveira and Geórgia C. Anselmo
Federal University of Ceará, Laboratory of Plant Virology, Fortaleza, Brazil

INTRODUCTION

Around the world, virus infections seriously affect the quality and quantity of agricultural products, including papaya (*Carica papaya* L.). Papaya is cultivated in tropical and subtropical areas, although it originated in Central America. It is known worldwide for its food and medicinal value (Manica 1982). Papaya is a fruit crop harvested all year around and is economically important to Brazil, which has a distinctive position in world production. Brazil has excellent conditions for producing tropical fruit crops, including papaya, and it is the largest producer of this crop in the world, with an estimated production of 1,890,000 tons in 2009 in a cultivated area close to 35,000 ha (IBGE 2012). Tropical fruit production has been the agribusiness activity with the greatest growth over the last 10 years in Northeastern Brazil, and papaya production has contributed the most to this intensive growth.

Unfortunately, papaya productivity is affected by several factors, mainly infectious diseases, especially those caused by viruses, which have been responsible for significant losses to the crop all over the country (Barbosa & Paguio 1982, Lima et al 1994, 2001, Lima & Lima 2002, Lima et al 2002, Santos et al 2003, Ventura et al 2004, Amaral et al 2006, Nascimento et al 2010). These virus diseases have been responsible for the most significant papaya losses around the world. The most important viruses that infect papaya in Northeastern

Brazil are *Papaya ringspot virus* (PRSV), family Potyviridae, genus *Potyvirus* (Lima et al 1994); *Papaya lethal yellowing virus* (PLYV), genus *Sobemovirus* (Silva et al 1997, Nascimento et al 2010); and Papaya meleira virus (PMeV), which is still being characterized taxonomically by the International Committee on Taxonomy of Virus (ICTV) (Marciel-Zambolim et al 2003).

PRSV causes a major disease in papaya, significantly reducing its production wherever it is cultivated (Lima & Gomes 1975, Conover 1976, Barbosa & Paguio 1982, Yeh & Gonsalves 1984, Lima & Camarço 1997, Gonsalves 1998). This is also true for Brazil, where the average number and weight of fruit per plant is reduced by 22% and 60%, respectively (Lima & Gomes 1975, Almeida & Carvalho 1978, Barbosa & Paguio 1982). The spectrum of PRSV systemic hosts is limited to members of the Caricaceae and Cucurbitaceae. According to their biologic properties, PRSVs are classified as type papaya (PRSV-P), which infects papaya and cucurbits, and type watermelon (PRSV-W), which infects only cucurbits (Purcifull et al 1984). PRSV-W was formerly named *Watermelon mosaic virus 1* (WMV-1), and it is serologically related to PRSV-P but not to *Watermelon mosaic virus* (WMV), another member of the genus *Potyvirus* (Purcifull & Hiebert 1979).

Symptoms in papaya are characterized by prominent mosaic and chlorosis on leaf lamina, leaf distortions, water-soaked oily streaks on the petioles and upper part of the trunk, and ring spots on the fruits, which are the basis for the common name of the disease. In severe symptoms, the distortion of the young leaves often results in the development of a shoestring appearance that resembles mite damage. Trees that are infected at a young stage remain stunted and do not produce economic crops (Barbosa & Paguio 1982, Purcifull et al 1984, Gonsalves 1998). The general properties of PRSV are similar to members of the genus *Potyvirus*, family Potyviridae. The virus is transmitted by mechanical inoculation and is not transmitted through seed. All types of PRSV and their isolates are transmitted by several species of aphid in a nonpersistent manner.

The genome of viruses of the genus *Potyvirus* is a unique molecule of positive-sense single-stranded RNA (ssRNA) of approximately 10 kb that contains a large open reading frame (ORF) for a polyprotein of about 3000 to 3300 amino acids; this polyprotein is processed by cleavage with several proteases, producing 12 functional proteins. Recent studies have demonstrated the presence of a short ORF called 'pretty interesting potyvirus' (PIPO) inserted in the P3 cistron and expressed by ribosomal frame-shifting to produce P3_PIPO; this has been shown to be essential for intercellular virus movement.

The virus particles are ca. 750–800 nm long and 12 nm wide, and the 5′ end of the ssRNA is linked to a virus protein (VPg). The 3′ end is polyadenylated (PoliA) (Shukla et al 1994, King et al 2012).

Producing papaya in isolated fields free from a source of virus has been recommended as a measure for controlling PRSV (Rezende & Costa 1993). However, other more effective and lasting control methods have been implemented lately, such as cross protection and the use of transgenic plants. PRSV strains HA5-1 and HA6-1, obtained from PRSV-P by induced mutation (Yeh & Gonsalves 1984), were introduced into Taiwan and showed considerable potential for controlling papaya ringspot by pre-immunization (Wang et al 1987, Yeh et al 1988). Genetically engineered papaya has also been used to successfully control the disease caused by PRSV in Hawaii (Gonsalves et al 2007, 2010).

The new disease called Meleira, or 'sticky disease,' caused by a virus (Marciel-Zambolim et al 2003) is causing serious damage to papaya production in several regions of Brazil, where it is spreading rapidly, reaching 100% incidence in some areas (Nakagawa et al 1987, Rodrigues et al 1989a,b). It is characterized by latex exudation from the fruits, which oxidizes, resulting in a sticky surface. Symptoms are also seen on the petiole and on the borders of young leaves, before the fruits are produced, which become necrotic after latex exudation. The fruits show irregular shape with yellow spots; this decreases their commercial value. The latex from fruits on infected plants is less viscous, and it cannot coagulate. The first symptoms appear when the plants are approximately 6 months old, mainly along the borders of the young leaves, resulting in a kind of burn and modification in the leaf's shape (Rodrigues et al 1989a,b). The presence of isometric virus particles 50 nm in diameter has been detected in leaves and latex from infected plants (Kitajima et al 1993). The virus particles were purified, making it possible to fulfill Koch's postulates for the disease by reproducing the disease symptoms in healthy plants inoculated with the purified virus preparation (Marciel-Zambolim et al 2003).

Lethal yellowing of papaya is a disease caused by *Papaya lethal yellowing virus* (PLYV) that occurs only in Northeastern Brazil. The symptoms are characterized by progressive leaf yellowing and greenish circular spots on the fruits. The virion is an isometric particle of ca. 30 nm, with an ssRNA genome of ca. 1.6×106 Da and a single-component coat protein of ca. 34.7 kDa. Although no biologic vector has been confirmed, the virus is spreading every year, probably by infected plantlets and contaminated tools. The virus infects only *C. papaya*, *Jacaratia heterophyla*, *J. spinosa*, *Vasconcellea cauliflora*, *V. quercifolia*, and *V. monoica*, all from the family Caricaceae. None of the 82 other species from 16 families are infected when inoculated with PLYV. The virus is very stable and can be detected in dried roots and leaves maintained in laboratory conditions for up to 120 days. The virus has the following physical properties: thermal inactivation point 80°C, longevity *in vitro* 60 days, and dilution end point 10^{-6}. High concentrations of virus particles can be purified from infected papaya, and good polyclonal antisera have been obtained. Phylogenetic analysis of the RNA-dependent RNA polymerase (RdRp) nucleotide sequences indicated that PLYV is a member of the genus *Sobemovirus*.

OCCURRENCE AND GEOGRAPHIC DISTRIBUTION OF *PAPAYA LETHAL YELLOWING VIRUS*

The disease caused by PLYV has been detected only in Northeastern Brazil. It was first identified in the State of Pernambuco (Loreto et al 1983, Nascimento et al 2010), followed by the States of Rio Grande do Norte (Oliveira et al 1989), Ceará (Lima & Santos 1991), Paraíba (Camarço et al 1998), and Bahia (Vega et al 1988), but it was never confirmed in the State of Bahia. The disease has become a serious problem for papaya producers in the region because it causes serious damage to crops—and because of its increasing dispersion throughout the orchards (Lima & Santos 1991, Kitajima et al 1992, Camarço et al 1996, Lima & Camarço 1997, Teixeira et al 1999, Lima & Lima 2002, Ramos et al 2008, Nascimento et al 2010). The virus is spreading throughout the region in an east-to-west direction because it was detected first in the State of Pernambuco, followed by Paraiba, Rio Grande do Norte, and Ceará. In the State of Ceará, the virus was first detected in the counties close to Rio Grande do Norte and has not yet been detected in the State of Piaui (West of Ceará) or in the Ceará counties close to the State of Piaui. Extensive and detailed surveys in the producing areas of Ceará and Rio Grande do Norte revealed the presence of PLYV to a lesser degree in the orchards, always occurring in small patches of infected plants within or near the edge of the orchards. This kind of disease distribution indicates that the virus is probably first introduced into an orchard by infected plantlets or contaminated tools. The first infected plants constitute the initial focus of the virus inside, or near, the edge of the orchard, and the virus spreads to neighboring plants by the grower's actions.

METHODS OF VIRUS DIAGNOSIS

Attempts to control plant diseases, including those caused by viruses, without sufficient information about their causal agents, their dissemination, and surviving properties, usually result in inadequate control and often in total failure. So, any attempt to establish a control program for a plant disease must be, always, preceded by a correct and accurate diagnosis.

Several methods can be used for arriving at a correct and definitive diagnosis of a plant disease caused by a virus; at the beginning of the study of plant virology, the symptoms represent important characteristics for identifying and characterizing the causal virus (Purcifull & Batchelor 1977, Almeida & Lima 2001, Naidu & Hughes 2001, Purcifull et al 2001, Astier et al 2007, Mulholland 2009, Lima et al 2012). Nevertheless, it is often impossible to diagnose plant virus infections by merely observing host symptoms. The symptoms alone are, usually, inadequate for a complete and correct diagnosis of a plant virus disease; this is because the symptoms caused by viruses vary according to the plant variety involved, the environmental conditions, the strain of the virus, the fact that sometimes different viruses can cause similar symptoms in the same plant species, and the fact that sometimes the disease could result from the synergistic effect of infection caused by two different viruses. However, a bioassay using a series of indicator plants remains an indispensable tool for detection and identification of plant viruses, and the original symptoms are still of great importance for plant virus denomination, as in the case of the disease caused by PLYV in papaya (Nascimento et al 2010). On the other hand, PLYV causes very typical and characteristic symptoms in the fruits (Fig. 11.1), which are distinguishable from those caused by the other virus diseases of papaya, mainly PRSV, which also causes symptoms on the fruits. Additionally, PLYV does not infect any virus indicator plants tested so far. So, symptoms and host range are very important in identifying a PLYV infection on papaya.

Several laboratory methods have been developed and adapted for virus identification, and serology is one of the most specific and easiest methods for obtaining a rapid and precise identification of plant viruses (Naidu & Hughes 2001, Purcifull et al 2001, Lima et al 2005, Astier et al 2007). Generally the methods that involve antigen–antibody reactions *in vitro* are simple and do not require sophisticated and expensive equipment.

Production of polyclonal antiserum

The most serious limitation of serology for plant virus identification and detection is the difficulty in producing a good virus-specific antiserum. Most antisera used for plant virus identification and detection are usually prepared by immunizing warm-blooded animals with purified plant virus or different types of viral protein. Most plant viruses, including PLYV, are good and effective antigens that, when artificially injected into a suitable warm-blooded animal, stimulate the production of specific antibodies that can be used in different

FIGURE 11.1 Papaya exhibiting symptoms caused by *Papaya lethal yellowing virus*. (A) A plant in a papaya orchard in the State of Ceará, Brazil, exhibiting the first symptoms of disease with a progressive leaf yellowing in the upper third portion of the plant canopy. (B) A plant showing an advanced stage of yellowing and death of the leaves throughout the entire plant canopy. (C) A papaya plant in an urban area with progressive leaf yellowing, acting as natural source of the virus. (D) Fruit with greenish and yellowish circular spots caused by PLYV.

serologic tests. Rabbits are commonly chosen for the production of polyclonal antiserum because they are easily housed and adapt well to being handled, but other animals, such as mice, goats, and chickens, can also be used. Several routes have been used to immunize rabbits with plant viruses, including the intravenous and intramuscular routes and through the foot pad.

The protocols of rabbit immunization vary greatly, but the following general immunization procedure has given satisfactory results for the preparation of good-titer plant virus antiserum, including for PLYV. The rabbits are immunized with purified virus preparation by three weekly injections, using, in each immunization, an aliquot of 500 μl from the purified virus preparation (0.5–1.0 mg ml^{-1}) emulsified with equal volume of Freund's incomplete adjuvant. Fifteen days after the last injection the rabbit can be bled for antiserum. Blood samples of 10 to 50 ml are taken by nicking the marginal ear vein of the animal and collecting blood in glass centrifuge tubes. The tubes with the blood samples are maintained in a water bath at 37°C for 1 hour, and the clear serum from a second centrifugation is collected, evaluated by indirect ELISA and/or Ouchterlony double-diffusion tests, and stored at −20°C. A high titer (1:512 000) PLYV polyclonal antiserum in indirect ELISA has been obtained from an immunized rabbit. Because of the high concentration of virus particles in infected tissues, and their high immunogenicity, polyclonal antiserum specific for PLYV was also obtained by oral immunization of rabbits with extracts from infected leaves or purified virus preparation.

Serologic methods used in virus detection

Several serologic techniques have already been developed for the identification and characterization of plant viruses, and the advent of ELISA has facilitated the use of serology for virus identification on a large scale (Van Regenmortel 1982, Hampton et al 1990, Naidu & Hughes 2001, Purcifull et al 2001, Lima et al 2005, Astier et al 2007, Nascimento et al 2010, Lima et al 2012). ELISA is a very specific and sensitive serologic technique introduced for the study and identification of plant viruses in the 1970s (Voller et al 1976, Clark & Adams 1977) in order to be able to detect virus particles at very low concentrations. Because of its adaptability, high sensitivity, and economic advantage in the use of reagents, ELISA is used in a wide range of situations, especially for indexing a large number of samples in a relatively short period of time. Although different variations of this serologic technique have been developed, the direct and the indirect ELISA are the most frequently used methods for the diagnosis of plant virus diseases (Clark & Bar-Joseph 1984, Almeida & Lima 2001, Lima et al 2012).

The following variations of the ELISA technique were successfully used for the detection of PLYV in infected plant tissues, contaminated soil, and water: indirect ELISA, immune precipitation ELISA (IP-ELISA), and a simple kit for plate-trapped antigen ELISA.

Indirect ELISA, or the plate-trapped antigen (PTA-ELISA) technique

As this serologic technique was initially developed to avoid the inconvenience and the difficulties of conjugating the enzyme with the IgG specific for each virus species to be used in the second layer of antibodies in direct ELISA, it requires antibodies produced in two different animal species.

Initially, in this method, the virus particles (the antigens) are trapped in the wells of the plate. Antibodies, raised in rabbits against viral antigens, are added to the wells. Use is then made of a 'universal IgG–enzyme conjugate,' which reacts with antibodies raised against all virus species, including PLYV. This so-called universal conjugate consists of IgGs, produced against the IgGs from the animal in which the virus antibodies had been raised, linked to a specific enzyme; the enzyme is selected according to the indicator substrate used. Alkaline phosphatase is the most common enzyme used; it acts on the substrate p-nitrophenol phosphate. As the PLYV antibodies were produced in rabbits, an anti-rabbit IgG—produced in goats—was used in the universal conjugate for all the indirect ELISA tests. The detecting antibody conjugate binds specifically to the primary virus-specific antibodies.

In this method, the wells of the ELISA plate are initially covered with extracts from PLYV-infected and healthy plant samples prepared in the proportion of 1:10 in carbonate buffer, pH 9.6. Then, the PLYV particles are covered with a layer of specific antibodies produced in rabbits. The antigen–antibody complexes (formed by combination of the viral antigens with specific antibodies) are then covered with the 'universal conjugate' consisting of anti-rabbit IgGs (produced in goats) linked to the enzyme alkaline phosphatase. The enzyme-linked anti-rabbit IgGs of the conjugate bind to the PLYV antibodies—which had reacted with the virus particles at the bottom of the ELISA plate wells; this binding is detected by colorimetric changes of the substrate p-nitrophenol phosphate that was added to the wells. After 20 and 40 minutes, the plates were analyzed in an ELISA plate reader using a filter for the 405 nm wavelength. Because a single universal antibody-conjugate can be used for detecting a wide range of plant viruses, the indirect ELISA technique is economic, practical, and suitable for virus detection in disease diagnosis and quarantine programs. Although the indirect ELISA technique is not very specific for plant virus strain or species identification, it can be used for virus species differentiation by antiserum cross absorption.

Similarly, a polyclonal antiserum can be absorbed with an extract from healthy plants to avoid back-cross reaction with plant proteins. For absorption of the polyclonal antiserum, one volume of the antiserum is mixed with two volumes of concentrated extract from healthy plants and the mixture is incubated at 37°C for 3 hours. The mixture of antiserum and extracts from healthy plants is centrifuged at 10,000 g for 10 minutes and the pellet is discharged. The absorbed polyclonal antiserum should not interfere with the results by reacting with extracts from healthy plants in all ELISA procedures.

Immune virus particle precipitation followed by ELISA (IP-ELISA)

Considering the problems with plant viruses where particles are not well adsorbed in the ELISA plate wells, this new ELISA technique (Fig. 11.2) was developed and validated for the detection of plant viruses from different families and genera, especially those from the genus *Comovirus* (Lima et al 2011b). This technique was successfully used to detect PLYV in infected plant tissue. As with the other ELISA procedures, approximately 2.0 g of PLYV infected plant tissues were ground in ELISA extraction buffer and 1.0 ml from the obtained extract was mixed with an equal volume of specific antiserum diluted to 1:1000 (v/v). The mixture of infected plant extract and the antiserum was incubated at 37°C for 3 hours and centrifuged at 5000 g for 10 minutes. The pellet containing the virus particles linked to the antibodies was resuspended in ELISA extraction buffer and used as in conventional indirect ELISA. The IP-ELISA was efficient for detecting PLYV and also for detecting virus species from different families and genera in different kinds of infected tissues. The use of IP-ELISA for detecting viruses was a sensitive and practical diagnostic technique for plant viruses, especially for *Cowpea severe mosaic virus* (CPSMV) and *Squash mosaic virus* (SQMV), genus *Comovirus* (Lima et al 2011b), in which the virus particles seem not to adsorb well in the bottom of the plate wells (personal observation).

A simple kit for plate-trapped ELISA

Companies dealing with immunobiologic products have developed practical kits for direct ELISA or double antibody sandwich (DAS-ELISA) but not kits suitable for indirect ELISA or PTA-ELISA. Considering the great problem of including infectious plant viruses in DAS-ELISA kits, a simple kit for PTA-ELISA was developed for plant virus identification, using PLYV as a model. Extracts from PLYV-infected papaya plant tissues were added to the ELISA plate wells, which were then sealed with plastic and kept in a refrigerator under laboratory conditions for different periods of time. At 10-day intervals, the plates were tested by the regular PTA-ELISA method, and after more than 150 days of incubation the plate-trapped antigen showed excellent results when used for PLYV detection, presenting absorption reading values over three times the values obtained for the respective controls with extracts of healthy plants. The PTA-ELISA kit also showed excellent results with five other virus species from the genera *Comovirus* (*Squash mosaic virus*, SqMV, and *Cowpea severe*

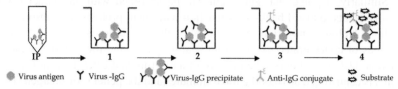

FIGURE 11.2 Immune precipitation enzyme-linked immunosorbent assay for the detection of *Papaya lethal yellowing virus* in infected papaya plants (diagrammatic).

mosaic virus, CPSMV), *Cucumovirus* (*Cucumber mosaic virus*, CMV) and *Potyvirus* (*Cowpea aphid-borne mosaic virus*, CABMV and *Zucchini yellow mosaic virus*, ZYMV), with ELISA absorption reading values for all the ELISA plate wells previously treated individually with PLYV, SqMV, CPSMV, CMV, CABMV, and ZYMV over three times the values obtained for the respective controls with extracts of healthy plants. The plate-trapped virus, together with its specific antiserum, could constitute a simple PTA-ELISA kit (Fig. 11.3), which permits the exchange of antisera between virologists without transferring infectious viruses from one laboratory to another to be used as control (Lima & Nascimento 2012).

Molecular techniques for virus detection

Although serology has been used extensively for plant virus identification on a large scale, the use of molecular techniques for plant virus identification and characterization is increasing throughout the world (Lima et al 2012). Several molecular methods have been developed for diagnosis and characterization of plant viruses, and the reverse transcription polymerase chain reaction (RT-PCR) has been shown to be a suitable method of research with RNA plant viruses (Ahlquist et al 1984, Mullis et al 1986, Mullis 1990, Lima et al 2012).

Polymerase chain reaction (PCR)

Because PCR is an *in vitro* method for amplifying target nucleic acid sequences, the speed, specificity, sensitivity, and versatility of this nucleic acid-based

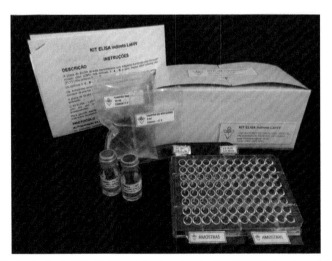

FIGURE 11.3   A simple PTA-ELISA kit developed at the Plant Virus Laboratory of the Federal University of Ceará. It is composed of an ELISA plate, with its wells treated with the antigen of the specific plant virus and extracts from a healthy plant; virus-specific antiserum; and buffers for dilution of the antiserum and preparation of plant samples.

detection system has made it suitable in many areas of research, including plant virology (Mullis et al 1986, Naidu & Hughes 2001, Lima et al 2012).

Reverse transcription polymerase chain reaction (RT-PCR)

Several variations of the PCR technique have been developed for identification of plant viruses with DNA genomes (Naidu & Hughes 2001, Lima et al 2012), and a method developed by Ahlquist et al (1984) is successfully used for the detection of RNA viruses by reverse transcription to produce a cDNA for amplification by PCR (RT-PCR). Because PCR and RT-PCR have the power to amplify a target nucleic acid present at an extremely low level in a complex mixture of heterologous sequences, they have become attractive and efficient methods for the diagnosis of plant virus diseases, including that caused by PLYV, which has been easily detected by RT-PCR using the specific primers PLYV-1: 5′ CTGAAGCGGATATTTCTGG 3′ and PLYV-2: 5′ GTGTATG-GCATA CAGTTATC 3′ (Silva et al 2000, Lima et al 2012).

Real-time quantitative PCR

A novel real-time quantitative PCR assay was developed for the detection and quantification of plant viruses (Mumford et al 2000), and the use of this technique has provided good results for the detection of PLYV in infected papaya tissue, demonstrating its potential for research with PLYV.

Immune precipitation reverse transcription polymerase chain reaction (IP-RT-PCR)

A variant form of RT-PCR, which combines the technical advantages of PCR with the practical advantages of serology, was developed for the detection of several different plant viruses, including PLYV (Lima et al 2011a). This new RT-PCR-based method involves the virus immune precipitation approach, previously described, and for this reason it is designated IP-RT-PCR (Fig. 11.4). This technique (Fig. 11.5) has also been validated for detecting the presence of four other virus species in different infected plant tissues: (a) cowpea (*Vigna unguiculata* subsp. *unguiculata*): *Cowpea severe mosaic virus*

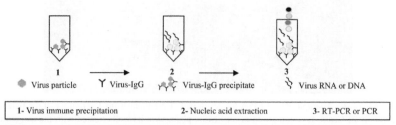

FIGURE 11.4 Immune precipitation polymerase chain reaction IP-RT-PCR for detection of plant viruses (diagrammatic).

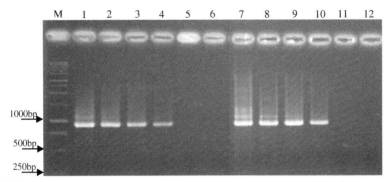

FIGURE 11.5   Results of reverse transcription polymerase chain reaction (RT-PCR) and immunoprecipitation PCR (IP-RT-PCR) of plant tissues infected with *Papaya lethal yellowing virus* at different dilutions of the plant extracts. Lane M: DNA ladder with standards of indicated length in kb; lanes 1–4: RT-PCR of PLYV-infected papaya in the dilutions of 1:10 (1), 1:100 (2), 1:1000 (3), and 1:10 000 (4); lanes 5, 6: RT-PCR of healthy papaya; lanes 7–10: IP-RT-PCR of PLYV-infected papaya at dilutions of 1:10 (7), 1:100 (8), 1:1000 (9), and 1:10 000 (10); and lanes 11, 12: IP-RT-PCR of healthy papaya. The gels were stained with ethidium bromide and analyzed under UV light.

(CPSMV), subfamily *Comovirinae*, genus *Comovirus*, and *Cucumber mosaic virus* (CMV), family Bromoviridae, genus *Cucumovirus*; (b) melon (*Cucumis melo*): *Squash mosaic virus* (SqMV), subfamily *Comovirinae*, genus *Comovirus*; and (c) watermelon (*Citrullus lanatus*): *Zucchini yellow mosaic virus* (ZYMV), family Potyviridae, genus *Potyvirus*. The PCR approach is very specific, simple, and practical; it minimizes problems with RNA extraction and combines the specificity of serology with the technical advantages of virus nucleic acid amplification.

VIRUS PROPERTIES, SYMPTOMS, AND HOST RANGE

Virions of the *Papaya lethal yellowing virus* (PLYV), a member of the genus *Sobemovirus*, are isometric particles ca. 30 nm in diameter (Fig. 11.6). A great number of isometric virus particles can be detected by electron microscopy in the cytoplasm and vacuoles of cells from leaves and fruits of infected plants. The virus genome is a unique fragment of ssRNA of ca. 1.6×10^6 Da, and its coat protein is composed by a single protein component of ca. 34.7 kDa. Characterization of the virus genome indicated that the RNA-dependent RNA polymerase (RdRp) cistron consists of 927 nt, and the cistron for the coat protein (CP) is 497 nt, with an overlapping region of 326 nt. The RdRp has two conserved motifs usually found in viral RdRp: GDD and FCSH. The CP has the conserved motif MPYTVGTWLRGVASNWSK found in all members of the genus *Sobemovirus* (Daltro et al 2012).

The origin of PLYV is unknown, but the virus could have come from a native wild host, where it could be surviving in natural conditions, or it could have resulted from a mutation of another plant virus that occurred in the region.

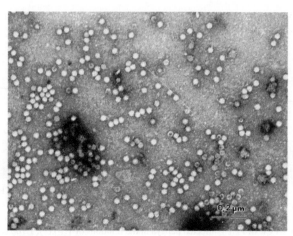

FIGURE 11.6 Electron micrograph of purified particles of *Papaya lethal yellowing virus.* *Courtesy of Professor Elliot W Kitajima.*

The PLYV causes serious symptoms that begin with a progressive leaf yellowing in the upper third part of the plant canopies, which wilt and finally die (Fig. 11.1). Greenish circular spots appear on the fruits, which turn yellowish when they are ripening (Fig. 11.1). Some studies demonstrated that the ripening process of fruits from infected plants is retarded and the pulp hardens, reducing the value of the fruits for marketing and especially for export. Inoculated young plants show mosaic, leaf distortion, and yellowing.

The host range of PLYV is restricted to plant species from the family Caricaceae, but the virus infects all commercial papaya types and varieties. Greenhouse experiments demonstrated that the virus also infects *Jacaratia heterophyla, J. spinosa, Vasconcellea cauliflora, V. quercifolia,* and *V. monoica,* all species from the family Caricaceae. The species *J. heterophyla* and *V. monóica* are not found growing naturally in Brazil, but *J. spinosa* and *V. quercifolia* are cultivated in the south and central west part of the country. Host-range studies indicated that PLYV does not infect any of the other 82 plant species from 16 different plant families that were inoculated with the virus, including the indicator plants *Chenopodium amaranticolor, C. murale, C. quinoa,* and *Nicotina benthamiana* (Amaral et al 2006, Nascimento et al 2010). The absence of symptoms and virus particles in the inoculated plants were confirmed by indirect linked immunosorbent assay (ELISA) and RT-PCR. All the plant virus species from the genus *Sobemovirus* have a narrow host range.

PLYV particles in high concentrations can be purified from infected papaya, with the isolation of ca. 300 mg of purified virus per kg of infected papaya leaves. The absorption spectrum of a purified preparation of PLYV has a maximum at ca. A258 nm (= 0.400 OD) and a minimum at A243 nm (= 0.255 OD); the ratio between the absorption at 260 and 280 (A260/A280) obtained for the purified virus preparation (1.80) was typical of a virus with polyhedral particles (Nordam 1973).

VIRUS TRANSMISSION AND STABILITY

Although no biologic vector has been confirmed, PLYV is spreading every year, probably by infected plantlets and contaminated agriculture tools. Transmission studies indicated that the virus is transmitted neither by *Myzus persicae* nor by *Diabrotica bivitulla* or *D. speciosa* (Lima & Santos 1991, Kitajima et al 1992, Silva 1996, Nascimento et al 2010), but it is readily transmitted mechanically and by human actions, including contaminated hands, agricultural tools, contaminated soil, and irrigation water (Camarço et al 1996, Lima et al 2001, Saraiva et al 2006, Nascimento et al 2010). The virus was efficiently transmitted by contaminated hands from infected to healthy papaya plants, even when the hands contaminated with extracts from infected plants were washed with tap water before being rubbed on the leaf surfaces of healthy plants (Table 11.1), demonstrating the high stability of the virus particles outside infected cells (Camarço et al 1998, Saraiva et al 2006, Nascimento et al 2010). Although the virus was not transmitted by seed from infected papaya fruits, it was serologically detected on the seed surface (Table 11.2). The virus was not serologically detected in germinated seeds and in plantlets raised from seeds from infected fruits (Table 11.2), nor was it detected in the embryos of seeds of infected fruits (Table 11.3). On the other hand, the virus was serologically detected on the seed *per se* and also in the tegument of seeds of infected fruits (Tables 11.2 and 11.3). Infective virus was also detected in soil from pots containing infected papaya plants and water used to irrigate infected plants. It was also demonstrated that healthy papaya plantlets can be infected if they are grown in contaminated soil (Camarço et al 1996, Nascimento et al 2010).

The use of virus-infected plantlets has contributed to the virus spreading among the producing areas and States, introducing the source of the virus into new orchards. The occurrence of the virus in small patches of infected plants

TABLE 11.1 Transmission of *Papaya lethal yellowing virus* by contaminated hands with extracts from infected papaya (*Carica papaya*) to healthy papaya plantlets determined by indirect enzyme-linked immune absorbent assay using antiserum for PLYV

Type of inoculation	Number of inoculated plantlets	Number of infected plantlets (% transmission)
Contaminated hands	104	13 (12.5%)
Contaminated and washed hands	116	6 (5.2%)
Control: hands with extracts from healthy plants	50	–

TABLE 11.2 Results of serologic tests with seeds, germinated seeds, and plantlets obtained from fruits of papaya (*Carica papaya*) naturally infected with *Papaya lethal yellowing virus*. In all cases the seeds were used with and without mucilage, and the samples were tested by indirect enzyme-linked immune absorbent assay using antiserum for PLYV

Type of papaya tissue	Number of samples tested		Number of samples infected	
	With mucilage	Without mucilage	With mucilage	Without mucilage
Seed *per se*	273	183	32	17
Germinated seeds	210	152	–	–
Plantlets	136	174	–	–

TABLE 11.3 The presence of *Papaya lethal yellowing virus* in the embryo and tegument of seeds from naturally infected papaya fruit, determined by indirect enzyme-linked immune absorbent assay (ELISA) using antiserum for PLYV

Type of papaya tissue	Number of samples tested	Number of samples with PLYV (% transmission)
Embryo from infected seeds	1128	– (–)
Tegument from infected seeds	670	112 (16.7%)
Embryo from healthy seeds	210	– (–)
Tegument from healthy seeds	230	– (–)

within or on the edge of the orchards indicates that the virus is probably introduced by infected plantlets or contaminated tools and spread by the grower's activities from those initial sources of virus to neighboring plants. For this reason, the production of plantlets in protected nurseries far from papaya commercial production orchards is an important control strategy for PLYV.

Several experiments have demonstrated that PLYV is a very stable virus. Infective virus was detected in dried roots and leaves from infected plants maintained under laboratory conditions for up to 120 days, demonstrating that debris from infected plants could function as a natural source of virus inside a papaya orchard. The following physical properties of the virus also demonstrate its high stability outside a host cell: thermal inactivation point 80°C, longevity *in vitro* 60 days, and dilution end point 10^{-6}. Nevertheless, the virus can be inactivated in leaves and roots removed from infected plants when they are subjected to solarization for a period of 12 days, but it maintained infectivity when the leaves and the roots were kept on the soil under natural conditions for 32 days, confirming that debris from infected plants could function as a natural source of virus inside a papaya orchard. The high stability of the virus particles could also facilitate their dissemination by human actions, which could explain the spreading of virus without a biologic vector (Camarço et al 1996, Lima & Lima 2002, Lima et al 2002, Saraiva et al 2006, Nascimento et al 2010). For this reason, it is important and necessary to practice rouging and elimination of old abandoned orchards as a preventive control measure to reduce the source of virus in the field (Lima & Lima 2002, Nascimento et al 2010). Considering that the high stability of PLYV in dried eradicated tissues from infected plants could interfere with the use of rouging for virus eradication in the field, solarization was shown to be an efficient agricultural practice for virus inactivation. The presence of PLYV was not detected by ELISA in papaya tissues 12 days after solarization, indicating possible virus degradation. On the other hand, PLYV was detected by ELISA in infected papaya tissues left over the soil and submitted to natural sun radiation 32 days after the tissue had been removed from the plants. The serologic results were confirmed by biologic tests when the tissues were inoculated into young healthy papaya plants. Although treatments with alcohol or neutral detergents were not efficient in eliminating infected PLYV from contaminated agricultural tools, the virus is inactivated with 10% sodium hypochlorite, indicating the efficiency of this agent for the treatment of contaminated tools.

VIRUS DISTRIBUTION INSIDE INFECTED PLANTS

Serologic studies with antiserum for PLYV, using equal amounts of tissue from each part of infected plants, demonstrated that the virus is uniformly distributed in the young, intermediate, and older leaves, stem, and roots. The absorption readings in ELISA tests for PLYV were very similar when an equal amount of each part of systemically infected plants, 30 days after virus inoculation, was tested by serology.

The presence of the virus in inoculated plants could be detected by indirect ELISA in the inoculated leaves 72 hours after inoculation; only 7 and 10 days later, the virus was detected in the stem and in the roots, respectively. Sixteen days after inoculation the virus was detected in the younger leaves, and the plant was systemically infected only 18 to 20 days after inoculation (Fig. 11.7). This is in agreement with what was proposed by Agrios (1997) in his textbook *Plant Pathology* (i.e., that virus moves from cell to cell, multiplies in most of them, and, after reaching the phloem cell, is rapidly transported over long distances within the plant). According to the results obtained, PLYV starts to replicate inside the inoculated cell as soon as it is

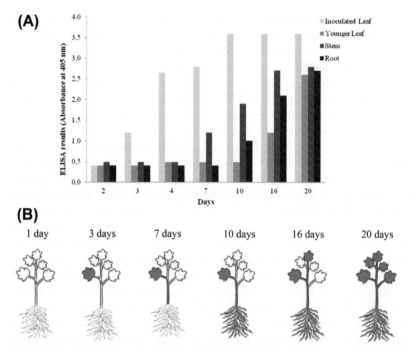

FIGURE 11.7   (A) Days after inoculation at which *Papaya lethal yellowing virus* was detected by indirect ELISA in the inoculated leaf, stem, roots, and upper leaves of a mechanically inoculated papaya plant (*Carica papaya*). (B) Schematic representation of the direction and rate of translocation of *Papaya lethal yellowing virus* in a mechanically inoculated papaya plant (*Carica papaya*). *Adapted from Agrios (1997).*

inoculated, but it takes approximately 18 to 20 days to systemically infect the entire plant, depending on the age and the size of the inoculated plant. Based on what is already known, the virus moves from cell to cell through the connecting plasmodesmata, and once the virus enters the phloem it is rapidly transported toward the growing regions.

INTERACTION WITH *PAPAYA RINGSPOT VIRUS*

Greenhouse studies with doubly inoculated papaya plants demonstrated a clear and severe synergistic interaction between PLYV and PRSV. The highly severe symptoms presented by the plants doubly infected with PLYV and PRSV showed yellowing, severe mosaic, leaf distortion, growth reduction, systemic necrosis, and the death of 50% of inoculated plants, indicating a highly synergistic effect between the viruses (Lima et al 1993). Plants with mixed infection, showing severe symptoms, were also found in natural conditions. This synergistic interaction indicates that the use of a mild strain should not be a practice recommended for controlling PRSV in Northern Brazil.

STRATEGIES FOR DISEASE CONTROL

Several strategies are recommended for controlling PLYV, the causal agent of papaya lethal yellowing disease, including the following measures: (a) virus-free certified plantlets; (b) eradication of virus-infected plants; (c) treatment of agricultural tools; and (d) additional precautions concerning virus transmission. Nevertheless, avoiding the virus is the most important strategy for controlling PLYV (Nascimento et al 2010).

Nurseries for the production of virus-free plantlets

Nurseries for producing papaya plantlets should be strategically isolated and located in areas free from the virus and distant from old papaya orchards. The use of virus-infected plantlets has contributed to viral dispersion among the producing areas and states, introducing a source of virus during the orchard's formation. Infected plantlets could introduce the primary source of virus into the orchards, and humans, through their agricultural practices, distribute the virus inside the orchards. For this reason, the production of plantlets in pro-tected nurseries far from papaya commercial production orchards represents an important control strategy for PLYV, especially when the orchards are located in a selected virus-free area. As a complementary action, the plantlets produced should be serologically indexed for PLYV and PRSV, with those plantlet lots that had negative results in the serologic tests receiving a virus-free certificate. Agricultural companies that produce tropical fruit and horticulture plantlets located in Northeastern Brazil produce papaya plantlets using good-quality seeds, free of viruses.

Eradication of the source of virus

Virus-infected plants exhibiting symptoms inside orchards should be eradi-cated and eliminated outside the fields. All the plants in old orchards that are no longer producing should be eliminated, even if they are not showing any symptoms. An efficient eradication program should involve the elimination of all sources of virus inside, and in the proximity of, areas where orchards will be established. Elimination of the source of virus in the fields should be guided and monitored by well-trained technicians who can recognize symp-toms of viral infection. In the case of doubts about virus infection, samples of suspicious plants should be serologically examined for the presence of a virus. The eradication programs need to include the participation of the producers and government institutions, and all papaya plants with virus symptoms should be eliminated from commercial orchards, including those from outside and from small backyards. In existing orchards, depending on the degree of virus incidence, rouging (eradication of initial sources of virus inside the orchards) should be practiced as a complementary control measure. The practice of roug-ing has been shown to be efficient in several papaya-producing areas, including for the control of PRSV, which is efficiently transmitted by aphids, in the State of Espirito Santo, the second highest papaya producer in Brazil. Although the success of the eradication program for controlling PRSV over a long period is questionable, especially because previous success was for a limited time in other regions, including Hawaii (Gonsalves 1998, Souza Jr & Gonsalves 1999), the eradication of infected plants to control PLYV should have a certain amount of success, considering that the virus does not have a biologic vector to disseminate it in the field. PLYV eradicated infected plant tissues should be submitted to solarization treatment for a 15-day period, which is enough to inactivate the virus.

Sterilization of agricultural tools

Considering the high stability of PLYV, its inactivation on the surface of contaminated agricultural tools is considered an important control measure to avoid virus dissemination inside an orchard. The agricultural tools used for trimming or harvesting practices in papaya orchards need to be treated by immersion in a solution of 10% sodium hypochlorite after use with each plant.

Additional precautions to avoid virus transmission

Considering the high stability of PLYV, precautions need to be taken to avoid virus dissemination inside an orchard through contaminated soil, irrigation water, seed coats from infected fruits, and contaminated agricultural tools.

REFERENCES

Agrios, G.N., 1997. Plant pathology, fourth ed. Academic Press San Diego, ch 13.

Ahlquist, P., French, R., Janda, M., Loesch-Fries, L.S., 1984. Multicomponent RNA plant virus infection derived from cloned viral cDNA. Proceedings of the National Academy of Sciences of the USA 81, 7066–7070.

Almeida, A.M.R., Carvalho, S.L.C., 1978. Ocorrência do vírus do mosaico do mamoeiro no estado do Paraná. Fitopatologia Brasileira 3, 220–225.

Almeida, A.M.R., Lima, J.A.A., 2001. Técnicas sorológicas aplicadas à fitovirologia. In: Almeida, A.M.R., Lima, J.A.A. (Eds.), Princípios e técnicas aplicados em fitovirologia, Edições Sociedade Brasileira Fitopatologia, Fortaleza, Ceará, Brazil, pp. 33–62.

Amaral, P.P., Resende, R.O., Souza Junior, M.T., 2006. Papaya lethal yellowing virus (PLYV) infects Vasconcellea cauliflora. Fitopatologia Brasileira 31, 517.

Astier, S., Albouy, J., Maury, Y., Robaglia, C., Lecoq, H., 2007. Principles of plant virology: genome, pathogenicity, virus ecology. Science Publishers, New Hampshire, USA.

Barbosa, F.R., Paguio, D.R., 1982. Vírus da mancha anelar do mamoeiro: incidência e efeito na produção do mamoeiro. Fitopatologia Brasileira 7, 365–373.

Camarço, R.F.E.A., Lima, J.A.A., Pio-Ribeiro, G., Andrade, G.P., 1996. Ocorrência do 'papaya lethal yellowing virus' no município de Santa Rita, Estado da Paraíba. (Abstr.). Fitopatologia Brasilera 29, 423.

Camarço, R.F.E.A., Lima, J.A.A., Pio-Ribeiro, G., 1998. Transmissão e presença em solo do 'papaya lethal yellowing virus'. Fitopatologia Brasileira 23, 453–458.

Clark, M.F., Adams, A.N., 1977. Characteristics of the microplate method of enzymelinked immunosorbent assay for the detection of plant viruses. Journal of General Virology 34 (3), 475–483.

Clark, M.F., Bar-Joseph, M., 1984. Enzyme immunosorbent assays in plant virology. In: Maramorosch, K., Koprowski, H. (Eds.), Methods in virology, Academic Press, New York, pp. 51–85.

Conover, R.A., 1976. A program for development of papaya tolerant to the distortion ringspot virus. Proceeding of the Florida State Horticultural Society 89, 229–231.

Daltro, C.B., Pereira, A.J., Cascardo, R.S., Alfenas-Zerbini, P., Beserra Jr., J.E.A., Lima, J.A.A., Zerbini, F.M., Zerbini, F.M., Andrade, E.C., 2012. Genetic variability of papaya lethal yellowing virus isolates from Ceará and Rio Grande do Norte states, Brazil. Tropical Plant Pathology 37, 37–43.

Gonsalves, D., 1998. Control of papaya ringspot virus in papaya: a case study. Annual Review of Phytopathology 36, 415–437.

Gonsalves, D., Suzuki, J.Y., Tripathi, S., Ferreira, S.A., 2007. *Papaya ringspot virus* (Potyviridae). In: Mahy, B.W.J., van Regenmortel, M.H.V. (Eds.), Encyclopedia of virology, Elsevier, Oxford, UK.

Gonsalves, D., Tripathi, S., Carr, J.B., Suzuki, J.Y., 2010. *Papaya ringspot virus*. Plant Health Instructor http://dx.doi.org/10.1094/PHI-I-2010-1004-01.

Hampton, R., Ball, E., De Boer, S., 1990. Serological methods for detection and identification of viral and bacterial plant pathogens. American Phytopathological Society, St Paul, MN, USA.

IBGE (Institudo Brasileiro de Geografia e Estatística). (2012). (accessed 19.12.2012) http://ibge.gov.br

King, A.N, Adams, N.J., Carstens, E.B. & Lefkowitz, E.J. (2012). Virus taxonomy. Ninth Report of the International Committee on Taxonomy of Viruses (ISBN 978-0-12-384684-6) 1, p. 727.

Kitajima, E.W., Oliveira, F.C., Pinheiro, C.R.S., 1992. Amarelo letal do mamoeiro solo no estado do Rio Grande do Norte. Fitopatologia Brasileira 17, 282–285.

Kitajima, E.W., Rodrigues, C., Silveira, J., Alves, F., Ventura, J.A., Aragão, F.J.L., Oliveira, L.H.R., 1993. Association of isometric viruslike particles, restricted to laticifers, with 'Meleira' ('sticky disease') of papaya (*Carica papaya*). Fitopatologia Brasileira 8, 118–122.

Lima, J.A.A., Camarço, R.F.E.A., 1997. Viruses that infect papaya in Brazil. Virus: Reviews & Research 2, 126–127.

Lima, J.A.A., Gomes, M.N.S., 1975. Identificação de 'papaya ringspot virus' no Ceará. Fitossanidade 2, 56–59.

Lima, J.A.A., Nascimento, A.K.Q., 2012. A simple kit of plate-trapped antigen enzyme-linked immunosorbent assay for identification of plant viruses. Virus Reviews and Research 17, 13. (Abstr.).

Lima, J.A.A., Santos, C.D.G., 1991. Isolamento de possível estirpe do vírus do amarelo letal do mamoeiro no Ceará. Fitopatologia Brasileira 16, 27. (Abstr.).

Lima, J.A.A., Marques, M.A.L., Camarço, R.F.E.A., 1993. Efeito sinérgico entre o vírus da mancha anelar e o vírus do amarelo letal do mamoeiro. Fitopatologia Brasileira 18, 289. (Abstr.).

Lima, J.A.A., Lima, A.R.T., Marques, M.A.L., 1994. Purificação e caracterização sorológica de um isolado do vírus do amarelo letal do mamoeiro 'solo' obtido no Ceará. Fitopatologia Brasileira 19, 437–441.

Lima, J.A.A., Lima, R.C.A., Gonçalves, M.F.B., 2001a. Production of policlonal antisera specific to plant viruses by rabbit oral immunization. Fitopatologia Brasileira 26, 774–777.

Lima, R.C.A., Lima, J.A.A., 2002. Viroses em mamoeiro e alternativas de controle/ Roberto Caracas de Araújo Lima. Secretaria da Agricultura Irrigada, Fortaleza, p. 40.

Lima, R.C.A., Lima, J.A.A., Souza Jr., M.T., Pio-Ribeiro, G., Andrade, G.P., 2001b. Etiologia e estratégias de controle de viroses do mamoeiro no Brasil. Fitopatologia Brasileira 26, 689–702.

Lima, R.C.A., Souza Jr., M.T., Pio-Ribeiro, G., Lima, J.A.A., 2002. Sequences of the coat protein gene from Brazilian isolates of *Papaya ringspot virus*. Fitopatologia Brasileira 27, 174–180.

Lima, J.A.A., Sittolin, I.M., Lima, R.C.A., 2005. Diagnose e estratégias de controle de doenças ocasionadas por vírus. In: Freire Filho, F.R., Lima, J.A.A., Silva, P.H.S., Ribeiro, V.Q. (Eds.), Feijão caupi: avanços tecnológicos, Embrapa Informação Tecnológica, Brasília, Brazil, pp. 404–459.

Lima, J.A.A., Nascimento, A.K.Q., Radaelli, P., Silva, A.K.F., Silva, F.R., 2011a. Immune precipitation polymerase chain reaction for identification of plant viruses. Virus Reviews and Research 16, 56. (Abstr.).

Lima, J.A.A., Nascimento, A.K.Q., Silva, F.R., Silva, A.K.F., Aragão, M.L., 2011b. An immune precipitation enzyme-linked immunosorbent (IP-ELISA) technique for identification of plant viruses. Virus Reviews and Research 16, 56. (Abstr.).

Lima, J.A.A., Nascimento, A.K.Q., Radaelli, P., Purcifull, D.E., 2012. Serology applied to plant virology. In: Molish-Al-Moslih, M. (Ed.), Serological diagnosis of certain human, animal and plant diseases, InTech, Rijekax, Croácia, pp. 71–94.

Loreto, T.J.G., Vital, A.F., Rezende, J.A.M., 1983. Ocorrência de um amarelo letal do mamoeiro solo no estado de Pernambuco. O Biológico 49, 275–279.

Manica, I., 1982. Fruticultura tropical: 3. Mamão. Agronômica Ceres, São Paulo, p. 255.

Marciel-Zambolim, E., Kunieda-Alonso, S., Matsuoka, K., Carvalho, M.G., Zerbini, F.M., 2003. Purification and some properties of *Papaya meleira virus*, a novel virus infecting papayas in Brazil. Plant Pathology 52, 389–394.

Mulholland, V., 2009. Immunocapture-PCR for plant virus detection. In: Robert, Burns (Ed.), Plant pathology: techniques and protocols, Scottish Agricultural Science Agency, ISBN: 1588297993, Edinburgh, UK, pp. 183–192.

Mullis, K.B., 1990. The unusual origin of the polymerase chain reaction. Scientific American 262, 56–65.

Mullis, K., Faloona, F., Scharf, S., Saiki, R., Horn, G., Erlich, H., 1986. Specific enzymatic amplification of DNA *in vitro*: the polymerase chain reaction. Cold Spring Harbor Symposia on Quantitative Biology 51, 263–273.

Mumford, R.A., Walsh, K., Barker, I., Boonham, N., 2000. Detection of *Potato mop top virus* and *Tobacco rattle virus* using multiplex real-time X fluorescent reverse-transcription polymerase chain reaction assay. Phytopathology 90, 448–453.

Naidu, R.A. & Hughes, J.d'A. (2001). Methods for the detection of plant virus diseases, In: Hughes J.d'A, Odu BO (eds) Plant virology in sub-Saharan Africa. Proceedings of a Conference Organized by IITA, International Institute of Tropical Agriculture, Nigeria, pp. 233–260.

Nakagawa, J., Takayama, Y., Suzukama, Y., 1987. Exudação de látex pelo mamoeiro. Estudo de Ocorrência em Teixeira de Freitas, BA. Congresso Brasileiro de Fruticultura, 9, Campinas. Anais, Campinas, SP, pp. 555–559.

Nascimento, A.K.Q., Lima, J.A.A., Nascimento, A.L.L., Beserra Jr., J.A.B., Purcifull, D.E., 2010. Biological, physical, and molecular properties of a *Papaya lethal yellowing virus* isolate. Plant Disease 94, 1206–1212.

Nordam, D., 1973. Identification of plant viruses. Methods and experiments. Centre for Agricultural Publishing and Documentation, Pudoc, Wageningen, Netherlands 88–102.

Oliveira, C.R.B., Ribeiro, S.G., Kitajima, E.W., 1989. Purificação e propriedades químicas do vírus do amarelecimento letal do mamoeiro isolado do Rio Grande do Norte. Fitopatologia Brasileira 14, 114. (Abstr.).

Purcifull, D.E., Batchelor, D.L., 1977. Immudiffusion tests with sodium dodecyl sulfate (SDS)-treated plant viruses and plant virus inclusions. University of Florida Agricultural Experimental Station Bulletin 788, p. 39.

Purcifull, D.E., Hiebert, E., 1979. Serological distinction of watermelon mosaic virus isolates. Phytopathology 19, 116–122.

Purcifull, D.E., Edwardson, J., Hiebert, E., Gonsalves, D., 1984. *Papaya ringspot virus*. Descriptions of plant viruses. CMI-AAB, Kew, 292, p. 8.

Purcifull, D.E., Hiebert, E., Petersen, M., Webb, S., 2001. Virus detection – serology. In: Maloy, O.C., Murray, T.D. (Eds.), Encyclopedia of plant pathology, John Wiley, pp. 1100–1109.

Ramos, N.F., Nascimento, A.K.Q., Goncalves, M.F.B., Lima, J.A.A., 2008. Presença dos vírus da mancha anelar e do amarelo letal em frutos de mamoeiro comercializados. Tropical Plant Pathology 33, 449–452.

Rezende, J.A.M., Costa, A.S., 1993. Controle do mosaico do mamoeiro por premunização: sucessos e dificuldades. Fitopatologia Brasileira 18, 258. (Abstr.).

Rodrigues, C.H., Alves, F.L., Marin, S.L.D., Maffia, L.A., Ventura, J.A., Gutierrez, A.S.D., 1989a. Meleira do mamoeiro no estado do Espírito Santo: enfoque fitopatológico. Selecta de Trabalhos sobre a Meleira do mamoeiro. EMCAPA, Linhares.

Rodrigues, C.H., Ventura, J.A., Maffia, L.A., 1989b. Distribuição e transmissão da Meleira em pomares de mamão no Espírito Santo. Fitopatologia Brasileira, 14,118.

Santos, H.P., Barbosa, C.J., Nickel, O., 2003. Doenças do Mamoeiro. In: Freire, F., das, C.O. (Eds.), Doenças de fruteiras tropicais de interesse agroindustrial, Embrapa Informação Tecnológica; Fortaleza, Embrapa Agroindústria Tropical, Brasília, pp. 391–434.

Saraiva, A.C.M., Paiva, W.O., Rabelo Filho, F.A.C., Lima, J.A.A., 2006. Transmissão por mãos contaminada e ausência de transmissão embrionária do vírus do amarelo letal do mamoeiro. Fitopatololgia Brasileira 31, 79–83.

Shukla, D.D., Ward, C.W., Brunt, A.A., 1994. The *Potyviridae*. CAB International, UK p. 116.

Silva, A.M.R., 1996. 'Papaya lethal yellowing virus': caracterização biológica e molecular. MS thesis. Federal University of Brasilia, Brasilia p. 122.

Silva, A.M.R., Kitajima, E.W., Souza, M.V., Resende, R., 1997. *Papaya lethal yellowing virus:* a possible member of the *Tombusvirus* genus. Fitopatologia Brasileira 22, 529–534.

Silva, A.M.R., Kitajima, E.W., Resende, R.O., 2000. Nucleotide and amino acid analysis of the polymerase and the coat protein genes of the papaya lethal yellowing virus. (Abstr.). Virus Reviews and Research 11, 196.

Souza JR., M.T., Gonsalves, D., 1999. Genetic engineering resistance to plant virus diseases: an effort to control papaya ringspot potyvirus in Brazil. Fitopatologia Brasileira 24, 485–502.

Teixeira, M.G.C., Lima, J.A.A., Sousa, A.E.B.A., Fernandes, E.R., 1999. Baixos graus de incidência do vírus do amarelo letal do mamoeiro em municípios do Rio Grande do Norte. Caatinga 12, 29–33.

Van Regenmortel, M.H.V., 1982. Serology and immunochemistry of plant viruses. Academic Press, New York ISBN-10: 0127141804.

Vega, J., Bezerra, J.L., Rezende, M.L.V., 1988. Detecção do vírus do amarelo letal do mamoeiro solo no estado da Bahia através de microscopia eletrônica. (Abstr.). Fitopatologia Brasileira 21, 147.

Ventura, J.A., Costa, H., Tatagiba, J.S., 2004. Papaya diseases and integrated control. In: Naqvi, S.A.M.H. (Ed.), Diseases of fruits and vegetables: diagnosis and management, vol. 2. Kluwer Academic Publishers, Dordrecht,, pp. 201–268.

Voller, A.A., Bartlett, A., Bidwell, D.E., Clark, M.F., Adams, A.N., 1976. The detection of viruses by enzyme-linked immunosorbent assay (ELISA). Journal of General Virology 33, 165–167.

Wang, H.L., Yeh, S.D., Chiu, R.J., 1987. Effectiveness of cross protection by mild mutants of papaya ringspot virus for control of ringspot desease of papaya in Taiwan. Plant Disease 71, 491–497.

Yeh, S.D., Gonsalves, D., 1984. Evulation of induced mutants of papaya ringspot virus for control by cross protection. Phytopathology 74, 1086–1091.

Yeh, S.D., Gonsalves, D., Wang, H.L., 1988. Control of papaya ringspot virus by cross protection. Plant Disease 72, 375–380.

Chapter 12

Establishment of endogenous pararetroviruses in the rice genome

Ruifang Liu and Yuji Kishima
Laboratory of Plant Breeding, Research Faculty of Agriculture, Hokkaido University, Sapporo, Japan

INTRODUCTION

Plant viruses with a DNA genome, which are present in host plant genomes, are divided into two categories: the single-stranded DNA (ssDNA) gemini-viruses (Kenton et al 1995, Bejarano et al 1996, Ashby et al 1997) and the double-stranded DNA (dsDNA) pararetroviruses (Harper et al 1999, Jakowitsch et al 1999, Ndowora et al 1999, Budiman et al 2000, Lockhart et al 2000, Mao et al 2000, Harper et al 2002, Staginnus & Richert-Poggeler 2006, Gambley et al 2008, Gayral et al 2008, Pahalawatta et al 2008).

Segments of the *Geminivirus* genome, including the viral replication origin and the adjacent *AL1* gene, have been found in the genomes of tobacco and related species (Bejarano et al 1996, Ashby et al 1997).

Pararetroviruses integrated into the host genome are referred to as 'endogenous pararetroviruses' (EPRVs) (Mette et al 2002, Staginnus et al 2009). Plant pararet-roviruses are classified in the family *Caulimoviridae*, which contains six genera (*Caulimovirus, Soymovirus, Cavemovirus, Petuvirus, Badnavirus*, and *Tungrovi-rus*) (Hull 2001, 2002, Fauquet 2005). EPRV-like sequences derived from four of these genera—*Cavemoviruses, Petuviruses, Badnaviruses*, and *Tungroviruses*—include *Banana streak virus* (BSV) in *Musa* spp. (Harper et al 1999, Ndowora et al 1999, Geering et al 2001, 2005, Harper et al 2005, Gayral et al 2008), *Petunia vein-clearing virus* (PVCV) in petunia (Richert-Poggeler & Shepherd 1997, Harper et al 2003, Noreen et al 2007), *Tobacco vein-clearing virus* (TVCV) in tobacco (Jakowitsch et al 1999, Lockhart et al 2000, Gregor et al 2004), and *Rice tungro bacilliform virus* (RTBV) in rice (Nagano et al 2000, 2002, Kunii et al 2004).

There are two forms of integrant: those that can form episomal viral infec-tions and those that cannot. Integrants of three pararetroviruses, BSV (Harper et al 1999, Ndowora et al 1999, Geering et al 2001, 2005, Harper et al 2005,

Gayral et al 2008, Gayral & Iskra-Caruana 2009), TVCV (Jakowitsch et al 1999, Lockhart et al 2000, Gregor et al 2004), and PVCV (Harper et al 2003, Richert-Poggeler et al 2003, Noreen et al 2007), can generate episomal infections in certain hybrid plant hosts in response to stress.

The DNA structure and localization of EPRVs within the host genome have been studied in detail in tobacco (Lockhart et al 2000), petunia (Richert-Poggeler et al 2003), banana (Ndowora et al 1999), and rice (Kunii et al 2004). Endogenous sequences represent linear forms of the original circular viral DNA. Only a few of the EPRV loci analyzed to date consist of continuous stretches of viral sequences. In most cases, the full-length viral genome had to be reconstructed from different segments, which were derived from several genomic loci.

Unlike retroviruses, integration into the host genome is not obligatory for the (dsDNA) pararetroviruses. Pararetroviruses do not possess integrase enzymes that are essential for the integration of retroviral DNA into the host DNA (Hull et al 2000, Peterson-Burch et al 2000, Harper et al 2002). Despite the functional inability of pararetroviruses to integrate into the host plant's genome, they have been found in various plant genomes over the past decade (Staginnus & Richert-Poggeler 2006). A few sequences were intact (Ndowora et al 1999, Lockhart et al 2000, Richert-Poggeler et al 2003), whereas the structures of EPRVs in genomic sequence data are often disrupted and truncated, with deletions, insertions, duplications, or inversions (Harper et al 1999, Jakowitsch et al 1999, Yang et al 2003, Kunii et al 2004, Hansen et al 2005, Bertsch et al 2009). Although nonhomologous end-joining has been considered to be the major mechanism for the integration of EPRVs, the details of this mechanism remain poorly understood (Hull et al 2000, Lockhart et al 2000, Harper et al 2002).

ENDOGENOUS *RICE TUNGRO BACILLIFORM VIRUS*-LIKE SEQUENCES ARE PREFERENTIALLY PRESENT BETWEEN AT DINUCLEOTIDE REPEATS IN RICE GENOMES

In South and Southeast Asia, *Rice tungro bacilliform virus* (RTBV), which is transmitted by green leafhoppers, causes one of the most serious diseases of rice with the assistance of *Rice tungro spherical virus* (RTSV) (Hibino et al 1979). Endogenous *Rice tungro bacilliform virus*-like sequences (ERTBVs) are embedded in the rice genome (Nagano et al 2000, 2002, Kunii et al 2004), although its currently active cognate, RTBV, has not been obtained from the genome. None of the ERTBVs are functionally intact as a virus, although consensus alignment provided a circular virus-like structure carrying two complete open reading frames (Fig. 12.1). At least 88 and 74 loci (including 13 unmapped segments) of ERTBV were identified in the japonica and indica databases, respectively (Liu et al 2012) (Fig. 12.2). Structural differences among the collected ERTBV segments appeared to be as a result of rearrangements of the segments, including deletions, insertions,

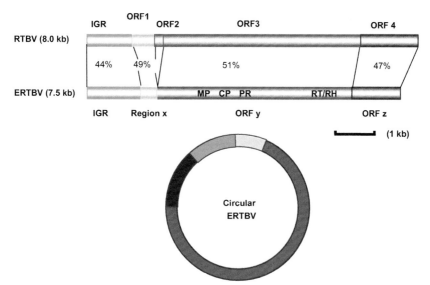

FIGURE 12.1 Deduced virus form of ERTBV that is assembled from genomic sequences. Comparison of the assembled ERTBV and RTBV is presented in the figure. Percentages indicate the nucleotide similarity of each of the corresponding segments or ORFs between ERTBV and RTBV. ERTBV lacks ORFs 1 and 2. An identical organization of ORFs and their orders is observed in their structures. ERTBV sequences consist of an intergenic region (IGR), Region x, ORF y, and ORF z. The nucleotide sequence for Region x corresponds with ORF 1, but ATGs for the initiation codon were not present. ORF 3 contains the movement protein (MP), the coat protein (CP), asparatic protease (PR), and RNase H (RT/RH). In the figure, different colors in the circle correspond to the above-mentioned segments. ERTBV is deduced to have been present in a circular structure; it is expressed in linear form for convenience.

inversions, and duplications. The structures varied among the segments, and they were unlikely to be active as viruses (Fig. 12.3). However, the nucleotide identities among ERTBV segments were more than 80% (Kunii et al 2004), so that each homologous part among the ERTBV segments was easily recognized. The ends of these ERTBVs were identified at 170 and 99 sites from the japonica and indica databases, respectively (Liu et al 2012). Most of the ERTBV segments were flanked by AT dinucleotide repeats (ATrs): 84% (143 sites) and 77% (76 sites) of sites for japonica and indica ERTBVs, respectively. The distances between ERTBV and ATr ranged from 0 bp to 759 bp, with an average of 145 bp. Thirty percent of the ends of ERTBVs were located less than 50 bp from an ATr. The probability of proximity for the two elements was extremely high, because the proportion of the ATrs in the rice genome (370 Mb) was calculated as about 0.2%. This estimation took into account complete ATrs that ranged from 10 bp to 232 bp, although a number of ATrs that were interrupted by other nucleotides were also observed in proximity to ERTBVs. Thus, ERTBVs have a high propensity to be localized adjacent to ATrs.

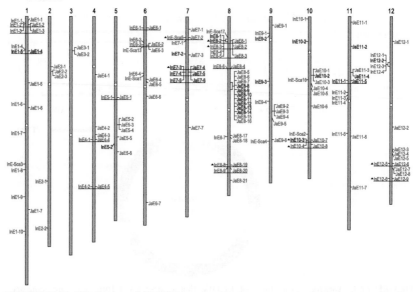

FIGURE 12.2 Distribution of ERTBVs on the 12 chromosomes in the japonica and indica genomes. Each vertical bar at the top indicates a chromosome number. A white dot on each chromosome bar shows the approximate position of the centromere. The japonica (Nipponbare) database contains 88 ERTBV sites (JaE: right side of the chromosome), and in the indica, the database contains 61 ERTBV sites (InE: left side). The chromosomal locations for 13 ERTBVs have not been identified. ERTBV segments that are flanked by ATrs are indicated with red letters. Underlines at ERTBV positions indicate the same sequence between japonica and indica; nine of them with solid triangles show incomplete homologies between the two ERTBVs.

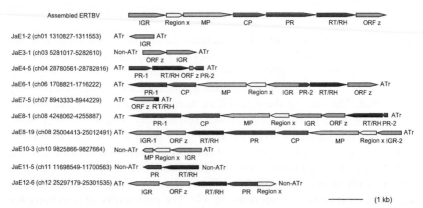

FIGURE 12.3 Rearrangement of japonica ERTBV segments. The assembled ERTBV consists of seven regions: intergenic region (IGR), Region x, movement protein (MP), coat protein (CP), asparatic protease (PR), and RNase H (RT/RH), which are shown in different colors. Ten japonica ERTBV segments (ERTBV no., chromosome no., and nucleotide positions) demonstrate the structural changes as represented by rearrangement of the segments. In addition, each segment contains a number of nucleotide changes, which are not shown in this figure. Arrows show the direction of sequence in the rice genome.

THE EXISTENCE OF AT DINUCLEOTIDE REPEATS PRIOR TO ENDOGENOUS *RICE TUNGRO BACILLIFORM VIRUS*-LIKE SEQUENCE INTEGRATION

A question arose as to whether ATrs were present before ERTBV insertion or whether ATr inserted coincidentally with ERTBVs. Comparison of ERTBVs in the japonica and indica genomes provided a means to address this question. Among the 88 and 61 ERTBV segments in the japonica and indica genomes, respectively, 22 segments were located at the same site in both genomes (Fig. 12.2). The rest of the ERTBV segments were considered to be present uniquely in either of the genomes and were absent at the same site in the other genome (i.e., empty donor sites) (Le et al 2000, Turcotte et al 2001). Database searches were able to identify 32 empty donor sites: 28 ERTBV sites in the japonica genome whose ERTBVs were absent in the indica genome and four ERTBV sites in the indica genome whose ERTBVs were absent in the japonica genome (Liu et al 2012). All the empty donor sites possessed ATrs at the corresponding ERTBV sites. These results are strong evidence that ATrs were present prior to the integration of ERTBVs, although the possibility cannot be excluded that ERTBVs had been excised to generate empty donor sites after the differentiation of japonica and indica.

Certain rice chromosomal segments have been duplicated in the course of the establishment of ancestral rice strains (Guyot & Keller 2004, Langham et al 2004), such that duplicated paralogous segments are found occasionally in the genome (Turcotte et al 2001). To validate the insertion of ERTBV into ATrs, a search was made for paralogous sequences of ERTBV flanking sequences. Five japonica ERTBVs (JaE 4-5, JaE 8-21, JaE 11-1, JaE 12-2, and JaE 12-9) showed ERTBV flanking sequences that matched ERTBV-unrelated segments at different positions in the japonica genome (Fig. 12.4). These five ERTBV-unrelated segments contained ATrs between both flanking sequences without ERTBV fragments (i.e., they were paralogous empty donor sites). Five pairs of the ERTBV and ERTBV-unrelated paralogous sites in the japonica genome also possess corresponding orthologous sites without ERTBV in the indica genome (Fig. 12.4). In each case, an ERTBV existed only at one site among the three or four sites (Fig. 12.4). The results led us to conclude that ATrs were present before the ERTBV insertions occurred and that the insertion of ERTBVs into the japonica or indica genomes was more plausible than the loss of ERTBVs.

POSSIBLE MECHANISMS OF INTEGRATION USING AT DINUCLEOTIDE REPEATS

AT-rich regions have been reported to be a favorable site for the insertion of transgenes into the *Arabidopsis* genome (Sawasaki et al 1998), SINE in *Brassica* (Tikhonov et al 2001), and Micropon in rice (Akagi et al 2001). These preferences are considered as consequences of active insertion of these donor elements. However, most of the DNA sequences in ATrs that were identified as insertions

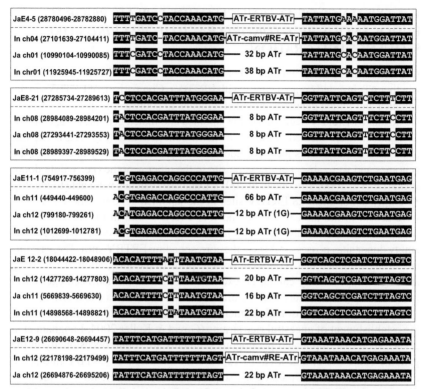

FIGURE 12.4 Five ERTBV sites in the japonica genome have paralogous and orthologous empty donor sites without ERTBV segments. Each ERTBV site (JaE) has three homologous sequences that share the same nucleotides (white letters in black) and AT-repeat (ATr) without the ERTBV sequence. These consist of paralogous, orthologous, and paralogous/orthologous sites in the japonica (Ja chromosome no. nuclotide positions) and indica genomes (In chromosome no. nucleotide positions). JaE12-9 has an orthologous site in indica and a paralogous site in japonica, but the third homologous (paralogous/orthologous) sequence in indica was not detected in the database. Among these sequences, two sites contained retrotransposoms instead of ERTBVs between ATrs.

did not connect to particular enzymatic systems for DNA integration. Even for the transposon-like sequences, few signs of target site duplications, which are the key indicators of insertions performed using their own transposases, were observed (Liu et al 2012). If such inserted DNA does not have its own enzymatic system for integration, insertion into the genome must be accomplished using host factors.

In higher eukaryotes, nonhomologous end joining (NHEJ), a DNA repair system without specific sequence homologies, could be the main mode of integration of various DNA segments through double-stranded breaks (DSBs) (Puchta 2005, Lieber 2010). Salomon & Puchta (1998) demonstrated that, in the tobacco genome, DSBs induced by the I-SceI restriction system gave rise to DNA insertions with a broad spectrum of unique and repeat genomic sequences at the breaks. These insertions, related to NHEJ, were explained by the synthesis-dependent strand annealing model that is an ectopic gap repair pathway using the nonspecific donor site in the chromatid as the template (Gloor et al 1991, Nassif et al 1994). The insertion sequences into ATrs are not always considered to be caused by episomal DNA itself but may also partially consist of the genomic sequence. On the other hand, the NHEJ pathway was also reported to be involved in integration events of external DNA segments, such as retroviruses in human cells (Taganov et al 2001, Tikhonov et al 2001), *Agrobacterium* T-DNA in plant cells (Kohli et al 2003, Somers & Makarevitch 2004), and long interspersed element (LINE) retrotransposition in animal cells (Suzuki et al 2009). Hence, NHEJ may be responsible for the mechanism of inserting episomal DNA and genomic sequences into the breaks at ATr sites. The ends of DSBs often induced nonhomologous recombination events with various filler DNAs at the breaks (Gorbunova & Levy 1997, Puchta 1999). If a single insertion event brought different segments into an ATr, recombination between the different segments might occur before the insertion. In the case where multiple insertion events occurred, the inserted sequences within an ATr might facilitate further independent insertions via homologous or non-homologous recombination events without DSBs.

AT DINUCLEOTIDE REPEATS ARE HOT SPOTS OF DOUBLE-STRANDED BREAKS

AT-rich regions in eukaryotic cells are often attached to the nuclear matrix, called scaffold/matrix attachment regions (S/MARs), which localize on the nuclear matrix or chromatin loop. Eukaryotic genomes are organized into loops fixed at S/MARs attached onto the nuclear matrix (Liebich et al 2002). Furthermore, AT-rich regions potentially give rise to a bending nature in DNA sequences that allows the groove to facilitate the binding of DNA-binding proteins (Carrera & Azorin 1994, Muller & Varmus 1994). Makarevitch & Somers (2006) demonstrated that T-DNA integration sites in the *Arabidopsis* genome were consistent with topoisomerase IIA cleavage sites, which have been frequently associated with S/MARs, because topoisomerase IIA can resolve topological problems caused by knotting and supercoiling at S/MARs. Therefore, ATrs cleaved by topoisomerase IIA might facilitate DSBs that have been utilized for DNA integration. Moreover, compared with the other SSRs, a greater number of ATrs were longer than 20 bp in the rice genome (Liu et al 2012). Therefore, this strongly suggested that ATrs are hot spots for DSBs in the rice genomes (Fig. 12.5).

DIFFERENCES IN *RICE TUNGRO BACILLIFORM VIRUS* AND ENDOGENOUS *RICE TUNGRO BACILLIFORM VIRUS*-LIKE SEQUENCE WITH REGARD TO GENOME INTEGRATION

ERTBV does not encode an integrase, so the machinery for integration might be dependent on host enzymes, as mentioned above. Although RTBV is currently

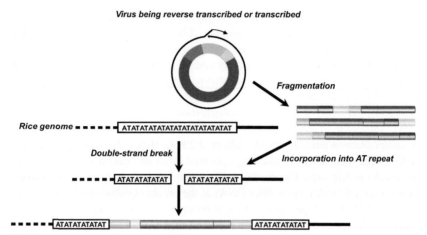

FIGURE 12.5 Processes by which ERTBV segments become incorporated into ATrs in the rice genome. DNA strands of the reverse transcription step of the virus are fragmented in the cells. At the same time, ATr in the rice genome causes disconnection due to a DNA double-strand break. Then, the fragmented virus segments are incorporated into the break sites. The processes from virus fragmentation to incorporation are accomplished by host machineries.

an active cognate of ERTBV, its integration has not been observed (Kunii et al 2004). The major structural difference between ERTBV and RTBV is open reading frame 2 (ORF2), which is present in RTBV but absent in ERTBV (Kunii et al 2004) (Fig. 12.1). The ORF2 protein was suggested to participate in RTBV capsid assembly through interaction with the coat protein in ORF3 (Herzog et al 2000). If RTBV lacks the ability to integrate into the rice genome, ORF2 is thought to be directly or indirectly involved in the suppression of incorporation of virus segments. Considering a number of the segments from ERTBV found in the rice genomes, ERTBV might have been frequently trapped in the rice genome when ERTBV was active as a virus. To facilitate incorporation into the rice genome, abundant fragments of ERTBV molecules might be suspended in the cells. We assume that ORF2 in RTBV suppresses fragmentation of the virus genome by completing capsid formation (Herzog et al 2000).

CONCLUSION

Each of the ERTBVs in the rice genome is highly rearranged, and none are functionally intact as a virus, although consensus alignment provided a circular virus-like structure carrying two complete open reading frames (Fig. 12.1). The two rice genomes decoded from the rice varieties japonica and indica allow us to examine integration events at ATrs through comparative analyses. A remarkable feature of the insertion sites of most ERTBVs is that they are flanked by ATrs (Kunii et al 2004, Liu et al 2012). The results suggest that ERTBVs, which lack their own integration enzymes, have been host-dependently trapped by ATrs in the genome and have not actively targeted ATrs (Fig. 12.5). Besides ERTBVs, AT-rich regions might be widely employed as a site to integrate external and internal DNAs into chromosomal DNA, acting as genomic dumping sites that drive evolutionary divergence (Liu et al 2012) (Fig. 12.5).

RTBV transmitted by green leafhoppers is the main causative agent for rice tungro disease (Hibino et al 1979). Kobayashi & Ikeda (1992) reported that the African rice species *Oryza glaberrima* and *O. barthii* showed much more severe systemic necrosis compared with the other rice species present in South and Southeast Asia after inoculation of both RTBV and RTSV. Kunii et al (2004) speculated a possible relationship between RTBV disease resistance and the copy number of ERTBV in the *Oryza* AA-genome species. To overcome tungro disease resistance, further analyses are necessary to focus on the copy number of ERTBV in the *Oryza* AA-genome species.

ACKNOWLEDGMENT

This work was supported by grants from the Ministry of Education, Culture, Sports, Science and Technology of Japan and International Research Network Program on Global Issues from the Ministry of Agriculture, Forestry and Fisheries.

REFERENCES

Akagi, H., Yokozeki, Y., Inagaki, A., Mori, K., Fujimura, T., 2001. Micron, a microsatellite-targeting transposable element in the rice genome. Molecular Genetics and Genomics 266, 471–480.

Ashby, M.K., Warry, A., Bejarano, E.R., Khashoggi, A., Burrell, M., Lichtenstein, C.P., 1997. Analysis of multiple copies of geminiviral DNA in the genome of four closely related *Nicotiana* species suggest a unique integration event. Plant Molecular Biology 35, 313–321.

Bejarano, E.R., Khashoggi, A., Witty, M., Lichtenstein, C., 1996. Integration of multiple repeats of geminiviral DNA into the nuclear genome of tobacco during evolution. Proceedings of the National Academy of Sciences of the USA 93, 759–764.

Bertsch, C., Beuve, M., Dolja, V.V., Wirth, M., Pelsy, F., Herrbach, E., et al., 2009. Retention of the virus-derived sequences in the nuclear genome of grapevine as a potential pathway to virus resistance. Biology Direct 4, 21.

Budiman, M.A., Mao, L., Wood, T.C., Wing, R.A., 2000. A deep-coverage tomato BAC library and prospects toward development of an STC framework for genome sequencing. Genome Research 10, 129–136.

Carrera, P., Azorin, F., 1994. Structural characterization of intrinsically curved AT-rich DNA sequences. Nucleic Acids Research 22, 3671–3680.

Fauquet, C., 2005. Virus taxonomy: classification and nomenclature of viruses. Eighth report of the International Committee on the Taxonomy of Viruses. Academic Oress, San Diego, CA.

Gambley, C.F., Geering, A.D.W., Steele, V., Thomas, J.E., 2008. Identification of viral and non-viral reverse transcribing elements in pineapple (*Ananas comosus*), including members of two new *Badnavirus* species. Archives of Virology 153, 1599–1604.

Gayral, P., Iskra-Caruana, M.L., 2009. Phylogeny of *Banana streak virus* reveals recent and repetitive endogenization in the genome of its banana host (*Musa* sp.). Journal of Molecular Evolution 69, 65–80.

Gayral, P., Noa-Carrazana, J.C., Lescot, M., Lheureux, F., Lockhart, B.E.L., Matsumoto, T., et al., 2008. A single *Banana streak virus* integration event in the banana genome as the origin of infectious endogenous pararetrovirus. Journal of Virology 82, 6697–6710.

Geering, A.D., Olszewski, N.E., Dahal, G., Thomas, J.E., Lockhart, B.E., 2001. Analysis of the distribution and structure of integrated *Banana streak virus* DNA in a range of *Musa* cultivars. Molecular Plant Pathology 2, 207–213.

Geering, A.D., Olszewski, N.E., Harper, G., Lockhart, B.E., Hull, R., Thomas, J.E., 2005. Banana contains a diverse array of endogenous badnaviruses. Journal of General Virology 86, 511–520.

Gloor, G.B., Nassif, N.A., Johnsonschlitz, D.M., Preston, C.R., Engels, W.R., 1991. Targeted gene replacement in *Drosophila* via P-element-induced gap repair. Science 253, 1110–1117.

Gorbunova, V., Levy, A.A., 1997. Non-homologous DNA end joining in plant cells is associated with deletions and filler DNA insertions. Nucleic Acids Research 25, 4650–4657.

Gregor, W., Mette, M.F., Staginnus, C., Matzke, M.A., Matzke, A.J.M., 2004. A distinct endogenous pararetrovirus family in *Nicotiana tomentosiformis*, a diploid progenitor of polyploid tobacco. Plant Physiology 134, 1191–1199.

Guyot, R., Keller, B., 2004. Ancestral genome duplication in rice. Genome 47, 610–614.

Hansen, C.N., Harper, G., Heslop-Harrison, J.S., 2005. Characterisation of pararetrovirus-like sequences in the genome of potato (*Solanum tuberosum*). Cytogenetic and Genome Research 110, 559–565.

Harper, G., Osuji, J.O., Heslop-Harrison, J.S., Hull, R., 1999. Integration of banana streak badnavirus into the *Musa* genome: molecular and cytogenetic evidence. Virology 255, 207–213.

Harper, G., Hull, R., Lockhart, B., Olszewski, N., 2002. Viral sequences integrated into plant genomes. Annual Review of Phytopathology 40, 119–136.

Harper, G., Richert-Poggeler, K.R., Hohn, T., Hull, R., 2003. Detection of petunia vein-clearing virus: model for the detection of DNA viruses in plants with homologous endogenous pararetrovirus sequences. Journal of Virological Methods 107, 177–184.

Harper, G., Hart, D., Moult, S., Hull, R., Geering, A., Thomas, J., 2005. The diversity of *Banana streak virus* isolates in Uganda. Archives of Virology 150, 2407–2420.

Herzog, E., Guerra-Peraza, O., Hohn, T., 2000. The rice tungro bacilliform virus gene II product interacts with the coat protein domain of the viral gene III polyprotein. Journal of Virology 74, 2073–2083.

Hibino, H., Saleh, N., Roechan, M., 1979. Transmission of two kinds of rice tungro-associated viruses by insect vectors. Phytopathology 69, 1266–1268.

Hull, R., 2001. Classifying reverse transcribing elements: a proposal and a challenge to the ICTV. International Committee on Taxonomy of Viruses. Archives of Virology 146, 2255–2261.

Hull, R., 2002. Matthew's plant virology. Academic Press, San Diego.

Hull, R., Harper, G., Lockhart, B., 2000. Viral sequences integrated into plant genomes. Trends in Plant Science 5, 362–365.

Jakowitsch, J., Mette, M.F., van Der Winden, J., Matzke, M.A., Matzke, A.J., 1999. Integrated pararetroviral sequences define a unique class of dispersed repetitive DNA in plants. Proceedings of the National Academy of Sciences of the USA 96, 13241–13246.

Kenton, A., Khashoggi, A., Parokonny, A., Bennett, M.D., Lichtenstein, C., 1995. Chromosomal location of endogenous geminivirus-related DNA sequences in *Nicotiana tabacum* L. Chromosome Research 3, 346–350.

Kobayashi, N., Ikeda, R., 1992. Necrosis caused by rice tungro viruses in *Oryza glaberrima* and *O. barthii*. Japanese Journal of Breeding 42, 885–890.

Kohli, A., Twyman, R.M., Abranches, R., Wegel, E., Stoger, E., Christou, P., 2003. Transgene integration, organization and interaction in plants. Plant Molecular Biology 52, 247–258.

Kunii, M., Kanda, M., Nagano, H., Uyeda, I., Kishima, Y., Sano, Y., 2004. Reconstruction of putative DNA virus from endogenous *Rice tungro bacilliform virus*-like sequences in the rice genome: implications for integration and evolution. BMC Genomics 5, 80.

Langham, R.J., Walsh, J., Dunn, M., Ko, C., Goff, S.A., Freeling, M., 2004. Genomic duplication, fractionation and the origin of regulatory novelty. Genetics 166, 935–945.

Le, Q.H., Wright, S., Yu, Z., Bureau, T., 2000. Transposon diversity in *Arabidopsis thaliana*. Proceedings of the National Academy of Sciences of the USA 97, 7376–7381.

Lieber, M.R., 2010. The mechanism of double-strand DNA break repair by the nonhomologous DNA end-joining pathway. Annual Review of Biochemistry 79, 181–211.

Liebich, I., Bode, J., Reuter, I., Wingender, E., 2002. Evaluation of sequence motifs found in scaffold/matrix-attached regions (S/MARs). Nucleic Acids Research 30, 3433–3442.

Liu, R., Koyanagi, K.O., Chen, S., Kishima, Y., 2012. Evolutionary force of AT-rich repeats to trap genomic and episomal DNAs into the rice genome: lessons from endogenous pararetrovirus. Plant Journal 72, 817–828.

Lockhart, B.E., Menke, J., Dahal, G., Olszewski, N.E., 2000. Characterization and genomic analysis of tobacco vein clearing virus, a plant pararetrovirus that is transmitted vertically and related to sequences integrated in the host genome. Journal of General Virology 81, 1579–1585.

Makarevitch, I., Somers, D.A., 2006. Association of *Arabidopsis* topoisomerase IIA cleavage sites with functional genomic elements and T-DNA loci. Plant Journal 48, 697–709.

Mao, L., Wood, T.C., Yu, Y.S., Budiman, M.A., Tomkins, J., Woo, S.S., et al., 2000. Rice transposable elements: a survey of 73,000 sequence-tagged-connectors. Genome Research 10, 982–990.

Mette, M.F., Kanno, T., Aufsatz, W., Jakowitsch, J., van der Winden, J., Matzke, M.A., et al., 2002. Endogenous viral sequences and their potential contribution to heritable virus resistance in plants. EMBO Journal 21, 461–469.

Muller, H.P., Varmus, H.E., 1994. DNA bending creates favored sites for retroviral integration: an explanation for preferred insertion sites in nucleosomes. EMBO Journal 13, 4704–4714.

Nagano, H., Oka, A., Kishima, Y., Sano, Y., 2000. DNA sequences homologous to *Rice tungro bacilliform virus* (RTBV) present in the rice genome. Rice Genetic Newsletter 17, 103–105.

Nagano, H., Kunii, M., Azuma, T., Kishima, Y., Sano, Y., 2002. Characterization of the repetitive sequences in a 200-kb region around the rice waxy locus: diversity of transposable elements and presence of veiled repetitive sequences. Genes & Genetic Systems 77, 69–79.

Nassif, N., Penney, J., Pal, S., Engels, W.R., Gloor, G.B., 1994. Efficient copying of nonhomologous sequences from ectopic sites via P-element-induced gap repair. Trends in Biochemical Sciences 14, 1613–1625.

Ndowora, T., Dahal, G., LaFleur, D., Harper, G., Hull, R., Olszewski, N.E., et al., 1999. Evidence that badnavirus infection in *Musa* can originate from integrated pararetroviral sequences. Virology 255, 214–220.

Noreen, F., Akbergenov, R., Hohn, T., Richert-Poggeler, K.R., 2007. Distinct expression of endogenous *Petunia vein clearing virus* and the DNA transposon dTph1 in two Petunia hybrida lines is correlated with differences in histone modification and siRNA production. Plant Journal 50, 219–229.

Pahalawatta, V., Druffel, K., Pappu, H., 2008. A new and distinct species in the genus *Caulimovirus* exists as an endogenous plant pararetroviral sequence in its host *Dahlia variabilis*. Virology 376, 253–257.

Peterson-Burch, B.D., Wright, D.A., Laten, H.M., Voytas, D.F., 2000. Retroviruses in plants? Trends in Genetics 16, 151–152.

Puchta, H., 1999. Double-strand break-induced recombination between ectopic homologous sequences in somatic plant cells. Genetics 152, 1173–1181.

Puchta, H., 2005. The repair of double-strand breaks in plants: mechanisms and consequences for genome evolution. Journal of Experimental Botany 56, 1–14.

Richert-Poggeler, K.R., Shepherd, R.J., 1997. *Petunia vein-clearing virus*: a plant pararetrovirus with the core sequences for an integrase function. Virology 236, 137–146.

Richert-Poggeler, K.R., Noreen, F., Schwarzacher, T., Harper, G., Hohn, T., 2003. Induction of infectious *Petunia vein clearing (pararetro) virus* from endogenous provirus in petunia. EMBO Journal 22, 4836–4845.

Salomon, S., Puchta, H., 1998. Capture of genomic and T-DNA sequences during double-strand break repair in somatic plant cells. EMBO Journal 17, 6086–6095.

Sawasaki, T., Takahashi, M., Goshima, N., Morikawa, H., 1998. Structures of transgene loci in transgenic *Arabidopsis* plants obtained by particle bombardment: junction regions can bind to nuclear matrices. Gene 218, 27–35.

Somers, D.A., Makarevitch, I., 2004. Transgene integration in plants: poking or patching holes in promiscuous genomes? Current Opinion in Biotechnology 15, 126–131.

Staginnus, C., Richert-Poggeler, K.R., 2006. Endogenous pararetroviruses: two-faced travelers in the plant genome. Trends in Plant Science 11, 485–491.

Staginnus, C., Iskra-Caruana, M.-L., Lockhart, B.E.L., Hohn, T., Richert-Poggeler, K.R., 2009. Suggestions for nomenclature of endogenous pararetroviral (EPRV) sequences in plants. Archives of Virology 154, 1189–1193.

Suzuki, J., Yamaguchi, K., Kajikawa, M., Ichiyanagi, K., Adachi, N., Koyama, H., et al., 2009. Genetic evidence that the non-homologous end-joining repair pathway is involved in LINE retrotransposition. PLoS Genetics 5, e1000461.

Taganov, K., Daniel, R., Katz, R.A., Favorova, O., Skalka, A.M., 2001. Characterization of retrovirus-host DNA junctions in cells deficient in nonhomologous-end joining. Journal of Virology 75, 9549–9552.

Tikhonov, A.P., Lavie, L., Tatout, C., Bennetzen, J.L., Avramova, Z., Deragon, J.M., 2001. Target sites for SINE integration in *Brassica* genomes display nuclear matrix binding activity. Chromosome Research 9, 325–337.

Turcotte, K., Srinivasan, S., Bureau, T., 2001. Survey of transposable elements from rice genomic sequences. Plant Journal 25, 169–179.

Yang, Z.N., Ye, X.R., Molina, J., Roose, M.L., Mirkov, T.E., 2003. Sequence analysis of a 282-kilobase region surrounding the citrus *Tristeza* virus resistance gene (Ctv) locus in *Poncirus trifoliata* L. Raf. Plant Physiology 131, 482–492.

Chapter 13

Volatile organic compounds and plant virus–host interaction

Y.L. Dorokhov and T.V. Komarova

*A N Belozersky Institute of Physico-Chemical Biology, Moscow State University, Moscow, Russia,
and N I Vavilov Institute of General Genetics, Russian Academy of Science, Moscow, Russia*

E.V. Sheshukova

N I Vavilov Institute of General Genetics, Russian Academy of Science, Moscow, Russia

INTRODUCTION

Plants, being rooted in the soil and therefore physically unable to change location, are subjected to a plethora of stress factors. These stresses can be either biotic or abiotic in nature. Biotic stress is caused by living organisms, including pathogens (bacteria, fungi, oomycetes, nematodes, viruses), herbivores, and parasitic plants. Conversely, unfavorable changes in environmental conditions, such as temperature, water availability, salinity, light, and nutrient availability, lead to abiotic stress. Therefore, to survive, plants must have a variety of defense mechanisms that effectively reduce the number of threats that they face and the impact of stress.

The plant protection system includes the organization of communication between plants and between different parts of the same plant. Plants use volatile organic compounds (VOCs) to communicate in the absence of physical contact (Holopainen 2004, Baldwin 2010, Holopainen & Blande 2012). In addition to simple compounds, such as oxygen, carbon dioxide, and water vapor, plants emit a large variety of different terpenes, fatty acid derivatives, benzenoids, phenylpropanoids, and amino-acid-derived metabolites (Pichersky & Gershenzon 2002, Pichersky et al 2006). Plant VOCs have multiple functions as internal plant hormones (e.g., ethylene, methyl jasmonate (MeJA), and methyl salicylate (MeSA)) that participate in communication with conspecific and heterospecific plants. VOCs also play important roles in plant–plant communication and communication with organisms of different trophic levels, such as herbivores, pollinators, and enemies of herbivores. Species-specific VOCs normally repel polyphagous herbivores and those that specifically feed on other plant species, but they may also attract specific herbivores and their natural enemies, which use VOCs as host location cues. The attraction of predators and parasitoids by

VOCs is considered to be an evolved indirect defense, whereby plants are able to reduce biotic stress caused by damaging herbivores (McCormick et al 2012). Plant volatile blends elicited by herbivores are quantitatively and qualitatively different from those released in response to mechanical damage alone. Therefore, host-plant-specific and herbivore-specific volatile release is a reliable cue that predators and parasitoids can use to find suitable prey. Different signaling pathways are induced by chewing insects, such as caterpillars, and piercing-sucking insects, such as aphids and whiteflies. Volatile release is also induced by infection with aphid-transmitted viruses. It has been suggested that viruses that are acquired rapidly by aphids induce volatile release to attract migratory aphids, but these viruses discourage long-term aphid feeding (De Vos & Jander 2010). We currently have a fairly complete understanding of the processes and metabolic pathways involved in the production of many VOCs (Dudareva & Pichersky 2008, Dudareva et al 2004, 2006), but we have an extremely limited understanding of how VOCs affect intercellular traffic and thus what impact VOCs have on the plant virus–host interaction. VOCs emitted by a damaged plant act on the plant's own leaves and on the organs of neighboring plants, modifying intercellular communication (Wenke et al 2010, Dorokhov et al 2012a). This modification of intercellular transport has a significant effect on pathogenic infections. Plant viruses, as opposed to bacteria, fungi, oomycetes, and nematodes, are more dependent on the state of intercellular transport because the life cycle of viruses includes the intercellular transport or cell-to-cell movement of the viral genetic material and long-distance spread throughout the plant.

In this chapter, we provide a review of the role of VOCs in the plant virus–host interaction, detail the current knowledge on VOC-induced factors involved in intercellular communication, and conclude with suggestions for future research in this field.

VOCs AS PRODUCTS OF PLANT METABOLIC PATHWAYS

Plant volatiles are the metabolites that plants release into the air. Plants emit an enormous volume of VOCs. Calculations have shown that 1/5th of the carbon dioxide fixed by plants re enters the atmosphere every day (Baldwin 2010). Today, over 1700 volatile compounds have been identified from more than 90 plant families, constituting approximately 1% of all currently known plant secondary metabolites (Pichersky & Gershenzon 2002, Pichersky et al 2006). Plant volatiles typically occur as a complex mixture of low-molecular-weight lipophilic compounds derived from different biosynthetic pathways. The biochemistry and molecular biology of plant volatiles is complex and involves the interplay of several biochemical pathways and hundreds of genes. From a chemical standpoint, VOCs belong to various classes of natural products, namely terpenoids (homo-, mono-, di-, sesquiterpenoids), fatty acid degradation products, phenylpropanoids, amino-acid-derived products, alkanes, alkenes, alcohols, esters, aldehydes, and ketones of various biogenetic origin (Holopainen 2004, Dudareva et al 2006). Many of these products are made more lipophilic before their release into the air by the removal or masking of hydrophilic functional groups through reduction, methylation, or acylation reactions. Many different metabolic pathways contribute to the volatiles that are released, and hence the volatile metabolome contains information about the plant's metabolic status.

To date, several compounds have been reported to function as between and within plant signals, including (i) the green leaf volatiles (GLVs), (ii) terpenes, (iii) the phytohormones, including MeJA, MeSA, and ethylene, and (iv) MeOH.

GLVs are associated with the smell of a freshly mown lawn. This class includes a range of C6 compounds, namely aldehydes [*trans*-2-hexenal], alcohols [*cis*-3-hexen-1-ol], and esters [*cis*-3-hexenyl acetate] derived from C18 fatty acids that are released from damaged membranes, dioxygenated by lipoxygenase enzymes, and cleaved by hydroperoxide lyases. These compounds are emitted upon mechanical damage of the plant as well as herbivore feeding (Yan & Wang 2006). Unlike terpenoids, GLVs are rapidly, immediately, and likely passively released from wounded leaves. In addition, the release of these C6 volatiles occurs not only locally at a wound site but also systemically in the distal leaves (Kessler & Baldwin 2002). These compounds are therefore indicative of any mechanical damage and could provide early signals to receiving plants.

Terpenoids are the largest group of secondary compounds, consisting of approximately 40,000 compounds, including at least 1000 monoterpenes and 6500 sesquiterpenes (Yu & Utsumi 2009). Terpenoids are able to provide rapid but also herbivore-damage-related signals to receiving plants. Terpenoids play a central role in generating the chemical diversity of plant volatiles and appear to have been under strong diversifying selection.

MeJA is a volatile derivative of jasmonic acid (JA), which is an integral component of the plant defense responses to insect feeding and mechanical damage of leaves. Application of MeJA to plant leaves has been shown to increase the production of proteinase inhibitors (Farmer & Ryan 1990) and endo-(1,3;1,4)-β-glucanase (Akiyama et al 2009) under laboratory conditions.

MeSA is synthesized from salicylic acid (SA), a non volatile chemical signal required for the establishment of systemic acquired resistance (SAR) (Vlot et al 2008). MeSA in leaf tissue plays a role similar to that of gaseous MeSA in the pathogen-induced defense response and in response to aphid feeding damage (Mann et al 2012). MeSA is emitted by tobacco in response to infection by the *Tobacco mosaic virus* (Shulaev et al 1997, Vlot et al 2008), but it is not released in response to mechanical wounding (Shulaev et al 1997, Dorokhov et al 2012a). Recently, it has been shown that the MeSA is a critical mobile signal for plant systemic acquired resistance, and it acts as a long-distance mobile signal (Park et al 2007, 2009, Carr et al 2010, Liu et al 2011).

Ethylene is one of three plant hormones that are emitted into the air in biologically active quantities. This hormone is derived from the oxidation of 1-amino-cyclopropane-1-carboxylic acid, which, in turn, is derived from the amino acid methionine. Ethylene contributes to plant resistance against necrotrophic pathogens and tolerance to submergence and drought stress. Ethylene regulates the growth and development of most plant parts, in particular flowers and fruits, controlling how plants use volatiles to advertise for pollinators and fruit dispersers, respectively (Pieterse et al 2009). Ethylene is the only plant volatile for which the molecular mechanisms of recognition by the ethylene receptor are understood in detail (Baldwin 2010).

MeOH and isoprene are quantitatively the most important plant volatiles after CO_2 (Seco et al 2007). MeOH is a product of the demethylation of pectin during cell wall (CW) formation by the CW enzyme pectin methylesterase (PME) (Micheli 2001, Pelloux et al 2007). Although emissions from volcanoes, generation from H_2 and CO_2 in sea-floor hydrothermal systems (Williams et al 2011), and combustion of biomass all contribute to terrestrial atmospheric MeOH, PME-mediated emissions from plants are likely the largest source of MeOH in the atmosphere (Razavi et al 2011). Mechanical damage in plants drastically increases MeOH emission. For example, in an alfalfa field, a significant MeOH flux was observed after plant cutting, and the emission of MeOH was enhanced during the 3 days that the alfalfa was drying (Warneke et al 2002).

Gaseous MeOH was traditionally considered to be a biochemical 'waste product' (Nemeček-Marshall et al 1995, Von Dahl et al 2006). Recently, however, the effects of PME-generated MeOH from plants ('emitters') on the defensive reactions of plants ('receivers') were studied (Dorokhov et al 2012a). Investigations demonstrated that increased MeOH emissions from *PME*-transgenic or mechanically wounded non-transgenic plants retarded the growth of the bacterial pathogen *Ralstonia solanacearum* in neighboring 'receiver' plants. This antibacterial resistance was accompanied by the upregulation of genes that control stress and cell-to-cell communication in the 'receiver.' These results led to the conclusion that MeOH is a signaling molecule that is involved in within-plant and plant-to-plant communication (Dorokhov et al 2012a).

LEAF WOUNDING IN VOC EMISSION AND PLANT VIRUS INFECTION

Plant viruses are a class of plant pathogens that specialize in movement from cell to cell and plant to plant. The limited amount of genetic information that plant viruses contain makes them dependent on the plant itself, including the agents and mechanisms developed by the plant to protect itself against pathogens. Through their role in plant communication, VOCs can influence not only the virus life cycle but also the virus–host–vector combination. This point of view is supported by studies on the role of plant wounding, VOC emission, and plant–virus infection.

In general, plants emit volatiles constitutively, and this is believed to be an adaptive mechanism for defense against abiotic and biotic stresses (Holopainen 2004, Holopainen & Blande 2012). Isoprene emission protects photosynthesis from transient heat stress (Laothawornkitkul et al 2009). Moreover, isoprenoids function as antioxidants in leaves and confer protection against O_3-induced oxidative stress and singlet oxygen accumulation during photosynthesis (Vickers et al 2009). MeOH accumulates in the intercellular air space or in the liquid pool at night, when the stomata close. In the morning, when the stomata open, a large MeOH release can be observed (Nemeček-Marshall et al 1995, Hüve et al 2007). MeOH metabolism in plants is often accompanied by a remarkable induction of biomass with a simultaneous increase of the photosynthetic efficiency and the more rapid development of some C3-plant species (Nonomura & Benson 1992, Gout et al 2000). Constitutively emitted VOCs also influence the behavior and physiology of some herbivores, having toxic, repellent, and deterrent effects (Holopainen 2004, Holopainen & Blande 2012). However, there are no data on the participation of constitutively emitted VOCs in the innate immunity of plants to viruses because the plant cell is bounded by a rigid CW that prevents direct contact between adjacent cells and virus particles. This physical barrier is a major bottleneck that plant viruses have to overcome to be able to spread from cell to cell for the subsequent systemic invasion of their host. Thus, plant wounding is an obligatory condition for virus entry. Mechanical damage to the CW provides the virus with an opportunity to enter the cell, where, after stripping, the released viral genetic material can initiate the synthesis of viral proteins and replication of the genome (Shaw 1999).

In the laboratory, to initiate reproduction of the *Tobacco mosaic virus*, the experimenter induces microtrauma of the leaf cuticle. In nature, virus penetration into the leaf tissue can occur via microdamage to the leaf cuticle, trichome, or CW due to damage caused by wind, rain, hail, or herbivore feeding. During mechanical inoculation in the laboratory and in nature, the virus enters a limited number of cells in the leaf. In a study of the process of so-called 'subliminal infection' with *Tobacco mosaic virus* (TMV) and a nonhost plant (Cheo 1970), the number of primary-infected cells was 1 cell per 50,000 to 150,000 uninfected cells (Sulzinski & Zaitlin 1982). The TMV particle is uncoated by a mechanism involving the ribosome and formation of a 'striposome,' referred to as cotranslational disassembly (Wilson 1984). A virus that replicates in the initially infected cell but fails to move from cell to cell is restricted to a single cell. The process of host translation of genomic and subgenomic RNA, which eventually leads to the formation of viral progeny, is accompanied by synthesis of the movement protein (MP) and transport of genomic RNA into neighboring cells. Cell-to-cell transport is necessary for the infection of larger areas of tissue. This transport takes place either as a viral ribonucleoprotein complex (Dorokhov et al 1983, Citovsky et al 1990) or as virus particles (Van Lent et al 1991), and it occurs via the plasmodesmata (PD) that connect adjacent cells. The cell-to-cell movement of viruses proceeds outwards from an initially infected cell. In the relatively slow process of cell-to-cell movement (one cell per 2 hours), the infection frontier advances only a small distance each day (Gillespie et al 2002). For systemic movement to other parts of the plant, the virus must move from the mesophyll cells to the leaf veins and vascular tissue (Vuorinen et al 2011). Similar pathways are employed by the plant host to traffic endogenous macromolecules, suggesting that viruses make use of host transport systems for their own movement (Leisner & Turgeon 1993). Once the virus enters the phloem, it moves at rates comparable to the rate of movement of photoassimilate. The speed of movement of photoassimilates is high. Photoassimilate synthesized in mature (source) leaves reaches the upper (sink) leaves within 60–100 minutes (Thorpe et al 2007). The direction and rate of photoassimilate transport depends on several factors, including the relative source and sink strengths, the proximity of the source to the sink, and the interconnections of the vascular system (Turgeon 1989). Phloem transport includes the loading (entry) of the virus into the phloem at source tissues, its circulation in the transport phloem, and its unloading (exit) from the phloem at the sink tissues. The rate of such movement of viral genetic material is hundreds of times higher than the rate of intercellular movement, and it has been estimated that TMV and *Potato virus X* (PVX) are transported at a rate of approximately 8 cm per hour (Capoor 1949). While it is assumed that most viruses move systemically through the phloem, xylem transport has been proposed for the beetle-transmitted *Sobemovirus* genus (Opalka et al 1998, Brugidou et al 2002). This assumption, however, does not apply to *Cocksfoot mottle sobemovirus* (Otsus et al 2012).

In contrast to the constitutive VOCs normally released from healthy intact plants, inducible volatiles are emitted after foliar damage or wounding, and these compounds are produced in larger quantities or in different ratios (Holopainen 2004, Holopainen & Blande 2012). In nature, leaf damage occurs as a result of exposure to wind, rain, and hail. Damage to the CW results in the release of GLV (Heil 2009), some terpenes (Piesik et al 2011), and MeOH (Von Dahl et al 2006, Körner et al 2009, Dorokhov et al 2012a). The emission of these compounds is very fast and can be detected immediately following mechanical damage (Dorokhov et al 2012a), which indicates the release of pre existing material (stored compounds). The MeOH emitted from wounded leaves is produced by two forms of PME: pre existing PME deposited in the CW before wounding, which allows rapid release of MeOH (Von Dahl et al 2006, Körner et al 2009), and PME that is synthesized *de novo* after wounding (Dorokhov et al 2012a), which likely involves generation of MeOH for an extended period (more than 8 hours). *De novo* synthesis of MeOH after leaf damage (i.e., only when needed) is more economical in terms of carbon usage and does not reduce plant fitness (Dicke & Van Loon 2000, Holopainen 2004).

MeOH emitted by a wounded plant attracts insects and bark beetles (*Hylurgops palliatus, Tomicus piniperda,* and *Trypodendron domesticum*), while longer-chain alcohols are not attractive (Byers 1992). Moreover, mice prefer the odor of MeOH to the odors of other plant volatiles in laboratory conditions, and MeOH exposure alters the accumulation of mRNA in the mouse brain (Dorokhov et al 2012b). This finding led to the conclusion that MeOH that is emitted by wounded plants may have a role in plant–animal signaling.

An attack by herbivorous insects is a complex event involving at least two different aspects, mechanical damage and chemical factors. The specific elicitors present in an herbivore's saliva are important for the induction of VOCs in some plant–insect systems (Wu et al 2008). Only the combination of both is able to induce the respective plant defenses. Feeding insects consume leaves by continually clipping off and ingesting small pieces of tissue. This process inflicts a series of mechanical wounds, usually accompanied by the simultaneous introduction of insect saliva and foregut secretions (oral secretions) into the damaged tissue (Wu et al 2008). The amounts of VOCs emitted by plants in response to phloem feeders are generally low and sometimes completely absent (Pareja et al 2007, Joó et al 2010), most likely because phloem feeders, unlike chewing herbivorous insects, inflict only minimal tissue damage on their host plants, circumventing stress signaling. Nevertheless, a study on *Myzus persicae* and peach cultivars demonstrated that even a moderate aphid attack can be sufficient to induce VOC emissions in plants. Aphid attacks induced emissions of mainly MeSA and terpenoids, whereas artificial wounding of leaves did not induce these compounds but resulted in the immediate emission of GLVs (Staudt et al 2010). Interestingly, even egg deposition by herbivorous insects induced the plant volatiles that can activate specific plant responses that significantly influence various members of higher trophic levels (Fatourus et al 2012). A variety of plant elicitors constitute the main mechanisms by which plants detect injury ('damaged-self recognition') and stimulate JA production (Heil 2009). When multiple enemies attack a plant, complex interactions between the SA, ethylene, and JA pathways can result in a shift from inhibitory to synergistic plant defensive responses to injury (Pieterse et al 2009).

VOC-MEDIATED PRIMING OF PLANTS AND ITS ROLE IN VIRUS INFECTION

Because VOCs move freely in the air, they may also affect neighboring plants and then mediate the phenomenon of 'plant–plant communication,' which has been found in taxonomically unrelated plants. VOCs emitted into the air as a result of mechanical damage of the leaf, or damage caused by an insect attack, may affect not only the leaves of adjacent levels on the same plant but may also affect the leaves of other plants. Theoretically, the release of VOCs as a signal of damage can lead to three possible responses in a neighboring plant. First, the neighboring plant does not respond to the signal. Second, after receiving a signal, the plant responds by creating conditions that promote plant virus infection (increasing sensitivity of plant, its sensitization). Finally, the plant responds to the signal by increasing its antiviral resistance (i.e., the creation of conditions that prevent virus infection). The VOC-mediated process of preparing plants for a putative attack can be referred to as 'priming' (i.e., the initiation of reactions before the impact of the pathogen). In general, the process of priming is related to plant immunity, whereby plants trigger their defenses in response to a signal or previous challenge so that they can react with increased severity (Holopainen & Blande 2012). Recent studies have shown that intact leaves neighboring a site of damage, and 'receiver' plants, detect VOCs as an alarm that allows them to prepare for a possible attack (Choudhary et al 2008, Frost et al 2008, Holopainen & Blande 2012). Priming via an airborne signal from an herbivore-damaged plant to an undamaged neighbor was observed in corn plants (Engelberth et al 2004). Plant–plant communication via VOCs is likely to be a common phenomenon in herbivore resistance, and similar volatile compounds can also mediate the beneficial effects that are caused by plant-growth-promoting rhizobacteria (Ryu et al 2004). Furthermore, exposure to VOCs such as *trans*-2-hexenal, *cis*-3-hexenal, or *cis*-3-hexenol enhanced the resistance of *Arabidopsis* against the fungal pathogen *Botrytis cinerea* (Kishimoto et al 2005), which indicates that VOCs may also induce disease resistance. Lima bean (*Phaseolus lunatus*) plants in a natural population became more resistant to a bacterial pathogen, *Pseudomonas syringae pv syringae*, when they were located close to conspecific neighbors in which systemic acquired resistance to pathogens had been chemically induced with benzothiadiazole (Yi et al 2009).

Priming of the plant, resulting in a change in phenotype, particularly the emergence of resistance, is apparently only one of many aspects of the process. The signals are received by plants, even though changes in phenotype are not observed. UV-C-irradiated plants can emit a volatile signal (MeSA and MeJA) in their immediate gaseous environment that leads to an increase in the frequency of homologous recombination in neighboring non irradiated plants (Yao et al 2011).

A signal from neighboring plants can change the biochemical pathways and transcriptional activity of the genome of the receiver plant. Changes in the transcription patterns of defense-related genes following exposure to VOCs have been described in several studies (Bate & Rothstein 1998, Arimura et al 2000, Farag et al 2005, Paschold et al 2006, Frost et al 2008). In 'receiver' plants, the emitted VOCs can upregulate *PR* genes, such as the basic type *PR-3* (chitinase), acidic type *PR-4* (thaumatin-like), *lipoxygenase* (LOX), *phenylalanine ammonia-lyase* (PAL), and *farnesyl pyrophosphate synthetase* (FPS). Other primed defense responses include accelerated production of trypsin-proteinase inhibitors in tobacco plants exposed to volatiles from damaged sagebrush (Kessler et al 2006). Multiple MeOH-inducible genes (MIGs) involved in defense and cell-to-cell trafficking have been detected (Dorokhov et al 2012a). The effects of the PME-generated MeOH released from wounded plants ('emitters') on the defensive reactions of neighboring 'receiver' plants can be explained on the basis of the activities of MIGs, most of which were classified as stress-response genes. The expression of the *LOX*, *PR-3*, *PR-4*, *FPS*, and *PAL* genes was increased slightly in MeOH-treated plants. Treatment with *cis*-3-hexen-1-ol stimulated the accumulation of *FPS* mRNA, but ethylene, MeSA, and MeJA treatment primarily upregulated *PAL* and *PR-4* with the accumulation of mRNAs (Dorokhov et al 2012a). In accordance with the data on the

MeOH-induced appearance of transcripts of genes involved in plant defense reactions, gaseous MeOH or vapors from wounded and *PME*-transgenic plants induced resistance to the bacterial pathogen *R. solanacearum* in the leaves of non-wounded neighboring 'receiver' plants (Dorokhov et al 2012a). Antibacterial resistance accompanied by MIG upregulation is likely to be related to induction of transcription of the *type II proteinase inhibitor (PI-II)* gene. Type II proteinase inhibitors are powerful inhibitors of serine endopeptidases in animals and microorganisms (Turra & Lorito 2011). The *PI-II* gene is not expressed in the leaves of healthy plants, but it is induced in leaves that have been subjected to different types of stress, including wounding and bacterial infection (Balandin et al 1995). *PME*-transgenic tobacco with high levels of *PI-II* expression demonstrated high resistance to *R. solanacearum* (Dorokhov et al 2012a). This finding supports the role of PI-II in the suppression of bacterial proteases.

Collectively, the results directly support a mechanism involving the VOC-induced priming and protection against bacteria, insects, fungi, and nematodes. However, when we consider the role of priming in the relationship of the virus and the plant host, the picture is ambiguous. The first study of the possible involvement of VOCs in plant priming and subsequent virus entry and reproduction was carried out 16 years ago (Shulaev et al 1997). Shulaev and his colleagues used tobacco plants (*Nicotiana tabacum* cv. *Xanthi* nc) carrying an *N* gene associated with a hypersensitive reaction at the site of TMV entry; they discovered that only TMV-inoculated plants produced gaseous MeSA. No MeSA was volatilized from healthy, mock-inoculated, or mechanically wounded tobacco plants. To establish whether the amount of emitted MeSA is sufficient to induce resistance in adjacent tobacco plants, the authors exploited a flow-through system and revealed moderate resistance to TMV in plants that received MeSA-containing air from TMV-inoculated plants. The authors concluded that TMV-infected plants emit MeSA, which is able to induce antivirus resistance in neighboring plants. Deng et al (2004) also reported that, in contrast to healthy intact controls, the leaves of tomato plants infected with TMV accumulated MeSA. Moreover, this MeSA accumulation was observed in tomato leaves treated with gaseous MeSA. However, no MeSA-mediated within-plant or plant-to-plant signaling was observed in experiments carried out in growth chambers, where control and test tobacco plants were intermingled and confined within a small space (Park et al 2009). It has been suggested that in nature, where conditions are not highly optimized for transmission of volatile signals and plants are generally less confined than in growth chambers, the MeSA signal is produced in a liquid rather than a gaseous form. Recent work by Attaran et al (2009) also demonstrated that in *A. thaliana*, MeSA is not required for systemic signaling and that SAR is established through *de novo* SA synthesis in systemic non infected leaves that are primed by the production of the mobile metabolite azelaic acid, a 9-carbon dicarboxylic acid.

Experiments with gaseous MeOH presented examples of priming in intact plants, leading to the creation of conditions conducive to viral infection

(Dorokhov et al 2012a). This effect could be explained by the activities of MIGs such as β-1,3-glucanase (BG) (Levy et al 2007a,b, Zavaliev et al 2011) and non-cell-autonomous pathway protein (NCAPP) (Lee et al 2003), enhancing cell-to-cell communication.

In accordance with these data, MeOH acts as a signal that facilitates cell-to-cell movement of 2×GFP as a reporter macromolecule for the movement through PD in different states of dilation. Moreover, BG or NCAPP may trigger PD dilation (gating) in leaves that have been agroinjected with binary vectors encoding BG or NCAPP (Dorokhov et al 2012a). A model proposing that MeOH-triggered PD dilation should enhance viral spread within the plant was confirmed in experiments where BG and NCAPP could enhance cell-to-cell communication and TMV RNA accumulation. Moreover, gaseous MeOH or vapors from wounded plants increased TMV reproduction in 'receivers' (Dorokhov et al 2012a).

Thus, MeOH has a contradictory effect on the sensitivity of the leaves of the 'receiver' plant to bacteria and viruses. The mechanisms that underlie this phenomenon are not clear, but we can consider the factors that may explain the inconsistency of the effects of MeOH. First, there is a fundamental difference between bacteria and viruses with respect to their modes of intercellular transport. Bacterial pathogens do not cross the plant CW boundaries because they inhabit the intercellular spaces (Lee & Lu 2011). In contrast, viral pathogens require cell-to-cell movement for local and systemic spread (Xu & Jackson 2010). Second, the most abundant MIGs can be divided into two groups according to their ability to participate in bacterial or viral pathogenesis. The first, including *PI-II* (Balandin et al 1995) and *PME* inhibitor (Volpi et al 2011), are involved in immunity against non viral pathogens. The second group of genes, which includes *NCAPP* (Lee et al 2003) and *MIG-21* (Dorokhov et al 2012a), is involved in the intercellular transport and the reproduction of the virus. The most abundant MIG, the *BG* gene, which encodes the basic pathogenesis-related 2 (PR-2) protein, is involved in antibacterial immunity. On the other hand, this protein accelerates the process of intercellular transport (Levy et al 2007a,b, Zavaliev et al 2011).

The role of MeOH in communication under natural conditions is unclear, and experiments in the field are necessary. In general, most of the experiments on the effects of VOCs on the pathogen–plant host interaction have been performed under laboratory conditions. In most reviews of the plant–herbivore–herbivore enemy relationship, additional experiments under natural conditions are needed to better understand the ecologic significance of the results obtained (Baldwin 2010; Holopainen & Blande 2012). This is also the case for investigations of plant virus transmission because emission patterns often differ between laboratory and field conditions. In this regard, it is important to estimate the distance at which a VOC can function as a signal. Sagebrush, *Artemisia tridentate*, has been used as a subject of numerous plant–plant communication studies conducted under field conditions. It has been shown that the distance over which interspecific communication occurred was 10 cm (Karban et al 2003). Intraspecific communication has also been demonstrated in sagebrush, and it was shown that communication occurred at distances of up to 60 cm between the clipped plants and the intact receivers (Karban et al 2006). Recently, Heil and Adame-Alvarez (2010) used wild-type lima bean plants to quantify the distances over which volatile signals move under field conditions, and they thereby determined whether these cues mainly trigger resistance in other parts of the same plant or in independent plants. Independent receiver plants exhibited airborne resistance to herbivores or pathogens at a maximum distance of 50 cm from a resistance-expressing emitter. If we apply these estimates to the natural situation of the virus–plant interaction, then VOCs will mostly act on the leaves of the same plant but not on those of neighboring plants. Given that the rate of VOC movement is much higher than the velocity of phloem compounds or the SAR signal, gaseous MeOH emitted by infected leaves would prime upper non-inoculated leaves for the arrival of infectious viral entities.

VOCs AND VIRUS-TRANSMITTED VECTORS

Volatile communication plays an important role in mediating the interactions between plants, aphids, and viruses in the environment (De Vos & Jander 2010). Aphids are a serious problem for agriculture despite being a relatively small insect group compared to others because they transmit numerous viruses. Nearly 50% of insect-borne viruses (275 out of 600) are transmitted by aphids. Many of the viruses transmitted by aphids cause diseases of major economic importance. The indirect damage that aphids cause through virus transmission often far exceeds their direct impact on crops. In response to aphid infestation, many plants initiate indirect defenses through the release of volatiles that are also induced by infection with aphid-transmitted viruses. From a virus-centric point of view, plant viruses rely on aphid transport not only to other parts of the same plant but also to more distant plants (Ng & Falk 2006, Dáder et al 2012). Therefore, it would be in the best interests of the virus to make the plant attractive for aphid probing but not necessarily more suitable for long-term aphid feeding.

There are two important aphid behaviors that affect the rate of virus transmission: (i) aphids feed preferentially on virus-infected tissue and (ii) aphids preferentially orient toward virus-infected plants (De Vos & Jander 2010). The particular plant–aphid–virus interaction determines the time required for virus acquisition. Plant viruses can be classified in two categories differing by the site at which the virus is retained by the vector and the retention period: non-circulative and circulative or persistent viruses, which frequently accumulate in the salivary glands.

Non-circulative, non persistent viruses

The optimal pattern of vector behavior for the transmission of non persistent viruses—attraction, probing, and rapid dispersal—appears to be quite different from that of the transmission of persistent viruses and may be expected to favor a different pattern of virus-induced changes in host–plant phenotypes. Viruses that are not persistently transmitted constitute a majority of plant viruses and represent many of those that cause the most severe economic losses in agricultural crops (Ng & Falk 2006). These viruses, which attach transiently to the aphid stylets, typically require shorter acquisition times than persistent, circulative viruses. Nonpersistent viruses are transmitted quickly, sometimes with a single aphid probe of a leaf, and they can make host plants less suitable for aphid feeding. Transmitted virions attach, through conserved protein–protein interactions, to specific regions within the aphid mouthparts during brief, exploratory probes of the epidermal cells of an infected plant and are transmitted effectively only if the vector disperses quickly (within minutes) to a new, susceptible plant. It has been suggested (De Vos & Jander 2010) that non persistent viruses might be expected to induce changes in the gustatory cues that repel aphids (after they have probed and acquired virions) rather than encouraging arrestment and colonization. Until recently, the limited evidence available suggests that aphid population growth is often reduced on plants infected by non persistently transmitted viruses, but it is not known how virus infection influences plant chemistry (e.g., volatile and contact cues) or vector behaviors relating directly to transmission. To address the lack of information about the disease ecology of non persistent plant viruses, Mauck et al (2010a) investigated the effects of infection by *Cucumber mosaic virus* (CMV), a non persistently transmitted virus, on plant chemistry and the interactions between cultivated squash plants (*Cucurbita pepo* cv. Dixie) and two generalist aphid vectors, *Aphis gossypii* and *M. persicae*. To explore its effects on plant–vector interactions, a series of field and greenhouse experiments was performed to assess aphid performance on healthy and infected plants. The results suggest that CMV-infected plants are poor hosts for *A. gossypii* and *M. persicae*: the populations reached higher levels on untouched and mock-inoculated plants than on CMV-infected plants. Greenhouse and field experiments showed that winged aphid colonizers emigrate more readily from infected plants. Gas chromatographic analysis of volatiles collected in the greenhouse and field from untouched, mock-inoculated, and CMV-infected *C. pepo* revealed that infected plants released significantly greater quantities of volatiles per gram of tissue than did healthy plants. In total, 38 compounds were released by squash plants. However a comparison of the overall blends indicated that no individual compound was responsible for these increased emissions; rather, infected plants released a blend that was qualitatively similar to that of healthy plants. The authors concluded that the attraction of aphids to the odors of CMV-infected plants may be explained by the elevated levels of volatile emissions that otherwise are similar to those of healthy plants.

Thus, these results reveal a pattern of interactions between CMV-infected plants and aphid vectors that is very different from that of persistently transmitted viruses and most likely more conducive to CMV's non persistent mode of transmission, which requires that aphids acquire the virus by probing infected plants but then disperse quickly, without colonizing the plant or initiating long-term feeding.

The conception that CMV-induced changes in host plant chemistry influence the behavior of vector herbivores is likely to also apply to non vector herbivores, such as the squash bug, *Anasa tristis*, which is a pest in this system. Mauck et al (2010b) found that adult *A. tristis* females preferred to oviposit on healthy plants in the field and that healthy plants supported higher populations of nymphs.

The congruence between the transmission pattern and the effects on host plant quality and attractiveness for vectors strongly suggests the possibility of a general mechanism, although additional plant–virus systems need to be explored. For example, examination of potato plants infected by *Potato virus Y* (PVY) or PVX revealed no apparent effects of infection on host-plant quality or vector attraction. PVX- and PVY-infected potatoes were neither better hosts nor were they preferentially colonized by *M. persicae,* the principal vector of the persistently transmitted virus, *Potato leafroll virus* (PLRV) (Castle et al 1998). PVY is a non-circulative virus transmitted in a non persistent manner by several aphid species. PVX does not require a vector, and it is typically transmitted mechanically. *M. persicae apterae* aggregated preferentially on PLRV-infected potatoes compared with uninfected plants or plants infected with PVX or PVY (Eigenbrode et al 2002). There were no essential differences in the amount and composition of volatiles from PLRV- and PVY- or PVX-infected plants.

A possible mechanism for the attraction of aphids to plants infected with CMV may involve the active viral protein 2b, which not only inhibits antiviral RNA silencing but also quenches the transcriptional responses of plant genes to JA, a key signaling molecule in the defense against insects. Ziebell et al (2011) found that infection of tobacco with a 2b gene deletion mutant (CMVΔ2b) induced strong resistance to *M. persicae* aphids, while CMV infection fostered aphid survival. Using an electrical penetration graph method, the authors found that higher proportions of aphids showed sustained phloem ingestion on CMV-infected plants than on CMVΔ2b-infected or mock-inoculated plants, although this did not increase the rate of growth of individual aphids. This indicates that the 2b protein could indirectly affect aphid-mediated virus transmission.

Persistent viruses

In contrast to non persistent viruses, persistent circulative viruses are taken into the aphid gut. The acquisition of persistent viruses requires sustained aphid feeding in the phloem of an infected plant over hours to days. This mode of transmission entails the ingestion of viruses present in plant phloem by feeding aphids and the subsequent movement of virions from the aphid gut through the body cavity to the salivary glands, where the virions reside but do not replicate. Once infected in this way, an aphid can transmit the virus through its saliva to new host plants persistently (i.e., for an extended period). This transmission appears to be favored by virus-induced changes in plants that encourage vector colonization and sustained feeding on infected plants. Furthermore, the enhanced quality of the plants infected with persistent viruses would lead to rapid growth of the aphid population, resulting in crowding and subsequent dispersal of aphids bearing the virus to new plants. Recent evidence confirms these suggestions and indicates that aphids are capable of responding behaviorally to differences in olfactory cues from plants infected with viruses and non infected control plants (Eigenbrode et al 2002, Jiménez-Martínez et al 2004). Among the best-documented examples of pathogen-induced effects on host odor cues are the induction of characteristic volatile emissions by two plant viruses, PLRV and *Barley yellow dwarf virus* (BYDV), that are more attractive to aphid vectors than emissions from healthy plants (Ngumbi et al 2007). In addition to inducing characteristic volatile blends in infected hosts, both of these pathogens improve the quality of the plants as hosts for vectors. The aphids *Rhopalosiphum padi* and *Schizaphis graminum* produce more offspring on BYDV-infected wheat and oats, respectively (Jiménez-Martínez et al 2004), and the main vector of PLRV, *M. persicae*, performs better on PLRV-infected potatoes. *M. persicae* and *R. padi* also preferentially arrest (i.e., remain) on virus-infected plants following exposure to tactile and gustatory cues. The results of studies on BYDV-infected wheat also suggest that responses of *R. padi* to BYDV-infected plants are caused by attraction rather than arrestment.

Thus, these viruses induce changes in the phenotypic traits of host plants that enhance vector attraction to, and arrestment on, infected plants.

The specific mechanisms that influence aphid responses to volatile compounds from virus-infected plants have not been investigated thoroughly. It is possible that one or several of the components whose levels are elevated by virus infection is a specific arrestant for the aphids. Alternatively, the entire blend released by the infected plants may be responsible. Eigenbrode et al (2002) reported that at least 11 headspace volatile components (*cis*-2-hexen-1-ol, *n*-nonane, nonanal, undecane, β-myrcene, (*R*)-(+)-limonene, (−)-transcaryophyllene, (+)-longifolene, α-pinene, β-pinene, and α-humulene) were elevated by PLRV infection but not by PVX or PVY infection or mock inoculation compared with non infected plants. Ngumbi et al (2007) followed up on these results to determine whether individual VOCs or blends of these compounds that are elevated as a result of PLRV infection elicit significant behavioral and physiologic activity in the aphids. It was shown that a blend of volatiles produced by PLRV-infected plants, which differs in its total concentration and relative composition from a blend from non infected potato plants, is an arrestant for the green peach aphid, whereas the individual components of the blend are not arrestants. This was the first study to demonstrate the importance of blends for *M. persicae* and the first to show the importance of blends vs. individual volatile compounds for eliciting behavioral responses in any aphid species.

CONCLUSION

Based on the available data, we can conclude that VOCs that are released into the air from damaged plants or plants compromised by herbivorous insects serve as an alarm to help neighboring plants or adjacent leaves to prepare for defense. It is not yet clear whether the purpose of VOC-transmitted signals is to pass the warning to adjacent plants or whether these neighbors just 'eavesdrop' on the signal that communicates with the other parts of the emitting plant. Regardless, VOC-mediated signals are transmitted to neighboring plants and promote the survival of the entire community. VOCs serve as an alarm signal that helps protect against herbivorous insects and plant pathogens such as bacteria, fungi, oomycetes, and nematodes (i.e., VOCs are unfavorable for pathogens and herbivores). However, considering the role of VOCs in the relationship between viruses and plants, we do not find a negative, or even a neutral, influence of VOCs on viruses. On the contrary, the findings described above indicate that VOCs sensitize the plant to allow the entry and spread of the virus within the plant and between plants by insect vectors. VOCs promote viral propagation. The favorable impact of VOCs on viruses may be explained by several factors, as follows.

The first possibility is that the virus, unlike the herbivorous insects, bacteria, fungi, oomycetes, and nematodes, does not represent a clear threat to the existence of the plant species. Such an explanation is quite controversial, and there is no convincing evidence to support this statement, although well-known disastrous consequences, such as potato blight caused by *Phytophthora infestans*, which struck Europe like 'a bolt from the blue' in the 1840s, are not related to a virus outbreak (Strange & Scott 2005). Little is known about viruses that cause the death of the plant host. The plant often sacrifices a portion of its cells (suicide or a phenomenon of hypersensitive reaction) to protect the whole plant, as in the case of TMV and *N*-gene-containing tobacco (Carr et al 2010).

The second possibility is related to the underlying differences in the genome structure and way of life of the plant viruses and other pathogens (bacteria, fungi, oomycetes, and nematodes). Plant viruses differ from other types of pathogens in a number of ways. First, the virus life cycle, including entry, replication, transcription, and progeny accumulation, takes place exclusively in the symplast (Xu & Jackson 2010). Furthermore, the survival of the virus depends on its ability to move from cell to cell in order to accumulate in sufficient levels and tissues to guarantee its survival. To this end, viruses exploit PD as cell-to-cell symplastic connections to achieve local spreading and systemic infection in their host plants. These channels provide access to most tissues, and their use leads to familiar patterns of infection. From the point of initial entry to the tissue, viruses spread from cell to cell through PD to create a local infection focus (cell-to-cell movement) until they encounter the tissues of the vasculature. At that point, they are loaded into the phloem elements for translocation to distal tissues (long-distance movement) (Vuorinen et al 2011). A virus must accomplish successful infection of a plant by utilizing a very limited amount of genetic material. Some plant virus genomes may encode as few as three proteins with an essential function. An example of such a component is movement protein. Because the physical size of the PD hinders the free movement of the virus or its genetic material, movement protein modifies the physical state of the symplastic tunnels, which leads to an increase of the size exclusion limit (SEL) of the PD. Due to their large population size and short generation time, viruses have a great potential to quickly evolve and adapt under the pressure of natural selection (Fraile & Garcia-Arenal 2010). The high incidence of mutation and recombination in viral genomes enhances the generation of new variants that, when the mutation results in a biologic advantage, quickly spread throughout the viral population (Moya et al 2004, Hanssen et al 2010). The other feature of many plant viruses that distinguishes them from other pathogens is the fact that a large majority of viruses depend on other organisms for dissemination from plant to plant (Ng & Falk 2006, Vuorinen et al 2011). Winged insects with sucking mouthparts (e.g., aphids, leafhoppers, and whiteflies) are responsible for the transmission of the majority of plant viruses and carry them efficiently over distance.

In summary, it is significant that the virus, with its small but highly variable genome, spends its entire life in the cell symplast, while other pathogens inhabit the apoplast. It should also be noted that as with apoplastic factors that are emitted into the air, VOCs affect the structure of the apoplast, including the CW.

The third possibility is that reinforced intercellular transport, which promotes viral replication, is also a necessary condition for the defense reactions, such as mechanisms of defense against bacterial pathogens using RNA interference. The symplast is not only the space in which viruses reproduce, it is also the site of RNA interference mechanisms that serve to eliminate foreign RNA. This specific degradation of RNA by RNA interference allows the host plant to effectively control viruses and other pathogens. It is known that the intracellular and intercellular transport of silencing factors is a necessary condition for effective RNA interference. Therefore, an MeOH-mediated increase of viral replication can be regarded as a compensation for the acquisition of plant antimicrobial resistance.

To conclude, our knowledge of phytogenic VOCs is still limited, and the role of these compounds in within-plant and plant–plant communication has not been fully elucidated. However, it is clear that globalization of agriculture has contributed to the spread of viruses. The cultural growth of plants (i.e., the cultivation of plants of the same species in the same place, where they are inevitably injured by mechanical processes) creates favorable conditions for the spread of viruses. The modern globalization of agriculture has led to the cultivation of plants in regions far from the places of origin of cultivated species and therefore away from viruses that they evolved with.

One option may be the use of transgenic plants as 'beacons' or 'disinfectants' with increased VOC emissions to trigger a protective response against pathogens and plant-eating insects (Dudareva & Pichersky 2008, Holopainen & Blande 2012). This suggestion is very attractive, but we must take into account the possibility that such a defense mechanism will favor the spread of the virus. Viral disease may be minimized by reduction of the virus's inoculums, inhibition of its virulence mechanisms, and promotion of genetic diversity in the crop.

REFERENCES

Akiyama, T., Jin, S., Yoshida, M., Hoshino, T., Opassiri, R., Cairns, K.J.R., 2009. Expression of an endo-(1,3;1,4)-beta-glucanase in response to wounding, methyl jasmonate, abscisic acid and ethephon in rice seedlings. Journal of Plant Physiology 166 (16), 1814–1825.

Arimura, G., Ozawa, R., Shimoda, T., Nishioka, T., Boland, W., Takabayashi, J., 2000. Herbivory-induced volatiles elicit defence genes in lima bean leaves. Nature 406 (6795), 512–515.

Attaran, E., Zeier, T.E., Griebel, T., Zeier, J., 2009. Methyl salicylate production and jasmonate signaling are not essential for systemic acquired resistance in *Arabidopsis*. Plant Cell 21 (3), 954–971.

Balandin, T., Van Der Does, C., Albert, J.M., Bol, J.F., Linthorst, H.J., 1995. Structure and induction pattern of a novel proteinase inhibitor class II gene of tobacco. Plant Molecular Biology 27 (6), 1197–1204.

Baldwin, I.T., 2010. Plant volatiles. Current Biology 20 (9), 392–397.

Bate, N.J., Rothstein, S.J., 1998. C6-volatiles derived from the lipoxygenase pathway induce a subset of defense-related genes. Plant Journal 16 (5), 561–569.

Brugidou, C., Opalka, N., Yeager, M., Beachy, R.N., Fauquet, C., 2002. Stability of *Rice yellow mottle virus* and cellular compartmentalization during the infection process in *Oryza sativa* (L.). Virology 297 (1), 98–108.

Byers, J.A., 1992. Attraction of bark beetles, *Tomicus piniperda, Hylurgops palliatus*, and *Trypodendron domesticum* and other insects to short-chain alcohols and monoterpenes. Journal of Chemical Ecology 18 (12), 2385–2402.

Capoor, S.P., 1949. The movement of tobacco mosaic viruses and potato virus X through tomato plants. Annals of Applied Biology 36 (3), 307–319.

Carr, J.P., Lewsey, M.G., Palukaitis, P., 2010. Signaling in induced resistance. Advances in Virus Research 76, 57–121.

Castle, S.J., Mowry, T.M., Berger, P.H., 1998. Differential settling by *Myzus persicae* (Homoptera: Aphididae) on various virus infected host plants. Annals of the Entomological Society of America 91 (5), 661–667.

Cheo, P.C., 1970. Subliminal infection of cotton by tobacco mosaic virus. Journal of Phytopathology 60, 41–46.

Choudhary, D.K., Johri, B.N., Prakash, A., 2008. Volatiles as priming agents that initiate plant growth and defence responses. Current Science 94 (5), 595–604.

Citovsky, V., Knorr, D., Schuster, G., Zambryski, P., 1990. The P-30 protein of tobacco mosaic virus is a single-strand nucleic acid binding protein. Cell 60 (4), 637–647.

Dáder, B., Moreno, A., Viñuela, E., Fereres, A., 2012. Spatio-temporal dynamics of viruses are differentially affected by parasitoids depending on the mode of transmission. Viruses 4 (11), 3069–3089. http://dx.doi.org/10.3390/v4113069.

Deng, C., Zhang, X., Zhu, W., Qian, J., 2004. Gas chromatography-mass spectrometry with solid-phase microextraction method for determination of methyl salicylate and other volatile compounds in leaves of *Lycopersicon esculentum*. Analytical and Bioanalytical Chemistry 378 (2), 518–522.

De Vos, M., Jander, G., 2010. Volatile communication in plant-aphid interactions. Current Opinion in Plant Biology 13 (4), 366–371.

Dicke, M., Van Loon, J.J.A., 2000. Multitrophic effects of herbivore-induced plant volatiles in an evolutionary context. Entomologia Experimentalis et Applicata 97 (3), 237–249.

Dorokhov, Y.L., Alexandrova, N.M., Miroshnichenko, N.A., Atabekov, J.G., 1983. Isolation and analysis of virus-specific ribonucleoprotein of tobacco mosaic virus-infected tobacco. Virology 127 (2), 237–252.

Dorokhov, Y.L., Komarova, T.V., Petrunia, I.V., Frolova, O.Y., Pozdyshev, D.V., Gleba, Y.Y., 2012a. Airborne signals from a wounded leaf facilitate viral spreading and induce antibacterial resistance in neighboring plants. PLoS Pathogens 8(4): e1002640. http://dx.doi.org/10.1371/journal.ppat.1002640.

Dorokhov, Y.L., Komarova, T.V., Petrunia, I.V., Kosorukov, V.S., Zinovkin, R.A., Shindyapina, A.V., Frolova, O.Y., Gleba, Y.Y., 2012b. Methanol may function as a cross-kingdom signal. PLoS ONE 7, e36122. http://dx.doi.org/10.1371/journal.pone.0036122.

Dudareva, N., Pichersky, E., 2008. Metabolic engineering of plant volatiles. Current Opinion in Biotechnology 19 (2), 181–189.

Dudareva, N., Pichersky, E., Gershenzon, J., 2004. Biochemistry of plant volatiles. Plant Physiology 135 (4), 1893–1902.

Dudareva, N., Negre, F., Nagegowda, D.A., Orlova, I., 2006. Plant volatiles: recent advances and future perspectives. Critical Reviews in Plant Sciences 25 (5), 417–440. http://dx.doi.org/10.1080/07352680600899973.

Eigenbrode, S.D., Ding, H., Shiel, P., Berger, P.H., 2002. Volatiles from potato plants infected with potato leafroll virus attract and arrest the virus vector, *Myzus persicae* (Homoptera: Aphididae). Proceedings of the Royal Society B: Biological Sciences 269 (1490), 455–460.

Engelberth, J., Alborn, H.T., Schmelz, E.A., Tumlinson, J.H., 2004. Airborne signals prime plants against insect herbivore attack. Proceedings of the National Academy of Sciences of the USA 101 (6), 1781–1785.

Farag, M.A., Fokar, M., Zhang, H., Allen, R.D., Paré, P.W., 2005. (Z)-3-hexenol induces defense genes and downstream metabolites in maize. Planta 220 (6), 900–909.

Farmer, E.E., Ryan, C.A., 1990. Interplant communication – airborne methyl jasmonate induces synthesis of proteinase-inhibitors in plant leaves. Proceedings of the National Academy of Sciences of the USA 87 (19), 7713–7716.

Fatouros, N.E., Lucas-Barbosa, D., Weldegergis, B.T., Pashalidou, F.G., Van Loon, J.J., Dicke, M., Harvey, J.A., Gols, R., Huigens, M.E., 2012. Plant volatiles induced by herbivore egg deposition affect insects of different trophic levels. PLoS ONE 7(8):e43607. http://dx.doi.org/10.1371/journal.pone.0043607.

Fraile, A., García-Arenal, F., 2010. The coevolution of plants and viruses: resistance and pathogenicity. Advances in Virus Research 76, 1–32.

Frost, C.J., Mescher, M.C., Dervinis, C., Davis, J.M., Carlson, J.E., De Moraes, C.M., 2008. Priming defense genes and metabolites in hybrid poplar by the green volatile cis-3-hexenyl acetate. New Phytologist 180 (3), 722–734.

Gillespie, T., Boevink, P., Haupt, S., Roberts, A.G., Toth, R., Valentine, T., Chapman, S., Oparka, K.J., 2002. Functional analysis of a DNA-shuffled movement protein reveals that microtubules are dispensable for the cell-to-cell movement of tobacco mosaic virus. Plant Cell 14 (6), 1207–1222.

Gout, E., Aubert, S., Bligny, R., Rebeille, F., Nonomura, A.R., 2000. Metabolism of methanol in plant cells. Carbon-13 nuclear magnetic resonance studies. Plant Physiology 123 (1), 287–296.

Hanssen, I.M., Lapidot, M., Thomma, B.P., 2010. Emerging viral diseases of tomato crops. Molecular Plant–Microbe Interactions 23 (5), 539–548.

Heil, M., 2009. Damaged-self recognition in plant herbivore defence. Trends in Plant Science 14 (7), 356–363.

Heil, M., Adame-Álvarez, R.M., 2010. Short signalling distances make plant communication a soliloquy. Biology Letters 6 (6), 843–845.

Holopainen, J.K., 2004. Multiple functions of inducible plant volatiles. Trends in Plant Science 9 (11), 529–533.

Holopainen, J.K., Blande, J.D., 2012. Molecular plant volatile communication. Advances in Experimental Medicine and Biology 739, 17–31.

Hüve, K., Christ, M.M., Kleist, E., Uerlings, R., Niinemets, Ü., Walter, A., Wildt, J., 2007. Simultaneous growth and emission measurements demonstrate an interactive control of methanol release by leaf expansion and stomata. Journal of Experimental Botany 58 (7), 1783–1793.

Jimenez-Martinez, E.S., Bosque-Perez, N.A., Berger, P.H., Zementra, R.S., Ding, H., Eigenbrode, S.D., 2004. Volatile cues influence the response of *Rhopalosiphum padi* (Homoptera: Aphididae) to *Barley yellow dwarf virus*-infected transgenic and untransformed wheat. Journal of Environmental Entomology 33 (5), 1207–1216.

Joó, É., Van Langenhove, H., Simpraga, M., Steppe, K., Amelynck, C., Schoon, N., Müller, J.-F., Dewulf, J., 2010. Variation in biogenic volatile organic compound emission pattern of *Fagus sylvatica* L. due to aphid infection. Atmospheric Environment 44, 227–234.

Karban, R., Maron, J., Felton, G.W., Ervin, G., Eichenseer, H., 2003. Herbivore damage to sagebrush induces resistance in wild tobacco: evidence for eavesdropping between plants. Oikos 100 (2), 325–332.

Karban, R., Shiojiri, K., Huntzinger, M., McCall, A.C., 2006. Damage-induced resistance in sagebrush: volatiles are key to intra- and interplant communication. Ecology 87 (4), 922–930.

Kessler, A., Baldwin, I.T., 2002. Plant responses to insect herbivory: the emerging molecular analysis. Annual Review of Plant Biology 53, 299–328.

Kessler, A., Halitschke, R., Diezel, C., Baldwin, I.T., 2006. Priming of plant defense responses in nature by airborne signaling between *Artemisia tridentate* and *Nicotiana attenuata*. Oecologia 148 (2), 280–292.

Kishimoto, K., Matsui, K., Ozawa, R., Takabayashi, J., 2005. Volatile C6-aldehydes and allo-ocimene activate defense genes and induce resistance against *Botrytis cinerea* in *Arabidopsis thaliana*. Plant and Cell Physiology 46 (7), 1093–1102.

Körner, E., Von Dahl, C.C., Bonaventure, G., Baldwin, I.T., 2009. Pectin methylesterase NaPME1 contributes to the emission of methanol during insect herbivory and to the elicitation of defence responses in *Nicotiana attenuata*. Journal of Experimental Botany 60 (9), 2631–2640.

Laothawornkitkul, J., Taylor, J.E., Paul, N.D., Hewitt, C.N., 2009. Biogenic volatile organic compounds in the Earth system. New Phytologist 183 (1), 27–51.

Lee, J.Y., Lu, H., 2011. Plasmodesmata: the battleground against intruders. Trends in Plant Science 16 (4), 201–210.

Lee, J.Y., Yoo, B.C., Rojas, M.R., Gomez-Ospina, N., Staehelin, L.A., Lucas, W.J., 2003. Selective trafficking of non-cell-autonomous proteins mediated by NtNCAPP1. Science 299 (5605), 392–326.

Leisner, S.M., Turgeon, R., 1993. Movement of virus and photoassimilate in the phloem: a comparative analysis. BioEssays 15 (11), 741–748.

Levy, A., Erlanger, M., Rosenthal, M., Epel, B.L., 2007a. A plasmodesmata-associated beta-1,3-glucanase in *Arabidopsis*. Plant Journal 49 (4), 669–682.

Levy, A., Guenoune-Gelbart, D., Epel, B.L., 2007b. Beta-1,3-Glucanases: plasmodesmal gate keepers for intracellular communication. Plant Signaling and Behavior 2 (5), 404–407.

Liu, P.P., Von Dahl, C.C., Klessig, D.F., 2011. The extent to which methyl salicylate is required for signaling systemic acquired resistance is dependent on exposure to light after infection. Plant Physiology 157 (4), 2216–2226.

Mann, R.S., Ali, J.G., Hermann, S.L., Tiwari, S., Pelz-Stelinski, K.S., Alborn, H.T., Stelinski, L.L., 2012. Induced release of a plant-defense volatile 'deceptively' attracts insect vectors to plants infected with a bacterial pathogen. PLoS Pathogens 8(3):e1002610. http://dx.doi.org/10.1371/journal.ppat.1002610.

Mauck, K.E., De Moraes, C.M., Mescher, M.C., 2010a. Deceptive chemical signals induced by a plant virus attract insect vectors to inferior hosts. Proceedings of the National Academy of Sciences of the USA 107 (8), 3600–3605.

Mauck, K.E., De Moraes, C.M., Mescher, M.C., 2010b. Effects of *Cucumber mosaic virus* infection on vector and non-vector herbivores of squash. Communicative and Integrative Biology 3 (6), 579–582.

McCormick, C.A., Unsicker, S.B., Gershenzon, J., 2012. The specificity of herbivore-induced plant volatiles in attracting herbivore enemies. Trends in Plant Science 17 (5), 303–310.

Micheli, F., 2001. Pectin methylesterases: cell wall enzymes with important roles in plant physiology. Trends in Plant Science 6 (9), 414–419.

Moya, A., Holmes, E.C., González-Candelas, F., 2004. The population genetics and evolutionary epidemiology of RNA viruses. Nature Reviews Microbiology 2, 279–288.

Nemeček-Marshall, M., MacDonald, R.C., Franzen, J.J., Wojciechowski, C.L., Fall, R., 1995. Methanol emission from leaves. Plant Physiology 108, 1359–1368.

Ng, J.C., Falk, B.W., 2006. Virus–vector interactions mediating nonpersistent and semipersistent transmission of plant viruses. Annual Review of Phytopathology 44, 183–212.

Ngumbi, E., Eigenbrode, S.D., Bosque-Pérez, N.A., Ding, H., Rodriguez, A., 2007. *Myzus persicae* is arrested more by blends than by individual compounds elevated in headspace of PLRV-infected potato. Journal of Chemical Ecology 33 (9), 1733–1747.

Nonomura, A., Benson, A., 1992. The path of carbon in photosynthesis: improved crop yields with methanol. Proceedings of the National Academy of Sciences of the USA 89 (20), 9794–9798.

Opalka, N., Brugidou, C., Bonneau, C., Nicole, M., Beachy, R.N., Yeager, M., Fauquet, C., 1998. Movement of *Rice yellow mottle virus* between xylem cells through pit membranes. Proceedings of the National Academy of Sciences of the USA 95 (6), 3323–3328.

Otsus, M., Uffert, G., Sõmera, M., Paves, H., Olspert, A., Islamov, B., Truve, E., 2012. *Cocksfoot mottle sobemovirus* establishes infection through the phloem. Virus Research 166 (1-2), 125–129.

Pareja, M., Moraes, M.C., Clark, S.J., Birkett, M.A., Powell, W., 2007. Response of the aphid parasitoid *Aphidius funebris* to volatiles from undamaged and aphid-infested *Centaurea nigra*. Journal of Chemical Ecology 33 (4), 695–710.

Park, S.W., Kaimoyo, E., Kumar, D., Mosher, S., Klessig, D.F., 2007. Methyl salicylate is a critical mobile signal for plant systemic acquired resistance. Science 318 (5847), 113–116.

Park, S.W., Liu, P.P., Forouhar, F., Vlot, A.C., Tong, L., Tietjen, K., Klessig, D.F., 2009. Use of a synthetic salicylic acid analog to investigate the roles of methyl salicylate and its esterases in plant disease resistance. Journal of Biological Chemistry 284 (11), 7307–7317.

Paschold, A., Halitschke, R., Baldwin, I.T., 2006. Using 'mute' plants to translate volatile signals. Plant Journal 45 (2), 275–291.

Pelloux, J., Rusterucci, C., Mellerowicz, E.J., 2007. New insights into pectin methylesterase structure and function. Trends in Plant Science 12 (6), 267–277.

Pichersky, E., Gershenzon, J., 2002. The formation and function of plant volatiles: perfumes for pollinator attraction and defense. Current Opinion in Plant Biology 5 (3), 237–243.

Pichersky, E., Noel, J.P., Dudareva, N., 2006. Biosynthesis of plant volatiles: nature's diversity and ingenuity. Science 311 (5762), 808–811.

Piesik, D., Pańka, D., Delaney, K.J., Skoczek, A., Lamparski, R., Weaver, D.K., 2011. Cereal crop volatile organic compound induction after mechanical injury, beetle herbivory (*Oulema* spp.), or fungal infection (*Fusarium* spp.). Journal of Plant Physiology 168 (9), 878–886.

Pieterse, C.M.J., Leon-Reyes, A., Van der Ent, S., Van Wees, S.C.M., 2009. Networking by small molecule hormones in plant immunity. Nature Chemical Biology 5 (5), 308–316.

Razavi, A., Karagulian, F., Clarisse, L., Hurtmans, D., Coheur, P.F., Clerbaux, C., Muller, J.F., Stavrakou, T., 2011. Global distributions of methanol and formic acid retrieved for the first time from the IASI/MetOp thermal infrared sounder. Atmospheric Chemistry and Physics 11, 857–872.

Ryu, C.M., Farag, M.A., Hu, C.H., Reddy, M.S., Kloepper, J.W., Pare, P.W., 2004. Bacterial volatiles induce systemic resistance in *Arabidopsis*. Plant Physiology 134 (3), 1017–1026.

Seco, R., Penuelas, J., Filella, I., 2007. Short-chain oxygenated VOCs: emission and uptake by plants and atmospheric sources, sinks, and concentrations. Atmospheric Environment 41 (12), 2477–2499.

Shaw, J.G., 1999. *Tobacco mosaic virus* and the study of early events in virus infections. Philosophical Transactions of the Royal Society of London, Series B, Biological Sciences 354 (1383), 603–611.

Shulaev, V., Silverman, P., Raskin, I., 1997. Airborne signalling by methyl salicylate in plant pathogen resistance. Nature 385, 718–721.

Staudt, M., Jackson, B., El-Aouni, H., Buatois, B., Lacroze, J.P., Poëssel, J.L., Sauge, M.H., 2010. Volatile organic compound emissions induced by the aphid *Myzus persicae* differ among resistant and susceptible peach cultivars and a wild relative. Tree Physiology 30 (10), 1320–1334.

Strange, R.N., Scott, P.R., 2005. Plant disease: a threat to global food security. Annual Review of Phytopathology 43, 83–116.

Sulzinski, M.A., Zaitlin, M., 1982. *Tobacco mosaic virus* replication in resistant and susceptible plants: in some resistant species virus is confined to a small number of initially infected cells. Virology 121 (1), 12–19.

Thorpe, M.R., Ferrieri, A.P., Herth, M.M., Ferrieri, R.A., 2007. 11C-imaging: methyl jasmonate moves in both phloem and xylem, promotes transport of jasmonate, and of photoassimilate even after proton transport is decoupled. Planta 226 (2), 541–551.

Turgeon, R., 1989. The sink-source transition in leaves. Annual Review of Plant Physiology and Plant Molecular Biology 40, 119–138.

Turra, D., Lorito, M., 2011. Potato type I and II proteinase inhibitors: modulating plant physiology and host resistance. Current Protein and Peptide Science 12 (5), 374–385.

Van Lent, J., Storms, M., Van Der Meer, F., Wellink, J., Goldbach, R., 1991. Tubular structures involved in movement of cowpea mosaic virus are also formed in infected cowpea protoplasts. Journal of General Virology 72, 2615–2623.

Vickers, C.E., Possell, M.P., Cojocariu, C., Velikova, V., Laothawornkitkul, J., Ryan, A., Mullineaux, P.M., Hewitt, C.N., 2009. Isoprene synthesis protects transgenic plants from oxidative stress. Plant, Cell and Environment 32 (5), 520–531.

Vlot, A.C., Klessig, D.F., Park, S.W., 2008. Systemic acquired resistance: the elusive signals. Current Opinion in Plant Biology 11 (4), 436–442.

Volpi, C., Janni, M., Lionetti, V., Bellincampi, D., Favaron, F., D'Ovidio, R., 2011. The ectopic expression of a pectin methyl esterase inhibitor increases pectin methyl esterification and limits fungal diseases in wheat. Molecular Plant–Microbe Interactions 24 (9), 1012–1019.

Von Dahl, C.C., Hävecker, M., Schlögl, R., Baldwin, I.T., 2006. Caterpillar-elicited methanol emission: a new signal in plant-herbivore interactions? Plant Journal 46 (6), 948–960.

Vuorinen, A.L., Kelloniemi, J., Valkonen, J.P., 2011. Why do viruses need phloem for systemic invasion of plants? Plant Science 181 (4), 355–363.

Warneke, C., Luxembourg, S.L., De Gouw, J.A., Rinne, H.J.I., Guenther, A.B., Fall, R., 2002. Disjunct eddy covariance measurements of oxygenated volatile organic compounds fluxes from an alfalfa field before and after cutting. Journal of Geophysical Research 107 (D8), 1–10.

Wenke, K., Kai, M., Piechulla, B., 2010. Belowground volatiles facilitate interactions between plant roots and soil organisms. Planta 231 (3), 499–506.

Williams, L.B., Holloway, J.R., Canfield, B., Glein, C.R., Dick, J.M., Hartnett, H.H., Shock, E.L., 2011. Birth of biomolecules from the warm wet sheets of clays near spreading centers. In: Golding, S.D., Glikson, M. (Eds.), Earliest life on earth, habitats, environments and methods of detection, Springer Science and Business Media, Dordrecht, Netherlands, Part 1, pp. 79–112.

Wilson, T.M., 1984. Cotranslational disassembly of tobacco mosaic virus in vitro. Virology 137 (2), 255–265.

Wu, J., Hettenhausen, C., Schuman, M.C., Baldwin, I.T., 2008. A comparison of two *Nicotiana attenuate* accessions reveals large differences in signaling induced by oral secretions of the specialist herbivore. Manduca sexta. Plant Physiology 146 (3), 927–939.

Xu, X.M., Jackson, D., 2010. Lights at the end of the tunnel: new views of plasmodesmal structure and function. Current Opinion in Plant Biology 13 (6), 684–692.

Yan, Z.G., Wang, C.Z., 2006. Wound-induced green leaf volatiles cause the release of acetylated derivatives and a terpenoid in maize. Phytochemistry 67 (1), 34–42.

Yao, Y., Danna, C.H., Zemp, F.J., Titov, V., Ciftci, O.N., Przybylski, R., Ausubel, F.M., Kovalchuk, I., 2011. UV-C-irradiated *Arabidopsis* and tobacco emit volatiles that trigger genomic instability in neighboring plants. Plant Cell 23 (10), 3842–3852.

Yi, H.-S., Heil, M., Adame-Alvarez, R.M., Ballhorn, D.J., Ryu, C.-M., 2009. Airborne induction and priming of plant defenses against a bacterial pathogen. Plant Physiology 151 (4), 2152–2161.

Yu, F., Utsumi, R., 2009. Diversity, regulation and genetic manipulation of plant mono and sesquiterpenoid biosynthesis. Cellular and Molecular Life Sciences 66 (18), 3043–3052.

Zavaliev, R., Ueki, S., Epel, B.L., Citovsky, V., 2011. Biology of callose (β-1,3-glucan) turnover at plasmodesmata. Protoplasma 248 (1), 117–130.

Ziebell, H., Murphy, A.M., Groen, S.C., Tungadi, T., Westwood, J.H., Lewsey, M.G., Moulin, M., Kleczkowski, A., Smith, A.G., Stevens, M., Powell, G., Carr, J.P., 2011. *Cucumber mosaic virus* and its 2b RNA silencing suppressor modify plant-aphid interactions in tobacco. Scientific Reports 1, 187.

Chapter 14

Diversity of latent plant–virus interactions and their impact on the virosphere

K.R. Richert-Pöggeler
Institute for Epidemiology and Pathogen Diagnostics, Julius Kühn-Institut, Braunschweig, Germany

J. Minarovits
University of Szeged, Faculty of Dentistry, Department of Oral Biology and Experimental Dental Research, Szeged, Hungary

INTRODUCTION

Viruses can be found in all kingdoms of life. The ninth report of the International Committee on Taxonomy of Viruses (ICTV, King et al 2012) lists 2284 virus and viroid species belonging to 349 different genera that have been identified in bacteria and archaea, invertebrates, plants, and vertebrates. Recent research has shown that the impact of viral entities on life goes far beyond their role as trigger of diseases. Curtis Suttle created the term 'virosphere,' which defines space where viruses are found and which is influenced by viruses. Viruses are the most abundant 'life'-forms on Earth. In one liter of seawater there are more viruses than there are people on Earth, and the total number of viruses found in the ocean outnumbers all the stars in the universe 10 million fold (Suttle 2007). Exploring the oceanic virosphere provides valuable insights into evolution and the origin of life, respectively. For example one viral entity, the *Mavirus*, was identified as a 'virophage,' parasitizing the giant *Cafeteria roenbergensis* virus. Sequence comparisons indicated that extant transposons of the Maverick/Polinton class may have originated from ancient relatives of *Mavirus* (Fischer & Suttle 2011).

The idea of 'virolution,' a term coined by Frank Ryan (2009), has emerged. It suggests that horizontal gene transfer facilitated by viruses and/or their endogenous counterparts plays a major role in coevolution of mutualistic symbionts. The occurrence and relevance of such events driven by viruses has been proposed for *Elysia chlorotica*, a mollusk that gained the ability of photosynthesis by such a process (Rumpho et al 2008).

Whereas viral latency has been a major field of research in bacterial, animal, and human viruses, with prominent representatives found among lysogenic bacteriophages, herpesviruses, and retroviruses, latent viruses in plants are only just starting to draw scientists' attention (Brüssow et al 2004, Minarovits 2006, Weiss 2006, Wang et al 2010).

One could speculate, however, that horizontal gene transfer mediated by mobile infectious entities, such as viruses, might have an impact during the evolution of plants, which have a sessile lifestyle. Besides natural selective pressure by changing environments, human interference by plant breeding and production can dramatically influence virus–plant coevolution, as is illustrated for selected ornamentals in this chapter. Furthermore, analogies as well as differences of virus latency in plant and animal hosts will be pointed out.

Diversity and biology of latent plant viruses and viroids

Historically, virus names often provide information about phenotypic changes and the symptoms they cause on the host from which they were first isolated. In the early years of virology no tools for imaging or analysis were available to provide information about particle structure and the organization of the viral genome. About one third of all classified viruses and viroids are represented by those infecting plants. However, current taxonomy is biased because 77% of recognized plant viruses derive from cultivated plants (Wren et al 2006). Knowledge about viruses in the natural flora needs to be expanded in order to give a realistic picture of the presence and distribution of viruses. Interestingly, that viruses can be present without causing symptoms was recognized early and is manifested in terms such as *latent, cryptic,* or *symptomless,* representing 7% and 4% of currently classified plant viruses and viroids, respectively. Viruses that cause asymptomatic infections can be found in 24 genera out of a total of 81 genera representing RNA and DNA viruses with various virion morphologies (Fig. 14.1; Table 14.1) and viroids in both *Pospiviroidae* and *Avsunviroidae* (King et al 2012 and references therein). The majority of the viruses listed in Table 14.1 have a narrow natural host range, indicating a high degree of adaptation to the host plant and coevolution. Regarding transmission, aphids play an important role, and evidently vertical transmission via seeds seems to be common to many latent viruses as well as viroids (Table 14.1). Persistent infections by endornaviruses constitute an extreme form of virus latency with a lack of encapsidation, cell-to-cell movement, and horizontal transmission (Roossinck et al 2011). The observed diversity in virus 'lifestyles' may also indicate that interactions causing latency are manifold and depend on the involved partners. *Potato virus M* (PVM), a carlavirus, is a common virus of potato occurring worldwide. Various strains of PVM have been isolated, and probably many more exist or are still evolving (Verhoeven et al 2006, Flatken et al 2008, Xu et al 2010). PVM infection can lead to yield

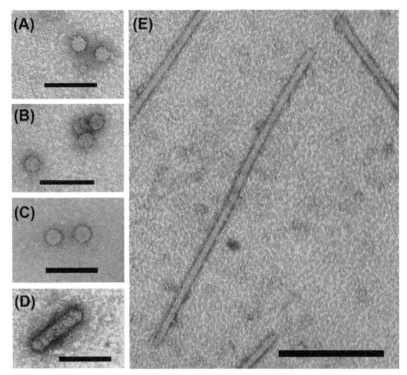

FIGURE 14.1 Virion morphology of selected latent plant viruses comprising isometric (A–C), bacilliform (D), and filamentous (E) particles. Negative contrast electron micrograph of virus preparations from crude sap stained with 1% uranyl acetate. (A) *Poinsettia mosaic virus*, 26–29 nm in diameter; (B) *Poinsettia latent virus*, 34 nm in diameter; (C) *Grapevine Algerian latent virus*, 32–35 nm in diameter; (D) badnavirus, 30 nm in width, modal length 130 nm; (E) carlavirus, 610–700 nm in length. Bar equals 100 nm in panels A–D and 200 nm in panel E.

losses of up to 45% depending on virus strain and potato cultivar (Xu et al 2010). In the ornamental *Solanum jasminoides* PVM does not cause symptoms (Fig. 14.2). Recently, two silencing suppressors have been identified for PVM that interfere with intracellular as well as systemic RNA silencing-based antiviral defense systems of the host (Senshu et al 2011). It would be interesting to see if the differences in virulence observed in the various PVM strains are correlated to the efficiency of the two viral suppressors blocking the RNA silencing machinery in potatoes and other hosts.

Interactions of latent plant viruses and viroids within the virosphere

The complexity of interactions within the virosphere is becoming more and more apparent. Biodiversity within the virosphere comprises direct virus–host, virus–vector, virus–virus, and virus–viroid interactions. In mixed infections, multiple

TABLE 14.1 Morphology, nucleic acid, host range, and modes of transmission from virus genera and viroid families with members featuring a latent phenotype

Name[a]	Morphology	Genome	Natural host range	Transmission
Alphacryptovirus	Isometric	dsRNA	Narrow	Cell division, seed, pollen
Aureusvirus, Pothos latent virus	Isometric	+ssRNA	Narrow	Soil-borne
Betacryptovirus	Isometric	dsRNA	Narrow	Cell division, seed, pollen
Cheravirus, Apple latent spherical virus	Isometric	+ssRNA	Wide or narrow	Seed, pollen
Bromovirus, Spring beauty latent virus	Isometric	+ssRNA	Narrow	Beetle
Ilarvirus, Spinach latent virus	Isometric	+ssRNA	Narrow	Mechanical, seed
Necrovirus, Olive latent virus 1	Isometric	+ssRNA	Narrow	Mechanical, soil
Nepovirus	Isometric	+ssRNA	Wide	Seed, nematode, mite
Polemo, Poinsettia latent virus	Isometric	+ssRNA	Narrow	Grafting, vegetative propagation
Strawberry latent ringspot virus	Isometric	+ssRNA	Wide	Seed, nematode
Tombusvirus, Grapevine Algerian latent virus	Isometric	+ssRNA	Narrow	Mechanical, soil
Tymovirus	Isometric	+ssRNA	Narrow	Mechanical
Caulimovirus, Horseradish latent virus	Isometric	dsDNA	Narrow	Aphid, mechanical
Oleavirus, Olive latent virus 2	Bacilliform	+ssRNA	Narrow	Mechanical
Carlavirus	Filamentous	+ssRNA	Wide or narrow	Aphid, mechanical
Foveavirus, Apricot latent virus	Filamentous	+ssRNA	Narrow	Mechanical, grafting

TABLE 14.1 Morphology, nucleic acid, host range, and modes of transmission from virus genera and viroid families with members featuring a latent phenotype—cont'd

Name[a]	Morphology	Genome	Natural host range	Transmission
Lolavirus, Lolium latent virus	Filamentous	+ssRNA	Narrow	Mechanical
Macluravirus	Filamentous	+ssRNA	Narrow	Aphid
Potyvirus	Filamentous	+ssRNA	Wide or narrow	Aphid, mechanical
Vitivirus, Heracleum latent virus	Filamentous	+ssRNA	Narrow	Aphid, mechanical
Hordeivirus	Rod-shaped	+ssRNA	Narrow	Seed, pollen
Tobamovirus	Rod-shaped	+ssRNA	Narrow	Mechanical
Endornavirus	No virions	dsRNA	Narrow	Seed
Avsunviroidae	No virions	Circular, ssRNA	Narrow	Mechanical, seed, pollen
Pospiviroidae	No virions	Circular, ssRNA	Wide or narrow	Mechanical, seed

When only one latent virus exists in the genus, corresponding features are listed.
[a]*Refers to virus genus or, if unassigned, to species and viroid family, respectively.*

FIGURE 14.2 *S. jasminoides* infected with *Potato virus M* and *Tomato apical stunt viroid* does not display symptoms.

partners are present and can consist of viruses belonging to the same or different genera, as well as viroids. Ornamentals can carry viruses and viroids that do not give rise to phenotypic changes, or cause damage, but which can act as a source of inoculum for related crop plants on which they have deleterious effects. In that respect, *Potato spindle tuber viroid* (PSTVd) infections of *S. jasminoides* were regarded as a potential threat for potato production and were eliminated according to EU legislation (Commission Decision 2007/410EC of June 2007). Interestingly, the vacant niche for a viroid in *S. jasminoides* seems to have been filled by another pospiviroid, *Tomato apical stunt viroid* (TASVd). It is assumed that TASVd-infected ornamentals are responsible for an outbreak of TASVd in tomato in the Netherlands (Verhoeven et al 2012). A mixed infection of a viroid and PVM does not induce symptoms in *S. jasminoides*, either (Verhoeven et al 2006; Fig. 14.2). For PSTVd it was proposed that the uneven viroid distribution in leaf tissue may support a dosage-dependent onset of host RNA silencing (Schwind et al 2009). Spacial separation—inhibiting virus–viroid encounters in the same cell—as well as low virus/viroid titers, may account for the maintenance of silencing control as observed in PVM/viroid infections. Future investigations should provide information on how viruses and viroids invade ornamentals, interact, and escape the surveillance system or serve as inocula for crop plants; information is also needed on what makes viruses/viroids stay asymptomatic in some solanaceous hosts while causing severe disease in others. The generation of chimeric DNA sequences has been reported for mixed infections consisting of a viroid-like element and *Carnation etched ring virus* (CERV), a caulimovirus, in carnation (Vera et al 2000). Replication of the double-stranded DNA virus, CERV, is mediated by a reverse transcriptase that uses an RNA template for DNA synthesis. Reverse transcription is also thought to generate DNA from the Carnation small viroid-like RNA and to be involved in the formation of extra chromosomal hybrid molecules comprising junctions of pararetrovirus and retroviroid-like DNA sequences (Vera et al 2000).

Transition from latency to the causation of symptoms

An even more complex system of interactions is man-made in the plant poinsettia, *Euphorbia pulcherrima*, that represents almost one third of potted plants produced in Germany (AMI report 2012). Commercially available poinsettia are infected with a phytoplasma to obtain a desirable 'free-branching' phenotype. Techniques such as meristem tip culture or heat therapy for virus elimination cannot be employed easily because they may also remove the phytoplasma from plants; hence, vegetatively propagated poinsettia often carry latent infections by *Poinsettia latent virus* (PnLV, Fig. 14.1), a polemovirus, and *Poinsettia mosaic virus* (PnMV, Fig. 14.1), a putative member of the *Tymoviridae* (aus dem Siepen et al 2005). The latter is asymptomatic when growing at a high temperature (Koenig et al 1986) due to a host control mechanism based on RNA silencing (Szittya et al 2003). Indeed, transgenic poinsettia carrying virus-derived hairpin RNA constructs were resistant to PnMV infection (Clarke et al 2008).

An additional level of complexity is reached when integrated, inducible virus sequences occur. Such interacting systems, which involve up to four different partners, are providing a new view on possible dynamic developments in open systems that react to changing environmental and endogenous conditions. Initial studies indicated the possibility of rapid evolution based on vertical as well as horizontal gene transfer and expression. One example exists in banana carrying integrated *Banana streak virus* (BSV) sequences in its genome as well as being infected with BSV horizontally transmitted by mealybugs (Iskra-Caruana et al 2010). Similarly to retroviruses, pararetroviruses of plants and animals also use reverse transcriptase during their replication cycle. Their double-stranded DNA genomes are present in episomal form in the host cells, similarly to herpesvirus genomes, but transcription of the genome is followed by reverse transcription of the viral transcript. Although pararetroviruses can replicate without integration of their genome into the cellular DNA, integration may occur, resulting, in plants, in infectious endogenous pararetrovirus (EPRV) sequences transmitted vertically (Richert-Pöggeler et al 2003, Gayral et al 2008). As in retroviral proviruses, EPRVs are subject to DNA methylation and may be induced by various means, including genome hybridization, tissue culture, abiotic stress, and wound stress (Harper et al 2002, Richert-Pöggeler et al 2003). In tomato, EPRV sequences that have lost the ability to induce virus infection appear to be controlled by a plant-specific RNA-mediated silencing mechanism resulting in DNA methylation at CHG and asymmetric CHH groups (where H is a base other than G). Accordingly, EPRV sequences were detected in heterochromatic regions, dispersed virtually on all chromosomes (Staginnus et al 2007). Besides DNA methylation, pericentromeric EPRV genomes in petunia were associated with histone H3 dimethylated at lysine 9 residues, representing inactive chromatin (Noreen et al 2007). Association of EPRV sequences with dimethylated lysine 9 as well as lysine 4 residues of histone H3—the latter representing active chromatin, as found in the petunia cultivar 'W138'—led to easier induction of EPRV sequences and detection of signaling molecules in the form of small interfering (si)RNAs. Therefore, epigenetic control of EPRV can be influenced by abiotic stress as well as host factors, resulting in the release of extrachromosomally replicating virus (Staginnus & Richert-Pöggeler 2006, Noreen et al 2007).

Latency in animal systems

Herpesviruses of humans, vertebrates, and invertebrate marine bivalve species are 'large genome' double-stranded DNA viruses defined by the morphology of the virion (for a review see McGeoch et al 2006). They are associated with a broad range of diseases due to productive, lytic replication in host cells, immunologic consequences of virus infection, and induction of malignant tumors. The genomes of extant herpesvirus strains coevolved with the genomes of their host species. In many cases infected individuals carry the viral genomes lifelong in their cells due to the epigenetic modification of the viral DNA and chromatin by cellular enzymes. DNA methylation and repressive marks (i.e., methylation of lysine residues in certain positions on the tails of histone H3 and H4) induce a closed chromatin configuration restricting the usage of viral promoters and preventing productive herpesvirus replication (reviewed by Minarovits 2006). It is interesting to note that the epigenetic mechanisms involved in the silencing of latent, episomal human herpesvirus genomes may differ depending on the host cell. It turned out that DNA methylation (i.e., modification of cytosines located in CpG dinucleotides) does not play a role in the regulation of herpes simplex virus type 1 (HSV-1) latent gene expression in neurons. In contrast, the nuclear matrix-attached episomal genomes of *Epstein–Barr virus* (EBV), a virus associated with a series of malignant tumors, are subject to DNA methylation in Burkitt's lymphoma and nasopharyngeal carcinoma cells. Histone modifications and protein–DNA interactions also affect the usage of latent EBV promoters (reviewed by Niller et al 2007).

Retroviruses (family: *Retroviridae*) infect a wide variety of species and replicate their RNA genomes with the help of reverse transcriptase (RT), a unique enzyme capable of converting the single-stranded viral RNA genome into double-stranded DNA (dsDNA). The dsDNA copy of the viral genome integrates into the cellular DNA, forming a provirus. Retroviruses infecting the somatic cells of their hosts are called exogenous retroviruses. The genomes of all vertebrate species studied so far, however, carry multiple copies of endogenous retroviral (ERV) genomes as well (reviewed by Weiss 2006). As a matter of fact these genetic elements represent the results of ancestral germ line infections by exogenous retroviruses. ERV proviruses can be inherited vertically from parents to offspring according to Mendelian laws. In contrast, exogenous retroviruses that may induce neoplasms or immunosuppression in their hosts are transmitted horizontally (Denner 2010). The regulatory sequences of the proviral genome are called long terminal repeats (LTRs). They are located at the 5′ and 3′ ends of the proviral genome and contain the retroviral promoter and enhancer elements. Proviral genomes are transcribed by the cellular RNA polymerase II enzyme that produces full-length viral genomes and viral mRNAs. In the cytoplasm, retroviral RNA genomes and translated viral proteins assemble to form viral particles that leave the cells by budding through the cell membrane (van Regenmortel et al 2000).

Integration of retroviral genomes into the host cell DNA permits spreading of the adjacent DNA methylation patterns and repressive or activating chromatin marks to the proviral domain. Thus, the cellular epigenetic regulatory machinery may affect the activity of retroviral promoters. Retroviral genomes are frequently silenced in various types of host cell. Recently a wide variety of silencing mechanisms were described that contribute to switching off the promoter of the *Human immunodeficiency virus* (HIV), the causative agent of acquired immunodeficiency syndrome (AIDS). Silencing of the HIV 5′ LTR involves CpG methylation that may inhibit the activity of the HIV promoter either directly, by blocking transcription factor binding, or indirectly, via attracting MBD2, a methyl-CpG binding protein. MBD2 may facilitate the buildup of a repressive chromatin structure through recruitment of histone deacetylases (Bednarik et al 1987, Blazkova et al 2009). Binding of transcriptional repressor proteins, histone methyltransferases, polycomb repressor complexes, and buildup of a nucleosomal structure blocking transcriptional initiation may also silence the HIV promoter (reviewed by Iglesias-Ussel and Romerio 2011). In addition, transcriptional interference by read-through transcripts originating at upstream host genes may inactivate the HIV LTR and render HIV proviruses dormant (Duverger et al 2009).

Epigenetic silencing of HIV LTR could be reverted by both DNA methyltransferase inhibitors and histone deacetylase inhibitors. In addition, inducers of the signal transducer NF-κB, such as tumor necrosis factor alpha (TNF-α) and Tat, the HIV-encoded transactivator, switched on the silent HIV promoter as well. Because latent HIV reservoirs prevent curative treatment of AIDS patients, recently a novel therapeutic approach, '*shock and kill*' therapy, was introduced using epigenetic drugs to reactivate latent HIV proviruses ('shock'), combined with an intensified highly active antiretroviral therapy (HAART, 'kill') to block the infection of new cells. Infected cells were expected to be destroyed by the reactivated virus or the immune system (Richman et al 2009, Deeks 2012, McNamara et al 2012).

CONCLUSION

In a world of globalization, virus spread becomes more efficient both with respect to space and time compared to former periods. Humans are promoting virus distribution either directly, by transport and exchange of virus- and/or viroid-infected material, or indirectly, by being contaminated with viruliferous vectors. This applies to both animal and plant systems.

The dogma that plants are immobile, in contrast to animals, may have to be revised. This is especially true for ornamentals that are traded on global markets. Furthermore, breeding, growth, and sale of ornamentals can occur in different places, exposing plants to changing climatic conditions and accompanying stresses. Further factors driving this change are growing markets, changing markets, economically challenging conditions, and a constant ambition to reduce energy costs during production. The latter is exemplified by reduced growing temperatures, which may even have adverse effects on plant virus surveillance systems (Szittya et al 2003).

Improved sequencing technologies revealed the omnipresence of viruses and their diversity (Kreuze et al 2009). Basic information regarding virus spread and invasion of hosts now has to be obtained. This would facilitate the analysis of driving forces determining viral coevolution. To meet this aim we need to explore the complex interactions that affect latency and the pathogenic or beneficial nature of virus–plant interactions. Knowing the trigger, it would be possible to evaluate the disease-causing potential of latent viruses. This is even more important for long-term relationships characteristic of latent virus infections in humans and perennial plants. It still is unresolved whether latent viruses are tolerated by the host because they can escape the surveillance system or whether the virus–host symbiosis is beneficial for the host during such a persistent relationship. For example, the viral coat protein of white clover cryptic virus seems to suppress nodulation in conditions of high nitrogen supply (Roossinck 2011).

There is an economic impact and relevance of studies on virus latency in plants. The European market for ornamentals, including non-EU countries, comprises 56% of the global market with a revenue for 2011 of more than 30 billion Euros. It is predicted that the market for ornamentals in the EU will further expand and grow to 37 billion Euros in the next three years (Swedish Chamber of Commerce 2011). For ornamentals, the place of production is often distinct from the markets and may even include different continents. International trade routes, and the global markets for ornamentals, may explain why the spectrum of virus and virus isolates, as well as viroids, may vary from year to year. The growing number of commercially available ornamental plant species and cultivars adds to the complexity when studying virus latency. The use of signaling molecules, such as small RNAs involved in the 'arms race' between host and virus, helped to design sensitive diagnostic tools, such as deep sequencing of siRNAs, to identify all partners involved in diseased as well as symptomless plants (Kreuze et al 2009). Last, but not least, latent virus-based vectors offer several advantages for studying gene functions in plants using virus-induced gene silencing (VIGS). *Apple latent spherical virus* (ALSV)-based vectors for VIGS did not induce obvious symptoms in most plants tested, and the induced silencing persisted throughout growth of the infected plant (Igarashi et al 2009). It is even more remarkable that vertical transmission of ALSV suggests viral invasion of meristematic tissue and may account for the ability of an ALSV-based vector to also silence a meristem gene (Igarashi et al 2009). Thus, latent viruses may also serve as useful tools for plant virus control using RNA-based vaccines that are currently studied in the European program 'Food and Agriculture COST Action FA0806.'

ACKNOWLEDGMENTS

We thank K Kobayashi, B Poeggeler, W Menzel, and T Kühne for stimulating discussions and valuable comments. We thank M Wassenegger for viroid analysis in *S. jasminoides*. We thank C Maaß and S Schuhmann for expert technical assistance, and C Maaß for photography of *S. jasminoides*.

REFERENCES

AMI (Agrarmarkt Informations-Gesellschaft mbH) Marktreport 2012. Anbauerhebung Zierpflanzen, Frühjahr 2012 – Produktions – und Wirtschaftstendenzen im Zierpflanzenbau, Bericht Nr. 1/2012.

aus dem Siepen, M., Pohl, J.O., Koo, B.-J., Wege, C., Jeske, H., 2005. *Poinsettia latent virus* is not a cryptic virus, but a natural polerovirus–sobemovirus hybrid. Virology 336, 240–250.

Bednarik, D.P., Mosca, J.D., Raj, N.B., 1987. Methylation as a modulator of expression of human immunodeficiency virus. Journal of Virology 61, 1253–1257.

Blazkova, J., Trejbalova, K., Gondois-Rey, F., Halfon, P., Philibert, P., Guiguen, A., Verdin, E., Olive, D., van Lint, C., Hejnar, J., Hirsch, I., 2009. CpG methylation controls reactivation of HIV from latency. PLoS Pathogens 5, e1000554.

Brüssow, H., Canchaya, C., Hardt, W.D., 2004. Phages and the evolution of bacterial pathogens: from genomic rearrangements to lysogenic conversion. Microbiology and Molecular Biology Reviews 68, 560–602.

Clarke, J.L., Spetz, C., Haugslien, S., Xing, S., Dees, M.W., Moe, R., Blystad, D.-R., 2008. *Agrobacterium tumefaciens*-mediated transformation of poinsettia, *Euphorbia pulcherrima*, with virus derived hairpin RNA constructs confers resistance to *Poinsettia mosaic virus*. Plant Cell Reports 27, 1027–1038.

Deeks, S.G., 2012. HIV: shock and kill. Nature 487, 439–440.

Denner, J., 2010. Endogenous retroviruses. In: Kurth, R., Bannert, N. (Eds.), Retroviruses, Caister Academic Press, Norfolk, UK, pp. 35–69.

Duverger, A., Jones, J., May, J., Bibollet-Ruche, F., Wagner, F.A., Cron, R.Q., Kutch, O., 2009. Determinants of the establishment of human immunodeficiency virus type 1 latency. Journal of Virology 83, 3078–3093.

Fischer, M.G., Suttle, C.A., 2011. A virophage at the origin of large DNA transposons. Science 332, 231–234.

Flatken, S., Ungewickell, V., Menzel, W., Maiss, E., 2008. Construction of an infectious full-length cDNA clone of *Potato virus M*. Archives of Virology 153, 1385–1389.

Gayral, P., Noa-Carrazana, J.-C., Lescot, M., Lheureux, F., Lockhart, B.E.L., Matsumoto, T., Piffanelli, P., Iskra-Caruana, M.-L., 2008. A single *Banana streak virus* integration event in the banana genome as the origin of infectious endogenous pararetrovirus. Journal of Virology 82, 6697–6710.

Harper, G., Hull, R., Lockhart, B., Olszewski, N., 2002. Viral sequences integrated into plant genomes. Annual Review of Phytopathology 40, 119–136.

Igarashi, A., Yamagata, K., Sugai, T., Takahashi, Y., Sugawara, E., Tamura, A., Yaegashi, H., Yamagishi, N., Takahashi, T., Isogai, M., Takahashi, H., Yoshikawa, N., 2009. Apple latent spherical virus vectors for reliable and effective virus-induced gene silencing among a broad range of plants including tobacco, tomato, *Arabidopsis thaliana*, cucurbits, and legumes. Virology 386, 407–416.

Iglesias-Ussel, M.D., Romerio, F., 2011. HIV reservoirs: the new frontier. AIDS Research 13, 13–29.

Iskra-Caruana, M.L., Baurens, F.C., Gayral, P., Chabannes, M., 2010. A four-partner plant-virus interaction: enemies can also come from within. Molecular Plant–Microbe Interactions 23, 1394–1402.

King, A.M.Q., Adams, M.J., Carstens, E.B., Lefkowitz, E.J., 2012. Virus taxonomy. Ninth report of the international committee on taxonomy of viruses. Academic Press, San Diego.

Koenig, R., Lesemann, D.-E., Fulton, R.W., 1986. *Poinsettia mosaic virus*. AAB descriptions of plant viruses no. 311, AAB.

Kreuze, J.F., Perez, A., Untiveros, M., Quispe, D., Fuentes, S., Barker, I., Simon, R., 2009. Complete viral genome sequence and discovery of novel viruses by deep sequencing of small RNAs: a generic method for diagnosis, discovery and sequencing of viruses. Virology 388, 1–7.

McGeoch, D.J., Rixon, F.J., Davison, A.J., 2006. Topics in herpesvirus genomics and evolution. Virus Research 117, 90–104.

McNamara, L.A., Ganesh, J.A., Collins, K.L., 2012. Latent HIV-1 infection occurs in multiple subsets of hematopoietic progenitor cells and is reversed by NF-κB activation. Journal of Virology 86, 9337–9350.

Minarovits, J., 2006. Epigenotypes of latent herpesvirus genomes. Current Topics in Microbiology and Immunology 310, 61–80.

Niller, H.H., Wolf, H., Minarovits, J., 2007. Epstein–Barr virus. In: Gonczol, E., Valyi-Nagy, T., Minarovits, J. (Eds.), Latency strategies of herpesviruses, Springer Science+Business Media, New York, pp. 154–191.

Noreen, F., Akbergenov, R., Hohn, T., Richert-Pöggeler, K.R., 2007. Distinct expression of endogenous petunia vein clearing virus and the DNA transposon dTph1 in two *Petunia hybrida* lines is correlated with differences in histone modification and siRNA production. Plant Journal 50, 219–229.

Richert-Pöggeler, K.R., Noreen, F., Schwarzacher, T., Harper, G., Hohn, T., 2003. Induction of infectious petunia vein clearing (pararetro) virus from endogenous provirus in petunia. EMBO Journal 22, 4836–4845.

Richman, D.D., Margolis, D.M., Delaney, M., Greene, W.C., Hazuda, D., Pomerantz, R.J., 2009. Challenge of finding a cure for HIV infection. Science 323, 1304–1307.

Roossinck, M.J., 2011. The good viruses: viral mutualistic symbiosis. Nature Reviews in Microbiology 9, 99–108.

Roossinck, M.J., Sabanadzovic, S., Okada, R., Valverde, R.A., 2011. The remarkable evolutionary history of endornaviruses. Journal of General Virology 92, 2674–2678.

Rumpho, M.E., Worful, J.M., Leeb, J., Kannan, K., Tylerc, M.S., Bhattacharya, D., Moustafa, A., Manhart, J.R., 2008. Horizontal gene transfer of the algal nuclear gene psbO to the photosynthetic sea slug *Elysia chlorotica*. PNAS 105, 17867–17871.

Ryan, F., 2009. Virolution. FPR-Books Ltd, UK.

Schwind, N., Zwiebel, M., Itaya, A., Ding, B., Wang, M.-B., Krczal, G., Wasseneger, M., 2009. RNAi-mediated resistance to *Potato spindle tuber viroid* in transgenic tomato expressing a viroid hairpin RNA construct. Molecular Plant Pathology 10, 459–469.

Senshu, H., Yamaji, Y., Minato, N., Shiraishi, T., Maejima, K., Hashimoto, M., Miura, C., Neriya, Y., Namba, S., 2011. A dual strategy for the suppression of host antiviral silencing: two distinct suppressors for viral replication and viral movement encoded by *Potato Virus* M. Journal of Virology 85, 10269–10278.

Staginnus, C., Richert-Pöggeler, K.R., 2006. Endogenous pararetroviruses: two-faced travelers in the plant genome. Trends in Plant Science 11, 485–491.

Staginnus, C., Gregor, W., Mette, M.F., Teo, C.H., Borroto-Fernández, E.G., Machado, M.L., Matzke, M., Schwarzacher, T., 2007. Endogenous pararetroviral sequences in tomato (*Solanum lycopersicum*) and related species. BMC Plant Biology 21 (7), 24.

Suttle, C., 2007. Marine viruses - major players in the global ecosystem. Nature Reviews in Microbiology 5, 801–812.

Swedish Chamber of Commerce, 2011. Market report. Focus on the EU and Swedish Market: Floricultural Products, May 2011.

Szittya, G., Silhavy, D., Molnar, A., Havelda, Z., Lovas, A., Lakatos, L., Banfalvi, Z., Burgyan, J., 2003. Low temperature inhibits RNA silencing-mediated defence by the control of siRNA generation. EMBO Journal 22, 633–640.

van Regenmortel, M.H.V., Fauquet, C.M., Bishop, D.H.L., Carstens, E.B., Estes, M.K., Lemon, S.M., Maniloff, J., Mayo, M.A., McGeoch, D.J., Pringle, C.R., Wicker, R.B. (Eds.), 2000. Retroviridae. Virus taxonomy, Academic Press, New York, pp. 369–387.

Vera, A., Daros, J.-A., Flores, R., Hernandez, C., 2000. The DNA of a plant retroviroid-like element is fused to different sites in the genome of a plant pararetrovirus and shows multiple forms with sequence deletions. Journal of Virology 74, 10390–10400.

Verhoeven, J.Th.J., Jansen, C.C.C., Roenhorst, J.W., 2006. First report of *Potato virus M* and *Chrysanthemum stunt viroid* in *Solanum jasminoides*. Plant Disease 90, 1359.

Verhoeven, J.Th.J., Botermans, M., Meekes, E.T.M., Roenhorst, J.W., 2012. *Tomato apical stunt viroid* in the Netherlands: most prevalent pospiviroid in ornamentals and first outbreak in tomato. European Journal of Plant Pathology 133, 803–810.

Wang, X., Kim, Y., Ma, Q., Hong, S.H., Pokusaeva, K., Sturino, J.M., Wood, T.K., 2010. Cryptic prophages help bacteria cope with adverse environments. Nature Communications 1, 147.

Weiss, R.A., 2006. The discovery of endogenous retroviruses. Retrovirology 3, 67.

Wren, J.D., Roossinck, M.J., Nelson, R.S., Scheets, K., Palmer, M.W., Melcher, U., 2006. Plant virus biodiversity and ecology. PLoS Biology 4, e80.

Xu, H., D'Aubin, J., Nie, J., 2010. Genomic variability in *Potato virus M* and the development of RT-PCR and RFLP procedures for the detection of this virus in seed potatoes. Virology Journal 7, 25.

Chapter 15

Viroid–insect–plant interactions

Noémi Van Bogaert
Department of Crop Protection, Faculty Bioscience Engineering, Ghent University, Ghent, Belgium, and Laboratory of Virology, Plant Sciences Unit—Crop Protection, Institute for Agricultural and Fisheries Research (ILVO), Merelbeke, Belgium

Guy Smagghe
Department of Crop Protection, Faculty Bioscience Engineering, Ghent University, Ghent, Belgium

Kris De Jonghe
Laboratory of Virology, Plant Sciences Unit—Crop Protection, Institute for Agricultural and Fisheries Research (ILVO), Merelbeke, Belgium

INTRODUCTION

What are viroids?

Viroids are nonprotein-encoding and highly structured, single-stranded RNA molecules, currently considered to be the smallest plant pathogens (Diener 1971, 2003, Flores et al 2005, Navarro et al 2012). The absence of a protein coat distinguishes viroids from viruses. Based on biochemical and structural characteristics, viroids are taxonomically divided into two families: *Avsunviroidae* and *Pospiviroidae* (Hadidi et al 2003). Rod like structures are typical for the pospiviroids, whereas more branched structures are typical of the avsunviroids (Codoner et al 2006). Another key difference between the two families is the location of replication. Avsunviroids replicate in the chloroplast, while pospiviroids reproduce within the nucleus (Flores et al 2005). The viroid first discovered—the type-species of the pospiviroids—is *Potato spindle tuber viroid* (PSTVd; Diener 1971, Fig. 15.1).

Viroids are notorious for causing plant diseases of considerable economic importance (Flores et al 2005). The induced symptoms depend largely on the host plant and the viroid in question but are usually characterized by diminished growth, stunting, leaf epinasty, necrosis, and flower and fruit deformations (Flores et al 2005, Owens & Hammond 2009). In potato crops, yield losses caused by PSTVd vary according to isolate, cultivar, and climatic conditions, with losses exceeding 60% (Pfannenstiel & Slack 1980). This is one of the

FIGURE 15.1 Schematic presentation of the secondary structure of PSTVd. Reproduced, with permission, from Góra-Sochacka 2004.

reasons why PSTVd has been listed as a quarantine organism in the European Union (EU) and many other countries, including Australia, Canada, Mexico, New Zealand, and the United States (de Hoop et al 2008).

DISTRIBUTION

During the last decade most of the European viroid research has focused on the pospiviroids. Their swift transmission, combined with the substantial risk they impose on economically important agricultural crops, has led to their listing as a quarantine organism in the EU. Currently, two pospiviroids, namely *Potato spindle tuber viroid* (PSTVd) and *Chrysanthemum stunt viroid* (CSVd) (Diener & Lawson 1972) are listed as A2 quarantine pest species by the European and Mediterranean Plant Protection Organisation (EPPO 2011). *Tomato apical stunt viroid* (TASVd) is the only pospiviroid to be included on the EPPO alert list, indicating that this viroid can possibly present a risk to EPPO member countries (EPPO 2011).

The pospiviroids are known to infect important economic crops, such as tomato (*Solanum lycopersicum*) and potato (*Solanum tuberosum*), inducing symptoms that vary with viroid strain, plant variety, and climatic conditions; however, the symptoms are generally characterized by reduced growth and chlorosis of the leaves (Flores et al 2005). During the past decade, however, various European surveys have revealed that many pospiviroids have also been *latently* present in ornamental plants belonging to the *Solanaceae* family (Verhoeven et al 2008a, 2010a, Luigi et al 2011). Subsequently, they can be potentially transmitted to major commercial crops belonging to the same family, such as tomato and potato (Verhoeven et al 2010c). These asymptomatic viroid-infected plants can act as reservoirs from which viroids may spread to cultivated species and induce diseases (Singh 2006, Matousek et al 2007, Verhoeven et al 2010c).

To illustrate this point, two large outbreaks of PSTVd and *Columnea latent viroid* (CLVd) were reported in tomato in Belgium in 2006 (Verhoeven et al 2007a). Although seed transmission was suspected, transmission from asymptomatic ornamentals and weeds through pruning tools, mechanical contact, or vectors could not be excluded. Since then, national surveys have shown that PSTVd, CLVd, CSVd, and TASVd have been repeatedly found in asymptomatic ornamental plants. *Citrus exocortis viroid* (CEVd) and *Tomato chlorotic dwarf viroid* (TCDVd) were also regularly reported.

Similar surveys in other European countries confirm the latent presence of viroids in different ornamental species. National surveys in Italy have found PSTVd in *Solanum jasminoides* and *Solanum rantonetti* (Di Serio 2007).

In 2005, *Petunia* hybrid plants from the United States were inspected after entering the post-entry quarantine station of the Plant Protection Service (PPS) in the Netherlands (Verhoeven et al 2007a). PPS found TCDVd to be present; this was the first report of this viroid in *P. hybrida* (Verhoeven et al 2007b). In 2008, CEVd was detected in *Verbena* sp., PSTVd in *Brugmansia* sp., and TASVd in *Cestrum* sp. (Verhoeven et al 2008b). All of these viroids were also found in *S. jasminoides* (Verhoeven et al 2008c).

Recently, Verhoeven et al (2012) demonstrated that the families *Gesneriaceae* (*Nematanthus* sp.), *Verbenaceae* (*Verbena* sp.), and *Apocynacea* (*Vinca* sp.) also include pospiviroid hosts. Additionally, this study indicated that ornamental species may act as sources of inoculum for pospiviroid outbreaks in tomato (Verhoeven et al 2012).

In the UK, the first report of a PSTVd outbreak in commercial tomatoes dates back to 2003 (Mumford et al 2003). Before this, PSTVd had been found only under controlled conditions in a UK potato germplasm collection (Mumford et al 2003). Since that date, many other pospiviroids have been detected in tomato and solanaceous ornamentals (FERA 2010). At least for PSTVd, worldwide distribution can now be assumed. In Peru, PSTVd has been detected in avocado (*Persea americana*), where infections often remain latent unless the tree is coinfected with ASBVd (Querci et al 1995). In New Zealand, PSTVd was reported to be associated with a new disease of glasshouse tomato and *Capsicum* crops (Lebas et al 2005).

In conclusion, new reports of pospiviroid members are emerging from all corners of the world. However, a global understanding of the prevalence and transmission of pospiviroid species remains far beyond the horizon.

TRANSMISSION

General mechanisms of virus and viroid transmission

The worldwide transmission of viroids is clearly related to human activity, mainly in the form of international trade. Vegetative propagation of plants and trafficking of commercial crops have been the main contributors to the global spread of these minute plant pathogens. The European Food Safety Authority (EFSA 2011) considers vegetative propagation of infected plant material to be the main source of viroid infection. In addition, both mechanical contact between infected and healthy plants and contamination with infected greenhouse materials play an important role (Verhoeven et al 2010b).

It has been hypothesized that the start of viroid epidemics in greenhouses is most commonly initiated by the presence of infected plants, with secondary spread being facilitated by insect activities (Singh & Singh 1998). Traditionally, this transmission through insect vectors is divided into three phases: acquisition, retention, and inoculation (Pirone & Blanc 1996). For plant viruses, it was estimated that more than 80% use arthropod vectors to move from one host to another (van den Heuvel et al 1999, Fereres & Moreno 2009). Most of these arthropods are of the insect order Hemiptera (Ng & Falk 2006). The best examples are aphids (*Aphidoidea*), whiteflies (*Aleyrodidae*), leafhoppers (*Cicadellidae*), and also thrips (order *Thysanoptera*) (van den Heuvel et al 1999). Aphids and whiteflies in particular seem to be very well adapted for virus transmission because their stylets recurrently pierce between plant cells to reach the phloem and/or to penetrate the actual cells without causing severe damage (Fereres & Moreno 2009). The latter organisms, and many other insect species, have been postulated as potential vectors of (pospi)viroids, but the importance of this form of vector transport has not yet been fully established.

Gray and Banerjee (1999) reviewed the most important molecular and cellular mechanisms by which viruses are transmitted between plants. For plant viruses, a distinction is made between *nonpersistent* viruses (i.e., those not retained by the insect vector for more than a few hours) and *persistent* viruses (i.e., those in a life-long association with the vector) (Gray & Banerjee 1999). Nonpersistent viruses are often called *stylet-borne* viruses because they are carried on the mouthparts of vectors and are lost once a vector has fed on a host (Power 2000). Similarly, those viruses retained in the foregut have been called *foregut-borne* viruses (Nault & Ammar 1989). *Cuticula-borne* viruses, comprising both stylet- and foregut-borne viruses, are those viruses that are carried on the cuticular lining of the vector feeding apparatus (Harris et al 1996).

Additionally, plant viruses have traditionally been classified into *circulative* and *non-circulative* viruses on the basis of whether they are being actively internalized into the vector's hemocoel or not (Gray & Banerjee 1999). Circulative viruses can be further divided into *propagative* viruses, which replicate in their arthropod vector as well as in their plant host, and *non-propagative* viruses, which replicate only in their plant hosts (Gray & Banerjee 1999). However, the majority of plant viruses are transmitted in a non-circulative, non-propagative way, being associated only with the cuticular linings of the mouthparts and the anterior part of the alimentary tract (van den Heuvel et al 1999).

It can be assumed that viroids, because they are not enveloped and do not encode any (movement) proteins, would have completely different ways of transmission as compared to plant viruses. To date, it is commonly believed that mechanical transmission through physical contact with insect parts and/ or products (e.g., pollen) is the most important transmission route. However, the importance of transencapsidation and transcomplementation also need to be investigated. Transmission through transencapsidation of a viroid into a virus capsid has received some, but not much, attention in the past in scientific studies (Querci et al 1997, Syller & Marczewski 2001). Transencapsidation can be defined as encapsidation of the nucleic acids of a virus or viroid into the virion of another virus. Earlier, transencapsidation was observed frequently for different luteo- and potyviruses (Falk et al 1995).

Transcomplementation is the phenomenon in which a viral protein—commonly a movement protein, an inhibitor of gene silencing, or a coat protein—enhances and/or supports the infection of a virus from a distinct species (Froissart et al 2002, Latham & Wilson 2008). Transcomplementation can be best exemplified by members belonging to the genus *Umbravirus*, which, unlike many other plant viruses, cannot be transmitted by aphids (Syller 2003). However, when the plant is coinfected with a suitable virus from the *Luteoviridae*, which acts as a helper, it does become aphid-transmissible (Taliansky et al 1996, Robinson et al 1999). During a mixed infection, the umbraviral RNA can be encapsidated by the capsid protein of the helper virus. The virion, accordingly assembled, is readily acquired by a luteovirid vector, which feeds on the infected plant and transmits the virus in a circulative fashion to the next plant (Syller 2003). The acquisition and transport of the virion through epithelial cell linings in the gut and salivary glands of the vector are supposed to happen by receptor-mediated endo- and exocytosis (van den Heuvel et al 1999). Other examples of trans-complementation include the *Potato aucuba virus* (PAMV) and the *Rice tungro bacilliform virus* (RTBV), for which vector transmission depends on a helper component (HC) produced by a coinfecting potyvirus and waïkavirus, respectively (Froissart et al 2002).

Figure 15.2 illustrates how transencapsidation and transcomplementation could be envisioned for viroids during non-circulative insect transmission.

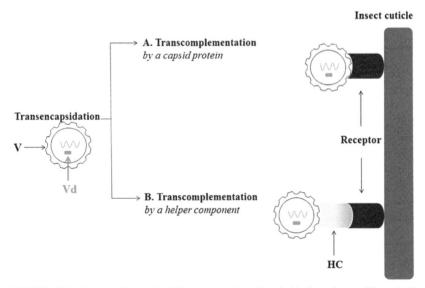

FIGURE 15.2   Scheme of non-circulative transmission of a viroid via an insect. Vd = viroid, V = icosahedral virus particle. Transencapsidation of a viroid into a virus capsid. Contact with the insect cuticle can be established through (A) transcomplementation by a capsid protein of the virus or (B) transcomplementation by a helper component (HC) of the virus. Modified from Froissart et al (2002).

Knowledge of *how* contact is made between the virion particle and the cuticular lining of the mouthparts or the anterior gut of an insect (i.e., the specific retention sites and receptors involved) is sparse (Fereres & Moreno 2009). In general terms, it is possible that the capsid protein of the virus in question is capable of binding a receptor on the cuticle (Froissart et al 2002). Alternatively, the virus might encode a helper component (HC) by forming a reversible 'bridge' between the virion and the receptor. This is referred to as the *bridge hypothesis* (Pirone & Blanc 1996).

Insect transmission of viroids

First attempts to transmit CSVd (Hollings & Stone 1973) and *Cucumber pale fruits viroid* (CPFVd) (van Dorst & Peters 1974) via aphids were unsuccessful. Additionally, these early experiments suggested that nematodes are not likely to act as viroid vectors (van Dorst & Peters 1974). However, the results of these early reports are strongly doubted (EFSA 2011). Conflicting reports may have been caused by the use of inaccurate assays, contrasting experimental designs, the use of visual readings, working in the field instead of in greenhouses, inaccurate detection of the viroid, contamination, etc. (Schumann et al 1980).

In 1980, Schumann et al successfully established the actual presence of PSTVd in potato plants by means of a gel electrophoresis assay. This study evaluated the transmission of PSTVd by six common insect pests of potato, all yielding negative results (Table 15.1). De Bokx and Piron (1981) investigated PSTVd transmission between tomato plants by three aphid species: foxglove aphid (*Aulacorthum solani* Kaltenbach), potato aphid (*Macrosiphum euphorbiae* Thomas), and green peach aphid (*Myzus persicae* Sulzer). As an inoculum source, infected tomato plants (cv. Sheyenne) and artificial diet solutions containing purified PSTVd were used. However, when aphids were allowed to feed for 20 seconds on the parafilm membrane enclosing the artificial diet, it seemed that they did not feed as successfully as they did on detached tomato leaves. The results showed that only *M. euphorbiae* transmitted PSTVd in a nonpersistent way (De Bokx & Piron 1981; Table 15.1).

A study by Galindo et al (1986) contradicted these results by showing highly efficient aphid transmission of *Tomato planta macho viroid* (TPMVd) by *M. persicae*. On the other hand, the cow pea aphid (*Aphis craccivora* Koch) transmitted *Tomato apical stunt viroid* (TASVd) with a low efficiency (Walter 1987).

Several years later, various studies demonstrated the transmission of PSTVd between potato plants by the aphid *M. persicae* after doubly infecting the plants with the viroid and *Potato leaf roll virus* (PLRV) (Querci et al 1997, Syller & Marczewski 2001). The latter virus belongs to the genus *Luteovirus* and is known to be transmitted by aphids (Goss 1930). Francki et al (1986) had already shown that PSTVd RNA can be transencapsidated by coat proteins of the *Velvet tobacco mottle virus* (VToMV). However, transencapsidation did not take place for *Potato virus Y* (PVY) (Singh et al 1992).

TABLE 15.1 Overview of insect transmission studies with PSTVd, TASVd, TCDVd, and TPMVd, showing + (Positive) and – (Negative) results (T = Transencapsidation)

Insect family	Species	Common name	PSTVd	TASVd	TCDVd	TPMVd
Aphididae	Aphis craccivora	Cowpea aphid		(4) –		
	Aulacortum solani	Foxglove aphid	(2) –			
	Macrosiphum euphorbiae	Potato aphid	(2) +			
	Myzus persicae	Green peach aphid	(1) –	(10) –		(3) +
			(2) –			
			(5) T			
			(6) T			
			(7) T			
Aleyrodidae	Bemisia tabaci	Tobacco whitefly		(10) –		
Thripidae	Frankliniella occidentalis	Western flower thrip	(10) –			
	Thrips tabaci	Onion thrip	(10) –			
Apidae	Apis mellifera	Honeybee	(10) –			
	Bombus terrestris	Bumblebee	(10) –	(8) +	(9) +	

Continued

TABLE 15.1 Overview of Insect Transmission Studies with PSTVd, TASVd, TCDVd, and TPMVd, showing + (Positive) and − (Negative) results (T = Transencapsidation) — cont'd

Insect family	Species	Common name	PSTVd	TASVd	TCDVd	TPMVd
Other	*Empoasca fabae*	Potato leafhopper	(1) −			
	Leptinotarsa decemlineata	Colorado potato beetle	(1) −			
	Lygus Lineolaris	Tarnished plant bug	(1) −			
	Melanoplus femur–rubrum	Redlegged grasshopper	(1) −			
	Prodenia eridania	Southern armyworm	(1) −			

References: (1) Schumann et al 1980, (2) De Bokx & Piron, 1981, (3) Galindo et al 1986, (4) Walter 1987, (5) Salazar et al 1995 (6), Querci et al 1997, (7) Syller & Marczewski, 2001, (8) Antignus et al 2007, (9) Matsuura et al 2010, (10) Nielsen et al 2012. Studies before 1980 are not listed.

In the experiments by Salazar et al (1995), where plants were doubly infected with PLRV and PSTVd, 100% transmission of PSTVd was achieved. No transmission was observed when source plants were infected with only the viroid (Salazar et al 1995). Following this research, Querci et al (1997) allowed apterous aphids to feed on either singly infected (PSTVd) or doubly infected (PSTVd + PLRV) source plants. Then, after a transmission access period (TAP) of 3 days, aphids were transferred to young uninfected potato plants. Inoculated plants were tested for PSTVd and/or PLRV after 15 and 45 days (after the TAP) by using a combination of nucleic acid spot hybridization (NASH) and enzyme-linked immunosorbent assay (ELISA) (Querci et al 1997). PSTVd was detected only in doubly infected plants, leading the authors to assume that transencapsidation of the viroid into the virus took place. To prove this hypothesis, different types of sample were treated, before RNA extraction, with *micrococcal nuclease*, an enzyme that degrades non-encapsidated PSTVd RNA with high efficiency. Samples treated with micrococcal nuclease, and exhibiting the presence of PSTVd after PCR (polymerase chain reaction), showed that PSTVd needs to be associated *within* the virus particle (Querci et al 1997).

Syller and Marczewski (2001) provide another confirmation of transencapsidation of PSTVd into PLRV. In this research, the response of potato plants to mixed infections with PSTVd and PLRV was strikingly more severe than infections with either pathogen alone.

Transencapsidation and subsequent transmission through insects (such as aphids) can potentially have important epidemiologic implications. A latently present viroid of a given crop can be incorporated into the capsid of a plant virus (e.g., into the icosahedral capsid of a *Luteovirus* sp.) and subsequently be transmitted by an insect vector (e.g., an aphid: Fig. 15.3). This pathway of trans-encapsidation, followed by vector-mediated transport, can result in the infection of another host plant (Fig. 15.3B) and can reveal the viroid symptoms that were not expressed in the former host (Francki et al, 1986). In Figure 15.3A, the viroid–virus association is taken up by an insect. However, the exact mechanism of this uptake and subsequent survival and transmission of the pathogens is not yet clear.

The research discussed above focused mainly on aphids as vectors of PSTVd. Over time, other insect and viroid species have also gained scientific attention. In 2007, Antignus et al (2007) investigated transmission of TASVd by silver-leaf whiteflies (*Bemisia tabaci*), green peach aphids (*M. persicae*), and bumblebees (*Bombus terrestris*). Whiteflies and aphids were introduced to TASVd-infected *Nicotiana rustica*, *Physalis floridensis*, and tomato source plants for 48 hours. Subsequently, they were transferred to individually caged healthy tomato plants for a 48-hour inoculation period and were tested for successful infection using Northern blot hybridization (Antignus et al 2007). Bumblebees were introduced into a 50-mesh screenhouse, where some of the tomato plants had been mechanically inoculated with TASVd. They concluded that no transmission of TASVd through (virus-free) *B. tabaci* or *M. persicae* took place

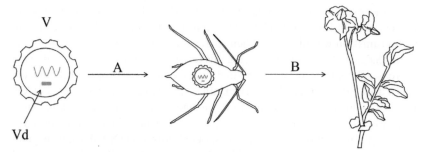

FIGURE 15.3   (A) Uptake of the viroid–virus system by an insect. (B) Insect-mediated transmission of the viroid–virus system between plants. Vd = viroid, V = icosahedral virus particle.

(Antignus et al 2007). Surprisingly, *B. terrestris* did seem capable of transporting TASVd from infected to viroid-free tomato plants (Antignus et al 2007). Therefore, the authors also suggested that transmission by bumblebees may be due to the wounding of the flowers during insect visits or by the introduction of infected pollen to the stigma of the flower.

The role of bumblebees as vectors in the transmission of viroids was also confirmed by Matsuura et al (2010) through experiments in greenhouses. Bumblebees (*Bombus ignitus*) were introduced into the greenhouses after mechanical inoculation of tomato plants with TCDVd (Matsuura et al 2010). After more than 1 month, TCDVd was detected by quantitative PCR (qPCR) in the non-infected plants (Matsuura et al 2010). The authors suggested that TCDVd is mechanically transmitted with crude sap via the insect mandibles. However, one should also consider horizontal transmission through viroid-contaminated pollen carried by bumblebees.

The results of Nielsen et al (2012) contradict the results of Antignus et al (2007) and Matsuura et al (2010). The study by Nielsen et al explored the transmission of PSTVd by thrips (*Frankliniella occidentalis* and *Thrips tabaci*), honeybees (*Apis mellifera*), and bumblebees (*B. terrestris*). Both intra- and interspecies transmission between ornamentals and crops of the *Solanaceae* were investigated, but no insect-mediated transmission could be recorded (Nielsen et al 2012). These authors emphasized that transmission of PSTVd by transencapsidation in PLRV particles was not considered in their experimental design but should certainly be taken into account in the future (Nielsen et al 2012). Additionally, we believe that foraging activities of honeybees and bumblebees should have been more closely monitored in order to yield evidence (if any) of their vectoring role.

CONCLUSION

The role and importance of insects as vectors in the transmission of viroids are still unclear. The limited numbers of studies exploring vector abilities of different species of insect have led to contradictory results. So far, only two species of aphids (*M. euphorbiae* and *M. persicae*) and bumblebees (*B. terrestris*) have been proven to play a role in viroid transmission. Furthermore, the exact mechanism and potential consequences of transencapsidation remain unresolved.

Differences in experimental design, test organisms, viroid isolates, and inoculation period hinder accurate comparison of the results. To address this problem, we recommend well-designed transmission experiments that use a broader range of viroid isolates and insect species. The latest molecular and visualization techniques can be used to detect and locate viroids within a host.

Further research is clearly needed to fill the aforementioned gaps in viroid epidemiology and to address the risks associated with the pathogenic nature of viroids. New information on viroid epidemiology is a prerequisite for the development and successful implementation of control strategies. For now, we recommend precautionary management measures until definitive answers become available.

ACKNOWLEDGMENTS

The authors acknowledge support for their research by the Institute for Agricultural and Fisheries Research (ILVO-Vlaanderen), the Fund of Scientific Research (FWO-Vlaanderen, Belgium), and the Special Research Fund of Ghent University (BOF-UGent).

REFERENCES

Antignus, Y., Lachman, O., Pearlsman, M., 2007. Spread of *Tomato apical stunt viroid* (TASVd) in greenhouse tomato crops is associated with seed transmission and bumble bee activity. Plant Disease 91 (1), 47–50, available from: ISI:000243011800007.

Codoner, F.M., Daros, J.A., Sole, R.V., Elena, S.F., 2006. The fittest versus the flattest: experimental confirmation of the quasispecies effect with subviral pathogens. PloS Pathogens 2 (12), 1187–1193, available from: ISI:000243209700009.

De Bokx, J.A., Piron, P.G.M., 1981. Transmission of potato spindle tuber viroid by aphids. Netherlands Journal of Plant Pathology 87, 31–34.

de Hoop, M.B., Verhoeven, J., Th, J., Roenhorst, J.W., 2008. Phytosanitary measures in the European Union: a call for more dynamic risk management allowing more focus on real pest risks. EPPO Bulletin 38, 510–515.

Diener, T.O., 1971. *Potato spindle tuber virus* 4. Replicating, low molecular weight RNA. Virology 45 (2), 411, available from: ISI: A1971K065800008.

Diener, T.O., 2003. Discovering viroids – a personal perspective. Nature Reviews Microbiology 1 (1), 75–80, available from: ISI:000220254100018.

Diener, T.O., Lawson, R.H., 1972. Chrysanthemum stunt, a viroid disease. Phytopathology 62 (7), 754, available from: ISI: A1972N043800083.

Di Serio, F., 2007. Identification and characterization of *Potato spindle tuber viroid* infecting *Solanum jasminoides* and *S-rantonnetii* in Italy. Journal of Plant Pathology 89 (2), 297–300, available from: ISI:000247707800020.

EFSA, 2011. Scientific opinion on the assessment of the risk of solanaceous pospiviroids for the EU territory and the identification and evaluation of risk management options. EFSA Journal 9 (8), 2330.

EPPO, 2011. EPPO A1 and A2 lists of pests recommended for regulation as quarantine pests. PM 1/2(20). OEPP/EPPO (European and Mediterranean Plant Protection Organization).

Falk, B.W., Passmore, B.K., Watson, M.T., Chin, L.S., 1995. The specificity and significance of heterologues encapsidation of virus and virus-like RNAs. In: Bills, D.D., et al. (Eds.), Biotechnology and plant protection, viral pathogenesis and disease resistance. Proceedings of the Fifth International Symposium, World Scientific, Singapore, pp. 391–415.

FERA, 2010. Emerging viroid threats to UK tomato production. The Food and Environment Research Agency, UK (S. Matthews-Berry).

Fereres, A., Moreno, A., 2009. Behavioural aspects influencing plant virus transmission by homopteran insects. Virus Research 141 (2), 158–168, available from: ISI:000265735600006.

Flores, R., Hernandez, C., de Alba, A.E.M., Daros, J.A., Di Serio, F., 2005. Viroids and viroid–host interactions. Annual Review of Phytopathology 43, 117–139, available from: ISI:000232286700006.

Francki, R.I.B., Zaitlin, M., Palukaitis, P., 1986. In vivo encapsidation of potato spindle tuber viroid by velvet tobacco mottle virus-particles. Virology 155, 469–473.

Froissart, R., Michalakis, Y., Blanc, S., 2002. Helper component-transcomplementation in the vector transmission of plant viruses. Phytopathology 92 (6), 576–579, available from: ISI:000175791400001.

Galindo, J., Lopez, M., Aguilar, T., 1986. Significance of *Myzus persicae* in the spread of tomato planta macha viroid. Fitopatologia Brasiliera 2, 400–410.

Góra-Sochacka, A., 2004. Viroids: unusual small pathogenic RNAs. Acta Biochimica Polonica 51 (3), 587–607, available from: ISI:000223774900002.

Goss, R.W., 1930. Insect transmission of potato virus diseases. Phytopathology 20, 136.

Gray, S.M., Banerjee, N., 1999. Mechanisms of arthropod transmission of plant and animal viruses. Microbiology and Molecular Biology Reviews 63 (1), 128, available from: ISI:000078999400005.

Hadidi, A., Flores, R., Randles, J.W., Semancik, J.S., 2003. Viroids. CSIRO, Collingwood, Victoria, Australia, p. 370.

Harris, K.F., PesicVanEsbroeck, Z., Duffus, J.E., 1996. Morphology of the sweet potato whitefly, *Bemisia tabaci* (Homoptera, Aleyrodidae) relative to virus transmission. Zoomorphology 116 (3), 143–156, available from: ISI: A1996VK58500005.

Hollings, M., Stone, O.M., 1973. Some properties of chrysanthemum stunt, a virus with the characteristics of an uncoated ribonucleic acid. Annuals of Applied Biology 74, 333–348.

Latham, J.R., Wilson, A.K., 2008. Transcomplementation and synergism in plants: implications for viral transgenes? Molecular Plant Pathology 9 (1), 85–103, available from: ISI:000251813900009.

Lebas, B.S.M., Clover, G.R.G., Ochoa-Corona, F.M., Elliott, D.R., Tang, Z., Alexander, B.J.R., 2005. Distribution of *Potato spindle tuber viroid* in New Zealand glasshouse crops of capsicum and tomato. Australasian Plant Pathology 34 (2), 129–133, available from: ISI:000229914400001.

Luigi, M., Luison, D., Tomassoli, L., Faggioli, F., 2011. Natural spread and molecular analysis of pospiviroids infecting ornamentals in Italy. Journal of Plant Pathology 93 (2), 491–495, available from: ISI:000293602400030.

Matousek, J., Orctova, L., Ptacek, J., Patzak, J., Dedic, P., Steger, G., Riesner, D., 2007. Experimental transmission of pospiviroid populations to weed species characteristic of potato and hop fields. Journal of Virology 81 (21), 11891–11899, available from: ISI:000250417400036.

Matsuura, S., Matsushita, Y., Kozuka, R., Shimizu, S., Tsuda, S., 2010. Transmission of *Tomato chlorotic dwarf viroid* by bumblebees (*Bombus ignitus*) in tomato plants. European Journal of Plant Pathology 126 (1), 111–115, available from: ISI:000272299700010.

Mumford, R.A , Jarvis, B., Skelton, A., 2003. The first report of *Potato spindle tuber viroid* (PSTVd) in commercial tomatoes in the U.K. New Disease Reports 8, 31.

Nault, L.R., Ammar, E., 1989. Leafhopper and planthopper transmission of plant-viruses. Annual Review of Entomology 34, 503–529, available from: ISI: A1989T004500023.

Navarro, B., Gisel, A., Rodio, M.E., Delgado, S., Flores, R., Di Serio, F., 2012. Viroids: how to infect a host and cause disease without encoding proteins. Biochimie 94 (7), 1474–1480, available from: ISI:000305502400007.

Ng, J.C.K., Falk, B.W., 2006. *Bemisia tabaci* transmission of specific *Lettuce infectious yellows virus* genotypes derived from in vitro synthesized transcript-inoculated protoplasts. Virology 352 (1), 209–215, available from: ISI:000239806800020.

Nielsen, S.L., Enkegaard, A., Nicolaisen, M., Kryger, P., Marn, M.V., Plesko, I.M., Kahrer, A., Gottsberger, R.A., 2012. No transmission of *Potato spindle tuber viroid* shown in experiments with thrips (*Frankliniella occidentalis*, *Thrips tabaci*), honey bees (*Apis mellifera*) and bumblebees (*Bombus terrestris*). European Journal of Plant Pathology 133 (3), 505–509, available from: ISI:000304445600002.

Owens, R.A., Hammond, R.W., 2009. Viroid pathogenicity: one process, many faces. Viruses-Basel 1 (2), 298–316, available from: ISI:000280337300013.

Pfannenstiel, M.A., Slack, S.A., 1980. Response of potato cultivars to infection by the *Potato spindle tuber viroid*. Phytopathology 70 (9), 922–926, available from: ISI: A1980KL09500024.

Pirone, T.P., Blanc, S., 1996. Helper-dependent vector transmission of plant viruses. Annual Review of Phytopathology 34, 227–247, available from: ISI: A1996VG88500014.

Power, A.G., 2000. Insect transmission of plant viruses: a constraint on virus variability. Current Opinion in Plant Biology 3 (4), 336–340, available from: ISI:000088052900011.

Querci, M., Owens, R.A., Vargas, C., Salazar, L.F., 1995. Detection of *Potato spindle tuber viroid* in avocado growing in Peru. Plant Disease 79 (2), 196–202, available from: ISI: A1995QJ18300020.

Querci, M., Owens, R.A., Bartolini, I., Lazarte, V., Salazar, L.F., 1997. Evidence for heterologous encapsidation of potato spindle tuber viroid in particles of potato leafroll virus. Journal of General Virology 78 (6), 1207–1211, available from http://vir.sgmjournals.org/content/78/6/1207. abstract.

Robinson, D.J., Ryabov, E.V., Raj, S.K., Roberts, I.M., Taliansky, M.E., 1999. Satellite RNA is essential for encapsidation of groundnut rosette umbravirus RNA by groundnut rosette assistor luteovirus coat protein. Virology 254 (1), 105–114, available from: ISI:000078622700011.

Salazar, L.F., Querci, M., Bartolini, I., Lazarte, V., 1995. Aphid transmission of potato spindle tuber viroid assisted by potato leafroll virus. Fitopatologia 30, 56–58.

Schumann, G.L., Tingey, W.M., Thurston, H.D., 1980. Evaluation of six insect pests for transmission of potato spindle tuber viroid. American Potato Journal 57, 205–211.

Singh, R.P., 2006. Viroids from ornamental plants – a potential threat to tomato and potato crops. Canadian Journal of Plant Pathology-Revue Canadienne de Phytopathologie 28 (2), 328–329, available from: ISI:000242391000056.

Singh, R.P., Singh, M., 1998. Specific detection of potato virus a in dormant tubers by reverse transcription polymerase chain reaction. Plant Disease 82 (2), 230–234, available from: ISI:000071671100017.

Singh, R.P., Boucher, A., Wang, R.G., Somerville, T.H., 1992. *Potato spindle tuber viroid* is not encapsidated *in vivo* by potato Virus-y particles. Canadian Journal of Plant Pathology-Revue Canadienne de Phytopathologie 14 (1), 18–21, available from: ISI: A1992JF35900003.

Syller, J., 2003. Molecular and biological features of umbraviruses, the unusual plant viruses lacking genetic information for a capsid protein. Physiological and Molecular Plant Pathology 63 (1), 35–46, available from: ISI:000187393000005.

Syller, J., Marczewski, W., 2001. *Potato leafroll virus*-assisted aphid transmission of potato spindle tuber viroid to potato leafroll virus-resistant potato. Journal of Phytopathology-Phytopathologische Zeitschrift 149 (3–4), 195–201, available from: ISI:000168494300011.

Taliansky, M.E., Robinson, D.J., Murant, A.F., 1996. Complete nucleotide sequence and organization of the RNA genome of groundnut rosette umbravirus. Journal of General Virology 77, 2335–2345, available from: ISI: A1996VF33000042.

van den Heuvel, J.F.J.M., Hogenhout, S.A., van der Wilk, F., 1999. Recognition and receptors in virus transmission by arthropods. Trends in Microbiology 7 (2), 71–76, available from: ISI:000080898400007.

van Dorst, H.J.M., Peters, D., 1974. Some biological observations on pale fruit, a viroid-incited disease of cucumber. Netherlands Journal of Plant Pathology 80, 85–96.

Verhoeven, J.T.J., Jansen, C.C.C., Roenhorst, J.W., Steyer, S., Michelante, D., 2007a. First report of potato spindle tuber viroid in tomato in Belgium. Plant Disease 91 (8), 1055, available from: ISI:000248198100032.

Verhoeven, J.T.J., Jansen, C.C.C., Werkman, A.W., Roenhorst, J.W., 2007b. First report of *Tomato chlorotic dwarf viroid* in *Petunia hybrida* from the United States of America. Plant Disease 91 (3), 324, available from: ISI:000244263500023.

Verhoeven, J.T.J., Jansen, C.C.C., Roenhorst, J.W., 2008a. First report of pospiviroids infecting ornamentals in the Netherlands: *Citrus exocortis viroid* in *Verbena* sp., *Potato spindle tuber viroid* in *Brugmansia suaveolens* and *Solanum jasminoides*, and *Tomato apical stunt* viroid in *Cestrum* sp. Plant Pathology 57 (2), 399, available from: ISI:000254193200091.

Verhoeven, J.T.J., Jansen, C.C.C., Roenhorst, J.W., 2008b. First report of pospiviroids infecting ornamentals in the Netherlands: *Citrus exocortis viroid* in *Verbena* sp., *Potato spindle tuber viroid* in *Brugmansia suaveolens* and *Solanum jasminoides*, and *Tomato apical stunt viroid* in *Cestrum* sp. Plant Pathology 57, 399.

Verhoeven, J.T.J., Jansen, C.C.C., Roenhorst, J.W., Steyer, S., Schwind, N., Wassenegger, M., 2008c. First report of *Solanum jasminoides* infected by *Citrus exocortis viroid* in Germany and the Netherlands and *Tomato apical stunt viroid* in Belgium and Germany. Plant Disease 92 (6), 973, available from: ISI:000256007700018.

Verhoeven, J.T.J., Botermans, M., Jansen, C.C.C., Roenhorst, J.W., 2010a. First report of *Tomato apical stunt viroid* in the symptomless hosts *Lycianthes rantonnetii* and *Streptosolen jamesonii* in the Netherlands. Plant Disease 94 (6), 791, available from: ISI:000277844300045.

Verhoeven, J.T.J., Huner, L., Marn, M.V., Plesko, I.M., Roenhorst, J.W., 2010b. Mechanical transmission of *Potato spindle tuber viroid* between plants of *Brugmansia suaveoles, Solanum jasminoides* and potatoes and tomatoes. European Journal of Plant Pathology 128 (4), 417–421, available from: ISI:000284596300001.

Verhoeven, J.T.J., Jansen, C.C.C., Botermans, M., Roenhorst, J.W., 2010c. Epidemiological evidence that vegetatively propagated, solanaceous plant species act as sources of *Potato spindle tuber viroid inoculum* for tomato. Plant Pathology 59 (1), 3–12, available from: ISI:000273477800002.

Verhoeven, J.T.J., Botermans, M., Meekes, E.T.M., Roenhorst, J.W., 2012. *Tomato apical stunt viroid* in the Netherlands: most prevalent pospiviroid in ornamentals and first outbreak in tomatoes. European Journal of Plant Pathology 133 (4), 803–810, available from: ISI:000305687000005.

Walter, B., 1987. Tomato apical stunt. In: Diener, T.O. (Ed.), The viroids, Plenum Press, New York, pp. 321–328.

Chapter 16

Engineering crops for resistance to geminiviruses

Akhtar Jamal Khan and Sohail Akhtar
Department of Crop Sciences, College of Agricultural and Marine Sciences, Sultan Qaboos University, Sultanate of Oman

Shahid Mansoor and Imran Amin
National Institute of Biotechnology and Genetic Engineering, Faisalabad, Pakistan

INTRODUCTION

Geminiviridae is a family of phytopathogenic viruses with a characteristic circular, single-stranded DNA (ssDNA) genome encapsidated in geminate particles (Goodman 1977, Harrison et al 1977, Gutierrez 2000, Hanley-Bowdoin et al 2000). Each geminate particle consists of 110 capsid protein subunits of 29–30 kDa and one molecule of ssDNA (virion-sense) of 2.5–3.0 kb (Fauquet et al 2000). The total number of nucleotides in one component varies from 2580 nt for *Maize streak virus* (Morris-Krisinich et al 1985) to 2993 nt for *Beet curly dwarf virus* (Stanley et al 1986). The four genera in this family are distinguished on the basis of insect vector, host specificity, and genome organization (Stanley et al 2005, Fauquet et al 2008). Begomoviruses constitute the largest genus; they are transmitted by the adult silver leaf whitefly (*Bemisia tabaci*) and infect a wide range of economically important dicotyledonous crops. Bipartite begomoviruses having two genomic components (DNA-A and DNA-B) are native to the New World (NW), but a small number occurs in the Old World (OW) as well. Monopartite begomoviruses with a single genomic component (homologous to DNA-A of bipartite begomoviruses) are indigenous to the OW only. Each genomic component is transcribed bidirectionally. OW monopartite viruses are often associated with satellite molecules, referred to as alpha- and beta-satellites. These satellites are half the size of the virus (~1.4 kb). Alpha-satellites can replicate on their own as they encode a replication-associated protein (Rep) (Mansoor et al 1999). Beta-satellites are entirely dependent on a helper virus for replication and encapsidation. They encode a single complementary sense strand gene called βC1 (Briddon et al 2003, Saeed et al 2005). βC1 is a pathogenicity determinant; it overcomes host defense responses and helps in the movement of the virus (Saeed et al 2005, 2007, Qazi et al 2007, Amin et al 2011). While no specific role for alpha-satellite is known (Briddon et al 2004, Cui et al 2004, Saunders et al 2004), recent evidence suggests that its replication-associated protein is a strong suppressor of post-transcriptional gene silencing (Nawaz-Ul-Rehman et al 2010). Begomovirus disease complexes are widespread in the Old World (OW), and their distribution is attributed mainly to the presence of their insect vector coupled with international trade.

Members of the family *Geminiviridae* have single-stranded (ss) DNA genomes and are classified into four genera, *Mastrevirus*, *Topocuvirus*, *Curtovirus*, and *Begomovirus*, based on their genome organization (Fig. 16.1), host range, and insect vector (Fig. 16.2) (Rybicki 1994, Briddon & Markham 1995, Hanley-Bow-doin et al 2000, Stanley et al 2005). The names of the genera of the geminiviruses were adopted from the abbreviations of their type members, such as *Begomovirus* from *Bean golden mosaic virus* (BGMV), *Mastrevirus* from *Maize streak virus* (MSV), *Curtovirus* from *Beet curly top virus* (BCTV), and *Topocuvirus* from *Tomato pseudo curly top virus* (TPCTV), respectively (Jeske 2009). Recently, ICTV has approved including three more genera, *Becurtovirus* (type species *Beat curly top Iran virus*), *Turncutovirus* (type species *Turnip curly top virus*), and *Eragrovirus* (type species *Eragrostic curvularia streak virus*), to the family *Geminiviridae* (Adam et al 2013; Fig. 16.2). The ssDNA circular genomes of

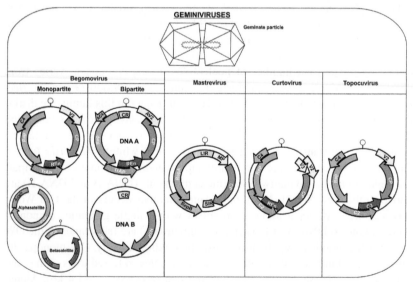

FIGURE 16.1 Genome organization of four genera belonging to the family *Geminiviridae* and satellites (beta-satellite and alpha-satellite) associated with *Begomovirus*. The position and orientation of virion-sense (V) and complementary-sense (C) genes and the predicted hairpin structure containing the TAATATTAC sequences are shown. Mastreviruses contain two noncoding regions, referred to as the large and small intergenic regions (LIR and SIR), respectively. The DNA-A and DNA-B components of bipartite begomoviruses are of same size, but DNA-A encodes six ORFs and DNA-B two ORFs.

geminiviruses replicate in the host nuclei of infected cells as double-stranded (ds), replicative intermediates by a rolling-circle mechanism (Hanley-Bowdoin et al 1999). The encapsidated form of the virus DNA is single-stranded and is referred to as the virion-strand. Rep binds to repeated elements (iterons) near the stem–loop structure and makes a site-specific nick at TAATATT↓AC in the loop region of the hairpin structure of the virion strand to initiate replication (Heyraud-Nitschke et al 1995). The rolling-circle DNA replication of the virion strand starts at the 3' adenine residue in the nonanucleotide sequence, which is designated as position 1 for numbering of the genomic sequence. The open reading frames (ORFs) are either encoded on the virion-sense strand (V) or the complementary-sense strand (C) and, for bipartite begomoviruses, A or B is used as a prefix for the genes encoded on the DNA-A or DNA-B component, respectively (Fauquet et al 2008).

The virion-sense strand of all geminiviruses encodes the only structural protein, coat protein (CP; ORF (A)V1), forming the distinctive geminate particles. The coat protein also interacts with the insect vector to facilitate transmission and is involved in movement in host plants (Hanley-Bowdoin et al 1999 (Fig. 16.1)). For New World begomoviruses V1 is the only virion-sense-encoded gene. However, the Old World begomoviruses encode an additional gene, (A) V2, which is involved in viral movement. In addition to the CP, the curtoviruses encode two additional genes: V2, responsible for regulating ssDNA/dsDNA levels, and V3, involved in virus movement. The genes encoded by TPCTV are not known yet. However, sequence similarities suggest that they reflect the functions of genes encoded by begomoviruses (Briddon et al 2003).

The dicot-infecting geminiviruses (with the exception of mastreviruses) encode four genes on the complementary-sense strand (Fig. 16.1). The replication-associated protein (Rep, encoded by ORF (A)C1) is a rolling-circle replication initiator protein with DNA nicking and joining activity. It is a sequence-specific

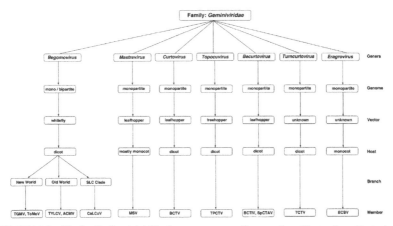

FIGURE 16.2   The family *Geminiviridae* has seven genera: *Begomovirus, Mastrevirus, Curtovirus, Topocuvirus, Becurtovirus, Turncurtovirus,* and *Eragrovirus,* which are distinguished by genome composition, vector, and host range. The *Begomovirus* genus can be divided into three branches. At the bottom of the figure some examples of members of the different genera are depicted.

DNA binding protein that nicks the virion-strand ahead of the 3' adenine to initiate rolling-circle replication and resolves new ssDNA circles by cutting and ligating. The protein recognizes small repeated sequences, known as 'iterons,' just upstream of the hairpin structure, which are, for the most part, species specific. The specificity of a particular Rep for its cognate iterons is such that, in most instances, the Rep from one virus species will not recognize the iterons of a second species to initiate replication.

Rep is also known for induction of host replication machinery, presumably to enable the virus to replicate in differentiated cells (Kong et al 2000, Egelkrout et al 2002). Rep binds with retinoblastoma related proteins (RBR) involved in the cell cycle that prevent cell entry into the S phase by sequestering transcription factors (Collin et al 1996). It has been shown that the Rep of TGMV binds to RBR through an 80-amino acid region that contains two predicted α-helices (Arguello-Astorga et al 2004). Host transcription is activated in mature leaves by relieving RBR/E2F repression. The ability of TGMV infection and RBR binding activity of Rep to overcome E2F-mediated blocking of the proliferating cell nuclear antigen (PCNA; Castillo et al 2004) promoter supports the fact that the geminivirus replication protein regulates host gene expression to some extent via the pRBR/E2F pathway. In this process, E2F binds to the PCNA promoter and recruits pRBR, which recruits chromatin remodeling activities, such as histone deacetylases and the SWI/SNF-like enzyme, to create a repressor complex (Zhang & Dean 2001). As a result an activation of host gene expression occurs and thus leads to the production of the required host DNA replication machinery. Rep also interacts with the replication factor C (RFC) complex, which helps in the transfer of PCNA to the replication fork (Luque et al 2002, Castillo et al 2004). Such interactions most probably occur in the early steps in the assembly of a DNA replication complex on the geminivirus origin.

The second complementary-sense gene (ORF (A)C2) encodes a protein that may act as a transcription factor that upregulates host-encoded genes and that may also be a suppressor of RNA interference (RNAi—a host defense mechanism that recognizes and degrades foreign and aberrant RNAs (reviewed by Baulcombe 2004)). For many begomoviruses the (A)C2 protein is also involved in upregulating the late (virion-sense) genes and is known as the transcriptional activator protein (TrAP). The replication enhancer protein (REn, ORF (A)C3) interacts with host factors, to create a cellular environment suitable for viral replication, and with Rep, to upregulate its activity. The (A)C4 protein (ORF C4) is, for many viruses, a pathogenicity determinant and in some cases a suppressor of RNAi.

The complementary-sense coding strategy of the mastreviruses is unusual among geminiviruses. The Rep is expressed from a spliced transcript of the RepA and RepB (ORFs C1 and C2, respectively) genes. The RepA protein, translated from a non-spliced transcript, interacts with host-encoded factors to create an environment suitable for virus replication and also is involved in upregulating the expression of late, virion-sense genes. The DNA-B component of bipartite begomoviruses encodes two additional genes: in the virion-sense the nuclear shuttle protein (NSP, ORF BV1) and in the complementary-sense the MP (ORF BC1). These act in a concerted manner to mediate cell-to-cell movement of the virus in plants (Stanley et al 2005).

The majority of the economically important geminiviruses are in the genus *Begomovirus*, which presently encompasses more than 200 species (Fauquet et al 2008). Begomoviruses infect only dicotyledonous plants and are transmitted by whiteflies (*Bemisia tabaci*), which can acquire the virus in less than 5 minutes and remain infectious for a lifetime (Czosnek 2007). Begomoviruses have monopartite or bipartite genomes (Lazarowitz et al 1992). All of the begomoviruses originating in North, Central, or South America (New World (NW)) have bipartite genomes (composed of DNA-A and DNA-B), while those originating in Europe, Africa, Middle East, India, China, and Japan (Old World (OW)) have either bipartite or monopartite genomes (Figs 16.1 and 16.2). Begomoviruses can be classified into three branches (OW, NW, and Squash leaf curl (SLC) clade) depending on their DNA sequence and place of origin (Fig. 16.2; Lazarowitz et al 1992). Some of the most prominent *Begomovirus* members intensively studied are *Cabbage leaf curl virus* (CaLCuV), *Tomato golden mosaic virus* (TGMV), *Bean golden yellow mosaic virus* (BGYMV), *African cassava mosaic virus* (ACMV), *Tomato yellow leaf curl virus* (TYLCV), *Mung bean yellow mosaic virus* (MYMV), and *Tomato mottle virus* (ToMoV).

There are three prominent examples of geminiviruses causing problems to agricultural crops that stand out for entirely different reasons. Cassava is a subsistence crop across much of sub-Saharan Africa and Southern India and is severely affected by bipartite begomoviruses that cause cassava mosaic disease (CMD) (Patil & Fauquet 2009). During the 1990s a particularly severe epidemic of CMD spread throughout Uganda, leading to food shortages and famine-related deaths (Thresh & Otim-Nape 1994). Recent reports have indicated that the CMD pandemic has since affected at least nine countries in East and Central Africa, covering an area of 2.6 million km^2 and causing an estimated annual economic loss of US$1.9–2.7 billion (Patil & Fauquet 2009). Tomato leaf curl disease affects tomato production throughout all tropical and subtropical areas and is caused by a number of distinct begomovirus species (Khan et al 2008). However, one particular species, the monopartite begomovirus *Tomato yellow leaf curl virus* (TYLCV), was first identified in the Middle East but has since spread throughout the Mediterranean region, the Americas, Australia, eastern China, and Japan (Cohen & Lapidot 2007). Begomoviruses are the most serious biotic limiting factor for agricultural production in Oman (Khan et al 2008, Idris et al 2011, Khan et al 2012a,b, 2013). This rapid spread has been attributed to the global trade in agricultural products. The tomato leaf curl disease is caused by a complex of monopartite begomoviruses that associate with a specific betasatellite (Khan et al 2008, Idris et al 2011).

Geminiviruses have spread around the world, causing significant losses in economically important crops and drawing the attention of the scientific community (Table 16.1). Their spread worldwide is due to climate change, changes

TABLE 16.1 A list of the economically important viral diseases caused by geminiviruses

Disease[a]	Virus genus	Host	Region or country	Yield loss	Reference
MSD	*Mastrevirus*	Maize	Sub-Saharan Africa	20–100%	Bosque-Perez 2000
CMD	*Begomovirus*	Cassava	Africa, India	15–20%	Legg & Fauquet 2004
CMD	*Begomovirus*	Cassava	Africa, India	Up to 90%	Patil et al 2005
CLCuD	*Begomovirus*	Cotton	Paskistan	30%	Mansoor et al 2011
BGMD/ BGYMD	*Begomovirus*	Bean	Florida, Central and South America	10–100%	Blair 1995, Faria & Maxwell 1999
YMD	*Begomovirus*	Legume	India	10–90%	Malathi et al 2005, Varma & Malathi 2003
TLCD/ TYLCD	*Begomovirus*	Tomato	Europe, Asia, Americas, Austria	20–80% Up to 100%	Moffat 1999, Moriones & Navas-Castillo 2000

[a]*The virus disease acronyms are: maize streak disease (MSD), cassava mosaic disease (CMD), cotton leaf curl disease (CLCuD), bean golden mosaic disease (BGMD), bean yellow golden mosaic disease (BGYMD), yellow mosaic disease (YMD), tomato leaf curl disease (TLCD), and tomato yellow leaf curl disease (TYLCD).*

in crop cultivation, increasing populations of insect vectors, increased movement of plants and people, and changes in agricultural practices such as the intensive use of insecticides and crop rotations (Anderson et al 2004, Morales & Jones 2004, Moriones & Navas-Castillo 2010). Many of these crops are staple foods in tropical and subtropical areas, making it economically and socially important to develop resistance strategies against geminivirus disease. The propagation of geminiviruses worldwide, in combination with their high evolution, recombination, and emergence rates, presents a major obstacle for managing geminivirus diseases. Strategies for controlling viruses of agricultural importance, including geminiviruses, are usually preventive by eradicating the vectors using chemical insecticides (Seal et al 2006). These strategies are effective in reducing the infectious potential of vectors (by lowering the number of vectors per plant), but often the virus has already been transmitted to the plant before the insect vector is killed. Moreover, because of their acute toxicity and adverse effects on the environment, concerns have been raised on the use of agrochemicals to control virus vectors. Due to the severity of viral epidemics, the difficulties of implementing efficient management practices, and the increasing demand for sustainable and environment-friendly agricultural practices, there is a great need to develop varieties that are resistant to pathogenic viruses. These factors necessitate that natural resistance be augmented with genetically engineered resistance that should provide more durable resistance to geminiviruses.

CONVENTIONAL DISEASE RESISTANCE STRATEGIES FOR GEMINIVIRUSES

Conventionally, viral diseases are controlled by controlling their insect vectors using either physical barriers or pesticides (Polston & Anderson 1997, Lapidot & Friedmann 2002, Pakniat-Jahromy et al 2010). Using pesticides to control insect vectors is expensive, environmentally hazardous, and requires frequent applications—often at higher doses than the recommended formulation, which has led to vector resistance (Morales 2001). Alternatively, physical barriers (Agryl® mesh screens and UV-absorbing plastic sheets) have been used to prevent exposure of the crops to the insect vectors (Cohen & Antignus 1994, Antignus 1998). However, the use of physical barriers adds to production costs and creates other problems such as shading, overheating, and poor ventilation (Lapidot & Friedmann 2002). As a consequence, vector control methods have not been successful in preventing or providing long-term control of geminivirus disease.

Conventional breeding is largely used to develop cultivars that are resistant to, or which tolerate, geminiviruses. The wild cultivars with endogenous viral resistance are used to cross with high-yielding susceptible cultivars to create a new resistant cultivar. The breeding for resistance method is time consuming and involves a difficult screening process that uses complex genetics (Lapidot & Friedmann 2002). Regardless of the repeated attempts to develop resistant or tolerant crop varieties by conventional breeding, success is limited because the tolerance is easily broken by the appearance of recombinant viruses or by an unfavorable environment with high virus diversity (Briddon & Stanley 2009).

NONCONVENTIONAL DISEASE RESISTANCE STRATEGIES FOR GEMINIVIRUSES

Transient control of ever evolving geminiviruses by conventional resistance approaches necessitated the emergence of nonconventional resistance strategies of plant transformation using tissue culture techniques. The objective is to make crop varieties inherently resistant to pathogen infection by introducing selected nucleic acid sequences into plants. The selected sequences actually determine the type of transgenic resistance obtained. If a susceptible plant is transformed with nucleic acid sequences derived from the pathogen itself, which may or may not encode a functional protein, it will be termed 'pathogen derived resistance' (PDR), and if the selected sequences are not of pathogen origin, it will be regarded as 'non-pathogen derived resistance' (NPDR).

The concept of PDR is based upon the phenomenon of cross protection (also known as 'mild strain interference'), in which plants infected with less virulent strains often develop resistance against highly virulent strains of that virus (Lecoq 1998). Although in many cases there appears to be no cross protection between geminiviruses, the interaction in most cases being either benign or synergistic (Vanitharani et al 2004), there is some evidence to suggest that the phenomenon might be useful for protection of cassava against begomoviruses that cause cassava mosaic disease in Africa (Owor et al 2004). Sanford and Johnson (1985) defined PDR as a strategy in which entire genes, or sequences of the pathogen's genome (either structural or nonstructural), are used to transform host plants so that protection against the pathogen from which the sequences were derived, or a range of closely related pathogens, is achieved.

A variety of geminivirus resistance strategies based on genetic engineering tools have been tested. They have focused mainly on pathogen-derived resistance, including the expression of viral proteins that interfere with virus infection or transcription of viral RNA to silence the expression of viral genes (Shepherd et al 2009). Analogous to CP-mediated resistance described for RNA viruses (Abel et al 1986), introduction of CP genes into plants has been used to create some level of resistance against monopartite geminiviruses that require CP for systemic infection (Briddon et al 1989, Rojas et al 2001). This strategy has not proven successful for bipartite begomoviruses, in which the NSP substitutes for the CP transport function (Ingham et al 1995, Pooma et al 1996, Azzam et al 1996, Frischmuth & Stanley 1998, Vanderschuren et al 2007). In contrast, the Rep protein is an excellent target for geminivirus resistance because of its essential nature, conserved sequence and function, and many protein–protein interactions. Expression of an N-terminally truncated Rep (T-Rep) protein from *Tomato yellow leaf curl Sardinia virus* (TYLCSV) in tomato conferred resistance to the homologous virus by repressing the viral promoter and affected a heterologous geminivirus by forming dysfunctional oligomers with its Rep protein (Lucioli et al 2003).

Other viral proteins have been expressed in transgenic plants and have yielded some level of disease resistance (Vanderschuren et al 2007). However, expressing viral proteins in plants can have detrimental effects on the plant's phenotype and provide specific resistance to a specific geminivirus genus or species. Transgenic *N. benthamiana* and tobacco expressing TrAP proteins from TGMV and BCTV had rather increased the susceptibility to these geminiviruses (Sunter et al 2001). Expression of the ACMV AC4 caused developmental abnormalities in *Arabidopsis* (Chellappan et al 2005).

Several examples of protein-mediated PDR designed for distinct plant-infecting RNA viruses in a range of plant species were successful even though the underlying mechanism of resistance remained undefined. Many of the geminivirus resistance strategies are highly specific for a given virus or set of viruses. In addition, they often require significant genomic sequence knowledge of the target virus/es. Geminiviruses rapidly adapt to new hosts and new environments, have high mutation and recombination rates, and often occur as mixed infections (Mansoor et al 2003, Amin et al 2010). These properties negatively impact the breadth and durability of many strategies. To overcome these challenges, peptide aptamer usage has emerged as a new broad-based strategy targeting the essential and conserved domains of a viral protein that cannot be altered without causing a severe decrease in virus fitness. Reports published using PDR or NPDR approaches to develop resistance to geminivirus are listed in Table 16.2 (PDR) and Table 16.3 (NPDR). Additionally the strategies investigated in efforts to obtain transgenic resistance against geminiviruses are summarized diagrammatically in Figure 16.3.

TABLE 16.2 A list of published reports using the pathogen-derived approach to develop genetic resistance against geminiviruses

Protein-mediated resistance			
PDR gene	Plant	Targeted virus[a]	Reference
MP of TGMV	N. benthamiana	ACMV	von Arnim & Stanley 1992
CP of TYLCV	S. lycopersicum	TYLCV	Kunik et al 1994
Truncated Rep of TYLCSV	N. benthamiana	TYLCSV	Noris et al 1996
Truncated Rep of TYLCSV	S. lycopersicum	TYLCSV	Brunetti et al 1997
Truncated MP of ToMoV	N. tabacum	ToMoV, CaLCV	Duan, et al 1997
Mutated Rep of ACMV	N. benthamiana	ACMV	Sangré et al 1999
Mutated Rep of BGMV	N. tabacum suspension cells	BGMV	Hanson & Maxwell 1999
Wild type/mutated NSP/MP of BDMV	S. lycopersicum	ToMoV	Hou et al 2000
Truncated Rep of TYLCSV	N. benthamiana	TYLCSV	Brunetti et al 2001
Truncated Rep of ToLCNDV	N. benthamiana	ToLCNDV, PHYVV, PYMV, ACMV	Chatterji et al 2001
Rep of ToMoV	S. lycopersicum	ToMoV	Polston & Hiebert 2001

Continued

TABLE 16.2 A list of published reports using the pathogen-derived approach to develop genetic resistance against geminiviruses—cont'd

Protein-mediated resistance			
PDR gene	Plant	Targeted virus[a]	Reference
Truncated Rep of TYLCV	N. benthamiana	TYLCSV, TYLCV	Lucioli et al 2003
Truncated Rep of TYLCV	S. lycopersicum	TYLCV	Antignus et al 2004
Mutated Rep of MSV	Zea mays	MSV	Shepherd et al 2005
Full-length Rep and truncated Rep of MYMV	N. tabacum	MYMV	Shivaprasad et al 2006
Truncated and mutated Rep of MSV	Digitaria sanguinalis and Zea mays	MSV	Shepherd et al 2007b
RNA interference			
Antisense Rep of TGMV	N. tabacum	TGMV	Day et al 1991
Antisense Rep of TGMV	N. tabacum	BCTV	Bejarano & Lichtenstein 1994
RNAi of CbLCV	Arabidopsis	CbLCV	Turnage et al 2002
RNAi of V2 gene of ToLCNDV	N. benthamiana	ToLCNDV	Mubin et al 2007
TGS of AC2/AL2 of TYLCCNV	N. benthamiana	TYLCCNV	Yang et al 2011
Truncated CP of ToMoV	N. tabacum	ToMoV	Sinisterra et al 1999
RNA silencing of AC2 of MYMV	A. thaliana	MYMV	Trinks et al 2005
siRNA of ACMV	M. esculenta	ACMV	Akbergenov et al 2006
Chemically synthesized siRNA to Rep of ACMV	N. tabacum protoplasts	ACMV	Vanitharani et al 2003
Rep of ACMV	Manihot esculenta	ACMV, EACMCV, SLCMV	Chellappan et al 2004

TABLE 16.2 A list of published reports using the pathogen-derived approach to develop genetic resistance against geminiviruses—cont'd

Protein-mediated resistance			
PDR gene	Plant	Targeted virus[a]	Reference
Antisense Rep of ToLCNDV	S. lycopersicum	ToLCNDV	Praveen et al 2005
Antisense Rep, TrAP and REn of ACMV	Manihot esculenta	ACMV	Zhang et al 2005
RNA silencing of CP of CMV and CBSV	Manihot esculenta	CMV, CBSV	Vanderschuren et al 2012

[a]The virus acronyms used are African cassava mosaic virus (ACMV), Bean golden mosaic virus (BGMV), Beet curly top virus (BCTV), Cabbage leaf curl virus (CaLCV), Cassava mosaic virus (CMV), Cassava brown streak virus (CBSV), East African cassava mosaic Cameroon virus (EAC-MCV), Maize streak virus (MSV), Mungbean yellow mosaic virus (MYMV), Pepper huasteco yellow vein virus (PHYVV), Potato yellow mosaic virus (PYMV), Sri Lankan cassava mosaic virus (SLCMV), Tomato golden mosaic virus (TGMV), Tomato leaf curl virus (ToLCV), Tomato leaf curl China virus (TLCCNV),Tomato leaf curl New Delhi virus (ToLCNDV), Tomato yellow leaf curl Sardinia virus (TYLCSV), Tomato yellow leaf curl virus (TYLCV), and Tomato mottle virus (ToMoV).

TABLE 16.3 A list of published reports using the non-pathogen-derived approach to develop genetic resistance against geminiviruses

Non-pathogen-derived resistance			
NPDR gene	Plant	Targeted virus[a]	Reference
Cell death induced by dianthin from Dianthus caryphyllus	N. benthamiana	ACMV	Hong & Stanley 1996
Artificial zinc-finger protein	A. thaliana	BSCTV	Sera 2005
Peptide aptamers inhibiting Rep	N. tabacum protoplasts	TGMV	Lopez-Ochoa et al 2006
Ribozyme-mediated resistance	Yeast	MYMIV	Ushasri et al 2007
Trapping of viral DNA by GroEL	S. lycopersicum	TYLCV	Akad et al 2007, Edelbaum et al 2009
Recombinant-antibody inhibiting Rep	N. benthamiana	TYLCV	Safarnejad et al 2009

[a]The virus acronyms used are African cassava mosaic virus (ACMV), Tomato golden mosaic virus (TGMV), Beet severe curly top virus (BSCTV), Tomato yellow leaf curl virus (TYLCV), and Mungbean yellow mosaic India virus (MYMIV).

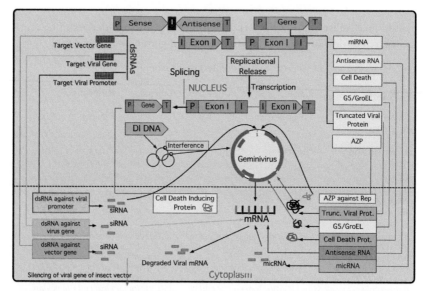

FIGURE 16.3 Different approaches to produce transgenic plants resistant to geminiviruses. microRNA, and antisense RNA, expressed from transgenes, are produced in the nucleus and are transported to the cytoplasm where they target the viral mRNAs. Cell-death-inducing protein mRNA, G5, GroEL, viral truncated proteins, and artificial zinc-finger (AZP) genes express their products in the cytoplasm, where they attack viral mRNA. In the 'InPAct' system, a circular DNA molecule with the gene of interest, promoter (P), terminator (T), and introns (I) are released by replication from the designed DNA structure integrated in the plant genome. The gene of interest, with introns and a hairpin structure in between, is expressed from the circular DNA molecule. The hairpin structure is removed, along with introns, during splicing, and this processed mRNA is transported to the cytoplasm, where it is translated into the desired protein, which performs its function. Hairpin RNAs produced from expression cassettes containing the virus gene sequence, in sense and antisense orientation, separated by an intron, are also transported to the cytoplasm, where they are processed to siRNAs, which enter the gene silencing pathway. Circular defective interfering DNA (DI-DNA) molecules of geminiviruses are produced from the partial tandem repeats, which are integrated into the plant genome, either by replicational release or recombination. These DI-DNAs replicate and compete with invading geminiviruses.

PATHOGEN-DERIVED RESISTANCE TO GEMINIVIRUSES

Coat-protein-mediated resistance

The first success with improved resistance to virus, using viral proteins, was reported by Abel et al (1986). In this study, transgenic plants expressing CP of TMV were developed. Since then CP-mediated resistance has been greatly exploited for different RNA viruses. Despite extensive studies, the molecular mechanisms governing CP-mediated resistance are not fully resolved. Monopartite begomoviruses use CP for systemic infection, and tomato plants expressing CP of TYLCV exhibited delayed disease symptoms (Briddon et al 1989, Rojas et al 2001). On the other hand, CP of bipartite begomoviruses is not essential for systemic infection because NSP on DNA-B substitutes the function (Ingham et al 1995, Pooma et al 1996). Therefore it is assumed that a CP-based resistance strategy will not generate resistance against bipartite begomoviruses. Consistent with this assumption, neither transgenic beans expressing the CP of *Bean golden mosaic virus* (BGMV) (Azzam et al 1996) nor *N. benthamiana* plants expressing ACMV CP (Frischmuth & Stanley 1998) displayed any resistance. Since CPs of geminiviruses are involved in transmission of viruses by interacting with the vector, they can be exploited to develop transgenic resistance (Briddon et al 1990, Noris et al 1998, Morin et al 2000). Systemic infection of ACMV was lost, coupled with reduced interaction with *Bemisia tabaci* when a CP-deficient ACMV clone was used for infection in cassava plants (Liu et al 1997). These studies suggest that vector specificity determinants reside in the CP of geminiviruses. Thus, by expressing a mutated nonfunctional CP, virus spread among its vectors can be obstructed in a crop field with geminivirus infection.

Movement-protein-mediated resistance

Geminivirus movement proteins are required for their cell-to-cell and long-distance movement and have been used to engineer resistance to various homologous and heterologous viruses. TGMV MP had a deleterious effect on the systemic movement of ACMV, whose MP has only 41% similarity with that of TGMV (von Arnim & Stanley 1992). Duan et al (1997) transformed *N. tabacum* with a mutated version of *Tomato mottle virus* (ToMoV) MP. Transgenic plants showed resistance to both ToMoV and CaLCuV. Using a similar approach, tomato plants were transformed with a mutated *Bean dwarf mosaic virus* (BDMV) MP gene, which showed resistance to ToMoV (Hou et al 2000). Nonfunctional MPs may compete for NSP interaction or oligomerization (Frischmuth et al 2004), and this could explain the resistance observed in plants expressing mutated MP. Though the use of MP transgenes seems to give broad-spectrum resistance, it is constrained by the fact that MP transgenes are often toxic when over-expressed in plant cells—and, in the case of begomoviruses, are known pathogenicity determinants. Therefore they need controlled expression to avoid undesirable effects on plant development (Covey & Al-Kaff 2000, Hou et al 2000).

Replicase-protein-mediated resistance

The multifunctional Rep protein of geminiviruses plays an important role in the regulation of viral gene transcription and the replication of the genome. As the genomes of ssDNA plant viruses do not encode any polymerase, it requires an interaction between the viral Rep protein and the host polymerase. *Geminivirus* Rep proteins have been extensively exploited for resistance development. Rep gene expression of ACMV was negatively regulated by its own protein product.

Repression of expression was achieved when a truncated Rep, comprising only the N-terminal 57 amino acids, was used in tobacco protoplasts (Hong & Stanley 1996). A similar approach was used to engineer resistance against TYLCSV in *N. benthamiana* (Noris et al 1996) and *Lycopersicum esculentum* (Brunetti et al 1997). Chatterji et al (2001) transiently expressed the oligomerization domain of the Rep of ToLCNDV in tobacco protoplasts and *N. benthamiana* plants. The protein not only inhibited the accumulation of homologous virus by 70–86% but reduced the accumulation of the heterologous viruses ACMV, *Pepper huasteco yellow vein virus* (PHYVV), and *Potato yellow mosaic virus* (PYMV) by 22–48%. The heterologous viruses used in this study had similar iteron sequences. Truncated Rep of TYLCSV was expressed in tomato plants, which conferred resistance to homologous virus by repressing its promoter activity. The formation of dysfunctional complexes reduced the accumulation of heterologous viruses as well. However, in some cases resistance was overcome due to transgene silencing (Lucioli et al 2003).

Not only the truncated but also full-size Rep proteins with function-abolishing mutations interfered with the replication of BGMV and ACMV in tobacco cell suspension and *N. benthamiana* plants, respectively (Hanson & Maxwell 1999, Sangaré et al 1999). Despite various success stories in using Rep proteins to generate resistance, some problems often come across due to interfering with the host cell cycle and transcription regulatory factors, thus disturbing the normal growth of transgenic plants (Antignus et al 2004). Hence, approaches were developed to combine truncation and mutation strategies in a way that the Rep interferes with the virus only and not with the host factors. Using such an approach, a success story against African maize pathogen 'MSV' came into being (Shepherd et al 2007a).

RNAi-mediated resistance

RNA silencing is a novel gene regulatory mechanism that reduces the transcript level by either suppressing transcription (transcriptional gene silencing (TGS)) or by activating a homology-dependent mechanism of RNA degradation (post-transcriptional gene silencing (PTGS) in plants (van Rij & Andino 2006), quelling in fungi (Escobar et al 2001), co-suppression or RNA interference (RNAi) in plants and RNAi in animals (Fire et al 1998)). The mechanism implicates the use of foreign double-stranded RNA, which is recognized and degraded by specialized protein complexes within many eukaryotic cells; RNAi is believed to be an evolutionarily conserved defense mechanism against viruses and transposable elements. It is a ubiquitous silencing mechanism present in all eukaryotes, including protozoa, animals, and plants (Hanon 2002, Baulcombe 2005). The power and utility of RNAi for specifically silencing the expression of any gene for which the sequence is available has driven its incredibly rapid adoption as a tool for reverse genetics in eukaryotic systems. RNA silencing is a preferred approach for triggering gene silencing in transgenic plants by the expression of dsRNAs homologous to viral sequences to get resistance (Fig. 16.4; Pooggin & Hohn 2004, Vanitharani et al 2005, Mansoor et al 2006, Aragão & Faria 2009, Shepherd et al 2009). The modern term RNAi was first coined while experimenting on a nematode, *Caenorhabditis elegans*, where dsRNA was established as a trigger for the gene silencing. In addition, sense and antisense RNA were also considered to be able to silence gene expression (Fire et al 1998, Hannon 2002). Actually, the phenomenon relates to a very back-dated work of Richard

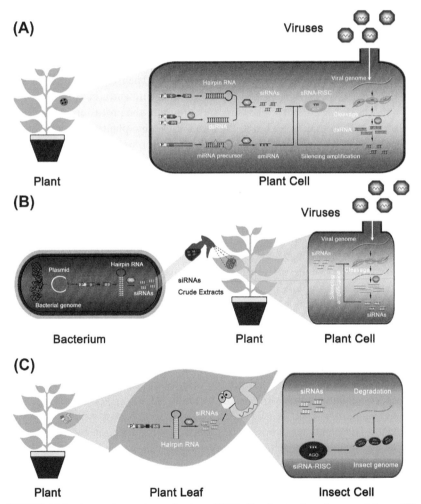

FIGURE 16.4 Approaches to the application of RNA silencing to plant disease resistance. (A) Expression of viral small RNA in host plants triggers antiviral silencing. (B) Sprayed bacterium-processed siRNAs confer resistance against virus. (C) Feeding on transgenic plants that carry RNAi constructs confers resistance against insects. As = antisense; P = promoter; s = sense. (Reproduced with permission from Figure 1 in Duan CG et al. Application of RNA silencing to plant disease resistance. Silence 2012;3:5, originally published by Biomed Central.)

Jorgensen, who had engineered transgenic plants of petunias to increase the pigmentation of flowers (Napoli et al 1990). The effect of exogenous transgenes proved contrary to expectation. Jorgensen was expecting deep-colored flowers with the introduction of a transgene, but he got variegated flowers in transgenic plants, and some plants were lacking pigment altogether. This indicated that not only were the transgenes themselves inactive but also that the added DNA sequences somehow affected expression of the endogenous loci (Napoli et al 1990). RNAi as a general defense mechanism was first understood by Fire et al (1998). Before this understanding, several studies had been performed targeting different viral genes using sense and antisense RNA constructs. Transgenic tobacco plants were produced by Day et al (1991) expressing an antisense (as)RNA construct containing sequences of the TGMV Rep, which were then challenged with infectious TGMV. The frequency of symptom development in transgenic plants was reduced significantly in comparison to the non-transformed control plants, correlating with the profusion of (as)RNA transcripts of the transgene. Later, the same transgenic plants imparted a good resistance response against BCTV as well because of high sequence homology of the targeted region of TGMV and BCTV Reps (Bejarano & Lichtenstein 1994). Asad et al (2003) targeted Rep, TrAP, and REn sequences of *Cotton leaf curl Khokhran virus* (CLCuKoV). Heritable resistance was noted in transgenic tobacco lines, which remained symptomless even upon exposure to viruliferous whiteflies. Tobacco plants transformed with an antisense RNA cassette targeting AV2 of *Tomato leaf curl New Delhi virus* (ToLCNDV) were resistant to the infection of this bipartite begomovirus (Mubin et al 2007). Similar reports regarding the usefulness of sense and antisense RNA technologies targeting different genes of geminiviruses are found, in which the transgenic plants show different resistance responses, such as being asymptomatic, the delayed appearance of symptoms, and a reduction in virus titer upon infection with the cognate virus (Bendahmane & Gronenborn 1997, Aragão et al 1998, Polston & Hiebert 2001).

Transgenic cassava plants expressing the full-length Rep gene of ACMV in sense orientation were produced and exhibited significant broad-spectrum resistance against a number of CMBs, including ACMV, *East African cassava mosaic Cameroon virus* (EACMCV), and *Sri Lankan cassava mosaic virus* (SCMV). The plants also showed resistance to dual infection by ACMV and EACMCV, which, in the field, results in a synergistic interaction that has been responsible for severe disease. Further analysis showed that the mechanism involved in the resistance was RNAi (Chellappan et al 2004). Similarly, Zhang et al (2005) produced transgenic cassava plants resistant to ACMV infection by expressing the Rep, TrAP, and REn genes of ACMV in antisense orientation. The transgenic lines showed reduced viral DNA replication. Vanderschuren et al (2007) reported that cassava plants expressing an intron-spliced hairpin construct containing CR sequences of ACMV showed accelerated recovery from infection upon challenge with ACMV and that recovery correlated with the appearance of transgene-derived siRNA. Overall, viral symptoms were attenuated and plants had less viral DNA accumulation compared with wild type plants. The authors suggested that the resistance was due to TGS but did not proceed to prove this.

Resistance using RNAi has also been achieved by targeting noncoding regions of geminiviruses. Transient expression of *Mungbean yellow mosaic virus* (MYMV) IR sequences as an intron-spliced hairpin resulted in complete recovery in blackgram plants infected with MYMV (Pooggin et al 2003). Similarly, an intron-spliced hairpin construct containing sequences of the IR conserved between TYLCV, TYLCSV, and *Tomato yellow leaf curl Malaga virus* yielded a broad-spectrum resistance when transiently expressed in tomato and *N. benthamiana* plants challenged with these viruses by *Agrobacterium*-mediated inoculation or whitefly transmission. No virus could be detected in plants, which were challenged with virus that had earlier been inoculated with the hairpin construct, using PCR; a positive correlation between resistance and the accumulation of TYLCV-specific siRNAs was observed in silenced plants (Abhary et al 2006). Bian et al (2006) proposed that a transgene carrying homology to the *Tomato leaf curl virus* (ToLCV) was silenced upon virus inoculation, while the virus managed to escape silencing possibly by *de novo* synthesized unmethylated DNA. In addition, plant viruses counteract plant defenses by encoding suppressors of gene silencing, which interfere at distinct steps of RNA-silencing pathways and thus break the resistance (Sharma & Ikegami 2008). The tomato inbred lines expressing Permease-I-like resistant gene (R gene from *Solanum habrochaites*) overexpressed severalfold following TYLCV inoculation (Eybishtz et al 2009). Silencing the Permease-I in resistant inbred lines using tobacco rattle virus-induced gene silencing led to a loss of the resistance in tomato. Furthermore, R-gene inbred lines developed viral symptoms typical of infected susceptible plants and accumulation of large amounts of virus (Eybishtz et al 2009).

Virus escaping from the silencing machinery due the presence of viral DNA-binding proteins, including the CP and NSP, which may serve as a shield for the viral DNA, is a major drawback in RNAi technology. Recently, RNAi technology has been successfully used to develop resistant transgenic common bean (*Phaseolus vulgaris*) against BGMV. These transgenic plants proved to be resistant against BGMV under field conditions (Aragão & Faria 2009).

miRNA-mediated resistance

More recently, researchers are using microRNA (miRNA) technology to confer resistance in plants against geminiviruses. miRNAs are another class of small RNAs of 21–24 nt, derived from transcripts with a distinctive RNA stem–loop secondary structure and best characterized for their role in developmental regulation (Bartel 2004). A large number of plant miRNAs has also been identified by computational and experimental approaches (Griffiths-Jones et al 2006,

Zhang et al 2006, Griffiths-Jones et al 2008). To date, over 2200 plant miRNAs have been discovered from over 30 plant species (Griffiths-Jones et al 2008).

miRNAs are derived from mature transcripts of inverted repeat precursor RNAs with partially double-stranded regions; therefore, the miRNA pathway does not require RDRs. miRNA genes are transcribed by RNA polymerase II (pol II), yielding a primary transcript. The primary transcripts, which are ultimately processed to produce miRNAs, are known as primary miRNAs (pri-miRNAs). Pri-miRNAs are processed initially by a nuclear RNAse-III like enzyme (equivalent to Drosha in animal) and yield pre-miRNAs. The downstream processing of pri-miRNAs involves the use of a protein named DAWDLE (DDL), which recruits predominantly DCL1 (Yu et al 2008). Pre-miRNAs are processed by Dicer together with HYL1 and SE (SERRATE) to form small RNA duplexes, which, after methylation by HEN1 at the 2'-OH of the 3' end, are transported to the cytoplasm with the help of an exportin homolog, HST (HASTY), to produce mature miRNAs (Katiyar-Agarwal & Jin 2010). Mature miRNAs are incorporated into RISCs, using AGO1 or AGO 2 as RNA slicer, and guide the complex either to promote the cleavage of target mRNAs or translational repression on the basis of sequence complementarity (Baumberger & Baulcombe 2005). In plants, in addition to mRNA cleavage or translational repression, DNA methylation is also induced by miRNAs (Bao et al 2004, Vrba et al 2013).

A number of miRNAs have been linked to biotic stress responses in plants. Pathogens of all types, such as bacteria, fungi, and viruses, are able to cause the upregulation or downregulation of certain miRNAs. Infection of *Turnip mosaic virus* (TuMV) causes upregulation of two miRNAs, bra-miR158 and bra-miR1885, in *Brassica rapa* (He et al 2008). miRNAs conferred resistance in transgenic *A. thaliana* plants against *Turnip yellow mosaic virus* (TYMV) and *Turnip mosaic virus* (TuMV) (Niu et al 2006). Recently, Amin et al (2011) documented the effects of diverse begomoviruses on the levels of a set of developmental miRNAs in *N. benthamiana*. The plants were infected with CLCuMV, ACMV, CaLCV, and TYLCV to note the levels of miR156, miR159, miR160, miR164, miR165, miR166, miR167, miR168, miR169, and miR170. The levels of most developmental miRNAs were increased after inoculation with these viruses. It has been shown that biogenesis of miRNA is not affected by the alteration of several nucleotides within an miRNA sequence. Therefore, it becomes possible to modify an miRNA sequence to target specific transcripts, which are not originally under the control of miRNA. Therefore, the backbone of an existing pre-miRNA can be used to generate artificial miRNAs with desired targets. The miRNA-mediated resistance approach is considered more effective in inhibiting viral infections in comparison to short hairpin siRNA technology. Although not many reports are available in the literature for the use of the miRNA strategy for geminivirus resistance, this approach can be exploited with the aim of developing broad-spectrum resistance (Niu et al 2006, Duan et al 2008, Ai et al 2011, Eamens et al 2011, Fahim & Larkin 2013).

DNA interference

In addition to genomic components, plants infected with geminiviruses often contain subgenomic DNAs (or subviral molecules) usually derived from partial deletion of the viral genome (Stenger et al 1992, Frischmuth & Stanley 1993). As they cause some interruption in helper virus replication and sometimes in movement, they are called defective interfering (DI) DNAs (Stanley et al 1990). This interference causes alterations in normal disease progression and results in amelioration of symptoms. Defective DNA molecules have been reported for the three families of DNA viruses of plants, namely *Geminiviridae*, *Nanoviridae*, and *Caulimoviridae*.

The use of DI-DNA as a resistance strategy started from the work of Stanley et al (1990). These authors transformed tobacco plants with a tandem repeat construct of a subgenomic defective DNA-B molecule derived from ACMV DNA-B. Transgenic plants showed ameliorated symptoms, compared to non-transformed plants, upon infection by ACMV. When these transgenic plants were inoculated with BCTV and TGMV, no sign of symptom amelioration was observed because these viruses were unable to amplify the subgenomic DNA, showing that the interaction is of a virus-specific nature. Later, Frischmuth et al (1997) linked the resistance phenotype of transgenic plants with the size of DI-DNA.

NON-PATHOGEN-DERIVED RESISTANCE TO GEMINIVIRUSES

As outlined earlier, silencing may compromise PDR that is protein-mediated, and virus-encoded suppressors can interfere with resistance based on the RNA silencing pathway. This suggests that, although RNAi is a valuable tool in generating resistance, greater efforts should be made to develop and utilize non-PDR strategies to confer resistance to viruses.

Virus-induced cell death

The hypersensitivity response (HR) is a defense mechanism used by plants against pathogens—including attack by viruses. In this mechanism, plants limit virus movement to the site of infection by inducing cell death of infected and neighboring tissue. Ribosome-inactivating proteins (RIPs) are naturally occurring plant toxins, which act by inhibiting protein synthesis by inactivating ribosomes (Narayanan et al 2005). Dianthin, a potent RIP, has been exploited to engineer transgenic resistance to ACMV in *N. benthamiana* (Hong et al 1996). In order to achieve this, dianthin was expressed from the ACMV virion-sense promoter. This strategy ensured the controlled expression of the toxin, facilitating the regeneration of phenotypically normal plants, and ensured that transgene expression was localized to virus-infected cells. When challenged with ACMV, transgenic plants produced atypical necrotic lesions on inoculated leaves, indicative of dianthin expression. Viral DNA accumulation was significantly reduced in these tissues, and plants exhibited attenuated systemic symptoms from which they recovered. Hussain et al (2005) found that tomato plants show HR for NSP of ToLCNDV, taking it as the target of host defense. Later on it was discovered that TrAP of ToLCNDV counters the aforementioned defense by suppressing HR (Hussain et al 2007), and the plants become unable to show HR. If such problems are solved, cell-death-inducing resistance mechanisms may prove to be worthwhile.

Peptide aptamer

One of the conserved features of geminiviruses is their Rep protein. Another approach for broad-spectrum resistance against geminiviruses is to target viral Rep using peptide aptamers. Peptide aptamers are small recombinant proteins usually consisting of approximately 20 amino acids. One of the characteristic features of peptide aptamers is that they are constrained within a scaffold protein. These recombinant proteins strongly bind to target proteins and thus interfere with their cellular functions (Hoppe-Seyler et al 2004, Baines & Colas 2006). Peptide aptamers were initially used to inactivate specific proteins *in vivo* in order to determine their functions (Colas et al 1996). Soon it was realized that that this strategy has applications in areas ranging from drug discovery to the identification of novel proteins involved in regulatory networks (Geyer et al 1999, Norman et al 1999, Baines & Colas 2006, Tomai et al 2006).

This strategy was first used to develop virus resistance in transgenic *N. benthamiana*, targeting the nucleoprotein (N) of *Tomato spotted wilt virus, Tomato chlorotic spot virus, Groundnut ring spot virus*, and *Chrysanthemum stem necrosis virus* (Uhrig 2003, Rudolph et al 2003). This study demonstrated that a certain peptide aptamer could confer broad-based viral resistance if an essential viral protein is targeted at conserved domains. The Rep protein is a good target for peptide aptamers, which, in turn, can inhibit geminivirus infection because Rep is essential for viral replication (Elmer et al 1988, Sunter et al 1990). Rep-specific aptamers were identified using a yeast two-hybrid system by using the N-terminal domain of TGMV Rep protein as bait. These aptamers were then used to engineer geminivirus resistance (Lopez-Ochoa et al 2006, 2009). Similarly, peptide aptamers were developed that bind to the Reps of diverse geminiviruses, including BCTV, CaLCuV, ACMV–Cameroon, and EACMV–Uganda (Lopez-Ochoa & Hanley-Bowdoin 2007), indicative of the potential of this strategy to develop a broad-spectrum resistance.

GroEL-mediated resistance

Under natural condition a virus–vector interaction is essential for efficient virus transmission. This interaction can be used for the benefit of the plants (Akad et al 2007). GroEL is a chaparonin protein produced by endosymbiotic bacteria from the whitefly vector *Bemisia tabaci*. The natural function of GroEL is to protect the virus during its passage through the hemolymph of the whitefly (Morin et al 1999). It was found that GroEL has a high affinity to bind to the CP of TYLCV; this property was exploited to develop transgenic resistance against TYLCV (Akad et al 2004). Due to the high affinity of GroEL proteins with a wide range of *Begomovirus* coat proteins, it was tested as a resistance transgene. The *B. tabaci* GroEL gene, expressed in transgenic tomatoes, produced phenotypic resistance against TYLCV infection. Transgenic tomato plants infected with TYLCV showed only very mild symptoms or no symptoms at all (Akad et al 2007). One of the drawbacks of this approach was that the viral loads in transgenic and non-transgenic plants were comparable (Akad et al 2007).

Artificial zinc-finger-protein-mediated resistance

A key feature in the use of DNA-binding proteins for resistance is the identification of virus-sequence-specific proteins that should not bind the host DNA. The use of transgenically expressed DNA-binding proteins to provide virus resistance relies on the identification of virus sequence-specific binding proteins that will not bind host DNA sequences. The Rep protein of geminiviruses has a high affinity for iterons (Rep-specific repeats) in the virion strand of geminiviruses. This property of Rep was exploited to develop artificial zinc-finger proteins (AZP) that can bind iterons (Sera & Uranga 2002, Sera 2005). The rationale of such an approach was that AZP would compete with the viral Rep due to its high affinity for dsDNA. This competition will lead to interference with viral replication, and thus viral replication can be blocked. This strategy has been successfully used for the development of resistance against TYLCV (Takenaka et al 2007, Koshino-Kimura et al 2009).

ssDNA-binding-protein-mediated resistance

Gene 5, commonly known as G5, is a small ssDNA-binding protein from *Escherichia coli* phage M13. It binds to DNA in a highly cooperative manner without pronounced sequence specificity. During synthesis of the viral ssDNA, it prevents the conversion of ssDNA into the dsDNA replicative form (van Duynhoven et al 1990, Padidam et al 1999). ToLCNDV DNA-A that lacked the CP gene was modified to express G5 under the control of a virion-sense promoter. The modified viruses led to the accumulation of wild type levels of ssDNA and high levels of dsDNA. The accumulation of ssDNA was apparently due to stable binding of G5 to viral ssDNA (van Duynhoven et al 1990). G5-expressing ToLCNDV did not spread efficiently in *N. benthamiana* plants, and inoculated plants developed only very mild symptoms. The authors proposed that the G5 expression interfered with the function of the NSP and thereby impaired virus spread. They claimed that the expression of G5 in transgenic plants may provide a novel way of controlling geminiviruses and that such resistance may be effective against all of them.

CONCLUSION

Mostly, model plants have been exploited to engineer resistance to geminiviruses with various genes. The multiple genetic resistance approaches should be tested in agronomic crops, bringing benefits to breeders and producers. The transgenic crops should exhibit an elevated level of resistance against multiple geminiviruses due to mixed infection and recombination, and such resistance should be durable in different agronomic conditions. Different strategies against geminivirus resistance should be exploited in agronomic crops, and best performing lines should be selected over several crop generations in natural conditions.

It was hypothesized that genes that pathogens used to parasitize their host can be used to engineer resistance in susceptible crops (Sanford & Johnson 1985). This hypothesis was soon tested and demonstrated for resistance against TMV (Abel et al 1986). Transgenic papaya cultivars resistant to *Papaya ring spot virus*, an RNA virus, were commercialized in 1998; that saved the papaya industry in Hawaii from devastation by PRSV (Gonsalves et al 1998). These RNA-resistant papayas are the third important transgenic crop grown commercially after herbicide and insect pests tolerance/resistant crops.

Despite the fact that the geminiviruses are one of the most destructive threats to agricultural crops in tropical and subtropical regions, no *Geminivirus*-resistant crops are grown commercially in farmers' fields. It is also a fact that despite the huge numbers of research articles describing various PDR and NPDR strategies for crop resistance to *Geminivirus* during the past two decades, until recently only one *Geminivirus*-resistant transgenic plant, a common bean, has been tested in field trials (Aragão & Faria 2009). It is unclear why transgenic resistance against geminiviruses has met with limited success despite the wide range of strategies that have been investigated. Homology-dependent strategies, such as RNAi, may not be effective due to the diversity of geminiviruses and their ability to rapidly evolve by mutation and recombination. The fact that their genomes encode multiple suppressors is also likely to play a part (Amin et al 2011).

The recombination rate of geminiviruses in a particular area is directly proportional to the diversity of these viruses in the crop plants grown there. In such crops, any single transgenic resistance mechanism would fail because of the overwhelming rate of recombination and evolution of new virus strains and species. This is the reason why a promising resistance strategy, 'RNAi,' has failed in the development of cotton resistant to geminiviruses in Asia and cassava in sub-Saharan Africa. The success of Aragão & Faria (2009) in maintaining transgenic resistance to BGMV in beans over several years of field trials is also attributable to a very low diversity of viruses affecting beans in that area.

In spite of that, the development of integrated multilayer resistance technology may offer a promising future to engineer resistance against geminiviruses. The foundation of any such strategy should be the best available host-plant resistance supplemented by at least two transgenic strategies of differing mechanisms of action—for example, an RNAi-based resistance mechanism supplemented with a non-PDR resistance such as GroEL or peptide aptamer. Genetic engineering offers a rapid and reliable tool to exploit *Geminivirus* resistance genes as compared to time-consuming traditional breeding used to introgress resistance genes. Traditionally bred cultivars offer medium resistance and often acquire undesirable traits. Genetic engineering can be used as an additional tool to implement *Geminivirus* resistance complementary to traditional breeding.

Currently, transgenic resistance is one of the most active research areas in plant protection studies. Identification of suitable target host genes, over or under expression of the selected genes, understanding the mechanisms implicated in the identification of pathogen effectors, and the mechanisms of maintenance of resistance over long periods of time, may be some of the future aspects of transgenic resistance strategies.

ACKNOWLEDGMENT

Financial support from The Research Council of Oman to AJK under grant number ORG/EBR/09/03 is gratefully acknowledged.

REFERENCES

Abel, P.P., Nelson, R.S., De, B., Hoffmann, N., Rogers, S.G., Fraley, R.T., Beachy, R.N., 1986. Delay of disease development in transgenic plants that express the tobacco mosaic virus coat protein gene. Science 232, 738–743.

Abhary, M.K., Anfoka, G.H., Nakhla, M.K., Maxwell, D.P., 2006. Post-transcriptional gene silencing in controlling viruses of the tomato yellow leaf curl virus complex. Archives of Virology 151, 2349–2363.

Adams, M.J., King, A.M.Q., Carstens, E.B., 2013. Ratification vote on taxonomic proposals to the International Committee on Taxonomy of Viruses (2013). Archives of Virology 158, 2023–2030.

Ai, T., Zhang, L., Gao, Z., Zhu, C.X., Guo, X., 2011. Highly efficient virus resistance mediated by artificial microRNAs that target the suppressor of PVX and PVY in plants. Plant Biology 13, 304–316.

Akad, F., Dotan, N., Czosnek, H., 2004. Trapping of *Tomato yellow leaf curl virus* (TYLCV) and other plant viruses with a GroEL homologue from the whitefly *Bemisia tabaci*. Archives of Virology 149, 1481–1497.

Akad, F., Eybishtz, D., Edelbaum, R., Gorovits, O., Dar-Issa, N., Czosnek, H., 2007. Making a friend from a foe: expressing a *GroEL* gene from the whitefly *Bemisia tabaci* in the phloem of tomato plants confers resistance to *Tomato yellow leaf curl virus*. Archives of Virology 152, 1323–1339.

Akbergenov, R., Si-Ammour, A., Blevins, T., Amin, I., Kutter, C., Vanderschuren, H., Zhang, P., Gruissem, W., Meins Jr., F., Hohn, T., Pooggin, M.M., 2006. Molecular characterization of geminivirus derived small RNAs in different plant species. Nucleic Acids Research 34, 462–471.

Amin, I., Ilyas, M., Mansoor, S., Briddon, R.W., Saeed, M., 2010. Role of DNA satellites in geminivirus disease complexes. In: Sharma, P. (Ed.), Emerging geminiviral diseases and their management, Nova Science Publishers, New York, pp. 209–234.

Amin, I., Hussain, K., Akbergenov, R., Yadav, J.S., Qazi, J., Mansoor, S., Hohn, T., Fauquet, C.M., Briddon, R.W., 2011. Suppressors of RNA silencing encoded by the components of the cotton leaf curl begomovirus-betasatellite complex. Molecular Plant–Microbe Interactions 24, 973–983.

Anderson, P.K., Cunningham, A.A., Patel, N.G., Morales, F.J., Epstein, P.R., Daszak, P., 2004. Emerging infectious diseases of plants: pathogen pollution, climate change and agrotechnology drivers. Trends in Ecology and Evolution 19, 535–544.

Antignus, Y., Lapidot, M., Hadar, D., Messika, Y., Cohen, S., 1998. Ultraviolet-absorbing screens serve as optical barriers to protect crops from virus and insect pests. Journal of Economic Entomology 91, 1401–1405.

Antignus, Y., Vunsh, R., Lachman, O., Pearlsman, M., Maslenin, L., Hananya, U., Rosner, A., 2004. Truncated Rep gene originated from *Tomato yellow leaf curl virus*-Israel (Mild) confers strain-specific resistance in transgenic tomato. Annals of Applied Biology 144, 39–44.

Aragão, F.J.L., Faria, J.C., 2009. First transgenic geminivirus-resistant plant in the field. Nature Biotechnology 27, 1086–1088.

Aragão, F.J.L., Ribeiro, S.G., Barros, L.M.G., Brasileiro, A.C.M., Maxwell, D.P., Rech, E.L., Faria, J.C., 1998. Transgenic beans (*Phaseolus vulgaris* L.) engineered to express viral antisense RNAs show delayed and attenuated symptoms to bean golden mosaic geminivirus. Molecular Breeding 4, 491–499.

Arguello-Astorga, G., Lopez-Ochoa, L., Kong, L.-J., Orozco, B.M., Settlage, S.B., Hanley-Bowdoin, L., 2004. A novel motif in geminivirus replication proteins interacts with the plant retinoblastoma-related protein. Journal of Virology 78, 4817–4826.

Asad, S., Haris, W.A., Bashir, A., Zafar, Y., Malik, K.A., Malik, N.N., Lichtenstein, C.P., 2003. Transgenic tobacco expressing geminiviral RNAs are resistant to the serious viral pathogen causing cotton leaf curl disease. Archives of Virology 148, 2341–2352.

Azzam, O., Diaz, O., Beaver, J. S., Gilbertson, R. L., Russell, D. R. Maxwell, D. P. (1996). Transgenic beans with the bean golden mosaic geminivirus coat protein gene are susceptible to virus infection. *Annual Report Bean Improvement Corporate*, *39*, 276–277.

Baines, I., Colas, P., 2006. Peptide aptamers as guides for small-molecule drug discovery. Drug Discovery Today 11, 334–341.

Bao, N., Lye, K.W., Barton, M.K., 2004. MicroRNA binding sites in *Arabidopsis* class III HD-ZIP mRNAs are required for methylation of the template chromosome. Developmental Cell 7, 653–662.

Bartel, D.P., 2004. MicroRNAs: genomics, biogenesis, mechanism, and function. Cell 116, 281–297.

Baulcombe, D., 2004. RNA silencing in plants. Nature 431, 356–363.

Baulcombe, D., 2005. RNA silencing. Trends in Biochemical Sciences 30, 290.

Baumberger, N., Baulcombe, D.C., 2005. *Arabidopsis* ARGONAUTE1 is an RNA Slicer that selectively recruits microRNAs and short interfering RNAs. Proceedings of the National Academy of Sciences of the USA 102, 11928–11933.

Bejarano, E.R., Lichtenstein, C.P., 1994. Expression of TGMV antisense RNA in transgenic tobacco inhibits replication of BCTV but not ACMV geminiviruses. Plant Molecular Biology 24, 241–248.

Bendahmane, M., Gronenborn, B., 1997. Engineering resistance against *Tomato yellow leaf curl virus* (TYLCV) using antisense RNA. Plant Molecular Biology 33, 351–357.

Bian, X.-Y., Rasheed, M.S., Seemanpillai, M.J., Rezaian, M.A., 2006. Analysis of silencing escape of *Tomato leaf curl virus*: an evaluation of the role of DNA methylation. Molecular Plant–Microbe Interactions 19, 614–624.

Blair, M.W., 1995. Occurrence of *Bean golden mosaic virus* in Florida. Plant Disease 79, 529.

Bosque-Perez, N.A., 2000. Eight decades of maize streak virus research. Virus Research 71, 107–121.

Briddon, R. W., Markhan, P. G. (1995). Geminiviridae. In F. A. Murphy et al (Ed.) Virus taxonomy: sixth report of the International Committee on Taxonomy of Viruses. New York: Springer-Verlag. pp. 158–165.

Briddon, R.W., Stanley, J., 2009. Geminiviridae. Encyclopedia of life sciences. Wiley, Chichester.

Briddon, R.W., Watts, J., Markham, P.G., Stanley, J., 1989. The coat protein of beet curly top virus is essential for infectivity. Virology 172, 628–633.

Briddon, R.W., Pinner, M.S., Stanley, J., Markham, P.G., 1990. Geminivirus coat protein replacement alters insect specificity. Virology 177, 85–94.

Briddon, R.W., Bull, S.E., Amin, I., Mansoor, S., Bedford, I.D., Dhawan, P., Rishi, N., Siwatch, S.S., Abdel-Salam, A.M., Markham, P.G., 2003. Diversity of DNA β: a satellite molecule associated with some monopartite begomoviruses. Virology 312, 106–121.

Briddon, R.W., Bull, S.E., Amin, I., Mansoor, S., Bedford, I.D., Rishi, N., Siwatch, S.S., Zafar, Y., Abdel-Salam, A.M., Markham, P.G., 2004. Diversity of DNA 1: a satellite-like molecule associated with monopartite begomovirus-DNA β complexes. Virology 324, 462–474.

Brunetti, A., Tavazza, M., Noris, E., Tavazza, R., Caciagli, P., Ancora, G., Crespi, S., Accotto, G.P., 1997. High expression of truncated viral Rep protein confers resistance to *Tomato yellow leaf curl virus* in transgenic tomato plants. Molecular Plant–Microbe Interactions 10, 571–579.

Brunetti, A., Tavazza, R., Noris, E., Lucioli, A., Accotto, G.P., Tavazza, M., 2001. Transgenically expressed T-Rep of *Tomato yellow leaf curl Sardinia virus* acts as a *trans*-dominant-negative mutant, inhibiting viral transcription and replication. Journal of Virology 75, 10573–10581.

Castillo, A.G., Kong, L.J., Hanley-Bowdoin, L., Bejarano, E.R., 2004. Interaction between a geminivirus replication protein and the plant sumoylation system. Journal of Virology 78, 2758–2769.

Chatterji, A., Beachy, R.N., Fauquet, C.M., 2001. Expression of the oligomerization domain of the replication-associated protein (Rep) of *Tomato leaf curl New Delhi virus* interferes with DNA accumulation of heterologous geminiviruses. Journal of Biological Chemistry 276, 25631–25638.

Chellappan, P., Masona, M.V., Vanitharani, R., Taylor, N.J., Fauquet, C.M., 2004. Broad spectrum resistance to ssDNA viruses associated with transgene-induced gene silencing in cassava. Plant Molecular Biology 56, 601–611.

Chellappan, P., Vanitharani, R., Fauquet, C.M., 2005. MicroRNA-binding viral protein interferes with *Arabidopsis* development. Proceedings of the National Academy of Sciences of the USA 102, 10381–10386.

Cohen, S., Antignus, Y., 1994. *Tomato yellow leaf curl virus*, a whitefly-borne geminivirus of tomatoes. Advances in Disease and Vector Research 10, 259–288.

Cohen, S., Lapidot, M., 2007. Appearance and expansion of TYLCV: a historical point of view. In: Czosnek, H. (Ed.), *Tomato yellow leaf curl virus* disease, Springer, Dordrecht, p. 3–12.

Colas, P., Cohen, B., Jessen, T., Grishina, I., McCoy, J., Brent, R., 1996. Genetic selection of peptide aptamers that recognize and inhibit cyclin-dependent kinase 2. Nature 380, 548–550.

Collin, S., Fernández-Lobato, M., Gooding, P.S., Mullineaux, P.M., Fenoll, C., 1996. The two nonstructural proteins from wheat dwarf virus involved in viral gene expression and replication are retinoblastoma-binding proteins. Virology 219, 324–329.

Covey, S.N., Al-Kaff, N.S., 2000. Plant DNA viruses and gene silencing. Plant Molecular Biology 43, 307–322.

Cui, X., Tao, X., Xie, Y., Fauquet, C.M., Zhou, X., 2004. A DNA ß associated with *Tomato yellow leaf curl China virus* is required for symptom induction. Journal of Virology 78, 13966–13974.

Czosnek, H., 2007. *Tomato yellow leaf curl disease*: management, molecular biology, breeding for resistance. Springer, Dordrecht ch 9.

Day, A.G., Bejarano, E.R., Buck, K.W., Burrell, M., Lichtenstein, C.P., 1991. Expression of an antisense viral gene in transgenic tobacco confers resistance to the DNA virus *Tomato golden mosaic virus*. Proceedings of the National Academy of Sciences of the USA 88, 6721–6725.

Duan, Y.-P., Powell, C.A., Webb, S.E., Purcifull, D.E., Hiebert, E., 1997. Geminivirus resistance in transgenic tobacco expressing mutated BC1 protein. Molecular Plant–Microbe Interactions 10, 617–623.

Duan, C.G., Wang, C.H., Fang, R.X., Guo, H.S., 2008. Artificial MicroRNAs highly accessible to targets confer efficient virus resistance in plants. Journal of Virology 82, 11084–11095.

Eamens, A.L., Agius, C., Smith, N.A., Waterhouse, P.M., Wang, M.B., 2011. Efficient silencing of endogenous microRNAs using artificial microRNAs in *Arabidopsis thaliana*. Molecular Plant 4, 157–170.

Edelbaum, D., Gorovits, R., Sasaki, S., Ikegami, M., Czosnek, H., 2009. Expressing a whitefly GroEL protein in *Nicotiana benthamiana* plants confers tolerance to *Tomato yellow leaf curl virus* and *Cucumber mosaic virus*, but not to *Grapevine virus A* or *Tobacco mosaic virus*. Archives of Virology 154, 399–407.

Egelkrout, E.M., Mariconti, L., Settlage, S.B., Cella, R., Robertson, D., Hanley-Bowdoin, L., 2002. Two E2F elements regulate the proliferating cell nuclear antigen promoter differently during leaf development. Plant Cell 14, 3225–3236.

Elmer, J.S., Brand, L., Sunter, G., Gardiner, W.E., Bisaro, D.M., Rogers, S.G., 1988. Genetic analysis of the tomato golden mosaic virus. II. The product of the AL1 coding sequence is required for replication. Nucleic Acids Research 16, 7043–7060.

Escobar, M.A., Civerolo, E.L., Summerfelt, K.R., Dandekar, A.M., 2001. RNAi-mediated oncogene silencing confers resistance to crown gall tumorigenesis. Proceedings of the National Academy of Sciences of the USA 98, 13437–13442.

Eybishtz, A., Peretz, Y., Sade, D., Akad, F., Czosnek, H., 2009. Silencing of a single gene in tomato plants resistant to *Tomato yellow leaf curl virus* renders them susceptible to the virus. Plant Molecular Biology 71, 157–171.

Fahim, M., Larkin, P.J., 2013. Designing effective amiRNA and multimeric amiRNA against plant viruses. Methods in Molecular Biology 942, 357–377.

Faria, J.C., Maxwell, D.P., 1999. Variability in geminivirus isolates associated with *Phaseolus* spp in Brazil. Phytopathology 89, 262–268.

Fauquet, C.M., Maxwell, D.P., Gronenborn, B., Stanley, J., 2000. Revised proposal for naming geminiviruses. Archives of Virology 8, 1743–1761.

Fauquet, C.M., Briddon, R.W., Brown, J.K., Moriones, E., Stanley, J., Zerbini, M., Zhou, X., 2008. Geminivirus strain demarcation and nomenclature. Archives of Virology 153, 783–821.

Fire, A., Xu, S., Montgomery, M.K., Kostas, S.A., Driver, S.E., Mello, C.C., 1998. Potent and specific genetic interference by double-stranded RNA in *Caenorhabditis elegans*. Nature 391, 806–811.

Frischmuth, S., Kleinow, T., Aberle, H.-J., Wege, C., Hülser, D., Jeske, H., 2004. Yeast two-hybrid systems confirm the membrane-association and oligomerization of BC1 but do not detect an interaction of the movement proteins BC1 and BV1 of *Abutilon* mosaic geminivirus. Archives of Virology 149, 2349–2364.

Frischmuth, T., Stanley, J., 1993. Strategies for the control of geminivirus diseases. Seminars in Virology 4, 329–337.

Frischmuth, T., Stanley, J., 1998. Recombination between viral DNA and the transgenic coat protein gene of *African cassava mosaic geminivirus*. Journal of General Virology 79, 1265–1271.

Frischmuth, T., Engel, M., Jeske, H., 1997. *Beet curly top virus* DI DNA-mediated resistance is linked to its size. Molecular Breeding 3, 213–217.

Geyer, C.R., Colman-Lerner, A., Brent, R., 1999. 'Mutagenesis' by peptide aptamers identifies genetic network members and pathway connections. Proceedings of the National Academy of Sciences of the USA 96, 8567–8572.

Gonsalves, D., Ferreira, S., Manshardt, R., Fitch, M., Slightom, J., 1998. Transgenic virus resistant papaya: new hope for control of papaya ringspot virus in Hawaii. APSnet Feature, American Pythopathological Society, www.apsnet.org/education/feature/papaya.

Goodman, R.M., 1977. Single-stranded DNA genome in a whitefly-transmitted plant virus. Virology 83, 171–179.

Griffiths-Jones, S., Grocock, R.J., Van Dongen, S., Bateman, A., Enright, A.J., 2006. miRBase: microRNA sequences, targets and gene nomenclature. Nucleic Acids Research 34, D140.

Griffiths-Jones, S., Saini, H.K., Van Dongen, S., Enright, A.J., 2008. miRBase: tools for microRNA genomics. Nucleic Acids Research 36, D154.

Gutierrez, C., 2000. Geminiviruses and the plant cell cycle. Plant Molecular Biology 43, 763–772.

Hanley-Bowdoin, L., Settlage, S.B., Orozco, B.M., Nagar, S., Robertson, D., 1999. Geminviruses: models for plant DNA replication, transcription, and cell cycle regulation. Critical Reviews in Plant Sciences 18, 71–106.

Hanley-Bowdoin, L., Settlage, S.B., Orozco, B.M., Nagar, S., Robertson, D., 2000. Geminiviruses: models for plant DNA replication, transcription, and cell cycle regulation. Critical Review in Biochemistry and Molecular Biology 35, 105–140.

Hannon, G.J., 2002. RNA interference. Nature 418, 244–251.

Hanson, S.F., Maxwell, D.P., 1999. *Trans*-dominant inhibition of geminiviral DNA replication by *Bean golden mosaic geminivirus rep* gene mutants. Phytopathology 89, 480–486.

Harrison, B., Barker, H., Bock, K.R., Guthrie, E.J., Meredith, G., Atkinson, M., 1977. Plant-viruses with circular single-stranded DNA. Nature 270, 760–762.

He, X.F., Fang, Y.Y., Feng, L., Guo, H.S., 2008. Characterization of conserved and novel micro-RNAs and their targets, including a TuMV-induced TIR-NBS-LRR class R gene-derived novel miRNA in Brassica. FEBS Letters 582, 2445–2452.

Heyraud-Nitschke, F., Schumacher, S., Laufs, J., Schaefer, S., Schell, J., Gronenborn, B., 1995. Determination of the origin cleavage and joining domain of geminivirus Rep proteins. Nucleic Acids Research 23, 910–916.

Hong, H., Saunders, K., Hartley, M.R., Stanley, J., 1996. Resistance to geminivirus infection by virus-induced expression of dianthin in transgenic plants. Virology 220, 119–127.

Hong, J., Stanley, J., 1996. Virus resistance in *Nicotiana benthamiana* conferred by *African cassava mosaic virus* replication-associated protein (AC1) transgene. Molecular Plant–Microbe Interactions 4, 219–225.

Hoppe-Seyler, F., Crnkovic-Mertens, I., Tomai, E., Butz, K., 2004. Peptide aptamers: specific inhibitors of protein function. Current Molecular Medicine 4, 529–538.

Hou, Y.M., Sanders, R., Ursin, V.M., Gilbertson, R.L., 2000. Transgenic plants expressing geminivirus movement proteins: abnormal phenotypes and delayed infection by *Tomato mottle virus* in transgenic tomatoes expressing the *Bean dwarf mosaic virus* BV1 or BC1 proteins. Molecular Plant–Microbe Interactions 13, 297–308.

Hussain, M., Mansoor, S., Iram, S., Fatima, A.N., Zafar, Y., 2005. The nuclear shuttle protein of *Tomato leaf curl New Delhi virus* is a pathogenicity determinant. Journal of Virology 79, 4434–4439.

Hussain, M., Mansoor, S., Iram, S., Zafar, Y., Briddon, R.W., 2007. The hypersensitive response to *Tomato leaf curl New Delhi virus* nuclear shuttle protein is inhibited by transcriptional activator protein. Molecular Plant–Microbe Interactions 20, 1581–1588.

Idris, A.M., Shahid, M.S., Briddon Khan, A.J., Zhu, J.K., Brown, J.K., 2011. An unusual alphasatel-lite associated with monopartite begomoviruses attenuates symptoms and reduces betasatellite accumulation. Journal of General Virology 92, 706–717.

Ingham, D.J., Pascal, E., Lazarowitz, S.G., 1995. Both bipartite geminivirus movement proteins define viral host range, but only BL1 determines viral pathogenicity. Virology 207, 191–204.

Jeske, H., 2009. Geminiviruses. Current Topics in Microbiology and Immunology 331, 185–226.

Katiyar-Agarwal, S., Jin, H., 2010. Role of small RNAs in host-microbe interactions. Annual Review of Phytopathology 48, 225–246.

Khan, A.J., Idris, A.M., Al-Saady, N.A., Al-Mahruki, M.A., Al-Subhi, A.M., Brown, J.K., 2008. A divergent isolate of *Tomato yellow leaf curl virus* from Oman with an associated DNAβ satellite: an evolutionary link between Asian and the Middle Eastern virus–satellite complexes. Virus Genes 36, 169–176.

Khan, A.J., Akhtar, S., Al-Shihi, A.A., Al-Hinai, F.M., Briddon, R.W., 2012a. Identification of *Cotton leaf curl Gezira virus* in papaya in Oman. Plant Disease 96, 1704.

Khan, A.J., Akhtar, S., Briddon, R.W., Ammara, U., Al-Matrushi, A.M., Mansoor, S., 2012b. Complete nucleotide sequence of watermelon chlorotic stunt virus originating from Oman. Viruses 4, 1169–1181.

Khan, A.J., Al-Matrushi, A.M., Fauquet, C.M., Briddon, R.W., 2013. The introduction of *East African cassava mosaic Zanzibar virus* to Oman harks back to 'Zanzibar, the capital of Oman', Virus Genes 46, 195–198.

Kong, L.J., Orozco, B.M., Roe, J.L., Nagar, S., Ou, S., Feiler, H.S., Durfee, T., Miller, A.B., Gruissem, W., Robertson, D., Hanley-Bowdoin, L., 2000. A geminivirus replication protein interacts with the retinoblastoma protein through a novel domain to determine symptoms and tissue specificity of infection in plants. EMBO Journal 19, 3485–3495.

Koshino-Kimura, Y., Takenaka, K., Domoto, F., Ohashi, M., Miyazaki, T., Aoyama, Y., Sera, T., 2009. Construction of plants resistant to TYLCV by using artificial zinc-finger proteins. Nucleic Acids Symposium Series 53, 281–282.

Kunik, T., Salomon, R., Zamir, D., Navot, N., Zeidan, M., Michelson, I., Gafni, Y., Czosnek, H., 1994. Transgenic tomato plants expressing the *Tomato yellow leaf curl virus* capsid protein are resistant to the virus. Biotechnology 12, 500–504.

Lapidot, M., Friedmann, M., 2002. Breeding for resistance to whitefly-transmitted geminiviruses. Annals of Applied Biology 140, 109–127.

Lazarowitz, S.G., Wu, L.C., Rogers, S.G., Elmer, J.S., 1992. Sequence-specific interaction with the viral AL1 protein identifies a geminivirus DNA replication origin. Plant Cell 4, 799–809.

Lecoq, H., 1998. Control of plant virus diseases by cross protection. In: Hadidi, A. (Ed.), Plant virus disease control, APS Press, Minnesota, pp. 33–40.

Legg, J.P., Fauquet, C.M., 2004. *Cassava mosaic geminiviruses* in Africa. Plant Molecular Biology 56, 585–599.

Liu, S., Bedford, I.D., Briddon, R.W., Markham, P.G., 1997. Efficient whitefly transmission of *African cassava mosaic geminivirus* requires sequences from both genomic components. Journal of General Virology 78, 1791–1794.

Lopez-Ochoa, L., Ramirez-Prado, J., Hanley-Bowdoin, L., 2006. Peptide aptamers that bind to a geminivirus replication protein interfere with viral replication in plant cells. Journal of Virology 80, 5841–5853.

Lopez-Ochoa, L., Hanley-Bowdoin, L. (2007). Use of peptide aptamers for broad-based geminivirus disease resistance. Conference abstract, American Society of Plant Biologists Plant Biology and Botany Joint Congress, Chicago, USA. http://abstracts.aspb. org/pb2007/public/M04/M04003.html.

Lopez-Ochoa, L., Nash, T.E., Ramirez-Prado, J., Hanley-Bowdoin, L., 2009. Isolation of peptide aptamers to target protein function. Methods in Molecular Biology 535, 333–360.

Lucioli, A., Noris, E., Brunetti, A., Tavazza, R., Ruzza, V., Castillo, A.G., Bejarano, E.R., Accotto, G.P., Tavazza, M., 2003. *Tomato yellow leaf curl Sardinia virus* rep-derived resistance to homologous and heterologous geminiviruses occurs by different mechanisms and is overcome if virus mediated transgene silencing is activated. Journal of Virology 77, 6785–6798.

Luque, A., Sanz-Burgos, A.P., Ramirez-Parra, E., M.M., C., Gutierrez, C., 2002. Interaction of geminivirus Rep protein with replication factor C and its potential role during geminivirus DNA replication. Virology 302, 83–94.

Malathi, V.G., Surendranath, B., Naghma, A., Roy, A., 2005. Adaptation to new hosts shown by the cloned components of mungbean yellow mosaic India virus causing cowpea golden mosaic in northern India. Canadian Journal of Plant Pathology 27, 439–447.

Mansoor, S., Khan, S.H., Bashir, A., Saeed, M., Zafar, Y., Malik, K.A., Briddon, R.W., Stanley, J., Markham, P.G., 1999. Identification of a novel circular single-stranded DNA associated with cotton leaf curl disease in Pakistan. Virology 259, 190–199.

Mansoor, S., Briddon, R.W., Bull, S.E., Bedford, I.D., Bashir, A., Hussain, M., Zafar, M.Y., Malik, K.A., Fauquet, C.M., Markham, P.G., 2003. *Cotton leaf curl disease* is associated with multiple monopartite begomoviruses supported by single DNA β. Archives of Virology 148, 1969–1986.

Mansoor, S., Zafar, Y., Briddon, R.W., 2006. Geminivirus disease complexes: the threat is spreading. Trends in Plant Science 11, 209–212.

Mansoor, S., Amin, I., Briddon, R.W., 2011. Geminiviral diseases in cotton. In: Oosterhuis, D.M. (Ed.), Stress physiology in cotton, The Cotton Foundation, Cordova, USA, pp. 125–148.

Moffat, A., 1999. Geminiviruses emerge as serious crop threat. Science 286, 1835.

Morales, F.J., 2001. Conventional breeding for resistance to *Bemisia tabaci* transmitted geminiviruses. Crop Protection 20, 825–834.

Morales, F.J., Jones, P.G., 2004. The ecology and epidemiology of whitefly-transmitted viruses in Latin America. Virus Research 100, 57–65.

Morin, S., Ghanim, M., Zeidan, M., Czosnek, H., Verbeek, M., van den Heuvel, J.F.J.M., 1999. A GroEL homologue from endosymbiotic bacteria of the whitefly *Bemisia tabaci* is implicated in the circulative transmission of *Tomato yellow leaf curl virus*. Virology 256, 75–84.

Morin, S., Ghanim, M., Sobol, I., Czosnek, H., 2000. The GroEL protein of the whitefly *Bemisia tabaci* interacts with the coat protein of trasmissible and nontransmissible begomoviruses in the yeast two-hybrid system. Virology 276, 404–416.

Moriones, E., Navas-Castillo, J., 2000. *Tomato yellow leaf curl virus*, an emerging virus complex causing epidemics worldwide. Virus Research 71, 123–134.

Moriones, E., Navas-Castillo, J., 2010. *Tomato yellow leaf curl disease* epidemics. In: Stansly, P.A., Naranjo, S.E. (Eds.), Bemisia: bionomics and managements of a global pest, Springer, Dordrecht, pp. 259–282.

Morris-Krisinich, B.A.M., Mullineaux, P.M., Donson, J., Boulton, M.I., Markham, P.G., Short, M.N., Davies, J.W., 1985. Bidirecional transcription of maize streak virus DNA and identification of the coat protein gene. Nucleic Acid Research 13, 7237–7256.

Mubin, M., Mansoor, S., Hussain, M., Zafar, Y., 2007. Silencing of the AV2 gene by antisense RNA protects transgenic plants against a bipartite begomovirus. Virology Journal 4, 10.

Napoli, C., Lemieux, C., Jorgensen, R., 1990. Introduction of a chimeric chalcone synthase gene into petunia results in reversible co-suppression of homologous genes in trans. Plant Cell 2, 279–289.

Narayanan, S., Surendranath, K., Bora, N., Surolia, A., Karande, A.A., 2005. Ribosome inactivating proteins and apoptosis. FEBS Letters 579, 1324–1331.

Nawaz-Ul-Rehman, M.S., Nahid, N., Mansoor, S., Briddon, R.W., Fauquet, C.M., 2010. Post-transcriptional gene silencing suppressor activity of two non-pathogenic alphasatellites associated with a begomovirus. Virology 405, 300–308.

Niu, Q.W., Lin, S.S., Reyes, J.L., Chen, K.C., Wu, H.W., Yeh, S.D., Chua, N.H., 2006. Expression of artificial microRNAs in transgenic *Arabidopsis thaliana* confers virus resistance. Nature Biotechnology 24, 1420–1428.

Noris, E., Accotto, G.P., Tavazza, R., Brunetti, A., Crespi, S., Tavazza, M., 1996. Resistance to tomato yellow leaf curl geminivirus in *Nicotiana benthamiana* plants transformed with a truncated viral C1 gene. Virology 224, 130–138.

Noris, E., Vaira, A.M., Caciagli, P., Masenga, V., Gronenborn, B., Accotto, G.P., 1998. Amino acids in the capsid protein of tomato yellow leaf curl virus that are crucial for systemic infection, particle formation, and insect transmission. Journal of Virology 72, 10050–10057.

Norman, T.C., Smith, D.L., Sorger, P.K., IMG, S.R.C., O'Rourke, S.M., Hughes, T.R., Roberts, C.J., Friend, S.H., Fields, S., Murray, A.W., 1999. Genetic selection of peptide inhibitors of biological pathways. Science 285, 591–595.

Owor, B., Legg, J.P., Okao-Okuja, G., Obonyo, R., Kyamanywa, S., Ogenga-Latigo, M.W., 2004. Field studies of cross protection with cassava mosaic geminiviruses in Uganda. Journal of Plant Pathology 152, 243–249.

Padidam, M., Sawyer, S., Fauquet, C.M., 1999. Possible emergence of new geminiviruses by frequent recombination. Virology 265, 218–225.

Pakniat-Jahromy, A., Behjatnia, S.A., Dry, I.B., Izadpanah, K., Rezaian, M.A., 2010. A new strategy for generating geminivirus resistant plants using a DNA betasatellite/split barnase construct. Journal of Virological Methods 170, 57–66.

Patil, B.L., Fauquet, C.M., 2009. *Cassava mosaic geminiviruses*: actual knowledge and perspectives. Molecular Plant Pathology 10, 685–701.

Patil, B.L., Rajasubramaniam, S., Bagchi, C., Dasgupta, I., 2005. Both *Indian cassava mosaic virus* and *Sri Lankan cassava mosaic virus* are found in India and exhibit high variability as assessed by PCR-RFLP. Archives of Virology 150, 389–397.

Polston, J.E., Anderson, P., 1997. The emergence of whitefly-transmitted geminiviruses in tomato in the western hemisphere. Plant Disease 81, 1358–1369.

Polston, J.E., Hiebert, E., 2001. Engineered resistance to tomato geminiviruses. In: Gilreath, P. (Ed.), Proceedings of the Florida Tomato Institute, University of Florida, pp. 19–22.

Pooggin, M., Hohn, T., 2004. Fighting geminiviruses by RNAi and vice versa. Plant Molecular Biology 55, 149–152.

Pooggin, M., Shivaprasad, P.V., Veluthambi, K., Hohn, T., 2003. RNAi targeting of DNA virus in plants. Nature Biotechnology 21, 131–132.

Pooma, W., Gillette, W.K., Jeffrey, J.L., Petty, I.T., 1996. Host and viral factors determine the dispensability of coat protein for bipartite geminivirus systemic movement. Virology 218, 264–268.

Praveen, S., Kushwaha, C.M., Mishra, A.K., Singh, V., Jain, R.K., Varma, A., 2005. Engineering tomato for resistance to tomato leaf curl disease using viral *Rep* gene sequences. Plant Cell, Tissue and Organ Culture 83, 311.

Qazi, J., Amin, I., Mansoor, S., Iqbal, M.J., Briddon, R.W., 2007. Contribution of the satellite encoded gene βC1 to cotton leaf curl disease symptoms. Virus Research 128, 135–139.

Rojas, M., Jiang, H., Salati, R., Xoconostle-Cazares, B., Sudarshana, M.R., Lucas, W.J., Gilbertson, R.L., 2001. Functional analysis of proteins involved in movement of the monopartite begomovirus, *Tomato yellow leaf curl virus*. Virology 291, 110–125.

Rudolph, C., Schreier, P.H., Uhrig, J.F., 2003. Peptide-mediated broad-spectrum plant resistance to tospoviruses. Proceedings of the National Academy of Sciences of the USA 100, 4429–4434.

Rybicki, E.P., 1994. A phylogenetic and evolutionary justification for three genera of *Geminiviridae*. Archives of Virology 139, 49–77.

Saeed, M., Behjatania, S.A.A., Mansoor, S., Zafar, Y., Hasnain, S., Rezaian, M.A., 2005. A single complementary-sense transcript of a geminiviral DNA β satellite is determinant of pathogenicity. Molecular Plant–Microbe Interactions 18, 7–14.

Saeed, M., Zafar, Y., Randles, J.W., Rezaian, M.A., 2007. A monopartite begomovirus-associated DNA β satellite substitutes for the DNA B of a bipartite begomovirus to permit systemic infection. Journal of General Virology 88, 2881–2889.

Safarnejad, M., Fischer, R., Commandeur, U., 2009. Recombinant antibody-mediated resistance against *Tomato yellow leaf curl virus* in *Nicotiana benthamiana*. Archives of Virology 154, 457–467.

Sanford, J.C., Johnson, S.A., 1985. The concept of parasite-derived resistance: deriving resistance genes from the parasites own genome. Journal of Theoretical Biology 115, 395–405.

Sangaré, A., Deng, D., Fauquet, C.M., Beachy, R.N., 1999. Resistance to *African cassava mosaic virus* conferred by a mutant of the putative NTP-binding domain of the Rep Gene (AC1) in *Nicotiana benthamiana*. Molecular Breeding 5, 95–102.

Saunders, K., Norman, A., Gucciardo, S., Stanley, J., 2004. The DNA beta satellite component associated with ageratum yellow vein disease encodes an essential pathogenicity protein (βC1). Virology 324, 37–47.

Seal, S.E., van den Bosch, F., Jeger, M.J., 2006. Factors influencing begomovirus evolution and their increasing global significance: implications for sustainable control. Critical Reviews in Plant Sciences 25, 23–46.

Sera, T., 2005. Inhibition of virus DNA replication by artificial zinc finger proteins. Journal of Virology 79, 2614–2619.

Sera, T., Uranga, C., 2002. Rational design of artificial zinc finger proteins using a non degenerate recognition code table. Biochemistry 41, 7074–7081.

Sharma, P., Ikegami, M., 2008. RNA-silencing suppressors of geminiviruses. Journal of General Plant Pathology 74, 189–202.

Shepherd, D.N., Martin, D.P., McGivern, D.R., Boulton, M.I., Thomson, J.A., Rybicki, E.P., 2005. A three-nucleotide mutation altering the *Maize streak virus* Rep pRBR-interaction motif reduces symptom severity in maize and partially reverts at high frequency without restoring pRBR-Rep binding. Journal of General Virology 86, 803–813.

Shepherd, D.N., Mangwende, T., Martin, D.P., Bezuidenhout, M., Kloppers, F.J., Carolissen, C.H., Monjane, A.L., Rybicki, E.P., Thomson, J.A., 2007a. *Maize streak virus*-resistant transgenic maize: a first for Africa. Plant Biotechnology Journal 5, 759–767.

Shepherd, D.N., Mangwende, T., Martin, D.P., Bezuidenhout, M., Thomson, J.A., Rybicki, E.P., 2007b. Inhibition of maize streak virus (MSV) replication by transient and transgenic expression of MSV replication-associated protein mutants. Journal of General Virology 88, 325–336.

Shepherd, D.N., Martin, D.P., Thomson, J.A., 2009. Transgenic strategies for developing crops resistant to geminiviruses. Plant Science 176, 1–11.

Shivaprasad, P.V., Thillaichidambaram, P., Balaji, V., Veluthambi, K., 2006. Expression of full-length and truncated Rep genes from *Mungbean yellow mosaic virus-Vigna* inhibits viral replication in transgenic tobacco. Virus Genes 33, 365–374.

Sinisterra, X.H., Polston, N.J.E., Abouzid, A.M., Hiebert, E., 1999. Tobacco plants transformed with a modified coat protein of tomato mottle begomovirus show resistance to virus infection. Phytopathology 89, 701–706.

Stanley, J., Markham, P.G., Callis, R.J., Pinner, M.S., 1986. The nucleotide sequence of an infectious clone of the geminivirus beet curly top virus. EMBO Journal 5, 1761–1767.

Stanley, J., Frischmuth, T., Ellwood, S., 1990. Defective viral DNA ameliorates symptoms of gemi-nivirus infection in transgenic plants. Proceedings of the National Academy of Sciences of the USA 87, 6291–6295.

Stanley, J., Bisaro, D. M., Briddon, R. W., Brown, J. K., Fauquet, C. M., Harrison, B. D., Rybicki, E. P., Stenger, D. C. (2005). Geminiviridae. In: Fauquet CM et al (eds) VIIIth Report of the International Committee on Taxonomy of Viruses. Virus Taxonomy. Academic Press: London, 301–326.

Stenger, D.C., Stevenson, M.C., Hormuzdi, S.G., Bisaro, D.M., 1992. A number of subgenomic DNAs are produced following agroinoculation of plants with beet curly top virus. Journal of General Virology 73, 237–242.

Sunter, G., Hartitz, M.D., Hormuzdi, S.G., Brough, C.L., Bisaro, D.M., 1990. Genetic analysis of tomato golden mosaic virus: ORF AL2 is required for coat protein accumulation while ORF AL3 is necessary for efficient DNA replication. Virology 179, 69–77.

Sunter, G., Sunter, J.L., Bisaro, D.M., 2001. Plants expressing tomato golden mosaic virus AL2 or beet curly top virus L2 transgenes show enhanced susceptibility to infection by DNA and RNA viruses. Virology 285, 59–70.

Takenaka, K., Koshino-Kimura, Y., Aoyama, Y., Sera, T., 2007. Inhibition of tomato yellow leaf curl virus replication by artificial zinc-finger proteins. Nucleic Acids Symposium Series 51, 429–430.

Thresh, J.M., Otim-Nape, G.W., 1994. Strategies for controlling African cassava mosaic geminivirus. In: Harris, K.F. (Ed.), Advances in disease vector research, Springer-Verlag, New York, pp. 215 236.

Tomai, E., Butz, K., Lohrey, C., von Weizsacker, F., Zentgraf, H., Hoppe-Seyler, F., 2006. Peptide aptamer-mediated inhibition of target proteins by sequestration into aggresomes. Journal of Biological Chemistry 281, 21345–21352.

Trinks, D., Rajeswaran, R., Shivaprasad, P.V., Akbergenov, R., Oakeley, E.J., Veluthambi, K., Hohn, T., Pooggin, M.M., 2005. Suppression of RNA silencing by a geminivirus nuclear protein, AC2, correlates with transactivation of host genes. Journal of Virology 79, 2517–2527.

Turnage, M.A., Muangsan, N., Peele, C.G., Robertson, D., 2002. Geminivirus-based vectors for gene silencing in Arabidopsis. Plant Journal 30, 107–114.

Uhrig, J.F., 2003. Response to Prins: broad virus resistance in transgenic plants. Trends in Biotechnology 21, 376–377.

Ushasri, C., Sunil, K.M., Deb, J.K., 2007. Intervention of geminiviral replication in yeast by ribozyme mediated downregulation of its Rep protein. FEBS Letters 581, 2675.

Vanderschuren, H., Stupak, M., Futterer, J., Gruissem, W., Zhang, P., 2007. Engineering resistance to geminiviruses–review and perspectives. Plant Biotechnology Journal 5, 207–220.

Vanderschuren, H., Moreno, I., Anjanappa, R.B., Zainuddin, I.M., Gruissem, W., 2012. Exploiting the combination of natural and genetically engineered resistance to Cassava mosaic and Cassava brown streak viruses impacting cassava production in Africa. PLoS ONE 7, e45277.

van Duynhoven, J.P.M., Folkers, P.J.M., Stassen, A.P.M., Harmsen, B.J.M., Konings, R.N.H., Hilbers, C.W., 1990. Structure of the DNA binding wing of the gene-V encoded single-stranded DNA binding protein of the filamentous bacteriophage M13. FEBS Letters 26, 1–4.

Vanitharani, R., Chellappan, P., Fauquet, C.M., 2003. Short interfering RNA-mediated interference of gene expression and viral DNA accumulation in cultured plant cells. Proceedings of the National Academy of Sciences of the USA 100, 9632–9636.

Vanitharani, R., Chellappan, P., Pita, J.S., Fauquet, C.M., 2004. Differential roles of AC2 and AC4 of cassava geminiviruses in mediating synergism and suppression of posttranscriptional gene silencing. Journal of Virology 78, 9487–9498.

Vanitharani, R., Chellappan, P., Fauquet, C.M., 2005. Geminiviruses and RNA silencing. Trends in Plant Science 10, 144–151.

van Rij, R.P., Andino, R., 2006. The silent treatment: RNAi as a defense against virus infection in mammals. Trends in Biotechnology 24, 186–193.

Varma, A., Malathi, V.G., 2003. Emerging geminivirus problems: a serious threat to crop production. Annals of Applied Biology 142, 145–164.

von Arnim, A., Stanley, J., 1992. Inhibition of *African cassava mosaic virus* systemic infection by a movement protein from the related geminivirus tomato golden mosaic virus. Virology 187, 555–564.

Vrba, L., Munoz-Rodriguez, J.L., Stampfer, M.R., Futscher, B.W., 2013. miRNA gene promoters are frequent targets of aberrant DNA methylation in human breast cancer. PLoS One 8, e54398.

Yang, X., Xie, Y., Raja, P., Li, S., Wolf, J.N., Shen, Q., Bisaro, D.M., 2011. Suppression of methylation-mediated transcriptional gene silencing by βC1-SAHH protein interaction during geminivirus-betasatellite infection. PLoS Pathogens 7, e1002329.

Yu, B., Bi, L., Zheng, B., Ji, L., Chevalier, D., Agarwal, M., Ramachandran, V., Li, W., Lagrange, T., Walker, J.C., 2008. The FHA domain proteins DAWDLE in *Arabidopsis* and SNIP1 in humans act in small RNA biogenesis. Proceedings of the National Academy of Sciences of the USA 105, 10073.

Zhang, B., Pan, X., Cannon, C.H., Cobb, G.P., Anderson, T.A., 2006. Conservation and divergence of plant microRNA genes. Plant Journal 46, 243–259.

Zhang, H.S., Dean, D.C., 2001. Rb-mediated chromatin structure regulation and transcriptional repression. Oncogene 20, 3134–3138.

Zhang, P., Vanderschuren, H., Futterer, J., Gruissem, W., 2005. Resistance to cassava mosaic disease in transgenic cassava expressing antisense RNAs targeting virus replication genes. Plant Biotechnology Journal 3, 385–397.

Cauliflower mosaic virus (CaMV) upregulates translation reinitiation of its pregenomic polycistronic 35S RNA via interaction with the cell's translation machinery

Mikhail Schepetilnikov and Lyubov Ryabova

Institut de Biologie Moléculaire des Plantes du CNRS, Université de Strasbourg, France

INTRODUCTION

Plant viruses have developed various strategies to express their genomes, including multiple forms of polycistronic translation, such as leaky scanning, frameshifting, read-through, and activated reinitiation/transactivation of polycistronic translation. The latter is the main focus of this chapter.

There is no evidence that plant viruses have evolved mechanisms that specifically shut down translation of cellular mRNAs. However, the use of non conventional translation mechanisms that compete with cellular mRNAs for the cell's translation machinery is a very common strategy among plant viruses. Such mechanisms include ribosomal shunting (Ryabova et al 2006) and internal initiation (Dreher & Miller 2006, albeit with only rare examples in plants) to avoid scanning through structural elements of the RNA leader that are essential for other aspects of the virus cell cycle. The use of host factors—elements of the cell translational machinery—is especially common. Cap-binding factors (eIF4E/eIFiso4E/eIF4G/eIFiso4G) are essential for several plant viruses to infect cells (Wang & Krishnaswamy 2012). Moreover, the targeting of cap-binding factors to mRNA to augment initiation efficiencies can be achieved not only via the cap structure but also via particular elements within 3′-untranslated regions of viral mRNAs (Miller et al 2007). However, caulimoviruses not only interact with the host translation machinery to exploit its properties but are able to modify its behavior via interaction with multiple host factors. Before introducing viral-activated mechanisms of reinitiation, we first describe the canonical translation initiation and reinitiation pathways, the host translation machinery, and the TOR signaling pathway.

CONTROL OF CELLULAR INITIATION AND REINITIATION OF TRANSLATION

Cap-dependent translation initiation in eukaryotes

The main translation initiation mechanism in eukaryotes (i.e., cap-dependent translation initiation) has many steps requiring numerous canonical translation initiation factors (eIFs; Hinnebusch & Lorsch 2012). During the first step, the 40S ribosomal subunit (40S) is loaded with a set of factors to form the 43S pre-initiation complex (43S PIC), which comprises the 40S ribosomal subunit, translation initiation factors (eIFs) eIF3, eIF1, eIF1A, eIF5, and the ternary complex (TC; eIF2xGTPxMet-tRNAi^{Met}). Cap-dependent initiation depends on loading of eIF4F—a complex of eIF4G, eIF4E, and eIF4A— onto the mRNA cap structure via eIF4E. eIF4A is a helicase that unwinds the mRNA leader together with eIF4B to trigger ribosomal scanning. eIF4G interacts with eIF4E and eIF4A, eIF4B, and with eIF3 and polyA-binding protein (PABP). The 43S PIC is brought into contact with the capped 5′-end of the mRNA via interaction between eIF4G, eIF4B, and eIF3, resulting in formation of the 48S PIC (Gallie 2002). Note that eIF3 is composed of 13 distinct subunits in humans and plants and orchestrates assembly of the 43S preinitiation complex (43S PIC) on mRNA (Browning et al 2001, Hinnebusch 2006). The 43S PIC scans along the mRNA until it encounters a suitable AUG start codon. After codon–anticodon base pairing at the first AUG in a favorable initiation context, eIF5 stimulates GTP hydrolysis and the release of eIF2-GDP from 40S ribosomes. A second GTPase, eIF5B, facilitates 60S subunit joining (Pestova et al 2000) to form the functional 80S ribosome to complete initiation of translation.

Canonical translation initiation factors and reinitiation-promoting factors (RPFs)

Translation reinitiation is strictly limited in eukaryotes. The efficiency of reinitiation depends on structural features of the mRNA, such as the size of the upstream ORF, and on the availability of initiation factors (Hinnebusch 1997, Kozak 2001). Generally, the length of the ORF, or rather the time required for its translation, is the main parameter strongly limiting reinitiation. The need for rapid uORF translation may be related to problems with *de novo* recruitment of initiator tRNA (Met-tRNAi^{Met}) within the ternary complex and the 60S ribosomal subunit required to accomplish the reinitiation event. eIF2 is the crucial translation initiation factor that brings TC to 40S and thus obviously needs to be reloaded for reinitiation. eIF3 is the main factor implicated in recruitment of the eIF2-GTP-Met-tRNAi ternary complex to the 40S ribosomal subunit and thus has been shown to increase reinitiation efficiency (Hinnebusch 1997, Park et al 2001). If factors necessary for reinitiation are shed from the ribosome as it translates longer ORFs, the time needed for ribosome loading with eIF2-GTP and Met-tRNA would be extended. This hypothesis is consistent with the demonstration that increasing the distance between long ORFs to beyond 200 nts can trigger low but detectable translation of the second ORF (Fütterer & Hohn 1991).

According to an existing model, eIFs required for resumption of scanning and/or recruitment of TC and 60S remain loosely associated with the elongating ribosome for a short time of a few elongation cycles in order to promote the following initiation event after termination of translation of a short ORF (sORF) (Kozak 2001, Pöyry et al 2004). During the long elongation event, eIFs dissociate from the translating ribosome, and reinitiation is precluded. Reinitiation-promoting factors (RPFs) include eIF3 and eIF4F (Park et al 2001, Pöyry et al 2004, Cuchalová et al 2010, Roy et al 2010, Munzarová et al 2011). Whether additional canonical eIFs are required for reinitiation needs to be clarified.

CONTROL OF TRANSLATION INITIATION BY THE TOR SIGNALING PATHWAY

TOR complexes in mammals

The target of rapamycin (TOR)—a serine/threonine kinase—is an evolutionary conserved 280-kDa Ser/Thr protein kinase that regulates cell growth and proliferation in response to cellular energy status, growth factors, hormones, and nutrient abundance (Ma & Blenis 2009). TOR was first described over 20 years ago as a target protein of the antifungal and immunosuppressant agent rapamycin (Heitman et al 1991). Rapamycin binds to its intracellular receptor FKBP12 and inhibits TOR-dependent signaling through direct binding to the FRB (FKBP12-Rapamycin-Binding) domain of TOR kinase. Mammalian TOR (mTOR) exists in two functionally and structurally distinct complexes: mTORC1 and mTORC2. mTORC1 mediates temporal control of cell growth by activating anabolic processes such as ribosome biogenesis, protein synthesis, transcription, and nutrient uptake and by inhibiting catabolic processes such as autophagy and ubiquitin-dependent proteolysis. Its core components are TOR, regulatory-associated protein of TOR (RAPTOR), and LST8 (mammalian lethal with SEC13 protein).

In plants, the cofactors, upstream effectors, and downstream targets of TOR are much less studied. In *Arabidopsis*, TOR is encoded by a single gene, disruption of which is lethal due to an early block in embryo development (Menand et al 2002). The current situation regarding rapamycin functionality in *Arabidopsis* is confusing—according to Xiong and Sheen (2012), rapamycin inhibits TORC1 function, while others find that functional FKBP12 (for example, yeast FKBP12) is required in addition (Menand et al 2002, Sormani et al 2007). Torin-1—another class of TOR inhibitors—can be used to selectively inactivate the TOR signaling pathway in plants (Schepetilnikov et al 2011). Torin-1 binds specifically within the ATP-binding pocket of the TOR kinase domain and blocks TOR activity and autophosphorylation in the ATP-competitive mode of action (Thoreen et al 2009). The *Arabidopsis* genome contains two *RAPTOR* (Anderson et al 2005, Deprost et al 2005) and LST8 (Moreau et al 2012) genes. Two homologs of the mammalian 40S ribosomal S6 (RPS6) protein kinase 1 (S6K1)—S6K1 and S6K2—are present in the *Arabidopsis* genome (Mahfouz et al 2006). Recent data suggest that TOR plays an important role in growth regulation (Deprost et al 2007, Sormani et al 2007), cell wall biogenesis (Leiber et al 2010), and regulation of autophagy in *Arabidopsis* (Liu & Bassham 2010).

TOR downstream targets in mammals and plants

Several steps of translation initiation are affected positively by mTORC1 via phosphorylation of several well-characterized substrates: eIF4E-binding proteins (4E-BPs), the 40S ribosomal protein S6 (RPS6) protein kinase 1 (S6K1), eIF4G, and elongation factor 2 (eEF2) kinase (reviewed by Caron et al 2010). In TOR inactivation conditions, 4E-BP is hypophosphorylated and bound to eIF4E. Thus 4E-BP precludes eIF4G interaction with eIF4E and abolishes cap-dependent translation initiation. When activated, TOR phosphorylates 4E-BP1, which trigger 4E-BP1 dissociation, complex formation between eIF4E, eIF4G, eIF4A, and eIF4B at the 5′ end of an mRNA, and restoration of translation (Ma & Blenis 2009). In plants, 4E-BPs have not yet been identified.

mTORC1 initiates activation of S6K1 at Thr 389 (Ma & Blenis 2009). Two main S6K1 phosphorylation sites—Thr229 in the catalytic loop and Thr389 in the hydrophobic motif close to carboxy-terminal kinase domain—have been shown to be essential for S6K1 activation, resulting in the formation of a docking site for phosphoinositide-dependent kinase 1 (PDK1), which then phosphorylates Thr229. In plants, S6K1 is phosphorylated by TOR at Thr449, which is the functional equivalent of Thr 389, in a Torin-1 responsive manner (Zhang et al 1994, Schepetilnikov et al 2011).

S6K phosphorylates its own set of downstream targets; many of these function in translation (Raught et al 2004). Recent data suggest a primary role for 4E-BPs and eIF4G in the TOR-dependent control of translation of TOP mRNAs that encode ribosomal proteins, elongation factors, and several other proteins associated with the assembly or function of the translational apparatus (Thoreen et al 2012).

TRANSLATION OF VIRAL POLYCISTRONIC Mrnas VIA REINITIATION

The virus-activated translation reinitiation strategy of plant pararetroviruses

The plant pararetroviruses—*Caulimoviridae*—include the icosahedral cau-limo-, soymo-, cavemo-, and petu viruses and the bacilliform badna- and tun-gro viruses. Members of the *Caulimoviridae* are distinct in some aspects of their genome arrangement and expression strategies. The plant pararetrovirus genome exists as episomal DNA in infected nuclei, as terminally redundant RNA in the cytoplasm, and as open circular dsDNA in virions. Capsid protein, protease, and reverse transcriptase/RNAse H genes are required for replica-tion, and an integrase is lacking in pararetroviruses (Rothnie et al 1994). Genes encoding a movement protein (MOV) and an aphid transmission factor (ATF) are used for intra- and inter plant spread together with a virion-associated protein (VAP) that is required for both movement (Stavolone et al 2005) and insect transmission (Leh et al 1999, 2001a, Plisson et al 2005; see Fig. 17.1).

Another class of genes, present in some but not all plant pararetroviruses, encode proteins that can transactivate post-transcriptional virus gene expression and/or have other post-translational functions. Finally, a non-conserved open reading frame VII (ORF VII) of unknown function is present at the beginning

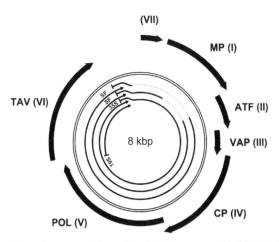

FIGURE 17.1 Schematic representation of the CaMV genome. Circular map of CaMV DNA (doubled circle). ORFs are indicated by thick black arrows and code for protein of unknown func-tion (VII), the cell-to-cell movement protein (MP, I), aphid transmission factors (ATF, II), virion-associated protein with MP and ATF functions (VAP, III), the precursor of the capsid proteins (CP, IV), the precursor of aspartic proteinase, reverse transcriptase and RNase H (POL, V), and an inclu-sion body protein/translational transactivator (TAV, VI). The inner circles represent the transcripts, the 35S RNA (35S) and 19S RNA (19S) transcripts, and two distinct spliced forms (SF).

of the viral sequence in most members of the *Caulimoviridae* and seems to be dispensable for viral function.

All genera of plant pararetroviruses produce a polycistronic RNA (except *Petunia vein clearing virus*, PVCV) with a single ORF encoding a single polyprotein that is then cleaved into the required viral proteins. The polycistronic caulimo-, soymo-, and probably cavemo virus RNAs have evolved a novel translation mechanism—virus-activated translation reinitiation—that is employed for translation of their polycistronic pregenomic RNAs via reinitiation. Translation reinitiation is unusual in eukaryotes. The mechanism of activated reinitiation is under the control of a viral reinitiation factor—transactivator/viroplasmin (TAV). TAV is encoded by ORF VI and is expressed from both the 35S pregenomic and the 19S subgenomic RNAs (Fig. 17.1). TAV is very abundant in infected cells and forms a dense matrix in the cytoplasm. It has many functions in the life cycle of the virus (Rothnie et al 1994). Most relevantly here, it is essential for reinitiation of translation of major ORFs on the 35S polycistronic RNA (Fütterer & Hohn 1991, Scholthof et al 1992).

Transactivation of polycistronic translation was first discovered for CaMV (Bonneville et al 1989, Fütterer & Hohn 1991, Scholthof et al 1992, Zijlstra & Hohn 1992) and soon after for *Figwort mosaic virus* (FMV, Gowda et al 1989) and *Peanut chlorotic streak virus* (PCSV; Maiti et al 1998). Indeed, reporter gene expression from different positions on the CaMV, FMV, and PCSV genomes in plant protoplasts proved a coupled translation of ORFs I through V under the control of TAV (Gowda et al 1989, Fütterer et al 1990, Scholthof et al 1992, Maiti et al 1998). In addition, TAV produced from the 19S RNA is able to activate TAV production from ORF VI on the 35S RNA (Driesen et al 1993). Importantly, TAV-activated polycistronic translation was proven to be essential for CaMV replication in a single cell (Kobayashi & Hohn 2003). CaMV is the most-studied plant pararetrovirus, and we concentrate the remainder of this chapter on CaMV translation strategies.

Ribosomal shunt on CaMV 35S RNA

CaMV uses reverse transcription for genome amplification (Rothnie et al 1994). It has two promoters that respectively ensure the production of the 35S pregenomic RNA and the 19S subgenomic RNA. The terminally redundant pregenomic 35S RNA is alternatively used as a replicative intermediate, a template for splicing, and as a polycistronic mRNA for expression of viral proteins. Several spliced versions of the 35S RNA have been identified, but all are polycistronic (Kiss-László et al 1995).

The polycistronic 35S RNA has a 600 nt long leader that contains several small uORFs and an extended hairpin structure, followed by tightly arranged long ORFs encoding all of the viral proteins (Fig. 17.1). RNA translation initiation on the 35S RNA is 5'-cap-dependent (Fütterer & Hohn 1991, Schmidt-Puchta et al 1997). The leader is loaded with complex secondary structure and multiple start codons that would be inhibitory to scanning but is bypassed by scanning ribosomes via a process known as ribosomal shunt. In the shunting process, linear scanning discontinues after the first sORF in front of a strong secondary-structure element and resumes downstream of the structured region and upstream of ORF VII (the mechanism is described in detail in Ryabova et al 2006). Two essential elements of CaMV shunting are the 5'-proximal small uORF (sORF A), which terminates in front of a stable structural element, and the structural element itself, which brings the long downstream ORF VII into close spatial proximity with sORF A. All known plant pararetroviruses contain a shunt structure (Pooggin et al 1999) with the exception of *Cestrum yellow leaf curl virus*, which does not contain uORFs within its relatively short leader (Stavolone et al 2003).

MECHANISM OF VIRUS-ACTIVATED REINITIATION OF TRANSLATION—INVOLVEMENT OF EIF3 AND RISP HOST FACTORS

Host factors in TAV-activated reinitiation

TAV-activated reinitiation has been studied extensively in CaMV. Interestingly, specific *cis*-sequence signals are not essential for transactivation of reinitiation as TAV can activate reinitiation after translation of the first ORF in an artificial bicistronic RNA containing two reporter ORFs (Fütterer & Hohn 1991, 1992). Accordingly, a stem structure at the cap-site inhibits expression of both reporters, while a stem between the two ORFs inhibits expression of only the second reporter (Fütterer & Hohn 1991). Unlike the case of GCN4 (see Hinnebusch 1997), TAV-mediated reinitiation is not dependent on the distance between the two ORFs and occurs equally efficiently either immediately after translation termination, when the two ORFs are linked by an AUGA quadruplet, or when the second ORF is located as far as 600 nt further downstream (Fütterer & Hohn 1991).

Taking into account that activated reinitiation is not much affected by the distance between the two consecutive ORFs, it was proposed that reinitiation-promoting factors, RPFs, do not have to be reacquired *de novo* but remain associated with the translating ribosome during the elongation step under conditions of TAV-activated reinitiation, and thus could be reused during the reinitiation event to promote 48S PIC formation and/or 60S recruitment (Park et al 2001). Our work revealed that the simple physical interaction of RPFs with the translating ribosome in the presence of TAV was not sufficient to overcome global cell barriers to reinitiation, and a change in the phosphorylation status of TAV interacting host factors to activate them rapidly was needed (Schepetilnikov et al 2011).

Several host factors have now been reported to interact with TAV (see Fig. 17.2). TAV can interact directly with the 60S ribosomal subunit via multiple ribosomal proteins—L24 (Park et al 2001), L18 (Leh et al 2001b) and L13 (Bureau et al 2004), and with eIF3 via subunit g (Park et al 2001).

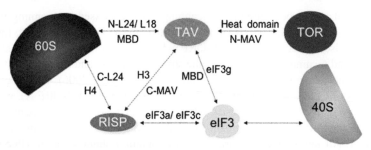

FIGURE 17.2 Protein–protein interactions between TAV and its partners. Interacting proteins are connected by thin interrupted lines. RISP can interact with 60S (via ribosomal protein L24), TAV, and eIF3 (via subunits a and c). TAV can bind 60S (via ribosomal proteins L18 and L24), RISP, and eIF3 (via the eIF3 subunit g). RISP and TAV are involved in complex formation with 40S via eIF3. TAV interacts with TOR. Proteins or protein domains mediating the above interactions are indicated. C-L24 (the L24 C-terminus), N-L24 (the L24 N-terminus), H4 (RISP α-helix 4), H3 (RISP α-helix 3), MBD (the multiple protein binding domain of TAV), MAV (the minimal transactivation domain of TAV), N-MAV (the MAV N-terminal domain), C-MAV (the MAV C-terminus), and Heat domain (the N-terminal heat repeat domain of TOR) are depicted.

Moreover, eIF3 was found to serve as a bridge between 40S and TAV (Park et al 2001) *in vitro*, suggesting complex formation between TAV and the eIF3-bound 40S ribosomal subunit. Later, two novel TAV interacting partners were identified and characterized: a reinitiation-supporting factor, RISP (Thiébeauld et al 2009), and a protein kinase, target-of-rapamycin (TOR; Schepetilnikov et al 2011). As can be seen in Figure 17.2, while TOR and RISP interact within the MAV domain (minimal domain of TAV), eIF3g and L24 bind MBD (multiple binding domain)—the two TAV domains shown to be essential for TAV transactivation function (de Tapia et al 1993, Park et al 2001). A novel plant factor—reinitiation supporting protein (RISP)—positively affects TAV function in reinitiation as well (Thiébeauld et al 2009). It seems that RISP plays the role of a TAV cofactor, whereby both can associate with the 60S ribosomal subunit via the same 60S ribosomal protein—L24 (RISP binds the C-terminus of L24, while TAV binds the L24 N-terminus; Fig. 17.2) and eIF3—via distinct subunits eIF3a and eIF3c (RISP) and eIF3g (TAV, Thiébeauld et al 2009). TAV and RISP mutants defective in their mutual interactions are less active, or inactive, in transactivation and viral amplification. Despite the observation of synergy between TAV and RISP in promoting reinitiation after long ORF translation, no translational enhancement is seen with RISP alone.

Characterization of eIF3 and RISP host factors in virus-activated reinitiation

As mentioned above, eIF3 is one of the important translation initiation factors, promoting nearly all steps of translation initiation (Hinnebusch 2006), suggesting a crucial role in reinitiation of translation. Evidently, eIF3a and eIF3g function in resumption of scanning of post-termination ribosomes in yeast (Cuchalová et al 2010, Munzarová et al 2011). In activated reinitiation, eIF3 can play a role in the resumption of scanning after termination of translation and/or recruitment of the TC during scanning.

One question to be addressed was whether TAV and its partners associate with actively translating ribosomes (polysomes). Analysis of polysomal content in extracts prepared from plants expressing a TAV transgene revealed drastic accumulation of eIF3 and RISP as well as TAV in polysomes (Park et al 2001, Thiébeauld et al 2009). This observation indicated that TAV can promote polysomal loading of eIF3/RISP and/or their stabilization on polysomes during elongation.

The data above suggest that TAV can enter polysomes via an eIF3 interaction. However, this does not appear to happen before the 60S-joining step. Although eIF4B was not found to interact with TAV, it was shown that it can out compete TAV for binding to eIF3 via subunit g and can preclude binding of TAV to 40S-bound eIF3 within the 48S PIC (Park et al 2004). Since eIF4B interactions with eIF3 can interfere with TAV binding, we have proposed that TAV associates with ribosome-bound eIF3 during or after the 60S-joining step, when eIF4B interactions with 40S-bound eIF3 are disrupted, while the link between 40S and eIF3 can be maintained further for a few elongation cycles (Pöyry et al 2004). This may explain the inability of TAV to affect the first round of initiation in plants (Park et al 2004). To summarize, TAV likely enters the cell translation machinery via interaction with eIF3g and as a complex with eIF3 and RISP binding 80S translating ribosomes.

This led to the current model as presented in Figure 17.4. TAV interacts with preinitiation complex somewhere prior to the 60S joining step and prevents dissociation of eIF3, RISP, and possibly other initiation factors, and therefore maintains the translating ribosome in a state competent for reinitiation immediately after termination of translation of the first ORF. We suggest that, in the presence of TAV, eIF3/RISP/TAV can travel with the elongating ribosome along the ORF on the solvent side of 60S through association with L18 in 60S (Park et al 2001), and thus without interfering with 80S binding of elongation factors. Interaction with L24, which is located close to the main factor binding site on 60S, is less likely—TAV binding there could interfere with canonical translation factor associations (Park et al 2001). After 60S release during translation termination, the TAV-eIF3-RISP complex would be transported back to 40S, where it can be used to reinitiate again.

Increasing evidence indicates that RISP forms part of two distinct complexes: (1) in the 43S PIC, where it associates with 40S as a complex with eIF3, and (2) in the 60S ribosomal subunit, where it binds the C-terminus of L24. These interactions are thought to provide a link between the 43S PIC and 60S and were implicated in 60S recruitment to the 48S PIC during the reinitiation step (see Thiébeauld et al 2009). Accordingly, RISP, TAV, and 60S co-localize in the epidermal cells of infected plants, and eIF3/TAV/RISP/L24 complex formation can be demonstrated *in vitro*. *In vitro* interactions between 60S and 43S-eIF3 can be mediated by RISP and strengthened by TAV.

Although the RISP interaction map within these two complexes is quite well established, the mechanism of RISP function in TAV-activated reinitiation is not clear. Recent studies revealed RISP phosphorylation at Ser 267 in response to TOR/S6K1 activation and the importance of RISP phosphorylation for its inter-action with TAV and function in TAV-activated reinitiation (Schepetilnikov et al 2011). RISP harbors the pattern RGRLES-267, which is found in many Akt or S6K1 substrates [phospho-Ser/Thr preceded by Lys/Arg at positions -5 and -3 (R/KxR/KxxS/T)] within its central domain. This motif was implicated earlier in TAV binding (Thiébeauld et al 2009, Schepetilnikov et al 2011). Thus, RISP activation is sensitive to TOR/S6K1 signaling activation. Although there is no direct evidence *in vivo* that phosphorylation of RISP is crucial for TAV-activated reinitiation of polycistronic translation, knockout of one (RISPa) of two genes encoding RISP results in a 3-fold decrease in TAV-activated reinitiation (Thiébeauld et al 2009). Thus, RISP can be considered as a novel component of the plant cell translation machinery; a mammalian ortholog of RISP has not yet been identified.

MECHANISM OF VIRUS-ACTIVATED REINITIATION OF TRANSLATION—INVOLVEMENT OF HOST FACTOR TOR

CaMV has the potential to activate the TOR signaling pathway *in planta* through direct TAV binding to TOR. A direct interaction between TAV and TOR was demonstrated by *in vitro* and *in planta* assays (Schepetilnikov et al 2011). The TOR binding site was located within the dsRNA-binding domain (dsR) of TAV (Fig. 17.2). TAV lacking the dsR domain is not able to interact with TOR *in vitro*, immunoprecipitate TOR, or upregulate TOR phosphorylation, and thus it is inactive in transactivation of polycistronic translation. Through-out our experiments, TOR phosphorylation was manifested by phosphorylation of S6K1 at TOR-specific Thr 449, and this phosphorylation was inhibited in response to Torin-1 application (Schepetilnikov et al 2013).

What is known about direct or indirect interactions between viruses and TOR signaling? Increasing evidence indicates that viruses may control key cel-lular signaling pathways to benefit the virus at the levels of protein synthesis, metabolism, growth, and survival. There are various ways that viruses activate and maintain the TOR pathway in favor of translational control, particularly via control of the phosphorylation status of 4E-BPs. Viruses that must main-tain cap-dependent translation (i.e., mammalian DNA viruses and many RNA viruses) try either to keep TORC1 active or maintain eIF4F complex integrity (Buchkovich et al 2008). Human cytomegalovirus (HCMV) infection induces TOR-dependent phosphorylation of 4E-BPs and eIF4G but not of S6K1, thus stimulating eIF4F complex assembly to increase cap-dependent translation initiation efficiency (Kudchodkar et al 2004). Two γ-herpesvirus proteins,

Epstein–Barr virus (EBV) LMP2A (Moody et al 2005), and *Kaposi's sarcoma herpes virus* (KSHV) G protein-coupled receptor vGPCR (Sodhi et al 2006) have also been implicated in activation of the TOR signaling pathway.

Interestingly, viruses have developed multiple mechanisms to activate TOR signaling in favor of viral translation. One such strategy results in stimulation of the PI3K-AKT pathway upstream of TOR kinase. Adenovirus protein E4-ORF1 stimulates PI3K, thus upregulating TOR signaling activation and viral replication (Gingras & Sonenberg 1997, O'Shea et al 2005). Remarkably, the herpes simplex virus (HSV-1) Us3 kinase mimics AKT-dependent phosphorylation of TSC2, which leads to TSC1/TSC2 complex inactivation and constitutive activation of TORC1 (Chuluunbaatar et al 2010). The human cytomegalovirus (HCMV) UL38 protein physically associates with TSC2, thus inactivating the TSC1/TSC2 complex (Moorman et al 2008). In contrast, binding of human papilloma virus (HPV) E6 oncoprotein to the tumor suppressor TSC2 results in TSC2 degradation (Lu et al 2004, Zheng et al 2008, Spangle & Münger 2010).

To sum up, although many viruses are known to affect the TOR pathway indirectly in order to stimulate cap-dependent translation initiation, the plant pararetrovirus CaMV has chosen a strategy that maintains high TOR phosphorylation status to upregulate reinitiation events on the same mRNA.

Is TAV a suitable tool for studying upstream regulation of TOR?

Despite a number of recent discoveries regarding the TOR pathway in plants, important aspects of TOR activation remain unresolved. In the mammalian system, insulin and growth factors regulate TORC1 activity via either the PI3K/Akt pathway or the Ras/MAPK pathway, which converge on the tuberous sclerosis heterodimeric complex (TSC1-TSC2) to inhibit the GTPase-activating function of the small GTPase Rheb (Ras homolog enriched in brain; Avruch et al 2009). Elimination or inactivation of the TSC complex results in an increase in GTP charging of Rheb and in constitutive activation of TORC1. Although plants contain many small GTPases, whether a Rheb homolog exists in *Arabidopsis* is still an open question. Thus, studies of TAV–TOR complexes might reveal a small GTPase or other protein factors participating in TOR activation in plants.

Accumulating data suggest that site-specific phosphorylation of mTOR can regulate its activation. Three phosphorylation sites (P-sites) in mTOR have been reported to date—the C terminus S2448 P-site, which is phosphorylated by S6K1 via a feedback loop; the S2481 autophosphorylation P-site; and the HEAT-repeat motif S1261 P-site used for TSC/Rheb signaling dependent phosphorylation (Acosta-Jaquez et al 2009). A Rheb-driven phosphorylation event at mTOR S1261 within the HEAT repeat domain was suggested to promote autokinase activity at S2481. Further work is required to identify and characterized *Arabidopsis* TOR phosphorylation sites.

S6K1 is phosphorylated by TAV-activated TOR within eIF3-containing preinitiation complexes

As mentioned above, S6K1 phosphorylation has been demonstrated within eIF3-containing initiation complexes in mammals, where eIF3 serves as a dynamic scaffold for binding of either phosphorylated TOR or inactive S6K1 (Holz et al 2005). According to the model proposed, in its inactive form, S6K1 associates with non-polysomal eIF3 complex, whereas TORC1 stays unbound. Upon activation, TORC1 is recruited to the eIF3 complex and phosphorylates S6K1. The TOR-mediated phosphorylation of S6K1 results in its dissociation from eIF3 and subsequent phosphorylation and activation by PDK1. According to the current model, the S6K1-eIF3/ TORC1-eIF3 complexes associate with the mRNA 5′ cap, bringing TORC1 and S6K1 into proximity with its other major targets. This could explain how TOR acts in translation initiation. A similar scenario was proposed for S6K1 phosphorylation by TOR in eIF3-containing preinitiation complexes in plants (Schepetilnikov et al 2013). Moreover, *At*TOR in its phosphorylated form can associate not only with eIF3-containing complexes but also with actively translating ribosomes (Schepetilnikov et al 2011), suggesting that eIF3-complexes and polysomes represent two platforms for S6K1 phosphorylation by TOR. Indeed, TOR, when activated, can associate with actively translating ribosomes, while in the inactive state polysomes are prebound by S6K1. Our current model states that active TOR is recruited to polyribosomes concomitantly with polysomal accumulation of eIF3 and RISP in a TAV-dependent manner to phosphorylate polysome-associated inactive S6K1 followed by RISP phosphorylation (see schematic presentation of corresponding phosphorylation events in Fig. 17.3).

The above hypothesis is consistent with the following experiment. In TAV transgenic plants, TAV, eIF3, RISP, and TOR associate with polysomes, and this process correlates with high RISP phosphorylation status. In contrast, in plants expressing TAVdsR mutants defective in TOR binding, TAVdsR, RISP, and eIF3 association with polysomes was affected only slightly, while TOR binding, and thus RISP phosphorylation, was abolished (Schepetilnikov et al 2011).

Current model of host factor functioning in TAV-activated reinitiation of translation

A putative model of how TOR signaling may contribute to TAV-mediated transactivation suggests that, upon overexpression, TAV binds to and activates TOR as well as entering polysomes via association with eIF3/ RISP prebound to 80S translating complexes at the beginning of elongation (Fig. 17.4).

During elongation, eIF3 and RISP could be transported by TAV to the rear side of 60S, where the eIF3-RISP-TAV complex can be stabilized via TAV binding to L18 (and/or L13) and not interfere with translation elongation. Activated TOR associates with polysomes, which, in turn, leads to activation of

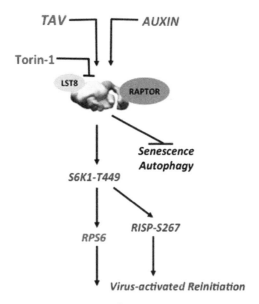

Translation/ Growth

FIGURE 17.3   A tentative model of the TOR signaling pathway in plant translation. The TOR pathway controlling translation and virus-activated reinitiation. Upon stimulation of TOR by either TAV or plant phytohormone auxin in a Torin-1 sensitive manner, S6K1 is phosphorylated at T449 and can phosphorylate at least two substrates: ribosomal protein S6 (RPS6) and RISP at S267. Activated RISP-TAV-eIF3 triggers virus-activated reinitiation.

S6K1 and phosphorylation of polysome-associated RISP. During termination of the first ORF, the reinitiation-competent eIF3/RISP-P/TAV complex, with or without TOR, is relocated back to the 40S subunit and begins to scan to the second ORF2 (in this case GUS ORF), recruiting the ternary complex and 60S on the way.

AUXIN AND TOR SIGNALING IN PLANTS

Our recent findings suggest that the TOR signaling pathway can be activated in response to the phytohormone auxin (Schepetilnikov et al 2013). Auxin plays a crucial role in a wide variety of plant morphogenetic and physiologic responses, and local auxin maxima represent signals for initiation of organ development (Benkova et al 2009, Lumba et al 2010). Although many of auxin's actions are mediated by transcription factors, auxin signaling plays a role in upregulation of translation as manifested by an increase in the level of actively translating ribosomes (Beltrán-Peña et al 2002, Turck et al 2004). This increase in translation level is likely the result of TOR signaling pathway activation (Schepetilnikov et al 2013). Our observation in plants— that TAV-activated TOR signaling mimics TOR activation in response to

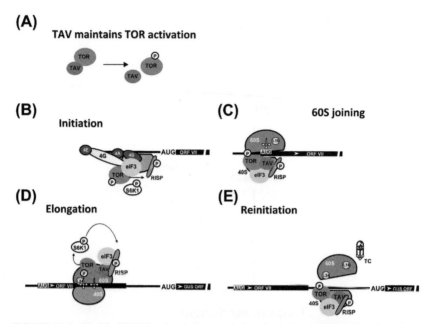

FIGURE 17.4   Model of TAV function during translation of polycistronic RNA. The 40S and 60S ribosomal subunits are depicted in gray. (A) TAV binds TOR and maintains TOR in a constitutively activated state. (B) eIF4B, in concert with eIF4F (4E/4G/4A) or eIFiso4F, interacts with eIF3 bound to RISP/S6K1 within the pre-initiation complex (eIF3-PIC). Activated TOR binds eIF3-PIC to trigger S6K1 and RISP phosphorylation. The complex scans until it encounters the first suitable start codon. (C) During the 60S subunit joining step eIF4B is displaced from the ribosome, while eIF3 is still associated with the solvent side of the 40S subunit. TAV binds to the eIF3/40S complex via eIF3 subunit g and RISP-P. ORF1 elongation begins. (D) During the elongation process, the TAV/eIF3/RISP/80S complex is stabilized, apparently in part by transfer of TAV/eIF3/RISP to the rear side of the 60S subunit through TAV interaction with L18 or/ and L13. When activated, TOR can bind polysomes and maintain the phosphorylation state of RISP. (E) The TAV/eIF3/RISP-P complex is relocated back to the 40S subunit via eIF3/40S interaction to assemble a reinitiation-competent 40S. eIF3, eIF4E (4E), eIF4G (4G), eIF4A (4A), eIF4B (4B), RISP, S6K1, TOR, and TAV are depicted.

auxin—suggests a novel role for auxin as an effector molecule in TOR pathway activation. It is interesting that, when activated in response to auxin, TOR can interact with S6K1-prebound polysomes, triggering S6K1 phosphorylation. We conclude that activated TOR displays the ability to interact with polyribosomes independently of the activating stimulus. The above data are consistent with the notion that TAV triggers TOR activation but is not required for active TOR binding to polysomes.

The discovery that TOR is an essential component of reinitiation after long ORF translation in the presence of the viral factor TAV might indicate that TOR can control reinitiation *per se*, including reinitiation after short ORF translation.

CONCLUSIONS AND PERSPECTIVES

The work described here provides new insights into how CaMV TAV has developed to capture the cell translation machinery and coerce it into performing functions that are normally restricted in eukaryotic cells. Information on the details of activation of TOR downstream targets as well as of TOR upstream effectors that has so far been missing in plants has also been expanded.

There is clearly a need to understand the mechanisms of RISP function in the TAV reinitiation pathway and in the cellular translation machinery, as well as the role of TAV partners and the mechanisms of TOR activation in plants in order to fully unravel the complexities of these highly tuned host/viral systems.

REFERENCES

Acosta-Jaquez, H.A., Keller, J.A., Foster, K.G., Ekim, B., Soliman, G.A., Feener, E.P., Ballif, B.A., Fingar, D.C., 2009. Site-specific mTOR phosphorylation promotes mTORC1-mediated signaling and cell growth. Molecular and Cellular Biology 29, 4308–4324.

Anderson, G.H., Veit, B., Hanson, M.R., 2005. The *Arabidopsis* AtRaptorgenes are essential for post-embryonic plant growth. BMC Biology 3, 12.

Avruch, J., Long, X., Ortiz-Vega, S., Rapley, J., Papageorgiou, A., Dai, N., 2009. Amino acid regulation of TOR complex 1. American Journal of Physiol Endocrinology and Metabolism 296, 592–602.

Beltrán-Peña, E., Aguilar, R., Ortíz-López, A., Dinkova, T.D., De Jiménez, E.S., 2002. Auxin stimulates S6 ribosomal protein phosphorylation in maize thereby affecting protein synthesis regulation. Plant Physiology 115, 291–297.

Benkova, E., Ivanchenko, M.G., Friml, J., Shishkova, S., Dubrovsky, J.G., 2009. A morphogenetic trigger: is there an emerging concept in plant developmental biology? Trends in Plant Science 14, 189–193.

Bonneville, J.M., Sanfaçon, H., Fütterer, J., Hohn, T., 1989. Posttranscriptional trans-activation in cauliflower mosaic virus. Cell 59, 1135–1143.

Browning, K.S., Gallie, D.R., Hershey, J.W.B., Hinnebusch, A.G., Maitra, U., Merrick, W.C., Norbury, C., 2001. Unified nomenclature for the subunits of eukaryotic initiation factor 3. Trends in Biochemical Science 26, 284.

Buchkovich, N.J., Yu, Y., Zampieri, C.A., Alwine, J.C., 2008. The TORrid affairs of viruses: effects of mammalian DNA viruses on the PI3K–Akt–mTOR signalling pathway. Nature Reviews Microbiology 6, 266–275.

Bureau, M., Leh, V., Haas, M., Geldreich, A., Ryabova, L., Yot, P., Keller, M., 2004. P6 protein of *Cauliflower mosaic virus*, a translation reinitiator, interacts with ribosomal protein L13 from *Arabidopsis thaliana*. Journal of General Virology 85, 3765–3775.

Caron, E., Ghosh, S., Matsuoka, Y., Ashton-Beaucage, D., Therrien, M., Lemieux, S., Perreault, C., Roux, P.P., Kitano, H., 2010. A comprehensive map of the mTOR signaling network. Molecular Systems Biology 6, 453.

Chuluunbaatar, U., Roller, R., Feldman, M.E., Brown, S., Shokat, K.M., Mohr, I., 2010. Constitutive mTORC1 activation by a herpesvirus Akt surrogate stimulates mRNA translation and viral replication. Genes & Development 24, 2627–2639.

Cuchalová, L., Kouba, T., Herrmannová, A., Dányi, I., Chiu, W.L., Valášek, L., 2010. The RNA recognition motif of eukaryotic translation initiation factor 3g (eIF3g) is required for resumption of scanning of posttermination ribosomes for reinitiation on GCN4 and together with eIF3i stimulates linear scanning. Mol Cell Biology 30, 4671–4686.

Deprost, D., Truong, H.N., Robaglia, C., Meyer, C., 2005. An *Arabidopsis* homolog of RAPTOR/KOG1 is essential for early embryo development. Biochemical and Biophysical Research Communications 326, 844–850.

Deprost, D., Yao, L., Sormani, R., Moreau, M., Leterreux, G., Nicolaï, M., Bedu, M., Robaglia, C., & Meyer, C. (2007). The *Arabidopsis* TOR kinase links plant growth, yield, stress resistance and mRNA translation. EMBO Reports, 8, 864–870.

de Tapia, M., Himmelbach, A., Hohn, T., 1993. Molecular dissection of the cauliflower mosaic virus translational transactivator. EMBO Journal 12, 3305–3314.

Dreher, T.W., Miller, W.A., 2006. Translational control in positive strand RNA plant viruses. Virology 344, 185–197.

Driesen, M., Benito-Moreno, R.M., Hohn, T., Futterer, J., 1993. Transcription from the CaMV 19 S promoter and autocatalysis of translation from CaMV RNA. Virology 195, 203–210.

Fütterer, J., Hohn, T., 1991. Translation of a polycistronic mRNA in the presence of the cauliflower mosaic virus transactivator protein. EMBO Journal 10, 3887–3896.

Fütterer, J., Hohn, T., 1992. Role of an upstream open reading frame in the translation of polycistronic mRNAs in plant cells. Nucleic Acids Research 20, 3851–3857.

Fütterer, J., Bonneville, J.M., Gordon, K., De Tapia, M., Karlsson, S., Hohn, T., 1990. Expression from polycistronic cauliflower mosaic virus pregenomic RNA. In: McCarthy, J.E.G., Tuite, M.F. (Eds.), Post-transcriptional control of gene expression, Springer, Berlin, pp. 347–357.

Gallie, D.R., 2002. Protein–protein interactions required during translation. Plant Molecular Biology 50, 949–970.

Gingras, A.C., Sonenberg, N., 1997. Adenovirus infection inactivates the translational inhibitors 4E-BP1 and 4E-BP2. Virology 237, 182–186.

Gowda, S., Wu, F.C., Scholthof, H.B., Shepherd, R.J., 1989. Gene VI of figwort mosaic virus (caulimo virus group) functions in posttranscriptional expression of genes on the full-length RNA transcript. Proceedings of the National Academy of Sciences of the USA 86, 9203–9207.

Heitman, J., Movva, N.R., Hall, M.N., 1991. Targets for cell cycle arrest by the immunosuppressant rapamycin in yeast. Science 253, 905–909.

Hinnebusch, A.G., 1997. Translational regulation of yeast GCN4. A window on factors that control initiator-tRNA binding to the ribosome. Journal of Biological Chemistry 272, 21661–21664.

Hinnebusch, A., 2006. eIF3: a versatile scaffold for translation initiation complexes. Trends in Biochemical Sciences 31, 553–562.

Hinnebusch, A.G., Lorsch, J.R., 2012. The mechanism of eukaryotic translation initiation: new insights and challenges. Cold Spring Harbor Perspectives in Biology 4, 10.

Holz, M.K., Ballif, B.A., Gygi, S.P., Blenis, J., 2005. mTOR and S6K1 mediate assembly of the translation preinitiation complex through dynamic protein interchange and ordered phosphorylation events. Cell 123, 569–580.

Kiss-László, Z., Blanc, S., Hohn, T., 1995. Splicing of *Cauliflower mosaic virus* 35S RNA is essential for viral infectivity. EMBO Journal 14, 3552–3562.

Kobayashi, K., Hohn, T., 2003. Dissection of cauliflower mosaic virus transactivator/viroplasmin reveals distinct essential functions in basic virus replication. Journal of Virology 77, 8577–8583.

Kozak, M., 2001. Constraints on reinitiation of translation in mammals. Nucleic Acids Research 29, 5226–5232.

Kudchodkar, S.B., Yu, Y., Maguire, T.G., Alwine, J.C., 2004. Human *Cytomegalovirus* infection induces rapamycin-insensitive phosphorylation of downstream effectors of mTOR kinase. Journal of Virology 78, 11030–11039.

Leh, V., Jacquot, E., Geldreich, A., Hermann, T., Leclerc, D., Cerutti, M., Yot, P., Keller, M., Blanc, S., 1999. Aphid transmission of cauliflower mosaic virus requires the viral PIII protein. EMBO Journal 18, 7077–7085.

Leh, V., Jacquot, E., Geldreich, A., Haas, M., Blanc, S., Keller, M., Yot, P., 2001a. Interaction between the open reading frame III product and the coat protein is required for transmission of cauliflower mosaic virus by aphids. Journal of Virology 75, 100–106.

Leh, V., Yot, P., Keller, M., 2001b. The cauliflower mosaic virus translational transactivator interacts with the 60S ribosomal subunit protein L18 of Arabidopsis thaliana. Virology 266, 1–7.

Leiber, R.M., John, F., Verhertbruggen, Y., Diet, A., Knox, J.P., Ringli, C., 2010. The TOR pathway modulates the structure of cell walls in *Arabidopsis*. Plant Cell 22, 1898–1908.

Liu, Y., Bassham, D.C., 2010. TOR is a negative regulator of autophagy in *Arabidopsis thaliana*. PLoS ONE 5, e11883.

Lu, Z., Hu, X., Li, Y., Zheng, L., Zhou, Y., Jiang, H., Ning, T., Basang, Z., Zhang, C., Ke, Y., 2004. Human papillomavirus 16 E6 oncoprotein interferences with insulin signaling pathway by binding to tuberin. Journal of Biological Chemistry 279, 35664–35670.

Lumba, S., Cutler, S., McCourt, P., 2010. Plant nuclear hormone receptors: a role for small molecules in protein–protein interactions. Annual Review of Cell and Developmental Biology 26, 445–469.

Ma, X.M., Blenis, J., 2009. Molecular mechanisms of mTOR-mediated translational control. Nature Reviews in Molecular Cell Biology 10, 307–318.

Mahfouz, M.M., Kim, S., Delauney, A.J., Verma, D.P., 2006. *Arabidopsis* TARGET OF RAPAMYCIN interacts with RAPTOR, which regulates the activity of S6 kinase in response to osmotic stress signals. Plant Cell 18, 477–490.

Maiti, I.B., Richins, R.D., Shepherd, R.J., 1998. Gene expression regulated by gene VI of caulimovirus: transactivation of downstream genes of transcripts by gene VI of peanut chlorotic streak virus in transgenic tobacco. Virus Research 57, 113–124.

Menand, B., Desnos, T., Nussaume, L., Berger, F., Bouchez, D., Meyer, C., Robaglia, C., 2002. Expression and disruption of the *Arabidopsis* TOR (target of rapamycin) gene. Proceedings of the National Academy of Sciences of the USA 99, 6422–6427.

Miller, W.A., Wang, Z., Treder, K., 2007. The amazing diversity of cap-independent translation elements in the 3'-untranslated regions of plant viral RNAs. Biochemical Society Transactions 35, 1629–1633.

Moody, C.A., Scott, R.S., Amirghahari, N., Nathan, C.A., Young, L.S., Dawson, C.W., Sixbey, J.W., 2005. Modulation of the cell growth regulator mTOR by Epstein-Barr virus-encoded LMP2A. Journal of Virology 79, 5499–5506.

Moorman, N.J., Cristea, I.M., Terhune, S.S., Rout, M.P., Chait, B.T., Shenk, T., 2008. Human cytomegalovirus protein UL38 inhibits host cell stress responses by antagonizing the tuberous sclerosis protein complex. Cell Host Microbe 3, 253–262.

Moreau, M., Azzopardi, M., Clément, G., Dobrenel, T., Marchive, C., Renne, C., Martin-Magniette, M.L., Taconnat, L., Renou, J.P., Robaglia, C., Meyer, C., 2012. Mutations in the *Arabidopsis* homolog of LST8/GβL, a partner of the target of Rapamycin kinase, impair plant growth, flowering, and metabolic adaptation to long days. Plant Cell 24, 463–481.

Munzarová, V., Pánek, J., Gunišová, S., Dányi, I., Szamecz, B., Valášek, L.S., 2011. Translation reinitiation relies on the interaction between eIF3a/TIF32 and progressively folded cis-acting mRNA elements preceding short uORFs. PLoS Genetics 7, e1002137.

O'Shea, C., Klupsch, K., Choi, S., Bagus, B., Soria, C., Shen, J., McCormick, F., Stokoe, D., 2005. Adenoviral proteins mimic nutrient/growth signals to activate the mTOR pathway for viral replication. EMBO Journal 24, 1211–1221.

Park, H.-S., Himmelbach, A., Browning, K.S., Hohn, T., Ryabova, L.A., 2001. A plant viral 'reinitiation' factor interacts with the host translational machinery. Cell 106, 723–733.

Park, H.-S., Browning, K.S., Hohn, T., Ryabova, L.A., 2004. Eucaryotic initiation factor 4B controls eIF3-mediated ribosomal entry of viral reinitiation factor. EMBO Journal 23, 1381–1391.

Pestova, T.V., Lomakin, I.B., Lee, J.H., Choi, S.K., Dever, T.E., Hellen, C.U., 2000. The joining of ribosomal subunits in eukaryotes requires eIF5B. Nature 403, 332–335.

Plisson, C., Uzest, M., Drucker, M., Froissart, R., Dumas, C., Conway, J., Thomas, D., Blanc, S., Bron, P., 2005. Structure of the mature P3-virus particle complex of cauliflower mosaic virus revealed by cryo-electron microscopy. Journal of Molecular Biology 346, 267–277.

Pooggin, M.M., Fütterer, J., Skryabin, K.G., Hohn, T., 1999. A short open reading frame terminating in front of a stable hairpin is the conserved feature in pregenomic RNA leaders of plant pararetroviruses. Journal of General Virology 80, 2217–2228.

Pöyry, T.A., Kaminski, A., Jackson, R., 2004. What determines whether mammalian ribosomes resume scanning after translation of a short upstream open reading frame? Genes & Development 18, 62–75.

Raught, B., Peiretti, F., Gingras, A.C., Livingstone, M., Shahbazian, D., Mayeur, G.L., Polakiewicz, R.D., Sonenberg, N., Hershey, J.W., 2004. Phosphorylation of eucaryotic translation initiation factor 4B Ser422 is modulated by S6 kinases. EMBO Journal 23, 1761–1769.

Rothnie, H.M., Chapdelaine, Y., Hohn, T., 1994. Pararetroviruses and retroviruses: a comparative review of viral structure and gene expression strategies. Advances in Virus Research 44, 1–67.

Roy, B., Vaughn, J.N., Kim, B.H., Zhou, F., Gilchrist, M.A., von Arnim, A.G., 2010. The h subunit of eIF3 promotes reinitiation competence during translation of mRNAs harboring upstream open reading frames. RNA 16, 748–761.

Ryabova, L.A., Pooggin, M.M., Hohn, T., 2006. Translation reinitiation and leaky scanning in plant viruses. Virus Research 119, 52–62.

Schepetilnikov, M., Kobayashi, K., Geldreich, A., Caranta, C., Robaglia, C., Keller, M., Ryabova, L.A., 2011. Viral factor TAV recruits TOR/S6K1 signalling to activate reinitiation after long ORF translation. EMBO Journal 30, 1343–1356.

Schepetilnikov, M., Mancera-Martinez, E., Dimitrova, M., Geldreich, A., Keller, M., Ryabova, L.A., 2013. TOR and S6K1 promote translation reinitiation of uORF-containing mRNAs via phosphorylation of eIF3h. EMBO Journal 32, 1087–1102.

Schmidt-Puchta, W., Dominguez, D., Lewetag, D., Hohn, T., 1997. Plant ribosome shunting in vitro. Nucleic Acids Research 25, 2854–2860.

Scholthof, H.B., Gowda, S., Wu, F.C., Shepherd, R.J., 1992. The full-length transcript of a caulimovirus is a polycistronic mRNA whose genes are trans-activated by the product of gene VI. Journal of Virology 66, 3131–3139.

Sodhi, A., Chaisuparat, R., Hu, J., Ramsdell, A.K., Manning, B.D., Sausville, E.A., Sawai, E.T., Molinolo, A., Gutkind, J.S., Montaner, S., 2006. The TSC2/mTOR pathway drives endothelial cell transformation induced by the Kaposi's sarcoma-associated herpesvirus G protein-coupled receptor. Cancer Cell 10, 133–143.

Sormani, R., Yao, L., Menand, B., Ennar, N., Lecampion, C., Meyer, C., Robaglia, C., 2007. Saccharomyces cerevisiae FKBP12 binds Arabidopsis thaliana TOR and its expression in plants leads to rapamycin susceptibility. BMC Plant Biology 7, 26.

Spangle, J.M., Münger, K., 2010. The human papillomavirus type 16 E6 oncoprotein activates mTORC1 signaling and increases protein synthesis. Journal of Virology 84, 9398–9407.

Stavolone, L., Kononova, M., Pauli, S., Ragozzino, A., de Haan, P., Milligan, S., Lawton, K., Hohn, T., 2003. *Cestrum yellow leaf curling virus* (CmYLCV) promoter: a new strong constitutive promoter for heterologous gene expression in a wide variety of crops. Plant Molecular Biology 53, 663–673.

Stavolone, L., Villani, M.E., Leclerc, D., Hohn, T., 2005. The virion-associated-protein (VAP) mediates *Cauliflower mosaic virus* (CaMV) movement from cell to cell. Proceedings of the National Academy of Sciences of the USA 102, 6219–6224.

Thiébeauld, O., Schepetilnikov, M., Park, H.-S., Geldreich, A., Kobayashi, K., Keller, M., Hohn, T., Ryabova, L.A., 2009. A new plant protein interacts with eIF3 and 60S to enhance virus-activated translation re-initiation. EMBO Journal 28, 3171–3184.

Thoreen, C.C., Kang, S.A., Chang, J.W., Liu, Q., Zhang, J., Gao, Y., Reichling, L.J., Sim, T., Sabatini, D.M., Gray, N.S., 2009. An ATP-competitive mammalian target of rapamycin inhibitor reveals rapamycin-resistant functions of mTORC1. Journal of Biological Chemistry 284, 8023–8032.

Thoreen, C.C., Chantranupong, L., Keys, H.R., Wang, T., Gray, N.S., Sabatini, D.M., 2012. A unifying model for mTORC1-mediated regulation of mRNA translation. Nature 485, 109–113.

Turck, F., Zilbermann, F., Kozma, S.C., Thomas, G., Nagy, F., 2004. Phytohormones participate in an S6 kinase signal transduction pathway in *Arabidopsis*. Plant Physiology 134, 1527–1535.

Wang, A., Krishnaswamy, S., 2012. Eukaryotic translation initiation factor 4E-mediated recessive resistance to plant viruses and its utility in crop improvement. Molecular Plant Pathology 13, 795–803.

Xiong, Y., Sheen, J., 2012. Rapamycin and glucose-target of rapamycin (TOR) protein signaling in plants. Journal of Biological Chemistry 287, 2836–2842.

Zhang, S.H., Lawton, M.A., Hunter, T., Lamb, C.J., 1994. atpk1, a novel ribosomal protein kinase gene from *Arabidopsis*. I. Isolation, characterization, and expression. Journal of Biological Chemistry 269, 17586–17592.

Zheng, L., Ding, H., Lu, Z., Li, Y., Pan, Y., Ning, T., Ke, Y., 2008. E3 ubiquitin ligase E6AP-mediated TSC2 turnover in the presence and absence of HPV16 E6. Genes Cells 13, 285–294.

Zijlstra, C., Hohn, T., 1992. *Cauliflower mosaic virus* gene VI controls translation from dicistronic expression units in transgenic *Arabidopsis* plants. Plant Cell 4, 1471–1484.

Chapter 18

Molecular mechanism of *Begomovirus* evolution and plant defense response

Vinutha T
Division of Biochemistry, IARI, New Delhi, India

Om Prakash Gupta[*]
Division of Quality and Basic Sciences, Directorate of Wheat Research, Karnal, India
[*]Corresponding author

Rama Prashat G
Division of Genetics, IARI, New Delhi, India

Veda Krishnan
Division of Biochemistry, IARI, New Delhi, India

Sharma P
Division of Crop Improvement, Directorate of Wheat Research, Karnal, India

INTRODUCTION

Due to the worldwide increase in population and the distribution of insect vectors and global movement of plant materials, *Begomovirus*-induced diseases have become a major constraint of crop production in tropical and subtropical regions (Rojas et al 2005, Seal et al 2006). Although viruses exhibit rapid sequence divergence over periods of time and mutations occur in variable positions of viral genomes (Strauss & Strauss 2001), genetic stability of plant viruses has been revealed over time, space, and host species separations (Kearney et al 1999); this agrees with the view that host-associated selection results in decreased diversity in viral genomes (Garcia-Arenal et al 2001). Studies on their geographic distribution has revealed that the begomoviruses affecting tomato in northern India are bipartite, while those affecting tomato in southern India are monopartite. These two groups of viruses are quite distinct in their biologic activity and genomic organization. Irrespective of their genomic nature, these viruses replicate in the host nuclei via double-stranded DNA intermediates using a rolling circle mechanism, whereas these viruses exist in single-stranded form during the infection cycle (Stenger et al 1995). *Begomo-virus* evolution is fastest in the wake of pandemics, which, in turn, reflects key changes in the factors affecting virus survival and spread. The probable genetic material among viruses with overlapping ranges would enable distinct viruses in the same region to evolve in a concerted manner. Moreover, some of the new variants will be preserved by geographic or biologic isolation and enhanced by various further genetic changes, to the point where new virus species can be considered to have evolved (Harrison & Robinson 1999).

It is becoming increasingly evident that increases in the intracellular concentrations of viral proteins have many consequences for host gene expression and metabolism. Some of these effects do not necessarily provide an advantage to the virus but nevertheless have effects on the host. One consequence of viral infection is the altered expression of host genes, which may lead to altered plant phenotypes. A major challenge has been to identify host genes with altered transcription profiles and to decipher how and why the changes are initiated. This information can then be used to investigate the functions of genes with altered

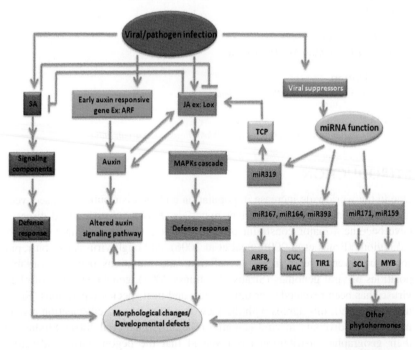

FIGURE 18.1 Host responses and altered gene expression associated with plant virus infections. Viral infections can also disrupt the functions of regulatory miRNAs (some of the miRNAs are taken as an example) and phytohormone signaling or biosynthesis, leading to developmental defects. Thus, some common ways in which a broad range of viruses may alter plant gene expression are outlined. Incompatible interactions between viruses and hosts have previously served as models for investigating host defense responses (modified from Whitham et al 2006).

expression profiles in plant–virus interactions (Whitham et al 2006). Figure 18.1 shows some common ways in which a broad range of viruses may alter plant gene expression by disrupting the functions of regulatory miRNAs, thereby altering phytohormone signaling or biosynthesis and leading to developmental defects.

EVOLUTION OF BEGOMOVIRUSES

The effects of recombination and pseudo-recombination on the evolution of begomoviruses and the epidemics of begomovirus-induced diseases have been extensively documented as compared to the mutational effects. Recombination is undoubtedly an important evolutionary mechanism in these viruses; it cannot create genetic variation *de novo*, so background mutation pressure must also be a key element in *Geminivirus* evolution (Ge et al 2007). Genetic diversity of a population of *Maize streak virus* (MSV), a *Mastrevirus* of the family *Geminiviridae* derived from a single isolate, demonstrated a quasispecies structure (Isnard et al 1998). There are numerous reports of the emergence of strains of *Geminivirus* with altered pathogenicity (Seal et al 2006), indicative of rapid genetic change that has been attributed to recombination or reassortment among different viral genomes.

The mutation frequency of a virus is determined by a combination of the intrinsic frequency of nucleotide mis-incorporation and the capability for mismatch repair, as well as the extent of specific selection or stochastic drift resulting from genetic bottlenecks imposed on the virus population. Unlike RNA viruses, begomoviruses replicate their genomes inside the nucleus by using the host replication machinery, presumably involving DNA polymerase α- and δ-like activities (Gutierrez 1999). Thus, these viruses were assumed to have higher replication fidelity and lower rates of mutation than RNA viruses (Rojas et al 2005). However, the large number of species (>100) (Fauquet et al 2003) and the continued reports of new species (Bull et al 2006), as well as the high degree of genetic diversity within a species (Stenger 1995, Patil et al 2005), suggest that begomoviruses have a high mutation rate and that they generate highly diverse populations in a short time. It was reported that DNA methylation inhibits the replication of *Tomato gold mosaic virus* in tobacco protoplasts (Brough et al 1992), implying that *Geminivirus* DNA may not be methylated and that the normal mechanisms for mismatch repair probably do not operate during, for example, the *Tomato gold mosaic virus* replication cycle (Inamdar et al 1992). Thus, it is possible that the mechanisms of mismatch repair may function differently during the replication of *Geminivirus* DNA and cellular DNA and that the lack of post-replication repair may be responsible for higher mis-incorporation in the *Geminivirus* progeny DNA (Seal et al 2006).

Mutations in the helix 4 motif of the AL1 gene of two distantly related begomoviruses revert at 100% frequency, suggesting that nucleotide substitutions occur with high incidence and are under strong selective pressure during *Geminivirus* infection (Seal et al 2006). Thus, in agreement with recent reports of high mutation rates for other ssDNA viruses infecting vertebrates and bacteria, nucleotide substitution events are likely to contribute to the diversity and rapid evolution of *Geminivirus* ssDNA genomes (Shackelton et al 2005).

CONSEQUENCES OF MUTATION AND RECOMBINATION

An important consequence of high rates of mutation and recombination is the continuous production of genetic variation in *Geminivirus* populations. This variability is balanced by a complex set of selection pressures, including those associated with intrinsic properties of the virus, such as the maintenance of essential nucleotide structures and replication signals and selection pressures to maintain crucial interactions with plant hosts and insect vectors (Astorga et al 2007). The evolutionary potential of geminiviruses needs to be considered in long-term control strategies, because any disease management effort will result in selective pressure on the virus population to adapt to new circumstances (McDonald & Linde 2002). A recent mathematical analysis of the potential impact of disease control strategies concluded that the use of resistant cultivars with reduced within-plant virus titers creates pressure on the target virus to evolve towards a higher multiplication rate (Van den Bosch et al 2006). The results reported here demonstrated experimentally that *Geminivirus* variants with residual replication capabilities are under strong selective pressure to generate variants that replicate to high titers. Given the large size and genetic heterogeneity of *Geminivirus* populations, and their capacity to rapidly change their genomes by recombination and mutation, it is necessary to devise resistance strategies that prevent virus replication and not simply reduce it because of the risk of generating more harmful variants that overcome resistance (Astorga et al 2007).

INTERACTION OF VIRAL SUPPRESSORS WITH THE SILENCING PATHWAY

Systemic infection by plant viruses frequently results in disease symptoms that resemble developmental defects, including loss of leaf polarity, loss of proper control of cell division, and loss of reproductive functions (Hull 2001). These and other phenotypes are frequently associated with virus-encoded pathogenicity factors, many of which are suppressors of RNA silencing (Voinnet et al 1999). RNA silencing suppressors from different plant viruses are structurally diverse. The RNA silencing mechanism functions as an adaptive immune response, which restricts the accumulation or spread of inducing viruses (Waterhouse et al 2001). Because they overcome this form of host defense, viral suppressors are also regarded as pathogenecity determinants, which attenuate host RNAi and also causes disease or developmental abnormalities. Suppressor proteins encoded by members of different virus families are distinct, suggesting that plant viruses evolved this counter-defensive mechanism independently on many occasions (Vaucheret et al 2001, Tijsterman et al 2002).

Suppressors from various viruses were shown to interfere not only with siRNA activities but also with microRNA (miRNA) activities and to trigger, as transgenes, an overlapping series of severe developmental defects in *Arabidopsis thaliana*. This suggests that interference with miRNA-directed processes may be a general feature contributing to pathogenicity of many viruses (Chellappan et al 2004). An intermediate in the miRNA biogenesis/RNA-induced silencing complex (RISC) assembly pathway causes accumulation of miRNA*s in the presence of suppressors (P1/HC-Pro, P21, or P19) that inhibit miRNA-guided cleavage of target mRNAs. Both P21 and P19, but not P1/HC-Pro, interact with miRNA/miRNA* complexes and hairpin RNA-derived short interfering RNAs (siRNAs) *in vivo*. In addition, P21 was shown to bind to synthetic miRNA/ miRNA* and siRNA duplexes *in vitro*. In this way several different suppressors act by distinct mechanisms to inhibit the incorporation of small RNAs into active RISCs (Chapman et al 2004). It was also shown that microRNAs (miRNAs) are involved in modulating plant viral diseases (Dunoyer et al 2004). microRNA-mediated gene silencing serves as a general defense mechanism against plant viruses (Lu et al 2008).

Most of the viral silencing suppressors, when overexpressed in transgenic plants, interfere with production and/or action of miRNAs, thus leading to various abnormalities of plant development, often resembling viral symptoms (Voinnet 2005). The AC4 protein encoded by the *African cassava mosaic virus Cameroon strain* is a silencing suppressor found to interact directly with the miRNA pathway. AC4 can bind to single-stranded miRNA or siRNA and inhibit miRNA-mediated negative regulation of gene expression in *Arabidopsis* plants; as a consequence, plants show developmental defects (Chellappan et al 2005). It was reported that viruses that produce most severe symptoms in plants (for example TMV and ToMV in tobacco) altered miRNA accumulation to a greater extent than viruses that produced mild symptoms (i.e., TEV and PVY). Furthermore, transgenic plants co-expressing movement protein and coat protein-based silencing suppressors exhibited similar abnormal development and phenotypic symptoms as compared to mutants of *A. thaliana*, in which miRNA pathways are altered. All of this evidence suggests that certain disease symptoms depend on alterations in miRNA levels in the plants during viral infection by interfering with miRNA-directed processes such as hormone signaling (Fig. 18.1), and thus this might be recognized as a general feature of virus pathogenicity (Bazzini et al 2007).

ROLE OF PHYTOHORMONES IN PLANT DEFENSE RESPONSES

Plant hormones play important roles in regulating developmental processes and signaling networks involved in plant responses to a wide range of biotic and abiotic stresses. The identification and characterization of several mutants affected in the biosynthesis, perception, and signal transduction of these hormones have been instrumental in understanding the role of individual components of each hormone signaling pathway in plant defense (Bari & Jones 2009). Host defense response through signaling molecules such as salicylic acid (SA), jasmonic acid (JA), and ethylene (ET) alters the plant developmental processes during defense responses against pathogens (Chung et al 2008). SA and JA/ET defense pathways are mutually antagonistic, and this suggests that the defense signaling network activated and utilized by the plant is dependent on the nature of the pathogen and its mode of pathogenicity (Fig. 18.1). Interactions between defense signaling pathways are an important mechanism for regulating defense responses against various types of pathogen. In recent years, several components regulating the cross-talk between SA, JA, and ET pathways have been identified. However, the underlying molecular mechanisms are not well understood. It was also shown that during pathogen infection, GH3 (early auxin responsive gene) is activated to modulate the auxin pathway, resulting in enhanced disease susceptibility through increasing IAA biosynthesis and de-repressing auxin signaling. Similarly, $GH_{3}5$ positively modulates the SA pathway to enhance the plant defense response through elevating SA biosynthesis, activating SA-induced genes, WRKYs, and basal defense-related genes. The host defense response, through signaling molecules such as SA, JA, and ET, alters the plant developmental processes during defense responses against pathogens (Chung et al 2008).

The abnormal growth forms of virus-infected plants have encouraged experiments that examine the effects of viruses on hormone levels in plants and vice versa (Jameson 2000). While it is difficult to formulate general rules about the effects of viruses on phytohormones, it is clear that abscisic acid, auxin, cytokinin, giberellin, and ethylene levels, alone or in combination, can all be perturbed, depending on the virus–host combination. Recently, direct links between auxin signaling and giberellin levels have been established for TMV and *Rice dwarf virus* (RDV), respectively. The helicase domain of the TMV 126- and 183-kDa replicase proteins was shown to interact with the Aux/indole 3-acetic acid (IAA) transcription factor IAA26 (Padmanabhan et al 2005), and this interaction contributed to the development of symptoms in plants (*Arabidopsis, N. benthamiana*); it also leads to the formation of nonfunctional IAA26 protein, causing phenotypic abnormalities, as was experimentally shown by silencing the IAA26 gene, which caused sililar phenotypic symptoms as infection with TMV. The ability of TMV to modulate symptoms through interaction with IAA26 was postulated to be due to interference with its normal role in forming heterodimers with auxin-response factors (ARF). It was also documented that *Tobacco mosaic virus* (TMV)-induced disease symptoms, including the loss of apical dominance, stunting, and leaf curling, are caused by the inappropriate expression of auxin-related genes, which is mediated through an interaction between TMV replicase and its specific target protein PAP1, a negative regulator of the auxin response factor (ARF) (Padmanabhan et al 2005). In the absence of auxin, Aux/IAA proteins, such as IAA26, bind to ARF transcription factors and prevent them from modulating the transcription of auxin-responsive genes.

However, in the presence of auxin, the Aux/IAA proteins are targeted for degradation, thus freeing the ARF proteins to modulate transcription of their target genes.

In addition to affecting auxin signaling events, it has been found that some viral proteins affect giberellic acid (GA) signaling. One such example is the P2 protein of RDV (*Rice dwarf virus*), which currently has no known role in the infection process except that it is a symptom determinant. Zhu and associates (2005) demonstrated—by a yeast two-hybrid assay—that P2 interacts with four different *ent*-kaurene oxidases or oxidase-like proteins. The *ent*-kaurene oxidases catalyze a step in the synthesis of GA, and *ent*-kaurene oxidase mutant rice plants are deficient in GA. In addition, this mutant rice possesses a phenotype similar to symptoms caused by RDV infection. The authors demonstrated that RDV infection results in decreased GA levels and symptoms that could be alleviated by GA application. Although host gene expression assays were not performed to confirm this, it is likely that GA-responsive genes are modulated by RDV infection. Taken together, emerging evidence suggests that auxin acts as an important component of hormone signaling networks involved in the regulation of defense responses and modulates defense and development responses. However, how auxin levels affect the balance of other hormones and fine-tune defense responses specific to different pathogens remains to be studied (Bari & Jones 2009). Numerous miRNAs have been predicted or validated to be involved in plant defense. For example miR-139 targets a gene encoding a mucin-like protein carrying a dense sugar coating against proteolysis, which is a pivotal step in pathogen invasion; miR160-3 acts on intracellular pathogenesis-related protein and miR408 provides defense through interaction with the genes encoding a copper ion-binding protein (Isam et al 2007). Overexpression of a plant miRNA (miR393) resulted in increased bacterial resistance (Navarro et al 2006). Therefore, it is thought that plant miRNA-directed RNAi or miRNA-specified mRNA destruction determines the balance in plant defense systems.

HOW DO PLANTS DEFEND THEMSELVES AGAINST THE EVOLUTIONARY POTENTIAL OF INVADING VIRUSES?

Plant defense systems are elegant examples of how nature can find highly efficient solutions to the problems it faces. Overall, it can be described as a co-evolution of defense and counter-defense mechanisms between the host plant and the invading virus (Lu et al 2008). A variety of defense responses have been reviewed recently (Carrington & Whitham 1998), but one of the most exciting areas of current research is post-transcriptional gene silencing (PTGS). RNA silencing suppression is a common property of plant viruses. Suppressor proteins are considered as pathogenicity determinants found in most viruses. Silencing suppressor proteins show a tremendous structural and sequence diversity that has been explained as an evolutionary convergence toward a common functional necessity (Li & Ding 2006).

Recent work on PTGS in plants has provided evidence that this mechanism functions as a general defense against virus invasion. Viral invasion can induce gene silencing and provide cross-protection against secondary virus infection (Ratcliff et al 1999, Ding 2000). At the same time, suppression of gene silencing is a general strategy used by a broad range of DNA and RNA plant viruses. Successful virus infection results from the ability of the virus to prevent PTGS-mediated degradation of its genome, either by directly incapacitating the plant's PTGS response or by moving through the plant more quickly than the PTGS response is initiated, or both (Waterhouse et al 1999). PTGS suppressors characterized so far include cucumovirus 2b, potyvirus HC-Pro, sobemovirus P1, tombusvirus P19, potex virus p25, and ACMV AC2 (TrAP) TYLCV-C C2 proteins (Carrington et al 2001, Voinnet 2001). There is much variation in the extent of suppression of PTGS by different viruses (Voinnet et al 1999), and individual suppressors can target different steps in the PTGS pathway (Anandalakshmi et al 2000, Voinnet et al 2000).

RNA-DIRECTED DNA METHYLATION AS A PLANT GENOME DEFENSE MECHANISM

High levels of dsRNA produced by viral infections or highly transcribed transgenes can provoke DNA methylation. For example, members of the geminiviridae are true DNA viruses that replicate circular, single-stranded DNA genomes in the nucleus by a rolling-circle mechanism that employs the host replication machinery (Hanley-Bowdoin et al 2004). The double-stranded DNA (dsDNA) intermediates that mediate both viral replication and transcription associate with cellular histone proteins to form mini-chromosomes (Pilartz & Jeske 2003). Transcripts produced from these mini-chromosomes are subject to PTGS, and geminiviruses and their associated satellites have been shown to encode a variety of proteins that can suppress this defense (Vanitharani et al 2004, Trinks et al 2005). In addition, given the role of RNA-directed methylation in silencing endogenous invasive DNAs, it is reasonable to propose that plants might also use methylation as a means to repress transcription and/or replication from a viral mini-chromosome (Bisaro 2006). A putative pathway (Fig. 18.2) for RNA-directed DNA methylation during *Geminivirus* infection suggests that viral genome targets may be transcribed by an RNA polymerase IVa complex (Pol IVa; containing NRPD1A and NRPD2). The resulting single-stranded RNA (ssRNA) is converted to dsRNA by complexes containing RDR2. The 24-nt siRNAs processed from dsRNA by DCL3 are loaded into complexes containing AGO4, which subsequently associates with Pol IVb (containing NRPD1B and NRPD2). The AGO4-associated siRNAs target the complex to homologous DNA sequences, where cytosine methyltransferases (e.g., DRM1/2) are recruited. Cytosine methyltransferases CMT3 and MET1 are primarily involved in methylation maintenance at CNG and CG sites, respectively. CNG methylation by CMT3 is also linked to H3K9 methylation carried out by KYP2 (Raja et al 2008).

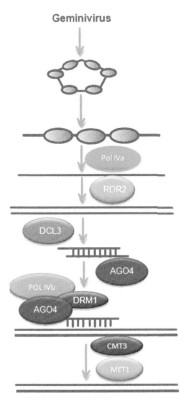

FIGURE 18.2   Putative pathway for RNA-directed DNA methylation during *Geminivirus* infection.

It was shown that *in vitro* methylation of *Geminivirus* DNA greatly reduces its ability to replicate in plant protoplasts (Brown et al 1992) and demonstrated that *Geminivirus* AL2 (also known as AC2 or C2) and L2 proteins can act as silencing suppressors by interacting with and inhibiting adenosine kinase (ADK) (Wang et al 2005). ADK is required for efficient production of the methyl group donor *S*-adenosylmethionine (SAM), and the primary defect of ADK-deficient yeast and plants is methylation deficiency (Moffatt et al 2002). Thus, it is possible that one role of the AL2 and L2 proteins is to counter a methylation-based defense. At least one *Geminivirus*-silencing suppressor protein has been hypothesized to counter this defense by inhibiting methylation reactions (Wang et al 2003).

CONCLUSION

Virus infection can disturb various developmental processes leading to symptom development by affecting host defense mechanisms. Plant hormones play a critical role in regulating plant developmental processes. In plants, microRNAs (miRNAs) target a wide range of mRNAs involved in various developmental processes, including hormone signaling. Virus-encoded pathogenecity factor/RNAi suppressor plays a significant role during infection, which leads to abnormal phenotypes in the plants—either through interfering with small RNA pathways, hormone signaling, or host defense pathways. This counter-defense response, in turn, helps viruses to establish themselves successfully in the plants. As a consequence, genetic diversity generated either by high mutation rates or by frequent recombination events allows the rapid evolution of viruses. Host-induced genome evolution allows for adaptation of plant viruses to their hosts, which is found to be either random or selected, depending on the virus and host species.

REFERENCES

Anandalakshmi, R., Marathe, R., Ge, X., Herr, J.M., Mau, C., Mallory, A., Pruss, G., Bowman, L., Vance, V.B., 2000. A calmodulin-related protein that suppresses post-transcriptional gene silencing in plants. Science 290, 142–144.

Astorga, G.A., Ascencio-Ibanez, J.T., Dallas, M.B., Orozco, B.M., Bowdoin, L.H., 2007. High-frequency reversion of *Geminivirus* replication protein mutants during infection. Journal of Virology 81 (20), 11005–11015.

Bari, R., Jones, J.D.G., 2009. Role of plant hormones in plant defence responses. Plant Molecular Biology 69, 473–488.

Bazzini, A.A., Hopp, H.E., Beachy, R.N., Asurmed, S., 2007. Infection and co-accumulation of tobacco mosaic virus proteins alter microRNA levels, correlating with symptom and plant development. Proceedings of the National Academy of Sciences *of the USA* 104, 12157–12162.

Bisaro, D.M., 2006. Silencing suppression by geminivirus proteins. Virology 344, 158–168.

Brough, C.L., Gardiner, W.E., Inamdar, N.M., Zhang, X.Y., Ehrlich, M., Bisaro, D.M., 1992. DNA methylation inhibits propagation of tomato golden mosaic virus DNA in transfected protoplasts. Plant Molecular Biology 18, 703–712.

Brown, J.K., Bird, J., 1992. White fly transmitted geminiviruses and associated disorders in Americas and the Carribbean Basin. Plant Disease 76, 220–225.

Bull, S.E., Briddon, R.W., Sserubombwe, W.S., Ngugi, K., Markham, P.G., Stanley, J., 2006. Genetic diversity and phylogeography of cassava mosaic viruses in Kenya. Journal of General Virology 87, 3053–3065.

Carrington, J.C., Whitham, S.A., 1998. Viral invasion and host defense: strategies and counter-strategies. Current Opinion in Plant Biology 1, 336–341.

Carrington, J.C., Kasschau, K.D., Johansen, L.K., 2001. Activation and suppression of RNA silencing by plant viruses. Virology 281, 1–5.

Chapman, E.J., Prokhnevsky, A.I., Gopinath, K., 2004. Viral RNA silencing suppressors inhibits the microRNA pathway at an intermediate step. Genes & Development 18, 1179–1186.

Chellappan, P., Vanitharani, R., Fauquet, C.M., 2004. Short interfering RNA accumulation correlates with host recovery in DNA virus-infected hosts, and gene silencing targets specific viral sequences. Journal of Virology 78, 7465–7477.

Chellappan, P., Vanitharani, R., Fauquet, C.M., 2005. MicroRNA-binding viral protein interferes with *Arabidopsis* development. Proceedings of the National Academy of Sciences *of the USA* 102, 10381–10386.

Chung, K.M., Kadunari, I., Naoyuki, U., Masao, T., 2008. New perspectives on plant defense responses through modulation of developmental pathways. Molecular Cells, OS, 107–112.

Ding, S.W., 2000. RNA silencing. Current Opinion in Biotechnology 11, 152–156.

Dunoyer, P., Lecellier, C.H., Parizotto, E.A., Himber, C., Voinnet, O., 2004. Probing the microRNA and small interfering RNA pathways with virus-encoded suppressors of RNA silencing. Plant Cell 16, 1235–1250.

Fauquet, C.M., Bisaro, D.M., Briddon, R.W., Brown, J.K., Harrison, B.D., Rybicki, E.P., Stenger, D.C., Stanley, J., 2003. Revision of taxonomic criteria for species demarcation in the family *Geminiviridae*, and an updated list of *Begomovirus* species. Archives of Virology 148, 405–421.

Garcia-Arenal, F., Fraile, A., Malpica, J.M., 2001. Variability and genetic structure of plant virus populations. Annual Review of Phytopathology 39, 157–186.

Ge, L.M., Zhang, J.T., Zhou, X.P., Li, H.Y., 2007. Genetic structure and population variability of *Tomato yellow leaf curl China virus*. Journal of Virology 81 (11), 5902–5907.

Gutierrez, C., 1999. *Geminivirus* DNA replication. Cellular and Molecular Life Science 56, 313–329.

Hanley-Bowdoin, L., Settlage, S., Robertson, D., 2004. Reprogramming plant gene expression: a prerequisite to geminivirus replication. Molecular Plant Pathology 5, 149–156.

Harrison, B.D., Robinson, D.J., 1999. Natural genomic and antigenic variation in whitefly-transmitted geminiviruses (*Begomoviruses*). Annual Review of Phytopathology 37, 369–398.

Hull, R., 2001. Matthews' plant virology, fourth ed. Academic Press, San Diego, CA.

Inamdar, N.M., Zhang, X.Y., Brough, C.L., Gardiner, W.E., Bisaro, D.M., Ehrlich, M., 1992. Transfection of heteroduplexes containing uracilguanine or thymine-guanine mispairs into plant cells. Plant Molecular Biology 20, 123–131.

Isam, F., Bjorn, V., Ralf, R., Wolfgang, R.H., Wolfgang, F., 2007. Evidence for the rapid expansion of microRNA-mediated regulation in early land plant evolution. Plant Biology 7, 13.

Isnard, M., Granier, M., Frutos, R., Reynaud, B., Petterschmitt, M., 1998. Quasispecies nature of three maize streak virus isolates obtained through different modes of selection from a population used to assess response to infection of maize cultivars. Journal of Genral Virology 79, 3091–3099.

Jameson, P., 2000. Cytokinins and auxins in plant–pathogen interactions – an overview. Plant Growth Regulation 32, 369–380.

Kearney, C.M., Thomson, M.J., Roland, K.E., 1999. Genome evolution of tobacco mosaic virus populations during long-term passaging in a diverse range of hosts. Archives of Virology 144, 1513–1526.

Li, F., Ding, S.W., 2006. Virus counter defense: diverse strategies for evading the RNA-silencing immunity. Annual Review of Microbiology 60, 503–531.

Lu, Y., Gan, Q., Chi, X., Qin, S., 2008. Roles of microRNA in plant defense and virus offense interaction. Plant Cell Reports 27, 1571–1579.

McDonald, B.A., Linde, C., 2002. Pathogen population genetics, evolutionary potential, and durable resistance. Annual Review of Phytopathology 40, 349–379.

Moffatt, B.A., Stevens, Y.Y., Allen, M.S., Snider, J.D., Pereira, L.A., Todorova, M.I., Summers, P.S., Weretilnyk, E.A., Martin-McCaffrey, L., Wagner, C., 2002. Adenosine kinase deficiency is associated with developmental abnormalities and reduced transmethylation. Plant Physiology 128 (3), 812–821.

Navarro, L., Dunoyer, P., Jay, F., Arnold, B., Dharmasiri, N., Estelle, M., Voinnet, O., Jones, J.D.G., 2006. A plant miRNA contributes to antibacterial resistance by repressing auxin signaling. Science 312, 436.

Padmanabhan, M.S., Goregaoker, S.P., Golem, S., Shiferaw, H., Culver, J.N., 2005. Interaction of the tobacco mosaic virus replicase protein with the Aux/IAA protein PAP1/IAA26 is associated with disease development. Journal of Virology 79 (4), 2549–2558.

Patil, B.L., Rajasubramaniam, S., Bagchi, C., Dasgupta, I., 2005. Both *Indian cassava mosaic virus* and *Sri Lankan cassava mosaic virus* are found in India and exhibit high variability as assessed by PCR-RFLP. Archives of Virology 150, 389–397.

Pilartz, M., Jeske, H., 2003. Mapping of Abutilon mosaic geminivirus minichromosomes. Journal of Virology 77, 10808–10818.

Raja, P., Bradley, C., Sanville, R., Buchmann, C., Bisaro, D.M., 2008. Viral genome methylation as an epigenetic defense against geminiviruses. Journal of Virology 82 (18), 8997–9007.

Ratcliff, F.G., MacFarlane, S.A., Baulcombe, D.C., 1999. Gene silencing without DNA: RNA-mediated cross-protection between viruses. Plant Cell 11, 1207–1215.

Rojas, M.R., Hagen, C., Lucas, W.J., Gibertson, R.L., 2005. Exploiting chinks in the plant's armor: evolution and emergence of geminiviruses. Annual Review of Phytopathology 43, 361–394.

Seal, S.E., Van-den-Bosch, F., Jeger, M.J., 2006. Factors influencing begomovirus evolution and their increasing global significance: implications for sustainable control. Critical Review Plant Science 25, 23–46.

Shackelton, L.A., Parrish, C.R., Truyen, U., Holmes, E.C., 2005. High rate of viral evolution associated with the emergence of carnivore parvovirus. Proceedings of the National Academy of Sciences of the USA 102, 379–384.

Stenger, D.C., 1995. Genotypic variability and occurrence of less than genome-length viral DNA forms in a field population of beet curly top geminivirus. Phytopathology 85, 1316–1322.

Strauss, J.H., Strauss, E.G., 2001. Virus evolution: how does an enveloped virus make a regular structure? Cell 105, 5–8.

Tijsterman, M., Ketting, R.F., Plasterk, R.H., 2002. The genetics of RNA silencing. Annual Review of Genetics 36, 489–519.

Trinks, D., Rajeswaran, R., Shivaprasad, P.V., Akbergenov, R., Oakeley, E.J., Veluthambi, K., Hohn, T., Pooggin, M., 2005. Suppression of RNA silencing by a geminivirus nuclear protein, AC2, correlates with transactivation of host genes. Journal of Virology 79, 2517–2527.

Van-den-Bosch, F., Akudibilah, G., Seal, S., Jeger, M., 2006. Host resistance and the evolutionary response of plant viruses. Journal of Applied Ecology 43, 506–516.

Vanitharani, R., Chellappan, P., Pita, J.S., Fauquet, C., 2004. Differential roles of AC2 and AC4 of cassava geminiviruses in mediating synergism and suppression of posttranscriptional gene silencing. Journal of Virology 78, 9487–9498.

Vaucheret, H., Béclin, C., Fagard, M., 2001. Post-transcriptional gene silencing in plants. Journal of Cell Science 114, 3083–3091.

Voinnet, O., 2001. RNA silencing as a plant immune system against viruses. Trends in Genetics 17, 449–459.

Voinnet, O., 2005. Induction and suppression of RNA silencing: insights from viral infections. Nature Review Genetics 6, 206–221.

Voinnet, O., Pinto, V.M., Baulcombe, D.C., 1999. Suppression of gene silencing: a general strategy used by diverse DNA and RNA viruses of plants. Proceedings of the National Academy of Sciences of the USA 96, 14147–14152.

Voinnet, O., Lederer, C., Baulcombe, D.C., 2000. A viral movement protein prevents spread of the gene silencing signal in Nicotiana benthamiana. Cell 103, 157–167.

Wang, G., Wei, L.N., Loh, H.H., 2003. Transcriptional regulation of mouse δ-opioid receptor gene by CpG methylation: involvement of Sp3 and a methyl-CpG-binding protein, MBD2, in transcriptional repression of mouse δ-opioid receptor gene in Neuro2A cells. Journal of Biological Chemistry 278, 40550–40556.

Wang, H., Buckley, K.J., Yang, X., Buchmann, R.C., Bisaro, D.M., 2005. Adenosine kinase inhibition and suppression of RNA silencing by geminivirus AL2 and L2 proteins. Journal of Virology 79, 7410–7418.

Waterhouse, P.M., Smith, N.A., Wang, M.B., 1999. Virus resistance and gene silencing: killing the messenger. Trends in Plant Science 4, 452–457.

Waterhouse, P.M., Wang, M.B., Lough, T., 2001. Gene silencing as an adaptive defense against viruses. Nature 411, 834–842.

Whitham, S.A., Yang, C., Goodin, M.M., 2006. Global impact: elucidating plant responses to viral infection. Molecular Plant–Microbe Interactions 19 (11), 1207–1215.

Zhu, S., Gao, F., Cao, X., Chen, M., Ye, G., Wei, C., Li, Y., 2005. The *Rice dwarf virus* P2 protein interacts with ent-kaurene oxidases in vivo, leading to reduced biosynthesis of gibberellins and rice dwarf symptoms. Plant Physiology 139, 1935–1945.

Chapter 19

Impact of the host on plant virus evolution

Xiao-fei Cheng
College of Life and Environmental Science, Hangzhou Normal University, Hangzhou, Zhejiang, P R China

Nasar Virk
Atta-ur-Rahman School of Applied Biosciences, National University of Sciences and Technology, Islamabad, Pakistan

Hui-zhong Wang
College of Life and Environmental Science, Hangzhou Normal University, Hangzhou, Zhejiang, P R China

INTRODUCTION

Plant viruses are obligate parasites that rely on the host cell for their survival and replication. The requirement of optimum conditions for propagation and the presence of host defense mechanisms force the viruses to continually coevolve with the host. Mutation is believed to be the most important way to achieve optimal adaptation to the host. It is known that both RNA and DNA viruses have high mutation rates (Holland et al 1982, Domingo 1997, Bonhoeffer & Sniegowski 2002, Domingo-Calap & Sanjuán 2011). In addition, viruses also use recombination, reassortment, and gene rearrangement to increase genome variability (Sztuba-Solinska et al 2011, Greenbaum et al 2012). In fact, the viral population produced during infection is composed of a group of complex variants termed viral quasispecies (Lauring & Andino 2010). All of these variants serve as a genetic reservoir and are selected by the virus itself and by pressure from its host. Selection results in specific footprints in the viral genome, such as genome composition, mutation bias, amino acid usage, synonymous codon usage, and dinucleotide usage. Detailed analysis of these footprints enables us to understand the evolutionary direction of the virus.

The mechanisms of viral evolution, such as mutation rate, quasispecies formation, recombination frequency, viral population variety, and viral phylogeny, have been studied extensively (Domingo et al 2008, Roossinck 1997, 2008). Instead of probing the aforementioned issues, the purpose of this chapter is to summarize the recent progress made in understanding how plants influence the evolution of viruses, particularly plant viruses. We divided our review into four parts: how plants affect genome stability, amino acid exchanges, synonymous codon usage, and dinucleotide bias.

IMPACT OF THE HOST ON VIRAL GENOME STABILITY

Due to the lack of proofreading of viral RNA polymerases, both plant and animal RNA viruses have mutation rates much higher than those of the host cell DNA, which is replicated via the DNA polymerase (Steinhauer et al 1992). The mutation rate for RNA viruses has been estimated to be within the range of 10^{-3} to 10^{-5} substitutions per nucleotide and per round of replication (Drake 1993). Plant DNA viruses, such as geminiviruses, can also evolve as quickly as their RNA counterparts (Arguello-Astorga et al 2007, Ge et al 2007, Duffy & Holmes 2008, 2009). Furthermore, plant RNA and DNA viruses can also produce vast numbers of indel (insertion and deletion) mutants during genome replication (Stanley et al 1990, Domingo 1992, Eigen 1996). All of these variants produced during viral replication form a huge genetic reservoir, termed 'viral quasispecies' (Lauring & Andino 2010). There is no doubt that among the viral population a proportion of the mutational variants might be beneficial for the virus, thus increasing its adaptability, while others could be lethal for the virus. However, all of these mutants will continue to survive in the host cell as long as the essential viral proteins for their replication are available. In some circumstances, the lethal mutants (e.g., the deletion mutants) will constitute the majority of the viral population; these viruses are shorter than the wild-type virus, and they replicate much faster (Zarling 1976, Shirako & Brakke 1984, de Oliveira Resende et al 1991, Moutailler et al 2011, Pu et al 2011). Most of the variants will be filtered out after a host alternation, because variants that contain one or more lethal mutations or lack essential proteins for their replication cannot survive in the new host. Actually, host alternations have been shown to play an important role in maintaining genome integrity and elimination of lethal mutants of viruses, especially arthropod-borne viruses, such as *Rice dwarf virus* (RDV) and *Tomato spotted wilt virus* (TSWV) (Sin et al 2005, Pu et al 2011). Interestingly, the distribution of mutational deletions in a given viral genome is not random; rather, it is concentrated in one or more mutational hot spots. For example, the major deletions within the TSWV genome generated by serial mechanical transmissions are localized in the gene encoding the precursor of glycoproteins (de Oliveira et al 1991), whereas the *3a* gene of RNA3 is the hot spot for *Cucumber mosaic virus* (CMV) deletion mutations generated during infection of different host plants (Graves & Roossinck 1995, Takeshita et al 2008).

Host plants may also have a role in increasing the genome complexity in some plant viruses, even promoting the emergence of new viral genes. For instance, the genome of viruses within the family *Closteroviridae* encodes two classes of genes. The first class includes the genes shared by all closteroviruses, termed 'conserved core genes'; these genes are involved in virus replication. Genes of the second class include species-specific genes (Karasev 2000, Martelli et al 2012). These latter genes show dramatic variation in their locations, numbers, and functions, and their products lack similarity with other proteins of closteroviruses. These results suggest that the latter genes originated separately in closteroviruses and underwent evolution recently. Functional analysis showed that the three non-conserved genes p33, p18, and p13 encoded by *Citrus tristeza virus* (CTV) have different roles in infecting various citrus hosts. All of these genes are dispensable for CTV replication in *Citrus macrophylla* and *Mexican lime* [*C. aurantifolia* (Christm.) Swing.] but are required for other citrus plants, including sour orange (*C. aurantium* L.), lemon [*C. limon* (L.) Burm.f.], grapefruit (*C. paradisi* Macf.) or calamondin (*C. mitis* Blanco) (Tatineni et al 2008, 2011). These results suggest that these genes in CTV evolved during adaptation to the new citrus hosts. Interestingly, the movement proteins encoded by all plant viruses are believed to have evolved during the adaptation of ancestral viruses to plant hosts as well (Mushegian & Koonin 1993, Melcher 2000, Lucas 2006).

It is noteworthy that viruses have also evolved several mechanisms to maintain their genome integrity, including evolving additional viral protein(s) to increase the fidelity of RNA-dependent RNA polymerase (RdRp), genome length checking, and restoring damaged genome termini (reviewed by Barr & Fearns 2010).

IMPACT OF THE HOST ON VIRAL AMINO ACID USAGE

As mentioned above, the need for adaptation to plants encourages viruses to quickly change their genomes; such changes include the emergence of new viral gene(s). In addition to this model of the emergence of new viral genes, plants are also able to influence the evolution of existing viral genes—especially those viral genes that encode proteins that directly interact with host proteins; this is because compatibility between viral proteins and host factors affects the efficiencies of viral replication, viral particle assembly, cell-to-cell movement, and other viral processes. Therefore, the interfaces of viral proteins that directly interact with host factors will be subjected to stronger selection pressure than other positions. For example, the plant eukaryotic translation initiation factors eIF4E and eIF4G interact directly with the potyviral genome-linked protein (VPg), which is essential for potyvirus multiplication. The eIF containing amino acid substitutions at the interaction interface will hamper the interaction between the viral VPg and eIFs, abolishing the replication of potyvirus and finally resulting in the emergence of recessive resistance against potyviruses in the host plant (Nieto et al 2011). In fact, a large portion of recessive resistance genes in crops against the potyviruses are eIFs (reviewed by Diaz-Pendon et al 2004, Truniger & Aranda 2009). Similarly, potyviruses are able to induce amino acid changes in their VPgs to overcome host eIF4E-mediated resistance (Truniger & Aranda 2009). These results suggest that the VPgs of potyviruses may have coevolved with the plant eIF4E (Charron et al 2008).

The point mutations introduced during viral replication can result in two possible consequences for the viral proteins: amino acid substitution (non-synonymous mutation) or only nucleotide substitution (synonymous mutation). Comparing the number of non-synonymous substitutions per non-synonymous site (d_N) to the number of synonymous substitution per synonymous site (d_S) in a given gene will give us an important indicator of the selective pressure acting on that gene (Hurst 2002). At the null hypothesis (no selection), the non-synonymous and synonymous substitution should take place at similar frequencies, and the value of d_N/d_S should approach 1.0. However, if a gene is under strong positive selection (also called Darwinian selection), the value of d_N/d_S should be higher than 1.0 and vice versa. Woelk & Holmes (2002) performed an analysis of the substitution patterns in the genes of animal RNA viruses; they found that the surface structural genes (e.g., envelope glycoprotein or outer capsid genes) of vector-borne RNA viruses have reduced rates of non-synonymous substitutions than those of nonvector-borne RNA viruses. This result is consistent with the different virus–host interactions between vector-borne and nonvector borne RNA viruses. The surface structural proteins of vector-borne viruses are under selection pressures from both their hosts and vectors, whereas the latter ones are only under the selection pressures of their hosts. Similarly, the nucleotide substitution model in the capsid genes of plant RNA viruses reflects the mode of transmission, suggesting the existence of host-specific selection pressures on certain viruses (Chare & Holmes 2004). In conclusion, there is no doubt that host plants affect the evolution of all viral proteins directly or indirectly during viral multiplication in their cells.

IMPACT OF THE HOST ON VIRAL SYNONYMOUS CODON CHOICE

Due to the degeneracy of genetic codons, all amino acids, except methionine and tryptophan, are encoded by more than one codon. Codons encoding the same amino acid are known as synonymous codons. The individual synonymous codons for a given amino acid are not used at similar frequencies in different genes or organisms, indicating a bias in codon usage (Grantham et al 1980). Synonymous codon usage is determined by many factors, such as translation selection, mutation pressure, gene transfer, amino acid conservation, RNA stability, hypersaline adaptation, and growth conditions (Ermolaeva 2001, Lynn et al 2002, Paul et al 2008). For viruses, the viral mutational preference is thought to be the most important factor that shapes viral synonymous codon usage (Jenkins & Holmes 2003, Adams & Antoniw 2004). However, the translational pressure due to tRNA availability, nucleotide acid abundance, and selection of CpG-suppressed clones in the host cell by the immune system, also affects viral synonymous codon usage and even determines the synonymous codon usage bias in some viral genes or particular viruses (Karlin et al 1990, Zhou et al 1999, Woo et al 2007, Lobo et al 2009, Aragonès et al 2010).

For example, Chantawannakul and Cutler (2008) found that the nucleotide composition at all codon positions and synonymous codon usage of viruses infecting the honeybee show a high degree of resemblance to that of the honeybee, suggesting that the long-term convergent evolution between honeybee and associated viruses results in the adaptation of virus synonymous codon usage to that of the host.

The first comprehensive analysis of the synonymous codon usage of plant viruses was carried out by Adams and Antoniw (2004). In this study, the synonymous codon usage bias of 385 plant viruses was measured with an effective number of codons (ENC), a simple method to quantify how far the codon usage of a gene departs from equal usage of synonymous codons (Wright 1990), and was correlated with the viral nucleotide composition, host type, and mode of transmission. They found that the ENC values of these viruses were positively correlated with those of viral GC contents in the third codon position but not with the host type they infect or the transmission model. As a result, they concluded that mutational bias, rather than translational selection, accounts for the observed variations in synonymous codon usage in plant viruses, and that there is no obvious impact of host translation selection in the viral synonymous codon usage. However, there are several pitfalls in their study. First, it is arbitrary to use only one indicator, the ENC value, to evaluate viral synonymous codon usage. Second, a direct comparison of the relative synonymous codon usage (RSCU) between the viruses and their respective hosts was not performed. Third, it is also arbitrary to categorize all the plant viruses based on the type of host plants they infect without considering features of the viral genome, for example, genome type (ssDNA, dsDNA, ssRNA, or dsRNA) and genome polarity (positive or negative), because viruses with different genome features were originated separately and may differ from each other greatly in many aspects, including the nucleotide composition and mutation bias. In fact, detailed analysis of the synonymous codon usage of begomoviruses (circular ssDNA viruses, *Geminiviridae*) showed that translational selection can be detected in the genomes of begomoviruses, especially in the highly expressed genes, although mutation bias appears to be the major determinant of the overall synonymous codon usage of begomoviruses (Xu et al 2008). Interestingly, we found a high degree of similarity of the synonymous codon usage between CTV and its citrus host (Cheng et al 2012). Additionally, the synonymous codon usage resemblance between woody plant-infecting closteroviruses and their woody hosts is higher than that between herbaceous plant-infecting closteroviruses and their herbaceous hosts (Cheng et al 2012). This result further confirms the influence of the host on synonymous codon usage in plant viruses. In another study, we also found that linear specific synonymous codon usage exists in viruses within the *Bunyviridae* and two phylogenetically related genera, *Tenuivirus* and *Emaravirus*, although the synonymous codon usage of most of these viruses shows a high degree of resemblance, suggesting that the mutational preference is the major factor influencing synonymous codon usage (our unpublished data).

In conclusion, several basic deductions can be drawn from the above studies. First, the synonymous codon usage of viruses within the same genus is always highly similar. In other words, mutational pressure is the major factor determining the overall synonymous codon usage. Second, translational pressure from the host also affects the viral synonymous codon usage, even if not in all plant viruses. Third, the influence of host translational pressure may be stronger in the genes that are highly expressed than in those expressed at lower levels. Fourth, the impact of host translational selection may be important in some particular plant viruses, such as those that coevolved with their plant hosts.

IMPACT OF THE HOST ON VIRAL DINUCLEOTIDE FREQUENCY

The dinucleotide frequency is the incidence of a given neighbor dinucleotide in a sequence (e.g., a gene or a genome). When all nucleotides are used randomly (no selection), the frequencies of the sixteen dinucleotide pairs should be similar. However, studies have shown that several dinucleotide pairs are always over-presented or underrepresented in the genomes tested (Kariin & Burge 1995, De Amicis & Marchetti 2000, Simmen 2008, Elango et al 2009), suggesting the existence of selection pressure(s). Actually, TpA was found to be repressed in almost all organisms tested, whereas CpG was always under-represented in the genomes of eukaryotic organisms (except in invertebrates) but not in prokaryotes (Kariin & Burge 1995). The depletion of TpA was thought to avoid nonsense mutation, to minimize improper transcription, and to reduce the risk of immune response (Karlin & Mrázek 1997, Forsbach et al 2008), whereas the repression of CpG in eukaryotic organisms was due to the cytosine methylation in these genomes, which is prone to mutate into thymine through spontaneous deamination, resulting in the dinucleotide TpG and the subsequent presence of a CpA in the opposite strand after DNA replication (Bird 1980). This result is consistent with the concomitant CpA and TpG overrepresentation in CpG-suppressed organisms. Similar research also has been carried out in viruses, including some plant viruses (Karlin et al 1994, Rima & McFerran 1997, Zsiros et al 1999, Tan et al 2004, Greenbaum et al 2008). Results showed that CpG was also predominantly repressed in viruses, especially in small eukaryotic RNA viruses (Karlin et al 1994, Rima & McFerran 1997). Interestingly, the CpG usages of viruses within the family *Flaviridae* and the picorna-like virus superfamily are consistent with that of their respective mammal, plant, and insect hosts (Jenkins et al 2001, Lobo et al 2009, Kapoor et al 2010), suggesting that the host has an important role in determining CpG dinucleotide usage.

The repression of CpG in DNA viruses is easily understood; however, the genomes of RNA viruses are not methylated, and DNA intermediates are not produced during the replication of the genomes. Thus, the CpG repression in RNA viruses was left as a mystery until the linkage between CpG dinucleotide and mammal innate antiviral immunity was recently discovered (Woo et al 2007, Greenbaum et al 2008, Jimenez-Baranda et al 2011). In the mammalian innate immune system, there are two types of receptor detecting abnormal or exogenous RNAs, such as viral RNAs: (i) the RIG-I-like RNA helicase receptors (RLH) and (ii) the Toll-like receptors (TLR) (Takeda et al 2003, Thompson & Locarnini 2007, Kawai & Akira 2008). The RLH contains three members, RIG-I, melanoma differentiation-associated gene 5 (MDA5), and LPG-2. These receptors recognize viral dsRNA without sequence specificity but differ from each other in their specificity for blunt- or sticky-ended dsRNA only. Within the 13 cloned TLRs, TLR3, TLR7/8, and TLR9 are specific in recognizing viral nucleic acids (Thompson & Locarnini 2007, Kawai & Akira 2008). Especially, TLR3 specifically recognizes viral dsRNAs with no obvious sequence specificity (Alexopoulou et al 2001), TLR9 binds to unmethylated viral DNA (Cornelie et al 2004), and TLR7/8 recognizes the ssRNAs that contain abnormal sequence motifs, including motifs of CpG in an AU-rich context (Diebold et al 2004, Heil et al 2004, Lund et al 2004, Jimenez-Baranda et al 2011). This result is further supported by the fact that the CpG dinucleotides were gradually lost in the genome of influenza virus B, since it shifted its host from swine to humans beginning in the 1820s (Greenbaum et al 2008, 2009).

As noted above, CpGs were also found to be repressed in some plant RNA viruses (Karlin et al 1994, Rima & McFerran 1997). However, a comprehensive analysis of the dinucleotide usage in plant RNA viruses has not yet been performed. Therefore, we downloaded all available genome sequences of plant RNA viruses deposited in the GenBank database (a total of 450 viruses) and calculated their CpG usages. Interestingly, the CpG dinucleotides were found to be underrepresented in the majority of plant RNA viruses, with the mean value of CpG odds ratio (CpG$_{O/E}$ value) of 0.74 ± 0.202 (Fig. 19.1). Moreover, the degrees of CpG repression in plant RNA viruses varies between different viral groups: the CpG was repressed greatly in plant ambisense ssRNA viruses (mean CpG$_{O/E}$ value is 0.27 ± 0.043), followed by negative and positive ssRNA viruses (mean CpG$_{O/E}$ values are 0.45 ± 0.066 and 0.75 ± 0.184, respectively), whereas it was almost normally distributed in plant dsRNA viruses (mean CpG$_{O/E}$ value is 0.85 ± 0.160). Interestingly, we also found host-related CpG usage in plant RNA viruses: the CpG was more repressed in dicot-infecting RNA viruses than in monocot-infecting RNA viruses (mean CpG$_{O/E}$ values are 0.72 ± 0.210 and 0.78 ± 0.162, respectively). These results indicate that the CpG usage of plant RNA viruses is also affected greatly by the host they infect. However, the source of the selection pressure is an interesting question to be addressed because no nucleic acid receptors involved in plant innate antiviral immune system have been discovered thus far.

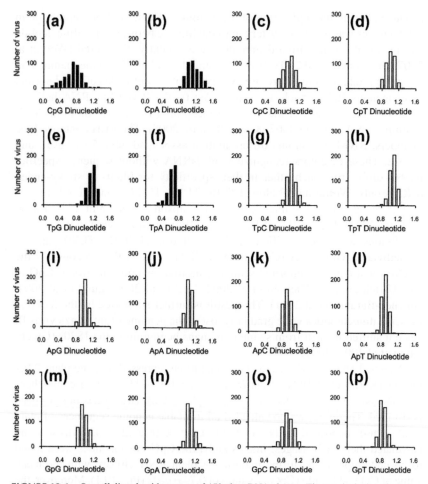

FIGURE 19.1 Overall dinucleotide usages of 450 plant RNA viruses. The y-axis depicts the number of viruses with the specific dinucleotide odds ratio values given on the x-axis. (a–p) The distribution patterns of CpG, CpA, CpC, CpT, TpG, TpA, TpC, TpT, ApG, ApA, ApC, ApT, GpG, GpA, GpC, and GpT, respectively. Four pairs of dinucleotides were found to be differentially represented (CpG, TpA, CpA, and TpG) in plant RNA viruses and are highlighted in black. Note that the distribution patterns of CpG and TpA are negatively deviated from 1.0, whereas the distribution patterns of TpG and CpA are positively deviated from 1.0, suggesting that CpG and TpA are repressed in most RNA viruses and that TpG and CpA are overrepresented in most RNA viruses.

CONCLUSION

In this chapter, we presented a summary of the evolutional pressure exerted by the plant host on viruses in the light of recent progress in the field. Based on the knowledge obtained thus far, it is clear that plants influence the evolution of the viruses infecting them in almost all aspects, including genome stability, protein emergence, amino acid usage, synonymous codon usage, and dinucleotide usage. However, the impact of the host on viral evolution is varied for different viruses and their hosts. Indeed, the evolutional direction of a given plant virus is determined by the virus itself and the environment, including the pressures from its host. In conclusion, tremendous progress has been made on the subject at hand in recent years—but a vast amount of research still needs to be undertaken because the battle with viruses is an eternal one and many unknown mysteries are still waiting to be uncovered.

REFERENCES

Adams, M.J., Antoniw, J.F., 2004. Codon usage bias amongst plant viruses. Archives of Virology 149, 113–135.

Alexopoulou, L., Holt, A.C., Medzhitov, R., Flavell, R.A., 2001. Recognition of double-stranded RNA and activation of NF-κB by Toll-like receptor 3. Nature 413, 732–738.

Aragonès, L., Guix, S., Ribes, E., Bosch, A., Pintó, R.M., 2010. Fine-tuning translation kinetics selection as the driving force of codon usage bias in the hepatitis A virus capsid. PLoS Pathogens 6, e1000797.

Arguello Astorga, G., Ascencio-Ibanez, J.T., Dallas, M.B., Orozco, B.M., Hanley-Bowdoin, L., 2007. High-frequency reversion of geminivirus replication protein mutants during infection. Journal of Virology 81, 11005–11015.

Barr, J.N., Fearns, R., 2010. How RNA viruses maintain their genome integrity. Journal of General Virology 91, 1373–1387.

Bird, A.P., 1980. DNA methylation and the frequency of CpG in animal DNA. Nucleic Acids Research 8, 1499–1504.

Bonhoeffer, S., Sniegowski, P., 2002. Virus evolution: the importance of being erroneous. Nature 420, 367–369.

Chantawannakul, P., Cutler, R.W., 2008. Convergent host-parasite codon usage between honeybee and bee associated viral genomes. Journal of Invertebrate Pathology 98, 206–210.

Chare, E.R., Holmes, E.C., 2004. Selection pressures in the capsid genes of plant RNA viruses reflect mode of transmission. Journal of General Virology 85, 3149–3157.

Charron, C., Nicolaï, M., Gallois, J.L., Robaglia, C., Moury, B., Palloix, A., Caranta, C., 2008. Natural variation and functional analyses provide evidence for co-evolution between plant eIF4E and potyviral VPg. Plant Journal 54, 56–68.

Cheng, X.F., Wu, X.Y., Wang, H.Z., Sun, Y.Q., Qian, Y.S., Luo, L., 2012. High codon adaptation in citrus tristeza virus to its citrus host. Virology Journal 9, 113.

Cornelie, S., Hoebeke, J., Schacht, A.M., Bertin, B., Vicogne, J., Capron, M., Riveau, G., 2004. Direct evidence that toll-like receptor 9 (TLR9) functionally binds plasmid DNA by specific cytosine-phosphate-guanine motif recognition. Journal of Biological Chemistry 279, 15124–15129.

de Amicis, F., Marchetti, S., 2000. Intercodon dinucleotides affect codon choice in plant genes. Nucleic Acids Research 28, 3339–3345.

de Oliveira, R.R., de Haan, P., de Avila, A.C., Kitajima, E.W., Kormelink, R., Goldbach, R., Peters, D., 1991. Generation of envelope and defective interfering RNA mutants of tomato spotted wilt virus by mechanical passage. Journal of General Virology 72, 2375–2383.

Diaz-Pendon, J.A., Truniger, V., Nieto, C., Garcia-Mas, J., Bendahmane, A., Aranda, M.A., 2004. Advances in understanding recessive resistance to plant viruses. Molecular Plant Pathology 5, 223–233.

Diebold, S.S., Kaisho, T., Hemmi, H., Akira, S., Reis e Sousa, C., 2004. Innate antiviral responses by means of TLR7-mediated recognition of single-stranded RNA. Science 303, 1529–1531.

Domingo, E., 1992. Genetic variation and quasi-species. Current Opinions in Genetics and Development 2, 61–63.

Domingo, E., 1997. Rapid evolution of viral RNA genomes. Journal of Nutrition 127, 958S–961S.

Domingo, E., Parish, C.R., Holland, J., 2008. Origin and evolution of viruses. Academic Press, New York.

Domingo-Calap, P., Sanjuán, R., 2011. Experimental evolution of RNA versus DNA viruses. Evolution 65, 2987–2994.

Drake, J.W., 1993. Rates of spontaneous mutation among RNA viruses. Proceedings of the National Academy of Sciences of the USA 90, 4171–4175.

Duffy, S., Holmes, E.C., 2008. Phylogenetic evidence for rapid rates of molecular evolution in the single-stranded DNA begomovirus tomato yellow leaf curl virus. Journal of Virology 82, 957–965.

Duffy, S., Holmes, E.C., 2009. Validation of high rates of nucleotide substitution in geminiviruses: phylogenetic evidence from *East African cassava mosaic viruses*. Journal of General Virology 90, 1539–1547.

Eigen, M., 1996. On the nature of virus quasispecies. Trends in Microbiology 4, 216–218.

Elango, N., Hunt, B.G., Goodisman, M.A., Yi, S.V., 2009. DNA methylation is widespread and associated with differential gene expression in castes of the honeybee, *Apis mellifera*. Proceedings of the National Academy of Sciences of the USA 106, 11206–11211.

Ermolaeva, M.D., 2001. Synonymous codon usage in bacteria. Current Issues in Molecular Biology 3, 91–97.

Forsbach, A., Nemorin, J.G., Montino, C., Muller, C., Samulowitz, U., Vicari, A.P., Jurk, M., Mutwiri, G.K., Krieg, A.M., Lipford, G.B., Vollmer, J., 2008. Identification of RNA sequence motifs stimulating sequence-specific TLR8-dependent immune responses. Journal of Immunology 180, 3729–3738.

Ge, L., Zhang, J., Zhou, X., Li, H., 2007. Genetic structure and population variability of *Tomato yellow leaf curl China virus*. Journal of Virology 81, 5902–5907.

Grantham, R., Gautier, C., Gouy, M., Mercier, R., Pavé, A., 1980. Codon catalog usage and the genome hypothesis. Nucleic Acids Research 8, 197.

Graves, M.V., Roossinck, M.J., 1995. Characterization of defective RNAs derived from RNA 3 of the Fny strain of cucumber mosaic cucumovirus. Journal of Virology 69, 4746–4751.

Greenbaum, B.D., Levine, A.J., Bhanot, G., Rabadan, R., 2008. Patterns of evolution and host gene mimicry in influenza and other RNA viruses. PLoS Pathogens 4, e1000079.

Greenbaum, B.D., Rabadan, R., Levine, A.J., 2009. Patterns of oligonucleotide sequences in viral and host cell RNA identify mediators of the host innate immune system. PLoS ONE 4, e5969.

Greenbaum, B.D., Li, O.T.W., Poon, L.L.M., Levine, A.J., Rabadan, R., 2012. Viral reassortment as an information exchange between viral segments. Proceedings of the National Academy of Sciences of the USA 109, 3341–3346.

Heil, F., Hemmi, H., Hochrein, H., Ampenberger, F., Kirschning, C., Akira, S., Lipford, G., Wagner, H., Bauer, S., 2004. Species-specific recognition of single-stranded RNA via toll-like receptor 7 and 8. Science 303, 1526–1529.

Holland, J., Spindler, K., Horodyski, F., Grabau, E., Nichol, S., VandePol, S., 1982. Rapid evolution of RNA genomes. Science 215, 1577–1585.

Hurst, L.D., 2002. The *Ka/Ks* ratio: diagnosing the form of sequence evolution. Trends in Genetics 18, 486–487.

Jenkins, G.M., Holmes, E.C., 2003. The extent of codon usage bias in human RNA viruses and its evolutionary origin. Virus Research 92, 1–7.

Jenkins, G.M., Pagel, M., Gould, E.A., de, A.Z.P.M., Holmes, E.C., 2001. Evolution of base composition and codon usage bias in the genus *Flavivirus*. Journal of Molecular Evolution 52, 383–390.

Jimenez-Baranda, S., Greenbaum, B., Manches, O., Handler, J., Rabadán, R., Levine, A., Bhardwaj, N., 2011. Oligonucleotide motifs that disappear during the evolution of influenza virus in humans increase alpha interferon secretion by plasmacytoid dendritic cells. Journal of Virology 85, 3893–3904.

Kapoor, A., Simmonds, P., Lipkin, W.I., Zaidi, S., Delwart, E., 2010. Use of nucleotide composition analysis to infer hosts for three novel picorna-like viruses. Journal of Virology 84, 10322–10328.

Karasev, A.V., 2000. Genetic diversity and evolution of *Closteroviruses*. Annual Review of Phytopathology 38, 293–324.

Kariin, S., Burge, C., 1995. Dinucleotide relative abundance extremes: a genomic signature. Trends in Genetics 11, 283–290.

Karlin, S., Mrázek, J., 1997. Compositional differences within and between eukaryotic genomes. Proceedings of the National Academy of Science of the USA 94, 10227–10232.

Karlin, S., Blaisdell, B.E., Schachtel, G.A., 1990. Contrasts in codon usage of latent versus productive genes of *Epstein–Barr virus*: data and hypotheses. Journal of Virology 64, 4264–4273.

Karlin, S., Doerfler, W., Cardon, L.R., 1994. Why is CpG suppressed in the genomes of virtually all small eukaryotic viruses but not in those of large eukaryotic viruses? Journal of Virology 68, 2889–2897.

Kawai, T., Akira, S., 2008. Toll-like receptor and RIG-1-like receptor signaling. Annals of the New York Academy of Sciences 1143, 1–20.

Lauring, A.S., Andino, R., 2010. Quasispecies theory and the behavior of RNA viruses. PLoS Pathogens 6, e1001005.

Lobo, F.P., Mota, B.E.F., Pena, S.D.J., Azevedo, V., Macedo, A.M., Tauch, A., Machado, C.R., Franco, G.R., 2009. Virus–host coevolution: common patterns of nucleotide motif usage in *Flaviviridae* and their hosts. PLoS ONE 4, e6282.

Lucas, W.J., 2006. Plant viral movement proteins: agents for cell-to-cell trafficking of viral genomes. Virology 344, 169–184.

Lund, J.M., Alexopoulou, L., Sato, A., Karow, M., Adams, N.C., Gale, N.W., Iwasaki, A., Flavell, R.A., 2004. Recognition of single-stranded RNA viruses by Toll-like receptor 7. Proceedings of the National Academy of Sciences of the USA 101, 5598–5603.

Lynn, D.J., Singer, G.A., Hickey, D.A., 2002. Synonymous codon usage is subject to selection in thermophilic bacteria. Nucleic Acids Research 30, 4272–4277.

Martelli, G.P., Agranovsky, A.A., Bar-Joseph, M., Boscia, D., Candresse, T., Coutts, R.H., Dolja, V., Hu, J., Jelkmann, W., Karasev, A.V., Martin, R.R., Minafra, A., Namba, S., Vetten, H.J., 2012. Closteroviridae. In: Andrew, M.Q.K. (Ed.), Virus taxonomy, Academic Press, San Diego, pp. 987–1001.

Melcher, U., 2000. The '30K' superfamily of viral movement proteins. Journal of General Virology 81, 257–266.

Moutailler, S., Roche, B., Thiberge, J.M., Caro, V., Rougeon, F., Failloux, A.B., 2011. Host alternation is necessary to maintain the genome stability of Rift valley fever virus. PLoS Neglected Tropical Diseases 5, e1156.

Mushegian, A.R., Koonin, E.V., 1993. Cell-to-cell movement of plant viruses. Insights from amino acid sequence comparisons of movement proteins and from analogies with cellular transport systems. Archives of Virology 133, 239–257.

Nieto, C., Rodríguez-Moreno, L., Rodríguez-Hernández, A.M., Aranda, M.A., Truniger, V., 2011. *Nicotiana benthamiana* resistance to non-adapted *Melon necrotic spot virus* results from an incompatible interaction between virus RNA and translation initiation factor 4E. Plant Journal 66, 492–501.

Paul, S., Bag, S., Das, S., Harvill, E., Dutta, C., 2008. Molecular signature of hypersaline adaptation: insights from genome and proteome composition of halophilic prokaryotes. Genome Biology 9, R70.

Pu, Y., Kikuchi, A., Moriyasu, Y., Tomaru, M., Jin, Y., Suga, H., Hagiwara, K., Akita, F., Shimizu, T., Netsu, O., Suzuki, N., Uehara-Ichiki, T., Sasaya, T., Wei, T., Li, Y., Omura, T., 2011. *Rice dwarf viruses* with dysfunctional genomes generated in plants are filtered out in vector insects: implications for the origin of the virus. Journal of Virology 85, 2975–2979.

Rima, B.K., McFerran, N.V., 1997. Dinucleotide and stop codon frequencies in single-stranded RNA viruses. Journal of General Virology 78, 2859–2870.

Roossinck, M.J., 1997. Mechanisms of plant virus evolution. Annual Review of Phytopathology 35, 191–209.

Roossinck, M.J., 2008. Plant virus evolution. Springer, Heidelberg.

Shirako, Y., Brakke, M.K., 1984. Spontaneous deletion mutation of soil-borne *Wheat mosaic virus* RNA II. Journal of General Virology 65, 855–858.

Simmen, M.W., 2008. Genome-scale relationships between cytosine methylation and dinucleotide abundances in animals. Genomics 92, 33–40.

Sin, S.H., McNulty, B..C., Kennedy, G.G., Moyer, J.W., 2005. Viral genetic determinants for thrips transmission of *Tomato spotted wilt virus*. Proceedings of the National Academy of Sciences of the USA 102, 5168–5173.

Stanley, J., Frischmuth, T., Ellwood, S., 1990. Defective viral DNA ameliorates symptoms of geminivirus infection in transgenic plants. Proceedings of the National Academy of Sciences of the USA 87, 6291–6295.

Steinhauer, D.A., Domingo, E., Holland, J.J., 1992. Lack of evidence for proofreading mechanisms associated with an RNA virus polymerase. Gene 122, 281–288.

Sztuba-Solinska, J., Urbanowicz, A., Figlerowicz, M., Bujarski, J.J., 2011. RNA–RNA recombination in plant virus replication and evolution. Annual Review of Phytopathology 49, 415–443.

Takeda, K., Kaisho, T., Akira, S., 2003. Toll-like receptors. Annual Review of Immunology 21, 335–376.

Takeshita, M., Matsuo, Y., Yoshikawa, T., Suzuki, M., Furuya, N., Tsuchiya, K., Takanami, Y., 2008. Characterization of a defective RNA derived from RNA 3 of the Y strain of cucumber mosaic virus. Archives of Virology 153, 579–583.

Tan, D.Y., Hair, B.M., Aini, I., Omar, A.R., Goh, Y.M., 2004. Base usage and dinucleotide frequency of infectious bursal disease virus. Virus Genes 28, 41–53.

Tatineni, S., Robertson, C.J., Garnsey, S.M., Bar-Joseph, M., Gowda, S., Dawson, W.O., 2008. Three genes of citrus tristeza virus are dispensable for infection and movement throughout some varieties of citrus trees. Virology 376, 297–307.

Tatineni, S., Robertson, C.J., Garnsey, S.M., Dawson, W.O., 2011. A plant virus evolved by acquiring multiple nonconserved genes to extend its host range. Proceedings of the National Academy of Sciences of the USA 108, 17366–17371.

Thompson, A.J.V., Locarnini, S.A., 2007. Toll-like receptors, RIG-I-like RNA helicases and the antiviral innate immune response. Immunology and Cell Biology 85, 435–445.

Truniger, V., Aranda, M.A., 2009. Recessive resistance to plant viruses. Advances in Virus Research 75, 119–159.

Woelk, C.H., Holmes, E.C., 2002. Reduced positive selection in vector-borne RNA viruses. Molecular Biology Evolution 19, 2333–2336.

Woo, P.C.Y., Wong, B.H.L., Huang, Y., Lau, S.K.P., Yuen, K.Y., 2007. Cytosine deamination and selection of CpG suppressed clones are the two major independent biological forces that shape codon usage bias in coronaviruses. Virology 369, 431–442.

Wright, F., 1990. The 'effective number of codons' used in a gene. Gene 87, 23–29.

Xu, X.Z., Liu, Q.P., Fan, L.J., Cui, X.F., Zhou, X.P., 2008. Analysis of synonymous codon usage and evolution of begomoviruses. Journal of Zhejiang University Science B 9, 667–674.

Zarling, D.A., 1976. Spontaneous mutation of RNA tumour viruses. La Ricerca in clinica e in laboratorio 6, 13–19.

Zhou, J., Liu, W.J., Peng, S.W., Sun, X.Y., Frazer, I., 1999. *Papillomavirus* capsid protein expression level depends on the match between codon usage and tRNA availability. Journal of Virology 73, 4972–4982.

Zsiros, J., Jebbink, M.F., Lukashov, V.V., Voute, P.A., Berkhout, B., 1999. Biased nucleotide composition of the genome of HERV-K related endogenous retroviruses and its evolutionary implications. Journal of Molecular Evolution 48, 102–111.

Virus-induced physiologic changes in plants

András Takács
University of Pannonia, Georgikon Faculty of Agricultural Sciences, Institute for Plant Protection, Keszthely, Hungary

József Horváth
University of Pannonia, Georgikon Faculty of Agricultural Sciences, Institute for Plant Protection, Keszthely, Hungary, and Kaposvár University, Department of Botany and Plant Production, Kaposvár, Hungary

Richard Gáborjányi
University of Pannonia, Georgikon Faculty of Agricultural Sciences, Institute for Plant Protection, Keszthely, Hungary

Gabriella Kazinczi
Kaposvár University, Department of Botany and Plant Production, Kaposvár, Hungary

CHLOROPLAST ULTRASTRUCTURE

Shortly after viral infection, investigation of a pathogen and the early events of host metabolism that lead to systemic infection in susceptible hosts (compatibility) or local infection in resistant plants (incompatibility) can begin. Historically, most research has focused on analyzing and elucidating the metabolic and ultrastructural changes of the hypersensitive reaction (HR) in order to unravel the phenomenon of plant resistance to viruses. Recently, more attention has been focused on the metabolism of systemically infected plants. However, only a few comparisons have been made at the ultrastructural level between these two different types of symptoms.

CHLOROPLAST DEGRADATION IN THE SUSCEPTIBLE HOSTS

Ultrastructural studies of infected plants have specifically focused on the degradation of chloroplasts in systemically infected plants in order to try to explain the decreased capacity of photosynthesis. The changes in plants that are associated with infections include a reduced number of chloroplasts, decreased chlorophyll content, large starch grains in the swollen and deformed plastids, and accumulation of osmiophilic plastoglobuli as a consequence of the disorganization of lamellar structures. It is generally accepted that the severity of macroscopic symptoms and cytologic alterations depend on the virus strain rather than on the host plant.

It was discovered that chloroplast degradation or decreased synthesis of chloroplast-associated proteins occurs in plants where the virus replication is associated with the chloroplasts, as in the case of *Barley stripe mosaic virus* (BSMV). However, abnormalities in chloroplast structure and function were reported in *Tobacco mosaic virus* (TMV)-infected plants too, where the virus replicated in the cytoplasm and not in the chloroplasts. Sometimes TMV virion-like structures have been found to aggregate in the stroma of chloroplasts. Later studies demonstrated that these pseudovirus-like structures were encapsidated forms of chloroplast DNA transcripts. According to Reneiro and Beachy (**19XX**), the TMV coat protein (CP) can directly impact chloroplast membranes and cause their instability, leading to chlorosis.

Ultrastructural changes of *Tomato spotted wilt virus* (TSWV) were followed in different stages of infection, from slight chlorosis to systemic necrosis. Studies focused on where membrane-bound virions form in the cell and on which genetic products are found in the viroplasm and fibrillar structures. Infected chloroplasts were found to be similar to healthy ones (see Fig. 20.1). Virus particles were detected only in the cytoplasm. The different mechanisms of virion maturation were also described. It was established that chloroplast ribosomes were not necessary for virus synthesis, but cytoplasmic ones were essential. As a secondary effect of viral infection, peripheral amoeboid extensions of chloroplast membranes were found in infected tissues. In the first stage, the thylakoid structure seemed to remain relatively intact.

Disorganized granal structures developed mainly in the second, late stage of infection (see Fig. 20.2). A similar structure of chloroplasts could be observed in the healthy but senescent leaves. Unexpectedly, cup-shaped chloroplasts were frequently seen in the acute phase, with mature virions present in their invaginations (see Fig. 20.3). The alterations in the fine structure of the TSWV-infected cells gradually led to necrosis; this was completely different from the rapid, hypersensitive necrosis induced by the *Tobacco necrosis virus* (TNV) (see Fig. 20.4).

CHLOROPLASTS IN THE RESISTANT OR NECROTIC HOST

In the resistant host, the first symptom of HR is the rapid change in the permeability of membranes. The oxidative burst of cells and the increase in peroxidase, polyphenol oxidase, and lipoxygenase activity lead to the breakup of membranes. Cytoplasmic invaginations that contain ribosomes in the chloroplasts and in the stroma are often found. The disorganization of chloroplasts and other organelles is a consequence of the rapid collapse of the cell structure (see

FIGURE 20.1   Chloroplasts from healthy tobacco leaves. Note the well-formed thylakoid membrane structures and grana.

Fig. 20.4). According to studies conducted by Weintraub and Ragetli (1964), Loebenstein (1972), and Almási et al (**XXXX**), the most characteristic events in the formation of necroses were the rapid collapse of cellular homeostasis and the specific changes responsible for the eventual blocking of the spread of the virus. Within a few hours of inoculation, no membrane-bound organelles were observed in the lesion area. Cytoplasmic and chloroplast membranes were disrupted and the thylakoid elements were dispersed. Cell death proceeded within a few hours. The discrete phases of necrobiosis could be followed only by observing cells adjacent to the dead ones.

The formation of local necrosis and the development of an 'active zone' around lesions have been discussed in detail by Israel and Ross (1967). The ultrastructure of local lesions was strikingly different from that of the encircling

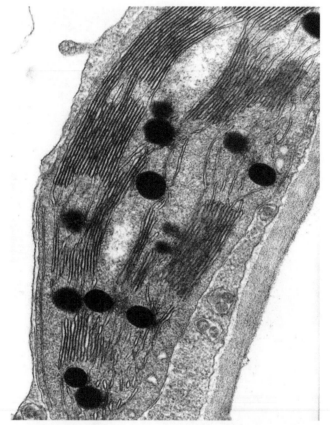

FIGURE 20.2 Slight disorganization of thylakoid membrane structure in chloroplasts from *Tomato spotted wilt virus* (TSWV)-infected tobacco leaves in the first stage of infection. Note the lipid plastoglobuli in the cytoplasm.

zone. Cells in the active zone encircling the mature lesions showed enhanced metabolic activity, containing large vacuoles and increased amounts of cytoplasm and ribosomes. The fine structure of these cells resembled the effect of induced juvenility. Chloroplasts in this area were intact or lens- or amoeboid-shaped. Sometimes dividing chloroplasts were also observed.

CHANGES IN CHLOROPHYLL–PROTEIN COMPLEXES AND CHLOROPLAST PROTEINS

Virus-infected cells contain reduced amounts of chlorophyll–protein complexes as compared to healthy plants. Koiwa et al (1992) proposed that TMV infection inhibits photosystem II (PSII) activity selectively by decomposing the light-harvesting antenna complex of PSII. However, there were no significant differences observed in the PSI reaction center.

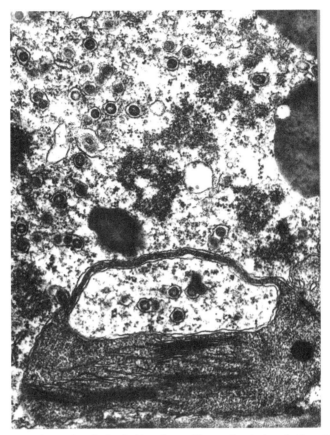

FIGURE 20.3 Abnormal vesiculated chloroplasts with cytoplasmic invagination at the acute phase of infection. Note the disrupted chloroplast with mature virions.

Among the stroma proteins, loss of activity of the small subunit of the enzyme ribulose-1,5-biphosphate-carboxylase-oxygenase (RuBisCO) was reported in the chlorotic tissues of *Cucumber mosaic virus* (CMV)-infected plants and in the chloroplasts of tobacco leaves infected with TSWV, together with a reduced level of 70S ribosomes. Paralleling the decrease in the chlorophyll content, a significant increase was measured in the enzyme activity of chlorophyllase and catalase. This was interpreted as a consequence of the release of the enzyme chlorophyllase bound to the chloroplast inner membrane following the disorder of the chloroplasts.

FLUORESCENCE EMISSION AND EXCITATION SPECTRA

The emission and excitation spectra of chlorophyll components provide good information about the status of chlorophyll–protein complexes. Emission spectra of control and TSWV-infected tobacco plants showed a slight increase in the 680–700 nm regions. This increase was more significant

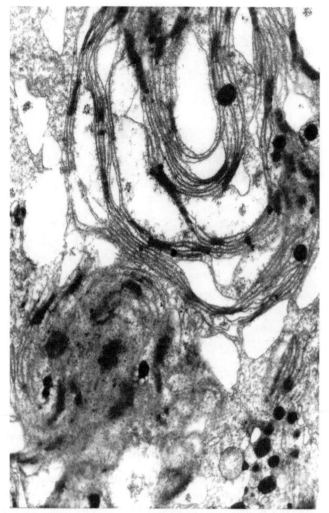

FIGURE 20.4   *Tobacco necrosis virus* (TNV)-infected cells with rapid cell disorganization.

in isolated chloroplasts (see Fig. 20.5). The excitation spectra of TSWV-infected leaves showed a characteristic decrease in the 690–730 nm regions. Similar but more pronounced changes were found in the spectra of isolated chloroplasts (see Fig. 20.6). In addition, the main peak shifted to the blue (678 nm). These changes show that chlorotic symptoms were connected to the ratio changes of chlorophyll–protein complexes of the photosynthetic apparatus.

Viral infection also influenced the activity of the photosynthetic electron transport chain. Half-time values of fluorescent decay after a single turnover flash showed that photosynthetic electron transport was significantly slower in virus-infected plants as compared to the healthy control.

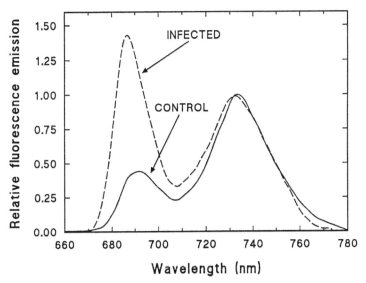

FIGURE 20.5 Emission spectra of chloroplasts isolated from *Tomato spotted wilt virus* (TSWV)-infected tobacco plants.

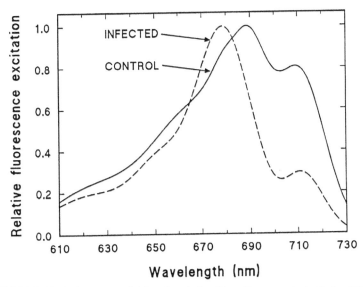

FIGURE 20.6 Excitation spectra of chloroplasts isolated from *Tomato spotted wilt virus* (TSWV)-infected tobacco plants.

INHIBITION OF CHLOROPHYLL BIOSYNTHESIS

Most plant viruses replicate in association with cellular or organelle membranes. Membranes play both structural and functional roles in the replication; however, little is known about the mechanism by which a virus converts membranes for its own use. Membrane reorganization and the *de novo* synthesis of intercellular membranes are mostly connected to the endoplasmic reticulum. Plastid envelope membranes can be involved in virus replication. Membrane fluidity depends on the presence of unsaturated fatty acids. The envelope is the site of galactolipid synthesis, and thus disturbed synthetic functions of envelope membranes may be responsible for the reduced amount of monogalactosyl-diacylglycerol (MGDG) and digalactosyl-diacylglycerol (DGDG).

Previous experiments have demonstrated the modified fine structure of chloroplasts but did not answer the question of whether the deterioration of chloroplast structure is a result of decomposition or the inhibition of chlorophyll biosynthesis. For this reason, Almási et al (**XXXX**) and Harsányi et al (**200X**) studied the effect of BSMV on the ultrastructure of etioplasts and the greening process of barley plants infected by seed transmission. The etioplasts of infected seedlings contained smaller prolamellar bodies (PLBs) with less regular membrane structure, while prothylakoid content was higher than in the control.

DECREASE IN POR QUANTITIES AND REDUCTION IN THE GALACTOLIPID CONTENT

The localization and ratio of nicotinamide adenine dinucleotide phosphate-oxidase (NADPH) to protochlorophyllide oxidoreductase (POR) were analyzed, and the galactolipid content and fatty acid distribution were measured in an attempt to understand the phenomenon by which infection of seed-transmitted BSMV alters membrane structures and inhibits chlorophyll biosynthesis of dark-grown barley plants. The immunolabeling intensity of POR in etioplasts of infected leaves was weaker, and the amount of the enzyme—measured by polyacrylamide gel elecrophoresis, immunoelectron microscopy, and Western blot—was lower than in noninfected plants. These results correlated with the earlier described reduction in the ratio of the photoactive 650 nm to nonphotoactive 630 nm-absorbing protochlorophyllide forms. The relative amount of galactolipids was lower in infected leaves. The amount of MGDG was reduced to 40%, and DGDG was reduced to 50% in control plants on a fresh weight basis. In the infected plants, the proportion of linolenic acid decreased in both galactolipids and total lipid content. The lower amount of highly unsaturated fatty acids and the reduced abundance of MGDG correlated with the previously detected reduction in the ratio of prolamellar body to prothylakoid membrane. The reduced amount of POR and the alteration in lipid composition resulted in a disturbed structure of PLBs. As a consequence, pigment synthesis in the greening process was inhibited in infected cells, explaining the appearance of chlorotic stripes in the infected tissues. These results show that the deteriorative effect of viral infection can be detected at a very early stage of leaf development.

CHANGES IN CO_2 FIXATION

In several host–virus interactions, disturbances in CO_2 fixation and related metabolic processes, or an alteration in the ratio of certain products, were reported. At the early stage of systemic TMV infection, more CO_2 molecules were assimilated by the photosynthetic process. As the infection progressed, this tendency inverted, and CO_2 uptake decreased in the infected tissues. The enzyme activity of phosphoenolpyruvate carboxylase increased in the leaves of Chinese cabbage infected by *Turnip yellow mosaic virus* (TYMV). In infected plants, amino acids translocated from the chloroplasts to the cytoplasm, so that CO_2 was built into amino acids instead of free sugars. The primary products of photosynthetic pathways transformed into glucose, amino acids, and organic acids, which are not able to translocate from the site of their production. Sugar and starch accumulation resulted from the disrupted translocation. Therefore, metabolites such as soluble carbohydrate could not get from the healthy leaves into infected ones. Many authors have investigated the relationship between chlorosis virus replication and starch accumulation; however, the severity of symptoms did not depend on the virus content or on the accumulation of starch in the chlorotic tissues.

Starch accumulation is a common feature of viral infection. Proliferation and enlargement of starch grains in chloroplasts in severely chlorotic tissues are often found. These cells have a reduced capacity for starch utilization. Doke and Hirai (1970**x**) found starch synthesis to be inhibited in tobacco systemically infected with TMV and maintained in light, although neither the rate of photosynthetic CO_2 fixation nor the activity of amylophosphorylase and amylase was affected. In the absence of light, starch degradation and amylase activity were stimulated.

EFFECT OF VIRUS REPLICATION PRODUCTS

The site of a virus's replication is seldom the chloroplast. However, products of viral replication, especially the coat protein (CP), may cause the inhibition of photosynthetic activity. Reneiro and Beachy (**XXXX**) described the accumulation of TMV CP and the correlation between CP concentration and the severity of symptoms. The CP was primarily attached to the thylakoid membrane. They suggested that the CP was bound to one of the chloroplast proteins, thus directly inhibiting the PSII system. Later studies confirmed the hypothesis that the CP is essential in the initiation of close physiologic interaction between a virus and its host plant.

Viral RNA was also detected in several viral infections. This is plausible in cases when the chloroplasts are sites of a virus's replication, for example in *Turnip yellow mosaic virus* (TYMV) in cabbage leaves, but not when replication takes place in the cytoplasm. It is theorized that pyruvate carboxylase (PC) or other viral gene products are translated by the chloroplast ribosomes from virus RNAs. In addition to CP, the different movement proteins (MPs) may also play a role in the inhibition of photosynthesis and the modification of symptoms. MPs affect photosynthetic activity in different ways—by altering the plasmodesmata size-dependent permeability of the palisade layer, by distribution of photosynthetic products, etc.

It may be concluded that viral infection leading to chlorosis or necrosis of infected leaf tissues induces a series of structural and metabolic changes that are related to the severity of symptoms and the type of host–parasite interaction. At the time of active virus multiplication, photosynthetic activity may increase with the stimulation of biosynthesis of amino acids, which are essential for virus multiplication. By the time systemic symptoms develop, a reduction in chlorophyll content, a decrease in photosynthetic activity, and a reduced rate of CO_2 assimilation have taken place. These changes are closely related both to chloroplast degeneration and inhibition of chloroplast biosynthesis. Chloroplasts may show signs of senescence, accompanied by a reduction in rRNA and ribosome function. Chlorosis is always accompanied by a lower amount of chlorophyll content, with alteration in the capacity of light absorbance leading to a decrease in light-dependent reactions. Reductions in photosynthetic phosphorylation and the Hill reaction, as well as inhibition of carbon dioxide assimilation, are common consequences of systemic viral infection. Finally, the overall rates of the photosynthetic process and photophosphorylation are affected. The lower energy supply of infected plants results in a lower growing capacity of the whole plant. On the other hand, in an incompatible host plant, the peroxidation of chloroplast and other inner membranes causes an ultimate reduction in photosynthesis. Accumulation of oxygen-free radicals is a direct consequence of the rapid tissue senescence and cell death.

REFERENCES

Almási, A., Ekés, M., Gáborjányi, R., 1996a. Comparison of ultrastructurral changes of *Nicotiana benthamiana* infected with three different viruses. Acta Phytopathologica et Entomologica Hungarica 31, 181–190.

Almási, A., Böddi, B., Szigeti, Z., Ekés, M., Gáborjányi, R., 1996b. Functional and structural destructions of chloroplasts in virus infected tobacco plants. X. Int. Congr. Virology, Jerusalem. Israel. Abstr, 24–039.

Almási, A., Apatini, D., Bóka, K., Böddi, B., Gáborjányi, R., 2000. BSMV infection inhibits chlorophyll biosynthesis in barley plants. Physiological and Molecular Plant Pathology 56, 227–233.

Almási, A., Harsányi, A., Gáborjányi, R., 2001. Photosynthetic alterations in virus infected plants. Acta Phytopathologica et Entomologica Hungarica 36, 15–29.

Doke, N., Hirai, T., 1970a. Effects of tobacco mosaic virus infection on photosynthetic CO_2 fixation and CO_2 incorporation into protein in tobacco leaves. Virology 42, 68–77.

Doke, N., Hirai, T., 1970b. Radioautographic studies on the photosynthetic CO_2 fixation in virus infected leaves. Phytopathology 60, 988–991.

Harsányi, A., Böddi, B., Bóka, K., Almási, A., Gáborjányi, R., 2002. Abnormal etioplast development in barley seedlings infected with BSMV by seed transmission. Physiologia Plantarum 114, 149–155.

Harsányi, A., Ryberg, M., Andersson, M.X., Bóka, K., László, L., Botond, G., Böddi, B., Gáborjányi, R., 2006. Alterations of POR quantity and lipid composition in etiolated barley seedlings infected by barley stripe mosaic virus (BSMV). Molecular Plant Pathology 7, 533–541.

Israel, H.W., Ross, A.F., 1967. The fine structure of local lesions induced by tobacco mosaic virus in tobacco. Virology 33, 272–286.

Koiwa, H., Kojima, M., Ikeda, T., Yoshida, Y., 1992. Fluctuations of particles on chloroplast thylakoid membranes in tomato plants infected with virulent or attenuated strain of tobacco mosaic virus. Annals of the Phytopathological Society of Japan 58, 58–64.

Loebenstein, G., 1972. Inhibition, interference and acquired resistance during infection. In: Kado, C.I., Agrawal, H. (Eds.), Principles and techniques in plant virology, Van Nostrand/Reinold, New York, pp. 32–61.

Reneiro, A., Beachy, R.N., 1986. Association of coat protein with chloroplast membranes in virus infected leaves. Plant Molecular Biology 6, 291–301.

Reneiro, A., Beachy, R.N., 1989. Reduced photosystem II activity and accumulation of viral coat protein in chloroplasts of leaves infected with tobacco mosaic virus. Plant Physiology 89, 111–116.

Weintraub, M., Ragetli, H.W., 1964. An electron microscope study of tobacco mosaic virus lesions in *Nicotiana glutinosa*. Journal of Cell Biology 23, 499–509.

FURTHER READING

Bailiss, K.W., 1970. Infection of cucumber cotyledons by cucumber mosaic virus and participation of chlorophyllase in the development of chlorotic lesions. Annals of Botany 34, 647–655.

Brakke, M.K., White, J.L., Samson, R.G., Joshi, J., 1988. Chlorophyll, chloroplast ribosomal RNA, and DNA are reduced by barley stripe mosaic virus systemic infection. Phytopathology 78, 570–574.

Esau, K., 1968. Viruses in plant hosts: form, distribution and pathogenic effects. University of Wisconsin Press, Madison, WI.

Fraser, R.S., 1987. Biochemistry of virus-infected plants. Wiley, Letchworth, UK.

Funayama, S., Sonoike, K., Terashima, I., 1997. Photosynthetic properties of leaves of *Eupatorium makinoi* infected by a geminivirus. Photosynthesis Research 53, 253–261.

Gáborjányi, R., Balázs, E., Király, Z., 1971. Ethylene production, tissue senescence and local virus infections. Acta Phytopathologica Acad. Sci. Hungarica 6, 51–55.

Gáborjányi, R., Fernandez, T.F., 1976. Induced alteration of peroxidase activities and the growth of peppers inoculated with tobacco etch virus. Acta Phytopathologica Acad. Sci. Hungarica 11, 277–281.

Goodman, R.N., Király, Z., Wood, R.K., 1986. The biochemistry and plant disease. University of Missouri, MO.

Goodman, R.N., Király, Z., Wood, K., 1991. A beteg növény biokémiája és élettana. Akadémiai Kiadó, Budapest, Hungary.

Gullner, G., Tóbiás, I., Fodor, J., Kőmíves, T., 1999. Elevation of glutathione level and activation of glutathione-related enzymes affect virus infection in tobacco. Free Radical Research 31, 155–161.

Hall, A.E., Loomis, R.S., 1972. An explanation of the difference in photsynthetic capabilities of healthy and beet yellows virus infected sugabeet (*Beta vulgaris* L.). Plant Physiology 50, 576–580.

Helms, K., Wardlaw, I.F., 1977. Effect of temperature of symptoms of tobacco mosaic virus and movement of photosynthate in *Nicotiana glutinosa*. Phytopathology 67, 344–350.

Ie, T.S., 1971. Electron microscopy of developmental stages of tomato spotted wilt virus in plant cells. Virolology 30, 468–479.

Kazinczi, G., Horváth, J., Takács, A., 2006. On the biological decline of weeds due to virus infection. Acta Phytopathologica et Entomologica Hungarica 41, 213–221.

Király, Z., Barna, B., Érsek, T., 1972. Hypersensitivity as a consequence, not the cause of plant resistance to infection. Nature 239, 456–458.

Király, Z., Barna, B., Kecskés, A., Fodor, J., 2002. Down-regulation of antioxidative capacity in a transgenic tobacco which fails to develop acquired resistance to necrotization caused by TMV. Free Radical Research 36, 981–991.

Kitajima, E.W., de Ávila, A.C., de Resende, O., Goldbach, R.W., Peters, D., 1992. Comparative cytological and immunogold labelling studies on different isolates of tomato spotted wilt virus. Journal of Submicroscopical Cytological Pathology 24, 1–14.

Kormelink, R., Kitajima, E.W., de Haan, P., Zudiema, D., Peters, D., Goldbach, R.W., 1991. The nonstructural protein (NS_s) encoded by the ambisense S RNS segment of tomato spotted wilt virus is associated with fibrous structures in infected plants. Virology 181, 459–468.

Lee, W.M., Ishikawa, M., Ahlquist, P., 2001. Mutation of host Delta 9 fatty acid desaturase inhibits brome mosaic virus RNA replication between template recognition and RNA synthesis. Journal of Virology 75, 2097–2106.

Magyarosy, A.C., Buchanan, B.B., Schürmann, P., 1973. Effect of a systemic virus infection on chloroplast function and structure. Virology 55, 426–438.

Matthews, R.E.F., 1981. Plant virology. Academic Press, New York.

Mohamed, N.A., 1973. Some effects of systemic infection by tomato spotted wilt virus on chloroplasts of *Nicotiana tabacum* leaves. Physiological Plant Pathology 3, 509–516.

Mohamed, N.A., Randles, J.W., 1972. Effect of tomato spotted wilt virus on ribosomes, ribonucleic acids and fraction I protein in *Nicotiana tabacum* leaves. Physiological Plant Pathology 2, 235–245.

Pogány, M., 1999. Plant viruses and plant regulators. In: Horváth, J., Gáborjányi, R. (Eds.), Plant viruses and virological methods, Mezőgazda Kiadó, Budapest, Hungary, pp. 284–340.

Rochon, D., Siegel, A., 1984. Chloroplast DNA transcripts are encapsidated by *Tobacco mosaic virus* coat protein. Proceedings of the National Academy of Science of the USA 81, 1719–1723.

Salomon, R., Sela, I., Soreq, H., Giveon, D., Littaur, U.Z., 1976. Enzymatic acylation of histidine to tobacco mosaic virus RNA. Virology 71, 74–84.

Southerland, M.W., 1991. The generation of oxygen free radicals during host plant responses to infection. Physiological Plant Pathology 39, 79–93.

Weststeijn, E.A., 1978. Permeability changes in the hypersensitive reaction of *Nicotiana tabacum* cv. Xanthi nc. after infection with tobacco mosaic virus. Physiological Plant Pathology 13, 253–258.

White, J.L., Brakke, M.K., 1983. Protein changes in wheat infected with streak mosaic virus qnd with barley stripe mosaic virus. Physiological Plant Pathology 22, 87–100.

Ziemiecki, A., Wood, K.R., 1976. Proteins synthesized by cucumber cotyledons infected with two strains of cucumber mosaic virus. Journal of Genetic Virology 31, 373–381.

Chapter 21

Virus–virus interactions

András Takács and Richard Gáborjányi
University of Pannonia, Georgikon Faculty of Agricultural Sciences, Institute for Plant Protection, Keszthely, Hungary

József Horváth
University of Pannonia, Georgikon Faculty of Agricultural Sciences, Institute for Plant Protection, Keszthely, Hungary, and Kaposvár University, Department of Botany and Plant Production, Kaposvár, Hungary

Gabriella Kazinczi
Kaposvár University, Department of Botany and Plant Production, Kaposvár, Hungary

CROSS PROTECTION

In the first three decades of the 20th century it was discovered that plants infected with a mild virus strain do not develop further symptoms when later inoculated with a strong strain. It was recognized early that interference occurred primarily between closely related viruses, and the term 'cross protection' was coined to indicate this relationship. Indeed, cross protection was used as a diagnostic test for relatedness between viral isolates. The potential for using a protective inoculation with a mild strain of virus as a disease-control measure against chance infection by a severe strain was recognized at an early stage. However, the potential for use of protective inoculations in crop protection was not rapidly taken up in crops under field conditions. Today, cross protection is used on only a few crops, and generally other methods are preferred if available.

Generally, the comparatively low uptake of cross protection in agricultural systems suggests that the disadvantages outweigh the advantages. There is a sensible reluctance to introduce viruses into the agricultural ecosystem because of possible deleterious consequences, and in general, cross protection has only been used when other methods, such as resistance, are not stable, where virus eradication has failed and the target virus has become endemic, or where the release could be carried out in controlled conditions, such as in greenhouse-grown crops.

A number of potential problems with cross protection have been identified. Even mild strains have often been shown to cause a decrease in crop yield.

Additionally, a cross-protecting virus might interact with other unrelated viral infections of a crop and produce synergistic damaging effects. There is also concern that a virus introduced to one crop for cross protection could spread to other species and possibly cause severe damage there. Numerous theories have been noted about the possible mechanisms of cross protection. Although there is strong evidence that coat protein (CP) plays a central role in crop protection, the mechanism is not confined solely to the inhibition of virus uncoating.

Despite concerns, there are many examples of crops and viruses in which effective cross protection has been demonstrated in laboratory or greenhouse experiments or field trials.

For example, a severe strain of *Cucumber mosaic virus* (CMV) originating from an infected tomato plant (Gastouni-Olympia, Greece) was isolated in tobacco (*Nicotiana tabacum* cv. Xanthi-nc) after three serial local lesion passages in *Chenopodium quinoa* and was designated CMV-G. CMV-G induces yellow mosaic (YM) symptoms in tobacco. In the *Solanaceae* family (tobacco, tomato, pepper), YM variants induced more severe symptoms than the mild mosaic (MM) variants. The YM and MM phenotypes were stable in tobacco for all seven passages attempted using the obtained YM and MM variants. Cross-protection experiments showed that an isolated MM variant was able to protect tobacco plants against a challenge infection by the YM variant.

To determine the pathogenicity domain and to apply cross protection, *Pepper mild mottle virus* (PMMoV) point mutations in the replicase (Rep) gene between the methyltransferase and helicase domains, and deletions truncating pseudoknots in the 3' non coding region (NCR), were constructed. Some mutants substituting a single amino acid in Rep residue 348 exhibited mild symptoms in *Nicotiana benthamiana* or pepper plants. Accumulation of these mutants was higher than that of other Rep mutants or wild-type PMMoV. Deletion mutants in the 3' NCR pseudoknot showed the lowest rate of viral replication and accumulation among the mutants tested. Six attenuated mutants, which combined 3' NCR deletions and single or double Rep substitution mutation, were constructed to investigate cross-protection effects on pepper plants. All six of the attenuated mutants showed milder symptom development than wild-type virus. These results suggest that Rep and the pseudoknot in the 3' NCR are major pathogenicity determinants of the virus, and engineered PMMoV-attenuated mutants can be useful for protection against the virus in pepper plants.

Papaya ringspot virus (PRSV) HA5-1, a mild mutant of type P Hawaii severe strain (PRSV P-HA), has been widely used to control PRSV type P strains in papaya but has not been shown to provide practical protection against PRSV type W strains in cucurbits. In order to widen the protection effectiveness against W strains, chimeric mild strains were constructed from HA5-1 to carry the heterologous 3' genomic region of a type W strain W-CI. Virus accumulation of recombinants and their cross-protection effectiveness against W-CI and P-HA were investigated. In horn melon and squash plants, the recombinant carrying both the heterologous CP-coding region and the 3' untranslated region (3' UTR), but not the heterologous CP-coding region alone, significantly enhanced protection against W-CI. Heterologous YUTR alone is critical for the enhancement of protection against W-C1 in horn melon but not in zucchini squash. In papaya, either the heterologous CP-coding region or the 3' UTR alone, but not together, significantly reduced the effectiveness of cross protection against P-HA. Our recombinants provide broader protection against both type W and P strains in cucurbits; however, protective effectiveness is also affected by virus accumulation, the organization of the 3' genomic region, and host factors.

SYNERGISTIC ANTAGONISTIC INTERACTIONS

When two unrelated plant viruses infect a plant simultaneously, synergistic viral interactions often occur, resulting in devastating diseases. The possible synergistic effect of a viral transgene on a superinfecting virus can enhance the symptoms of the superinfecting virus. Such synergy between viruses is well known. Transgenic and non-transgenic plant lines should be inoculated with viruses that they are expected to encounter in the field, and the symptoms produced should be compared. More severe symptoms in transgenic lines may be an indication of a synergistic effect. However, they may also be caused by somaclonal variation resulting from the transformation process.

The use of ultrastructural paracrystalline arrays composed of coinfecting viruses (referred to as mixed virus particle aggregates [MVPAs]) were noted in the majority of the mixed infections. When the flexuous rod-shaped potyvirus particles involved in MVPAs were sectioned transversely, specific geometric patterns were noted within some doubly infected cells. Although similar geometric patterns were associated with MVPAs of various virus combinations, unique characteristics within patterns were consistent in each mixed infection. Centrally located virus particles within some MVPAs appeared swollen (e.g., *Southern bean mosaic virus* [SBMV] mixed with *Blackeye cowpea mosaic virus* [BlCMV], CMV mixed with BlCMV, and *Sunn hemp mosaic virus* [SHMV] mixed with *Soybean mosaic virus* [SoMV]). These arrays showed mixed infection of plant viruses by adding the additional dimension of visualizing the interactions between the coinfecting viruses.

There were 32- and 64-fold increases in RNA accumulation in *Sweet potato mild mottle virus* (SPMMV) and *Sweet potato feathery mottle virus* (SPFMV), respectively, in mixed infection with *Sweet potato chlorotic stunt virus* (SPCSV) in sweetpotato cv. Tanzania plants. However, accumulation of SPCSV in mixed infection with SPMMV or SPFMV was reduced by 2- to 4-fold, indicating an antagonistic interaction. Data indicated that SPMMV and SPFMV had an additive effect on synergy with SPCSV because symptom severity was further increased in triple infection of the viruses, whereas SPMMV and SPFMV showed no detectable mutual interaction or synergy when they coinfected sweet potato cultivars in the absence of SPCSV. These data indicate that SPMMV and SPFMV, both of which are members of the family *Potyviridae*, are beneficiaries of a synergistic interaction with SPCSV.

Pepper huasteco virus (PHV) and *Pepper golden mosaic virus* (PepGMV) are found in mixtures in many horticultural crops in Mexico. This combination constitutes an interesting, naturally occurring model system that can be used to study several aspects of virus–virus interactions. Possible interactions between PHV and PepGMV were studied at four levels: symptom expression, gene expression, replication, and movement. In terms of symptom expression, the interaction was shown to be host-dependent because antagonism was observed in pepper, whereas synergy was detected in tobacco and *N. benthamiana*. PHV and PepGMV did not generate viable pseudorecombinant viruses; however, their replication increased during mixed infections. An asymmetric complementary movement was observed because PHV was able to support the systemic movement of PepGMV A, whereas PepGMV did not support the systemic distribution of PHV A. Heterologous transactivation of both CP promoters also was detected. Several conclusions can be drawn from these experiments. First, viruses that coinfect the same plant can interact at several levels (replication, movement) and in different manners (synergy, antagonism); some interactions might be host-dependent. Additionally, natural mixed infections may be a potential source of geminivirus variability, generating viable tripartite combinations that can facilitate recombination events.

Several plant DNA viruses produce significant quantities of deleted versions of their DNA in infected plants, which is generally correlated with a slowing down of the replication process of the viral DNA. These deleted versions of the viral DNA are called defective-interfering (DI) DNA because of their inhibitory effect on the helper virus. The sizes of the DI-DNA for different plant viruses can vary from one-tenth of the size of the viral genome to one-half, and most are encapsidated. Sequence analysis suggests that DI-DNA is formed by deletion, duplication, inversion, rearrangement, and sometimes by insertion of non viral DNA sequences involving the viral genome and its satellites. The role of the host plant in the formation of DI-DNA is also important, as DI-DNA is readily formed in experimental hosts rather than their natural hosts. Symptom modulation by DI-DNA is believed to occur through competition for essential viral and host factors, which reduces the levels of the helper virus. There is also new evidence pointing to the role of DI-DNA in activating PTGS in the plant against viral transcripts, which is also likely to contribute to symptom amelioration. The possibility of DI-DNA playing a role in the integration of pieces of viral DNA into plant genomes also exists. Most importantly, DI-DNA has the potential to act as a tool in developing novel control strategies against viruses in crop plants and to act as gene expression or silencing vectors.

RECOMBINATION

RNA viruses in humans, animals, and plants evolve rapidly, accumulating mutations and RNA recombination. Three sorts of recombination have been identified. Homologous recombination occurs with crossovers between related RNA at precisely matched sites, aberrant homologous recombination occurs with crossovers between related RNA at non-corresponding sites, and non-homologous recombination occurs with crossovers between unrelated RNA at non-corresponding sites. There is considerable evidence of extensive recombination of RNA viruses, and it is likely that all three mechanisms have been involved at one time or another. It is generally believed that recombination plays an important role in the evolution of RNA viruses. Forthcoming is recombination between superinfecting viral RNA and RNA expressed from a transgene through the aberrant homologous recombination mechanism. The finding of recognizable host RNA sequences within viral RNA is suggestive of non-homologous recombination.

The commercialization of transgenic plants containing virus-derived sequences for disease resistance promises both economic and environmental benefits through improvements in crop productivity and quality, concurrent with a reduction in the use of chemical pesticides and other agricultural inputs. The mass cultivation of these crops, however, also poses potential risks. One of the more controversial risks is the potential for environmental impacts arising as a result of genetic exchange between naturally occurring plant viruses and virus-derived sequences deployed in some genetically modified crop plants. This is because there is a finite probability that virus–virus recombination and virus–transgene recombination could give rise to a new (chimeric) virus that is capable of spreading virulent disease in the environment.

Although viral recombination is a natural phenomenon that has occurred many times in the past, the use of virus transgenes in plants is a very recent development, and many people have voiced significant concern about the potential problems these may pose.

The severe economic consequences of emerging plant viruses highlight the importance of studies of the evolutions of plant viruses. One issue of particular relevance is the extent to which the genomes of plant viruses are shaped by recombination. A phylogenetic survey of recombination frequency in a wide range of positive-sense RNA plant viruses was conducted using 975 capsid gene sequences and 157 complete genome sequences. In total, 12 of the 36 RNA virus species analyzed showed evidence of recombination, comprising 17% of the capsid gene sequence alignments and 44% of the genome sequence alignments. It could be argued that recombination is a relatively common process in some plant RNA viruses, most notably the potyviruses.

It has also been established that host ribonucleases and host-mediated viral RNA turnover play major roles in RNA virus recombination and evolution.

The recombination of the genomes of geminiviruses has been studied using a statistical procedure developed to detect gene conversions. Complete nucleotide sequences of geminiviruses were aligned, and recombination events were detected by searching for pairs of virus sequences that were significantly more similar than expected based on random distribution of polymorphic sites. The analysis revealed that recombination is very frequent and occurs between species and within and across genera. Tests identified 420 statistically significant recombinant fragments distributed across the genome. These results suggest that recombination is a significant contributor to geminivirus evolution. The high rate of recombination may be contributing to the recent emergence of new geminivirus diseases.

HETEROENCAPSIDATION

Heteroencapsidation involves the superinfection of a plant expressing the CP of a virus via an unrelated virus. Heteroencapsidation is the encapsidation of the genome of the superinfecting virus by the CP of the other virus. The main property of CP under consideration is the vector transmission characteristic. CP is involved in long-distance viral movement around infected plants, and heteroencapsidation could enhance the movement of a superinfecting virus that does not normally move systemically.

However, the introduction of CP genes into plants presents the potential risk of encapsidating a superinfecting viral genome in the transgenic protein, an event that could change the epidemiology of a disease. Because CP is involved in the interactions of the virus particle with its vector, the release in the field of such transgenic plants could alter the transmission properties of some important viruses. CP-transformed transgenic plants bear the potential risk of releasing genetically engineered plants into the environment. There are several examples of heteroencapsidation in transgenic plants, both between viruses of the same group and between unrelated viruses.

The expression of CP in transgenic plants has been shown to be very effective in protecting plants from viruses. To detect the potential heterologous encapsidation of the CMV genome by *Alfalfa mosaic virus* (AMV) CP expressed in transgenic tobacco plants, a system of immunocapture (IC) and amplification by the polymerase chain reaction (PCR) was optimized. This was highly sensitive and provided a reliable selection of the heterologously encapsidated CMV genome in the presence of natural CMV particles. As little as 2 pg of virus could be detected via the IC/PCR technique. Evidence of heterologous encapsidation in the CMV genome was found in 11 of the 33 transgenic plants tested 2 weeks after CMV inoculation. This demonstrates a significant rate of heterologous encapsidation events between two unrelated viruses in transgenic plants.

Transgenic *N. benthamiana* plants expressing the CP of an aphid-transmissible strain of *Plum pox virus* (PPV) were infected by a non-aphid-transmissible strain of *Zucchini yellow mosaic virus* (ZYMV-NAT), in which the CP had a D-T-G amino acid triplet instead of the D-A-G triplet essential for aphid transmission. The aphid vector *Myzus persicae* could acquire and transmit ZYMV-NAT from these plants but not from infected *N. benthamiana* control plants that were not transformed or that were transformed but not expressing the PPV coat protein. The aphid-transmitted ZYMV subcultures were shown to be still non-aphid-transmissible from plants not expressing PPV CP, which indicated that their transmission was not due to RNA recombination or reversion to the aphid-transmissible type. In immunosorbent electron microscopy experiments using the decoration technique, virus particles in the infected control plants could be coated only with ZYMV antibodies, whereas virus particles in the infected transgenic plants expressing the PPV coat protein could be coated not only with ZYMV antibodies but also in part with PPV antibodies. This suggests that aphid transmission of ZYMV-NAT occurred through heterologous encapsidation.

GENE SILENCING

The notion that the introduction of alien RNA into an organism can cause silencing of endogenes and transgenes in plants came to light during the last decades of the 20th century. It was based on the discoveries of virus-induced gene silencing (VIGS) and the protection against pathogenic viruses by preinfection with less pathogenic plant viruses or components of such viruses, as well as on cosuppression phenomena. Gene silencing was first detected in plants in 1990. The breakthrough in RNA silencing research was the discovery by Mello, Fire, and associates that double-stranded RNA (dsRNA) can silence specifically homologous genes in the nematode *Caenorhabditis elegans*. This discovery in *C. elegans*, published in 1998, immediately initiated studies in protozoa, metazoa, fungi, and plants. Similar RNA-silencing mechanisms, albeit with some notable differences, were subsequently identified in almost all eukaryotic organisms in which they were sought. Investigators dealing with the different organisms were aware of each others' results and a very active field of study emerged within a few years. Investigators of plant RNA silencing benefited from the findings in other organisms, especially in *C. elegans*, *Drosophila*, and mammals, where the protein complexes involved in RNA silencing, such as the Dicer complex and the RNA-induced silencing complex (RISC), were studied intensively. The study of RNA silencing in plants followed two avenues. In one avenue, the process of initiation of endogenous dsRNA was followed; in addition, the fate and impact of dsRNA introduced into plant cells were investigated. It was discovered that this dsRNA is cut into similar 21-nt fragments, and the derived ssRNA may guide the RISC to cleave specific mRNA sequences. In the other avenue, the formation of 'hairpin' or 'stem loop' RNA sequences from transcripts of genomic sequences were investigated. The development of these RNA structures into mature microRNA was studied and the possible roles of endogenously formed and introduced microRNAs in the regulation of expression of plant genes were gradually revealed.

RNA silencing in plants was discovered to be a mechanism whereby invading nucleic acids, such as transgenes and viruses, are silenced through the action of small (20–26 nt) homologous RNA molecules. These mechanisms of RNA silencing have evolved to defend plants against viral infection, as well as to regulate gene expression for growth and development. RNA silencing can reduce the expression of specific genes through post-transcriptional gene silencing, the microRNA pathway, and also through transcriptional gene silencing. Post-transcriptional gene silencing also acts as an antivirus mechanism. By suppressing this antivirus defense mechanism, viruses affect all three silencing pathways in addition to the intercellular signaling mechanism that transmits RNA-based messages throughout the plant. Productive viral infection may therefore disrupt the normal gene expression patterns in plants, resulting, at least in part, in a symptomatic phenotype. However, viruses counteract this antiviral defense by expressing silencing suppressor proteins, which are potent weapons in the 'arms race' between plants and invading viruses. There are several cellular silencing pathways in addition to those involved in defense. Endogenous silencing pathways have important roles in gene regulation at the transcriptional, RNA stability, and translational levels. They share a common core of small RNA generator and effector proteins with multiple paralogs in plant genomes, some of which have acquired highly specialized functions. These proteins efficiently inhibit RNA silencing by interacting with various steps of the different silencing pathways, and these mechanisms of suppression are being unraveled progressively. Cosuppression of transgenes and their homologous viral sequences by RNA silencing is a powerful strategy for achieving high-level viral resistance in plants. Gene-silencing strategies can be applied to protect horticultural and field crops from viral infection and to provide results of field tests. The effectiveness and stability of RNA-mediated transgenic resistance are assessed, taking into account the effects of viral, plant, and environmental factors.

Geminiviruses are ssDNA viruses that infect a range of economically important crop species. A pathogen-derived transgenic approach has been developed to generate high levels of resistance against these pathogens in a susceptible cultivar of cassava (*Manihot esculenta*). Integration of the apolipoprotein C2-linked (ACl) gene (which encodes the replication-associated protein) from *African cassava mosaic virus* imparted resistance against a homologous virus and provided strong cross protection against two heterologous species of cassava-infecting geminiviruses. Short-interfering RNAs specific to the ACl transgene were identified in the two most resistant transgenic plant lines prior to virus challenge. Levels of ACl mRNA were suppressed in these plants. When challenged with geminiviruses, viral DNA was reduced by up to 98% as compared to controls, providing evidence that integration of ACl initiates protection against viral infection via a post-transcriptional gene-silencing mechanism. The robust cross resistance reported has important implications for field deployment of transgenic strategies to control germiniviruses.

VIGS is used to analyze gene function in dicotyledonous plants but is used less in monocotyledonous plants (particularly rice and corn), partially due to the limited number of virus expression vectors available. It is reported that the cloning and modification for VIGS of a virus from *Festuca arundinacea* caused systemic mosaic symptoms in barley, rice, and a specific cultivar of maize (Va35) under greenhouse conditions. Through sequencing, the virus was determined to be a strain of the *Brome mosaic virus* (BMV). The virus was named F-BMV, and genetic determinants that controlled the systemic infection of rice were mapped to RNAs 1 and 2 of the tripartite genome. cDNA from RNA 3 of the Russian strain of BMV (R-BMV) was modified to accept inserts from foreign genes. Coinoculation of RNAs 1 and 2 from F-BMV and RNA 3 from R-BMV expressing a portion of a plant gene to leaves of barley, rice, and maize plants resulted in visual silencing-like phenotypes. The visual phenotypes were correlated with decreased target host transcript levels in the corresponding leaves. The VIGS visual phenotype varied from that maintained during silencing of actin 1 transcript expression to transient with incomplete penetration through affected tissue during silencing of phytoene desaturase expression. F-BMV RNA 3 was modified to allow greater accumulation of the virus while minimizing virus pathogenicity. The modified vector C-BMVA/G was shown to be useful for VIGS.

FURTHER READING

Candelier, H.P., Hull, R., 1993. *Cucumber mosaic-virus* genome is encapsidated in alfalfa mosaic-virus coat protein expressed in transgenic tobacco plants. Transgenic Research 2, 277–285.

Chare, E.R., Holmes, E.C., 2006. A phylogenetic survey of recombination frequency in plant RNA viruses. Archives of Virology 151, 933–946.

Chellappan, P., Masona, M.V., Vanitharani, R., Taylor, N.J., Fauquet, C.M., 2004. Broad spectrum resistance to ssDNA viruses associated with transgene-induced gene silencing in cassava. Plant Molecular Biology 56, 601–611.

Cheng, C.P., Serviene, E., Nagy, P.D., 2006. Suppression of viral RNA recombination by a host exoribonuclease. Journal of Virology 80, 2631–2640.

Ding, X.S., Schneider, W.L., Chaluvadi, S.R., Mian, M.A., Nelson, R.S., 2006. Characterization of a *Brome mosaic virus* strain and its use as a vector for gene silencing in monocotyledonous hosts. Molecular Plant-Microbe Interaction 19, 1229–1239.

Foster, G.D., Taylor, S.C., 1998. Plant virology protocols: from virus isolation to transgenic resistance. Humana Press, New Jersey, USA.

Harrison, B.D., Robinson, D.J., 2005. Another quarter century of great progress in understanding the biological properties of plant viruses. Annals of Applied Biology 146, 15–37.

Lecoq, H., Ravelonandro, M., Wipfscheibel, C., Monsion, M., Raccah, B., Dunez, J., 1993. Aphid transmission of a non-aphid-transmissible strain of *Zucchini yellow mosaic* potyvirus from transgenic plants expressing the capsid protein of *Plum pox potyvirus*. Molecular Plant-Microbe Interactions 6, 403–406.

Martin, E.M., Cho, J.D., Kim, J.S., Goeke, S.C., Kim, K.S., Gergerich, R.C., 2004. Novel cytopathological structures induced by mixed infection of unrelated plant viruses. Phytopathology 94, 111–119.

Méndez-Lozano, J., Torres-Pacheco, I., Fauquet, C.M., Rivera-Bustamante, R.F., 2003. Interactions between geminiviruses in a naturally occurring mixture: *Pepper huasteco virus* and *Pepper golden mosaic virus*. Phytopathology 93, 270–277.

Mukasa, S., 2004. Genetic variability and interactions of three sweetpotato infecting viruses. Doctoral Thesis, Uppsala 2004. Acta Universitatis Agriculturae Sueciae Agraria 477.

Napoli, C., Lemieux, C., Jorgensen, R., 1990. Introduction of a chimeric chalcone synthase gene into petunia results in reversible co-suppression of homologous genes in trans. Plant Cell 2, 279–289.

Padidam, M., Sawyer, S., Fauquet, C.M., 1999. Possible emergence of new geminiviruses by frequent recombination. Virology 265, 218–225.

Patil, B.L., Dasgupta, I., 2006. Defective interfering DNAs of plant viruses. Critical Review of Plant Science 25, 47–64.

Sclavounos, A.P., Voloudakis, A.E., Arabatzis, C., Kyriakopoulou, P.E., 2006. A severe Hellenic CMV tomato isolate: symptom variability in tobacco, characterization and discrimination of variants. European Journal of Plant Pathology 115, 163–172.

Yoon, J.Y., Il Ahn, H., Minjea, K., Tsuda, S., Ryu, K.H., 2006. *Pepper mild mottle virus* pathogenicity determinants and cross protection effect of attenuated mutants in pepper. Virus Research 118, 23–30.

You, B.J., Chiang, C.H., Chen, L.F., Su, W.C., Yeh, S.D., 2005. Engineered mild strains of *Papaya ringspot virus* for broader cross protection in cucurbits. Phytopathology 95, 533–540.

Index

Note: Page numbers with "f" denote figures; "t" tables; "b" boxes.

Printed and bound by CPI Group (UK) Ltd, Croydon, CR0 4YY

14/05/2025

01871861-0001